The Humongous Book of SAT Math Problems

650 math problems with comprehensive solutions, including the math sections from three full practice SATs

Translated for People who don't Speak Math!!

ALPHA
A member of Penguin Group (USA) Inc.

by W. Michael Kelley

ALPHA BOOKS

Penguin Group (USA) Inc., 375 Hudson Street, New York, New York 10014, USA • Penguin Group (Canada), 90 Eglinton Avenue East, Suite 700, Toronto, Ontario M4P 2Y3, Canada (a division of Pearson Penguin Canada Inc.) • Penguin Books Ltd., 80 Strand, London WC2R 0RL, England • Penguin Ireland, 25 St. Stephen's Green, Dublin 2, Ireland (a division of Penguin Books Ltd.) • Penguin Group (Australia), 250 Camberwell Road, Camberwell, Victoria 3124, Australia (a division of Pearson Australia Group Pty. Ltd.) • Penguin Books India Pvt. Ltd., 11 Community Centre, Panchsheel Park, New Delhi—110 017, India • Penguin Group (NZ), 67 Apollo Drive, Rosedale, North Shore, Auckland 1311, New Zealand (a division of Pearson New Zealand Ltd.) • Penguin Books (South Africa) (Pty.) Ltd., 24 Sturdee Avenue, Rosebank, Johannesburg 2196, South Africa • Penguin Books Ltd., Registered Offices: 80 Strand, London WC2R 0RL, England

International Standard Book Number: 978-1-61564-271-7

Library of Congress Catalog Card Number: 2013942517

15 14 13 8 7 6 5 4 3 2 1

Interpretation of the printing code: The rightmost number of the first series of numbers is the year of the book's printing; the rightmost number of the second series of numbers is the number of the book's printing. For example, a printing code of 13-1 shows that the first printing occurred in 2013.

Printed in the United States of America

Contents

Introduction

Are you studying for the SAT? Are you looking for a little practice or to brush up on your basic math skills to earn the highest possible SAT math score you can? Do the very letters S, A, and T (in that order) give you shortness of breath or cause a flop sweat? Is a friend or loved one struggling to prepare for the SAT?

If you answered YES to any of these questions, then you NEED this book. Here's why:

Fact #1: You need advice from someone who has beaten the SAT at its own game.

I used to teach SAT preparatory classes, and one day a student of mine challenged me. "You only know the answers to these problems because you've got the teacher's edition!" he said. "If you ever took the real SAT like us, our scores would beat yours easily."

Challenge accepted. I signed up for the next SAT test, and (even though I don't usually like to toot my own horn, in this case I have to) I got a perfect score. Suddenly, all of my students wanted me to help them study for the SAT!

Fact #2: Not all SAT prep books, products, and companies have your best interests in mind.

SAT prep is a big market—lots of students take the test, so there's plenty of money to be made. I have known and worked with a lot of people in the SAT market, and I can tell you one thing for sure: There are a LOT of people who are more interested in taking your money than they are in producing a quality product. This is sad but so very true. This book is different. It is designed to help you understand the math you need to master.

Fact #3: SAT books LOVE to give you tons of problems but HATE explaining how to work them out.

Time and time again, you reach the hardest problem at the end of a math section, and you get completely stumped. The problem might as well be written in another language. Frustrated, you flip to the answers, and this is the answer explanation you get: "C." THAT'S IT!? It just lists the correct answer? How in the world are you supposed to learn from that?

Even the books that include answers feel smug in their explanations. They have the nerve to say things like "The answer is obviously D because of triangle properties." Obvious? Obvious to WHOM? The whole reason you're buying a book to get help is because the answers aren't all obvious. Why does no one understand this?

Fact #4: Reading lists of facts is fun for a while, but then it gets old. Enough with this list—let's cut to the chase.

This book is (figuratively) dripping with SAT practice problems. After all, it is HUMONGOUS! It includes everything the College Board says you should study to prepare for the test. To be honest, if 650 problems aren't enough, then you've got some kind of crazy math hunger, my friend, and I'd seek professional help if I were you.

All of my notes are off to the side like this and point to the parts of the book I'm trying to explain.

This practice book was good at first, but to make it GREAT, I went through and worked out all the problems and took notes in the margins when I thought something was confusing or needed a little more explanation. In the skill-building chapters of the book, I also drew little skulls next to the hardest problems, so you'd know not to freak out if they were too challenging. After all, if you're working on a problem and you're totally stumped, isn't it better to know that the problem is SUPPOSED to be hard? It's reassuring, at least for me. (I left the skulls out of the practice SAT tests, because I want them to look and feel just like real SAT tests, which don't usually come with skull sketches on them—that might be a little unsettling.)

I think you'll be pleasantly surprised by how detailed the answer explanations are, and I hope you'll find my little notes helpful along the way. Call me crazy, but I think that people who WANT to prepare for the SAT and are willing to spend the time drilling their way through practice problems should actually be able to figure the problems out and learn as they go, but that's just my 2¢.

Good luck, and make sure to come visit my website at www.calculus-help.com. If you feel so inclined, drop me an e-mail and give me your 2¢. (Not literally, though—real pennies clog up the Internet pipes.)

—Mike Kelley

Acknowledgements

Special thanks to Alpha Books, who continues to support The Humongous Books. Also, thanks to Lisa Kelley and Rob Halstead, who proofread and double-checked everything I write, unsung heroes behind the scenes.

Trademarks

Dedication

As always, this book is dedicated to the four most important people in my life, my family.

For my wife Lisa, the most giving woman I know. She works endlessly, maintains countless sports, work, publishing, proofreading, after-school activity, and other family schedules and asks for nothing in return. She is the first in line to help someone in need, even when her own life is packed too full of stuff. You are my inspiration to be a better man and the reason our kids are as awesome as they are.

For my son Nicholas, who is growing up far too quickly. You are the kindest, most compassionate, and yet most intense little boy I know, and I am amazed by the strength of your character. You always make the wise choice and you never give up, thereby figuring out the secret of life at the age of 10.

For my daughter Erin, who loves classic cars and the wind in her hair. You are my confident child, my bold adventurer, a girl who knows who she is and who she wants to be. Keep taking chances, because I will always be here to pick you up if you fall.

For my daughter Sara, the tenderhearted caregiver who lavishes love on stuffed animals, hamsters, and basically anything furry. Since birth you have wanted to be a grown-up just so you can be a mom and have a little Chub-Chub of your own. If they're half as kind as you, they'll be lucky.

Part One

GET TO KNOW THE SAT

Including the kinds of questions you'll face

Before you get neck-deep in math review, you should take a moment to learn about the SAT itself. Your goal is to attack and vanquish the SAT, but to be a worthy combatant, you must first learn about your opponent. What makes it tick? What sorts of crafty tricks does it have up its proverbial sleeves? Does it follow any sort of predictable patterns that you can exploit to give yourself an edge?

Resist the temptation to skip over this part! It contains all sorts of incredibly useful strategies and tips. Do you know when you should guess? What should you do when you are completely stuck on a question? How much time should you spend on each problem? You'll find answers to all of these questions (and more).

Chapter 1 focuses on the structure of the SAT and general strategies you can apply to all questions, including how to pace yourself and why an SAT solution is usually not the same sort of solution you'd turn in for a math class. The next two chapters focus on specific strategies for the two different types of math questions you'll encounter: multiple-choice (also known as "selected response") questions and student-produced response ("grid-in") questions.

Chapter 1

GENERAL SAT STRATEGIES

SAT FAQs

Have you ever met people who are almost supernaturally good at tests? They brag that no matter how little they study, they can ace any sort of exam—and much to your chagrin, they're usually not exaggerating. If you're like me, you envy the voodoo-like power they exert over tests, because you always need to spend time stressing and sweating as you study while they float gently by you, happy and serene, on a cloud of self-satisfaction.

This chapter represents your first step on a journey, the goal of which is to become one of those obnoxiously talented test takers. You begin by dissecting the SAT so that you know exactly what to expect when you crack open the exam book on test day. Next, you tackle one of the biggest problems students face on the test: pacing themselves. Finally, you explore strategies that help you break through mental blocks and misconceptions that stand in the way of you and your dream score.

Take a deep breath and buckle up. The first leg of your SAT preparation journey begins now, as you compile a toolbox full of SAT tips, tricks, techniques, strategies, and maybe a couple of stray Allen wrenches. Some of the questions in this chapter are purely informational, reviewing categories of questions, explaining how the test is scored, and describing how each individual mathematics section is organized. However, a few sample SAT-style questions are mixed in as well, to help you start warming up your mental muscles.

Structure of the Test

How is the SAT designed?

1.1 Each SAT contains nine separate sections. Of these, how many contain math problems, and how many problems appear in each mathematics section?

An SAT contains three scored mathematics sections. Keep in mind that one section of every official SAT test is not scored—it is present to equate scores between tests so that no single version of the SAT is any easier or harder than any other version. Therefore, you may encounter four mathematics sections when you take the test, but one of them will not count toward your score.

The unscored section varies, so an individual test may contain an unscored mathematics section, critical reading section, or writing multiple-choice section.

One mathematics section contains 20 multiple-choice questions, and you will have 25 minutes to complete it. Another section contains 18 questions, and you are also allotted 25 minutes to finish it. The remaining section contains 16 questions and a time limit of 20 minutes. Although the sections may appear in any order, they usually appear in the order described here. In other words, the 20-question section usually appears early in the test, the 18-question section usually appears toward the middle of the test, and the 16-question section is often located near the end of the test. If any of the sections repeat—for example, if your test includes two 18-question mathematics sections—one of those sections is unscored.

1.2 How many of the mathematics questions are multiple choice and how many are student-produced (grid-in) questions? Do the grid-in questions always appear in the same section or does it vary from test to test?

The mathematics sections of the SAT contain a total of 54 questions, 44 of which are multiple-choice and 10 of which are grid-ins. The grid-in questions always appear at the end of the 18-question mathematics section. Thus, the first eight questions of that section are multiple-choice and the final 10 questions are grid-ins.

1.3 Four major mathematical content areas are tested on the SAT. Name the content areas and construct a circle graph that illustrates approximately how many questions of each type appear on the SAT.

In other words, a pie chart

The four mathematical content areas tested on the SAT are:

- Numbers and operations
- Algebra and functions
- Geometry and measurement
- Data analysis, statistics, and probability

Do not assume that each content area is equally represented on the test. According to the College Board, the company that designs and delivers the SAT, the 54 math questions on the test are categorized as follows:

- Numbers and operations: 11–13 questions (approximately 22 percent of the total)
- Algebra and functions: 19–21 questions (approximately 37 percent of the total)

- Geometry and measurement: 14–16 questions (approximately 28 percent of the total)
- Data analysis, statistics, and probability: 6–7 questions (approximately 13 percent of the total)

This data is reflected in the circle graph below.

To learn how to construct and interpret circle graphs, see Problems 10.11–10.13.

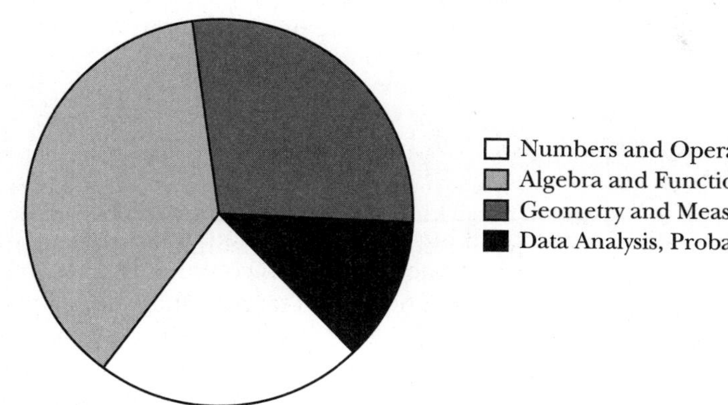

1.4 How are the questions within each mathematics section organized? Identify the single exception to this organizational system.

The questions in each mathematics section are roughly organized in order of difficulty. That does *not* mean that the first question is the easiest, the second question is slightly harder than the first, the third question is slightly harder than the second, and so on. Instead, the first cluster of questions in each section should be significantly easier than the final cluster of questions in the same section. Bookended between the easy and hard questions is a cluster of questions of medium difficulty—not super easy but not too hard either.

Why does the SAT place the hard questions at the end of a section? Encountering a difficult question early discourages students ("If I can't even get the first question right, there's no way I am going to do well") and compromises the test's ability to assign a fair score accurately.

There is one exception to this rule. On the 18-question section that contains both multiple-choice and grid-in questions, each part is organized separately. In other words, the 8 multiple-choice questions range from easy to hard, and then the 10 grid-in questions start over with easy questions and progress to hard.

1.5 On how many of the mathematics sections can you use a calculator? What types of calculators are prohibited on the SAT? Can you use a calculator application on your cell phone?

You may use an approved calculator on all three mathematics sections of the SAT, but calculators cannot be used on any of the other sections. You may choose to bring a basic four-function calculator (that only adds, subtracts, multiplies, and divides), a scientific calculator, or a graphing calculator. The College Board reserves the right to decide what models of calculators are approved on the SAT, so you should refer to their website.

Most calculators are allowed, with a few notable exceptions: Calculators that have a full alphabetic keyboard, require an electrical cord to operate, or make noise are prohibited. You should bring spare batteries and make sure you have practiced with your calculator so you do not waste precious time trying to figure out how your calculator works during the test.

Although a wealth of excellent calculator applications are available for your smart phone, they are not allowed either, because you may not use your cell phone at any time during the exam. At the test center, you will be required to turn your phone fully off; if it rings or vibrates during the test, you run the risk of being escorted from the facility and having your scores nullified. If the proctor suspects you of cheating, your phone may even be confiscated, so you should probably leave your phone at home or at least in your car on test day.

1.6 Are you allowed to bring a sheet of formulas to the SAT? What formulas, if any, are provided for you on test day?

You may not bring notes of any kind to the test. However, the following formulas and facts are printed in the test book at the beginning of every mathematics section:

- Area of a circle: $A = \pi r^2$
- Circumference of a circle: $C = 2\pi r$ (or $C = \pi d$)
- Area of a rectangle: $A = l \cdot w$
- Area of a triangle: $A = (1/2)\ bh$
- Volume of a box: $V = l \cdot w \cdot h$
- Volume of a cylinder: $V = \pi r^2 h$
- Pythagorean theorem: $c^2 = a^2 + b^2$
- Leg lengths of 30°–60°–90° triangle: x, $2x$, and $x\sqrt{3}$
- Leg lengths of 45°–45°–90° triangle: s, s, $s\sqrt{2}$
- The arc of a circle measures 360 degrees
- The angles within a triangle have measures that sum to 180 degrees

To review these triangles (and to figure out what in the world this means), see Problem 8.24.

1.7 Statistically, SAT questions are categorized according to one of three levels of difficulty: easy, medium, and hard. Construct a circle graph that illustrates approximately how many questions of each category appear on a standard SAT.

The SAT has been administered for a lengthy period of time, and its designers are constantly analyzing the data produced when students test. Therefore, they can predict how a student population will perform on specific questions. The designations "easy," "medium," and "hard" are based on the number of students that will likely answer a question correctly. If they anticipate that a majority of test takers will answer a question correctly, it is placed in the cluster of easy questions.

The number of easy, medium, and hard questions varies from section to section and from test to test. Generally, approximately 35 percent of the questions are easy, 40 percent of the questions are medium, and 25 percent of the questions are hard, as illustrated in the following figure.

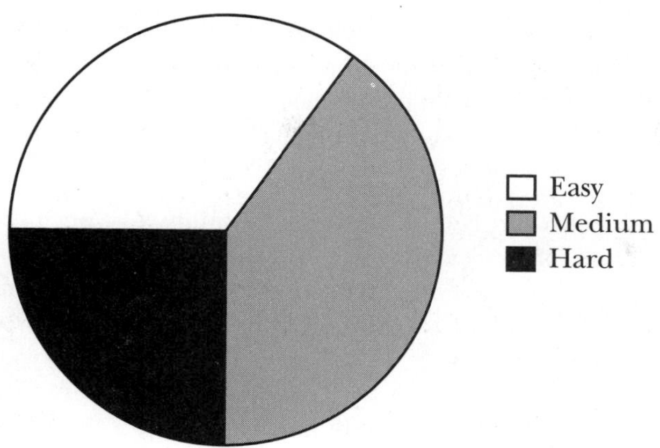

1.8 Most, if not all, students will need paper to complete calculations, draw graphs, construct diagrams, and perform other tasks in order to solve problems on the SAT. How many sheets of paper are you allowed to bring with you on the day of the test?

Commonly referred to as "scratch paper"

You may not bring paper into the testing facility with you—not even blank sheets of paper. All of your calculations should be completed in the test booklet itself; plenty of empty space is provided for you. Make sure to transfer all of your answers from the test booklet to the answer sheet. If you circle the correct answer in the test booklet but do not mark it in on the answer sheet, you will not receive credit.

If you prefer finishing the problems first and then transferring your solutions to the answer sheet at the end of the section, make sure to leave enough time. Test proctors will not give you extra time to bubble your answers once time expires.

1.9 Do you begin the SAT with a perfect score and lose points as you answer questions incorrectly or, alternately, begin with a zero score and earn points as you answer questions correctly?

You begin the SAT with a score of zero, earn points as you answer questions correctly, and (in some cases) lose points as you answer questions incorrectly.

1.10 Explain how raw scores are calculated for the mathematics sections of the SAT.

You earn one point for each question you answer correctly. You receive zero points for incorrect answers. Furthermore, you lose one-fourth of a point for each *multiple-choice* question that you answer incorrectly. There is no penalty for answering grid-in questions incorrectly (other than losing the opportunity to earn a point for a correct answer). Because the SAT contains 54 math questions, the highest possible mathematics raw score is 54.

1.11 What is the difference between a raw score and a scaled score?

Once your raw score has been calculated, it undergoes a complex statistical analysis, part of which is called "normalization." This process ensures that different forms of the SAT, which contain different questions with potentially different difficulty levels, report consistent results. The result of this process is a scaled score, the familiar value ranging between 200 and 800 points that you report to college admission offices.

Raw scores map to scaled scores differently for each test. For example, a raw score of 54 on the mathematics section will always map to a scaled score of 800, but if the questions in a particular version of the SAT were more difficult than usual, you might be able to earn a scaled score of 800 with a raw score of 51.

1.12 A certain student recently took the SAT. An analysis of her performance on the mathematics sections revealed that she answered 41 questions correctly. She omitted two multiple-choice questions, completed all of the grid-in questions, and answered 11 questions incorrectly (3 grid-ins and 8 multiple-choice). What was her mathematics raw score?

She answered 41 questions correctly, earning one raw score point for each correct answer. Therefore, she has an initial raw score of 41 before you account for her omitted and incorrect answers. There is no penalty for omitting questions, so the two multiple-choice questions she did not answer do not affect her initial raw score of 41.

However, she answered 11 questions incorrectly. There is no penalty for incorrect grid-in answers, but one-fourth of a point is subtracted for each incorrect multiple-choice answer. Multiply 8 by 0.25 (or 1/4, the same value in fraction form) to calculate the total penalty she incurs.

If she had answered nine multiple-choice questions incorrectly, the penalty would have been $0.25(9) = 2.25$. Decimal penalties are rounded to the nearest whole number. In this case, 2.25 is rounded down to 2, so the penalty for missing 8 questions is the same as the penalty for missing 9 questions.

$$0.25(8) = 2.00 = 2$$

Subtract 2 from 41 to calculate her final raw score: $41 - 2 = 39$.

Pacing and Practicing

Testing the way you test

1.13 How long should you spend on each problem?

In Problem 1.1, you identify the number of questions in each mathematics section and the length of time you have to complete those questions. Divide the length of time given for each section by the number of questions in that section to calculate the average time you have to complete each question.

First section: 25 minutes $\div$ 20 questions ≈ 1.25 minutes per question

Second section: 25 minutes $\div$ 18 questions ≈ 1.39 minutes per question

Third section: 20 minutes $\div$ 16 questions ≈ 1.25 minutes per question

You have slightly longer than one minute to complete each question. Because the second section contains grid-in questions (which require you to bubble more than one oval to record your answer on your answer sheet), you are allotted slightly more time per question in that section.

If you strive to spend one minute or less on easy and medium questions of each section, you will have extra time to spend on the hard questions and to check your answers.

1.14 How can you pace yourself to ensure that you have a chance to answer each question without the pacing strategy, itself, requiring extra time?

As Problem 1.13 explains, your goal should be to spend no more than one minute on each easy and medium question. Easy and medium questions comprise approximately 75 percent of each section.

You are allowed to bring a watch to the SAT to help pace yourself. Rather than check the watch after every question, get in the habit of checking the time every fifth question. In other words, after you complete the fifth question, no more than five minutes should have passed. After 10 questions, no more than 10 minutes should have passed. If you follow this strategy, you should have plenty of time on the hard questions at the end of each section.

When you stop to check the time, you should also check to make sure you are bubbling the answer sheet correctly. Did you skip a line? Don't wait until the end of the section to find out you're recording the answer to #20 in the space for #21, with no time left to figure out what went wrong.

Remember, on the second mathematics section, the multiple-choice and grid-in questions are organized separately, from easy to hard. Therefore, questions 7 and 8 (the final two multiple-choice questions in the section) are probably hard, and you should save them until the end of the section. Return to them *after* the first seven or eight grid-in questions, and make sure you leave the answer sheet spaces blank that correspond with the questions you skip on the test.

1.15 When should you skip a question?

Problem 1.14 outlines an excellent pacing strategy—one minute per question until the questions get hard. If you find yourself stuck on any question for longer than one minute, you should skip it and return to it later. *Every student gets stuck—even students who receive perfect scores on the SAT.* However, you cannot let that slow you down. If you get stuck, move on without fretting or agonizing over it.

You have to suppress the thought that skipping a question means you're giving up on it. Many students simply refuse to skip a question, wasting a lot of time to get one correct answer when they could have gotten three or four other correct answers in the same time period.

Remember, no single question is worth more than any other question. The easiest math problem at the beginning of every section is worth exactly the same as the hardest, most ridiculous, cringe-worthy problem at the end of the section. Your goal should be to sail through the questions you are sure to get correct before getting bogged down in the tricky ones.

Note that you cannot return to a question once the section is over. If you plan on returning to the question, you have to do so before the time in that section runs out.

1.16 The image below is a reproduction of a student's work in an SAT test booklet. The student originally skipped this question, marking in the test booklet before he continued. Explain how the marks he made saved him time when he returned to this question.

4. Which set of numbers below <u>does not</u> represent the lengths of the sides of a valid triangle?

(A) 1, 2, 3
~~(B) 3, 4, 5~~
(C) 5, 10, 12
~~(D) 7, 24, 25~~
(E) 70, 100, 150

This student employed two key strategies when he omitted the question. First, he circled the question number, making it easy to find in a test booklet that is quickly covered in notes, calculations, scribbles, and diagrams as the test progresses. Second, he crossed out two of the choices that he determined were definitely incorrect before he moved on. This saved him time, because he did not need to start over again and attempt to re-think his way through the problem when he returned to it.

This may sound fairly obvious to you, but if you skip a question, make sure that you skip the corresponding row on the answer sheet. In this case, the student skips question 4, so he needs to leave question 4 on the answer sheet blank as he moves on to question 5. Keep your eyes open when you take the SAT—you will almost always spot a student madly erasing a large chunk of the answer sheet just before time runs out in one of the sections, trying to fix errors caused by omitting questions and bubbling wrong.

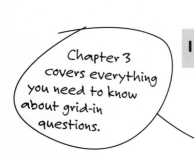

This is a good reason to circle correct answers in the test booklet. If you make an error on the answer sheet and have to bubble all of your answers for the section again, it helps to know what the answers are. Otherwise, you'll have to work from memory or redo all of the problems.

1.17 Are the grid-in questions easier or harder than the multiple-choice questions?

Compared to the multiple-choice questions, the grid-in questions tend to contain fewer easy questions. It is not clear whether the content of the grid-in questions, the lack of answer choices from which to begin, the mechanics of recording a numerical answer in a different format, or the inability to guess causes this disparity in difficulty, but generally speaking, students do not perform as well when they have to produce their own answers instead of selecting them from a list of options.

1.18 Which of the mathematics sections is the hardest?

Recall that your primary goal is to answer all of the easy and medium questions correctly. From this perspective, the sections present approximately the same difficulty because no single section consistently contains more hard questions. However, as Problem 1.17 explains, grid-in questions tend to be more difficult than multiple-choice questions. As a result, the 18-question mathematics section tends to contain fewer easy questions and more medium questions than the multiple-choice only sections.

Chapter 3 covers everything you need to know about grid-in questions.

1.19 What is the best way to practice for the SAT?

First, you need to make sure you have mastered all of the content you need to know. That means working through the chapters of this book and getting a firm grasp on the mathematical concepts the SAT is evaluating. Once you have a basic grasp on the content, it is time to practice, practice, practice.

The best way to prepare for the testing experience is to replicate that experience as closely as possible when you take practice tests. In other words, you should take the practice test under the same time constraints as the actual SAT, complete all of the sections in order, and do so in a single sitting. Would you consider running a marathon if you had never actually run 26.2 miles all at once?

After you finish the test, take the time to analyze your performance. Did you find specific areas of weakness—for example, did you get a lot of questions wrong that contained geometric diagrams? Did you make careless mistakes? Did you pace yourself well? Every student is different, and your path to SAT success is unique to you. Diagnose your struggles and use the lessons you learn to improve your performance on the next practice test.

Breaking Mental Blocks

What to do when you get stuck

Note: Problems 1.20–1.21 refer to the SAT question below, which is of medium difficulty.

In a certain numerical sequence, each term is exactly *d* larger than the term before it. If the tenth term is 43 and the third term is 8, what is the first term?

(A) –2
(B) 0
(C) 1
(D) 3
(E) 5

1.20 The sequence described in this SAT question is an arithmetic sequence, because the difference between consecutive terms is a constant value, *d*. The *n*th term (a_n) of an arithmetic sequence is defined as $a_n = a_1 + (n-1)d$, such that a_1 is the first term of the sequence. Attempt to apply this formula to answer the question, and explain why it is not necessarily the best approach for this question on the SAT.

You may be wondering why this question is asking you to apply a formula that does not work well. The reason is simple: Too many students believe that memorizing piles of formulas will help them on the SAT. The writers of the test are clever; they can find ways of asking questions that make memorizing formulas an utter waste of time. You can answer many of the math questions without any complex formulas or theorems!

For example, in this case, you are given the tenth term and the third term: $a_{10} = 43$ and $a_3 = 8$. You are asked to calculate the first term (a_1), but the value of d is not given (nor is it obvious). Now, *could* you figure out d if you had to? Sure. You know that $a_n = 43$ when $n = 10$. Substitute those values into the arithmetic sequence formula.

$$\begin{aligned} a_n &= a_1 + (n-1)d \\ a_{10} &= a_1 + (10-1)d \\ 43 &= a_1 + 9d \end{aligned}$$

That equation contains two variables, so you cannot solve it just yet. However, you also know that $a_n = 8$ when $n = 3$, so you can plug those values into the sequence formula as well.

$$\begin{aligned} a_n &= a_1 + (n-1)d \\ a_3 &= a_1 + (3-1)d \\ 8 &= a_1 + 2d \end{aligned}$$

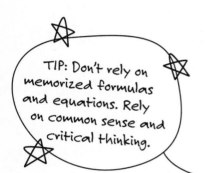

TIP: Don't rely on memorized formulas and equations. Rely on common sense and critical thinking.

Now you have two equations written in terms of two unknowns ($43 = a_1 + 9d$ and $8 = a_1 + 2d$). You can solve this system of equations using substitution or elimination.

Pause. Time out. This is a lot of work. It is very mathematically sound and your math teacher would be proud of you, but all of this is going to take much longer than the one minute you are supposed to spend on medium difficulty questions. Besides, what happens if you forget the formula? You are out of luck!

Don't misunderstand. You will need to know some basic algebra formulas, and in some cases memorizing extra formulas might help. However, you'll feel much more comfortable and confident on the SAT if you have faith in your ability to think logically rather than putting your faith in a heap of memorized formulas you may or may not need.

In the next problem, you answer this SAT question in a more logical manner.

Note: Problems 1.20–1.21 refer to the SAT question in Problem 1.20.

1.21 Answer the question without applying the formula presented in Problem 1.20. Think critically and analytically; it may help to construct a diagram.

Break down the situation. A sequence is a list of terms that follows some sort of pattern. In this sequence, each term is some amount d larger than the previous term. For example, if $d = 2$, then each term in the sequence is 2 larger than the term before it. However, this question does not tell you what d is.

The following diagram summarizes what you know about the sequence, which isn't much. You are only given two terms of the sequence—the third and the tenth—and you are directed to calculate the first term based on that information alone.

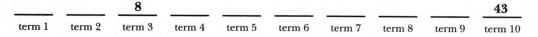

		8							43
term 1	term 2	term 3	term 4	term 5	term 6	term 7	term 8	term 9	term 10

Before you panic, take a deep breath. You actually know more than what is presented in the diagram. You may not know exactly what the ninth term is, but you know that it is d less than the tenth term. Therefore, you can say that the ninth term is $a_9 = 43 - d$. Similarly, the eighth term is d less than the ninth term: $a_8 = 43 - d - d$. Combine the like terms $-d$ and $-d$ to neaten up the expression: $a_8 = 43 - 2d$. Repeat the process: $a_7 = 43 - 3d$, $a_6 = 43 - 4d$, $a_5 = 43 - 5d$, etc. and fill in the blanks of the diagram.

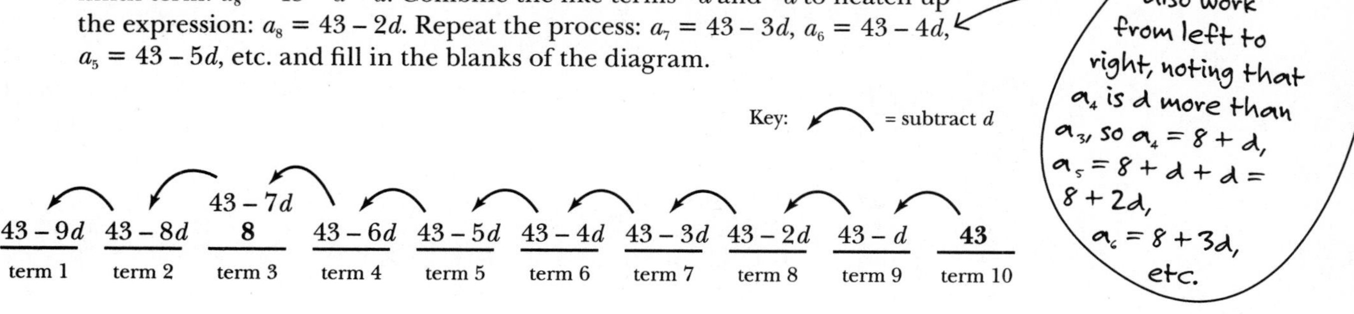

You now know that the third term is equal to $43 - 7d$, and you already knew that the third term was equal to 8. Therefore, you can set those values equal to create an equation that calculates d.

$$\begin{aligned} 43 - 7d &= 8 \\ -7d &= 8 - 43 \\ -7d &= -35 \\ d &= \frac{-35}{-7} \\ d &= 5 \end{aligned}$$

Each term of the sequence is 5 more than the term before it. According to the diagram, the first term is equal to $43 - 9d$. Substitute $d = 5$ into that equation to answer the question.

$$\begin{aligned} a_1 &= 43 - 9d \\ &= 43 - 9(5) \\ &= 43 - 45 \\ &= -2 \end{aligned}$$

You conclude that the correct answer is (A). This may not have been your solution strategy, but hopefully it inspires you. Algebraic logic is the only thing standing between you and a higher mathematics score on the SAT—formulas are rarely the answer.

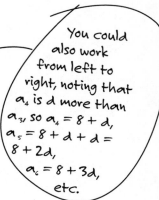

Note: Problems 1.22–1.24 refer to the SAT question below.

If $2|x+1|>3$, which of the following is NOT a value of x?

(A) -4

(B) -2

(C) $\frac{2}{3}$

(D) $\frac{3}{4}$

(E) 1

1.22 The solution to an absolute value inequality $|x+a|>b$ is the union of the solutions of the linear inequalities $x+a>b$ and $x+a<-b$. Apply these formulas to determine which of the answer choices is correct.

As Problems 1.20–1.21 demonstrate, the approach you take to solving math problems on the SAT is not always the approach you are encouraged to take in a traditional math classroom. In a classroom, you are expected to know and explain *why* you got an answer correct, showing your work and justifying your conclusions. On the SAT, you are only asked for the correct answer. Any shortcut you can take on your way to that answer is time in the bank for later questions. That said, here is another example of a traditional versus streamlined SAT approach.

In a traditional classroom, you may be asked to find the solution set of the absolute value inequality $2|x+1|>3$. You would begin by isolating the absolute value expression on the left side of the inequality symbol, dividing each side by 2.

$$\frac{\cancel{2}|x+1|}{\cancel{2}}>\frac{3}{2}$$

$$|x+1|>\frac{3}{2}$$

Now you would rewrite the absolute value inequality as two separate non-absolute value inequalities, as directed by the problem.

$$x+1>\frac{3}{2} \quad \text{or} \quad x+1<-\frac{3}{2}$$

Solve each of the new inequalities for x.

$$x>\frac{3}{2}-1 \qquad\qquad x<-\frac{3}{2}-1$$

$$x>\frac{3}{2}-\frac{2}{2} \quad \text{or} \quad x<-\frac{3}{2}-\frac{2}{2}$$

$$x>\frac{1}{2} \qquad\qquad x<-\frac{5}{2}$$

This inequality is true for all x-values that are either greater than (but not equal to) 1/2 or less than (but not equal to) –5/2. The correct answer is (B), because –2 is neither greater than 1/2 nor less than –5/2; instead, –2 lies between those values on the number line. While this approach does, indeed, produce the correct answer, it is not the quickest technique. Additionally, it requires you to memorize a set of transformation equations that allow you to rewrite the absolute value inequality as a pair of non-absolute value inequalities.

Note: Problems 1.22–1.24 refer to the SAT question in Problem 1.22.

1.23 Rather than applying the technique described in Problem 1.22, answer the question by back-substituting, evaluating the given absolute value inequality for each of the answer choices until the inequality is not true.

The book introduces the concept of back-substitution here, because it's a prime example of a time-saving, not-usually-ok-in-a-math-classroom-but-fine-on-the-SAT technique. Refer to Problems 2.1–2.5 for more information about back-substitution.

Instead of identifying the solution set of all possible answers (the approach in Problem 1.22), you should take the most direct route to the solution. Simply plug each answer choice into the inequality until one of them produces a false statement. Start with $x = -4$, choice (A).

$$\begin{aligned} 2|x+1| &> 3 \\ 2|-4+1| &> 3 \\ 2|-3| &> 3 \\ 2(3) &> 3 \\ 6 &> 3 \quad \text{True} \end{aligned}$$

That value makes the statement true, so move on to choice (B), $x = -2$.

$$\begin{aligned} 2|-2+1| &> 3 \\ 2|-1| &> 3 \\ 2(1) &> 3 \\ 2 &> 3 \quad \text{False} \end{aligned}$$

The inequality is false when $x = -2$, so the answer is (B). Compared to the solution in Problem 1.22, this approach required very little prior knowledge and minimal effort. All you needed to do was evaluate some expressions for a specific x-value, something you learn to do in the first few weeks of any algebra class.

Note: Problems 1.22–1.24 refer to the SAT question in Problem 1.22.

1.24 What word in the question immediately makes this problem tricky? What simple test booklet marking strategy helps you avoid errors related to this word?

The word "NOT" makes this question tricky. Most math classes focus on finding solutions, whereas this question asks you to find something *other than* a solution. Students moving quickly through the SAT tend to focus on equations and numbers, but not always the words of the problem that provide the context. If you didn't see "NOT" in the problem, and you decided to apply the back-substitution technique presented in Problem 1.23, you would have stopped when answer choice (A) made the inequality true.

The SAT realizes that negation words in a question need to be highlighted. That is why the words are either underlined, written in all capital letters, or bold faced. Still, students moving too quickly miss them. In other cases, students pick up on the negative word but then forget about it as they work their way through the problem. By the time they reach the answer choices, it has totally slipped their minds.

TIP: Circle negation words like NOT and EXCEPT in a question, and make sure you take them into consideration when checking your answers.

The best way to keep focused on those negative words when they pop up is to circle them in your answer booklet whenever you see them. Then, when you are finished the section and are checking your answers before time runs out, the negative word will grab your attention.

1.25 A portion of a student's SAT test booklet is reproduced below. Because of an error, the student's answer is not listed among the answer choices.

If $2(x - 1) - 3(x + 4) = -13$, which of the following is the value of x?

(A) -15
(B) -1
(C) 4
(D) 13
(E) 27

$2(x-1)-3(x+4)=13$
$2(x)+2(-1)-3(x)-3(4)=13$
$2x-2-3x-12=13$
$-1x - 14 = 13$
$-x=13+14$
$-x=27$
$x=-27$
??

Identify the student's error and explain how to avoid making that error yourself.

The student's work is nearly perfect. One big problem: The first line is copied incorrectly. The right side of the equation should be –13, but the student miscopied the number, omitting the negative sign. There is a lesson here. When you check your work, make sure that you copied the problem correctly, or try not to copy problems at all.

TIP: Any time you copy an equation from the test booklet, you should double-check to make sure you did so accurately.

1.26 Answer the following SAT question using a calculator.

If $64^{\frac{1}{x+1}} + 16^{\frac{1}{x-3}} = 6$, which of the following is the value of x?

(A) 2
(B) 3
(C) 4
(D) 5
(E) 6

See Problem 4.7 for more information.

Attempting to solve this equation would be very time consuming. Rather than investing that time and (most likely) discovering that you cannot solve for x, back-substitute each of the answer choices into the equation. Notice that the equation contains rational (fractional) exponents, which are radical expressions in disguise.

If you cannot remember how to deal with rational exponents, don't panic! Use your calculator to substitute the values of x into the left side of the equation. When you get an answer of 6 (the value on the right side of the equation), you have identified the correct value of x. Spoiler alert: The correct answer to this question is (D), $x = 5$.

The most commonly used calculators on the SAT test are graphing calculators made by Texas Instruments (TI). If you own a TI calculator, you type the expression below to evaluate the left side of the given equation for $x = 5$. Parentheses are used to enclose the exponents and the denominators because they are expressions.

If you type 64^1/(5+1), the calculator will follow the order of operations, raising 64 to the power of 1 and then dividing by 6. You must use parentheses to tell the calculator that the exponent is the entire expression (1/(5+1)).

64^(1/(5+1))+16^(1/(5–3))

To reduce the number of parentheses, you could simplify the denominators of the fractions: $5 + 1 = 6$ and $5 - 3 = 2$.

64^(1/6)+16^(1/2)

The result is 6, so the correct answer is (D), $x = 5$.

1.27 The beginning of an SAT question is duplicated below. Use the information given to label the diagram as completely as possible.

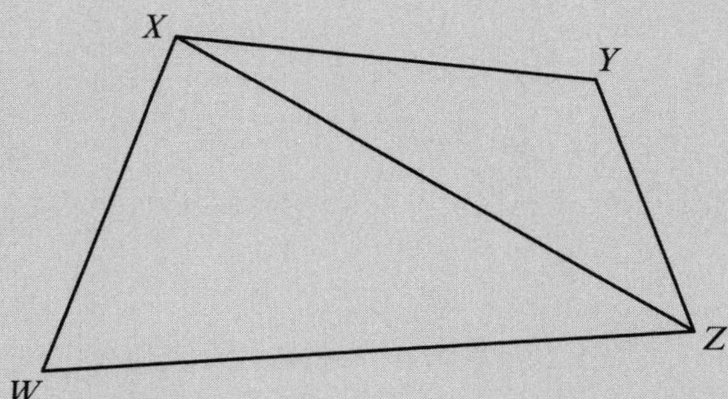

Note: Figure not drawn to scale.

This means that you can't make judgments based on how large the angles or segments appear in the diagram. For all you know, WXYZ might be a rectangle and angle W might actually be larger than angle Y.

In the diagram above, angle WZX has twice the measure of angle YXZ, triangle WXZ is isosceles, angle WXZ has twice the measure of angle YZX, and the measure of angle YZX is 50°.

Remember, you can write all over the test booklet, so label the diagrams in the test booklet. There's no reason to recopy the diagrams before you label them—this takes valuable time and you run the risk of copying incorrectly (as Problem 1.25 explains).

You know that angle YZX measures 50°, so write that on the diagram. Angle WXZ measures 100°, twice the measure of angle YZX. You also know that angle WZX is twice the size of angle YXZ, but you don't know how large either of those angles is, so let a represent the measure of YXZ and $2a$ represent the measure of WZX. Finally, triangle WXZ is isosceles, which means it has two sides of equal length and two angles of equal measure. No triangle can contain two 100° angles—the sum of all three angles in a triangle has to equal 180°, which means two angles that account for 200° are too large—so angle W must be congruent to angle WZX; both of those angles have measure $2a$.

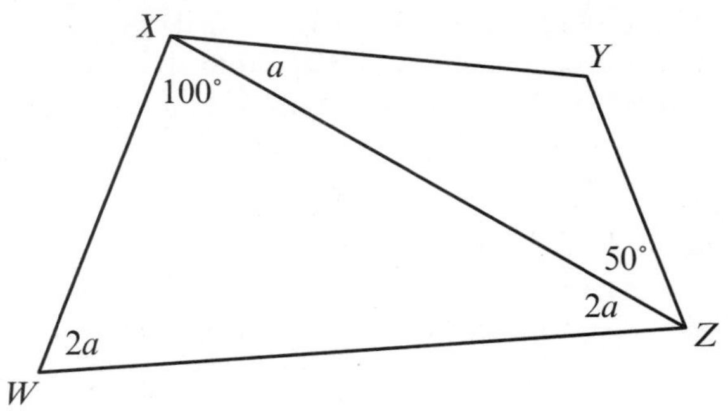

If you can't figure out why, don't worry. There's plenty of geometry review to come.

If you already possess mad geometry skills, you could squeeze even more information out of this diagram, figuring out that $a = 20$, which means that $\overline{YZ}$ and $\overline{WZ}$ are actually perpendicular because they form a right angle. That figure definitely is not drawn to scale.

1.28 Draw a diagram to solve the SAT question below.

Points A, B, C, D, and E lie on the same line in the same plane. E is the midpoint of $\overline{AC}$, D is the midpoint of $\overline{CE}$, and A is the midpoint of $\overline{BC}$. If the length of segment $\overline{BC}$ is 16, which of the following is the length of $\overline{AD}$?

(A) 2

(B) 4

(C) 6

(D) 8

(E) 10

Start by drawing a line and marking points A and C on the line. E is the midpoint of $\overline{AC}$, so place E at the halfway point between A and C. Point D should be placed at the halfway point between C and E. Finally, because A is the midpoint of $\overline{BC}$, place B left of point A such that the distances AB and AC are equal.

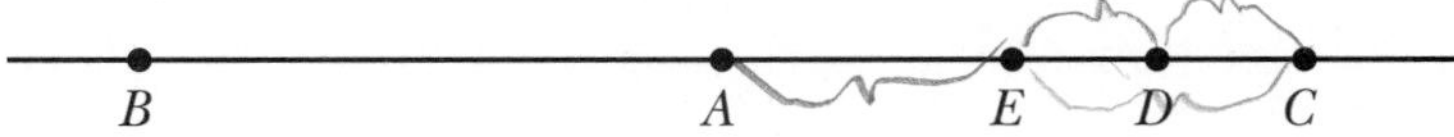

The question states that $BC = 16$. Point A divides the segment in half, so $AB = AC = 8$. Point E divides $\overline{AC}$ in half, so $AE = CE = 4$. Point D divides $\overline{CE}$ in half, so $CD = DE = 2$.

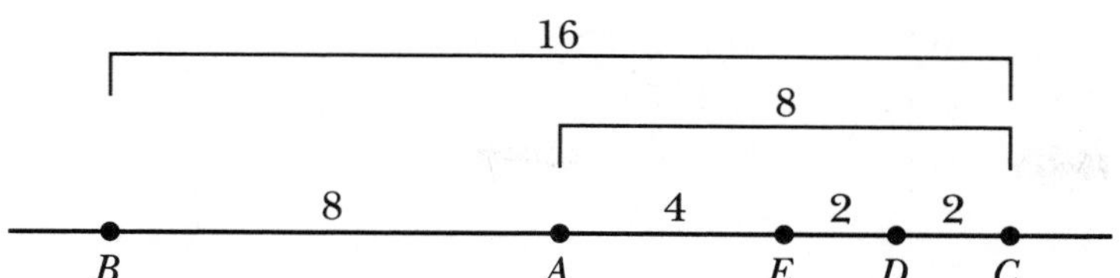

You are asked to calculate *AD*. Notice that $AD = AE + DE$. According to the diagram, $AE = 4$ and $DE = 2$, so $AD = 4 + 2 = 6$. The correct answer is (C).

1.29 Look for a factoring shortcut to answer the following SAT question.

If $x^2 - y^2 = 10$ and $x - y = 2$, which of the following is the value of x in terms of y?

(A) $x = 5y$

(B) $x = \frac{1}{y-5}$

(C) $x = \frac{1}{5-y}$

(D) $x = y - 5$

(E) $x = 5 - y$

"x in terms of y" means x should appear by itself on one side of the equation, and the other side of the equation is some expression containing y's (but no x's).

Recognizing common patterns on the SAT saves you a lot of time. For example, this question contains $x^2 - y^2$, which is called a difference of perfect squares. That difference can be factored into $(x + y)(x - y)$. Rewrite the first equation of the question in factored form.

$$x^2 - y^2 = 10$$
$$(x+y)(x-y) = 10$$

The question also states that $x - y = 2$, so replace $x - y$ in the equation above with the equivalent value 2.

$$(x + y)(2) = 10$$

Solve the equation for x.

$$x + y = \frac{10}{2}$$
$$x + y = 5$$
$$x = 5 - y$$

The correct answer is (E).

1.30 Answer the following SAT question. Identify the most commonly selected incorrect answer choice and explain why more students select it than any of the other incorrect answer choices.

If the ratio of 1 to t is twice the ratio of t to 6, what is the value of t^4?

(A) $\frac{1}{3}$

(B) $\frac{\sqrt{3}}{3}$

(C) $\sqrt{3}$

(D) 3

(E) 9

Ratios can be expressed as fractions. For example, the ratio of 1 to t is the fraction $1/t$. According to the question, $1/t$ equals two times the ratio $t/6$.

$$\frac{1}{t} = 2\left(\frac{t}{6}\right)$$

Simplify the right side of the equation.

$$\frac{1}{t} = \frac{2}{1}\left(\frac{t}{6}\right)$$
$$\frac{1}{t} = \frac{2t}{6}$$
$$\frac{1}{t} = \frac{t}{3}$$

Cross-multiply to solve the proportion.

$$1 \cdot 3 = t^2$$
$$3 = t^2$$
$$\pm\sqrt{3} = t$$

There's a lot of algebra in this example, but don't worry if some of it is over your head. The review is still coming. The whole point of this problem is to remind you to read the directions carefully.

Notice that the question asks you to identify the value of t^4, not the value of t. Many test takers stop here and select answer (C), which is incorrect. Raise $\pm\sqrt{3}$ to the fourth power.

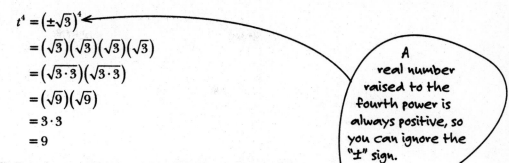

$$\begin{aligned} t^4 &= \left(\pm\sqrt{3}\right)^4 \\ &= \left(\sqrt{3}\right)\left(\sqrt{3}\right)\left(\sqrt{3}\right)\left(\sqrt{3}\right) \\ &= \left(\sqrt{3\cdot 3}\right)\left(\sqrt{3\cdot 3}\right) \\ &= \left(\sqrt{9}\right)\left(\sqrt{9}\right) \\ &= 3\cdot 3 \\ &= 9 \end{aligned}$$

A real number raised to the fourth power is always positive, so you can ignore the "±" sign.

The correct answer is (E). Pay close attention to what the question is actually asking.

Chapter 2

MULTIPLE-CHOICE STRATEGIES

Eeny, meeny, miny, moe

In Chapter 1, you get to know the SAT, spend some time thinking about how to pace yourself on a mind-bendingly long test, and then explore some general strategies that you can apply to many of the questions on the SAT. This chapter focuses specifically on strategies for the multiple-choice questions, which represent slightly more than 80 percent of the test.

Some students ignore the answer choices, acting as though they don't even exist until they have come up with an answer of their own. Other students rely too heavily on the answer choices and wind up getting fooled by the clever SAT designers, who know how to create choices that are tantalizing but, unfortunately, incorrect.

There is no single strategy that works for every kind of multiple-choice question, but there are a few ways to make those answer choices work for you.

Before I took the SAT, someone told me that I should always guess (C) if I didn't know an answer. Why? According to this test "guru," lengthy statistical analyses of the SAT revealed that answer (C) occurs more commonly than any other answer. This was amazing information! Unfortunately, it was also wrong. There are no tricks to sussing out the right answer. Administering the SAT is, after all, a business with a reputation, and it is in the best interest of the College Board to make sure there are no back doors to correct answers.

That doesn't mean there are NO strategies that will help you answer multiple-choice questions. In fact, some of the time-saving tricks you'll learn in this chapter are so useful, so essential, that the SAT developers expect you to know them.

Back-Substitution

Plugging and chugging

2.1 How are the answer choices in a multiple-choice question listed? Why might this be useful information?

Answer choices are almost always listed in order, from least to greatest or greatest to least. This sometimes works to your advantage when you apply the back-substitution technique—substituting the answer choices into the question to get an answer. Sometimes, plugging in the answer choices in order produces results that are also in order! In these cases, you should start by plugging answer choice (C) into the question. If the result is too high, then the answer must be (A) or (B); if it is too low, the answer is (D) or (E).

Note that this strategy does not always work, because the equations in the questions are not always well behaved, but when it does work, it is a real time-saver. One other note: Because "back-substitution" sounds unnecessarily formal (and no fun at all), from this point forward, this book refers to the technique as the "plug and chug" method. Plug the answers into the question, chug your way through the math, and see if the answer works. See? You're having more fun already.

If you have no idea what this means, you skipped Problem 2.1. Caught you!

2.2 Plug and chug your way through the SAT question below.

If four consecutive positive integers have a product of 32,760, which of the following represents the least of those four integers?

(A) 8
(B) 9
(C) 10
(D) 11
(E) 12

The problem is referring to the product of four consecutive integers, which are four integers in a row. If 8 is the least of those integers, the product would be $8 \cdot 9 \cdot 10 \cdot 11$. That product is smaller than the product when 9 is the least integer: $9 \cdot 10 \cdot 11 \cdot 12$. The larger the answer choice, the larger the result when you plug and chug. Not only is this a prime plug and chug opportunity, it is also a perfect time to start with choice (C). If the product is too large, you only have to test (A) and (B); if it is too small, you only have to test (D) and (E).

Use your calculator to compute the product when 10 is the smallest of four consecutive integers: $10 \cdot 11 \cdot 12 \cdot 13 = 17{,}160$. This product is not as large as 32,760, so move on to choice (D): $11 \cdot 12 \cdot 13 \cdot 14 = 24{,}024$. Still not large enough. That means that answer *must* be (E), but you might as well check to make sure—it only takes a few button presses on your calculator: $12 \cdot 13 \cdot 14 \cdot 15 = 32{,}760$. The answer is definitely (E).

2.3 Plug and chug your way through the SAT question below.

If $\sqrt{4x} = \frac{x}{2}\sqrt{2}$, what is the value of x?

(A) 12
(B) 8
(C) 4
(D) 2
(E) 1

The question contains x's, and the answer choices are values of x, so this is a perfect opportunity to plug and chug. However, starting with answer choice (C) is not clearly beneficial here. It does not matter how large the values are—it only matters that the two sides of the equation are equal. You might as well start with answer choice (A) and work your way through the choices until one value of x satisfies the equation. Substitute $x = 12$ into the equation to determine if it makes the equation true.

$$\begin{aligned}\sqrt{4x} &= \frac{x}{2}\sqrt{2}\\ \sqrt{4(12)} &= \frac{12}{2}\sqrt{2}\\ \sqrt{48} &= 6\sqrt{2}\\ \sqrt{16\cdot 3} &= 6\sqrt{2}\\ 4\sqrt{3} &\neq 6\sqrt{2}\end{aligned}$$

Those equal signs are a little misleading, because you don't find out until the last step, in a surprise twist ending, that the sides of the equation are not equal after all. Move on to (B): $x = 8$.

$$\begin{aligned}\sqrt{4(8)} &= \frac{8}{2}\sqrt{2}\\ \sqrt{32} &= 4\sqrt{2}\\ \sqrt{16\cdot 2} &= 4\sqrt{2}\\ 4\sqrt{2} &= 4\sqrt{2}\end{aligned}$$

No disappointing surprise this time; $x = 8$ actually worked. That means the answer is (B). Should you plug the rest of the answers in as well, to make sure they don't work? Probably not. You may want check your calculations to ensure you did not make any mistakes, but you can confidently move on to the next problem without obsessively plugging the x-values of (C), (D), and (E) into the equation.

For more practice on weird, made-up functions and operations like this, see Problems 6.78 and 6.80.

2.4 Plug and chug your way through the SAT question below.

For all integers a and b, let $a \square b = \dfrac{4b - a^2}{ab}$. What is the value of $-3 \square 2$?

(A) $-\dfrac{17}{6}$

(B) $-\dfrac{8}{3}$

(C) $-\dfrac{4}{3}$

(D) $\dfrac{1}{6}$

(E) $\dfrac{8}{3}$

If you've never seen a math operation that looks like a square before and translates into a crazy fraction, don't worry. There is no operation like this in real life; it is invented for this problem only. Instead of plugging the answers into the question, this time you plug the numbers $a = -3$ and $b = 2$ into the expression.

$$a \square b = \frac{4b - a^2}{ab}$$

$$-3 \square 2 = \frac{4(2) - (-3)^2}{(-3)(2)}$$

$$-3 \square 2 = \frac{8 - (9)}{-6}$$

$$-3 \square 2 = \frac{-1}{-6}$$

$$-3 \square 2 = \frac{1}{6}$$

Plug -3 into a because they both appear on the left side of the square. Plug 2 into b because they are on the right side of the square.

The correct answer is (D).

2.5 Plug and chug to solve the difficult SAT question below.

If $g(x) = 2x + 1$ and $f(x) = \dfrac{[g(x) - 6]^2}{g(x-3)}$, for which of the following values of x is $f(x) = 9$?

(A) 7
(B) 6
(C) 3
(D) –2
(E) The answer cannot be determined from the information given

You are told that this problem is from the hard section of the test, so it is either one of the last problems in a section or one of the last multiple-choice problems before the grid-in questions. Here's a good rule of thumb: If the question is supposed to be difficult, the answer is *probably not* going to be "the answer cannot be determined."

Why? If the question is meant to be hard, most students will struggle to answer it, assume it is not possible, and tend to select (E). This question would be useless! Hard questions are meant to help separate the high scorers from very high scorers, so if everyone guesses (E), this question has not helped differentiate between those two groups. Therefore you can almost immediately disregard (E) as a possible answer.

Good tip: The answers to HARD questions are almost NEVER the answers that look obviously correct, or it would break the SAT.

That said, it is time to plug and chug. In this problem, you have to plug into the function $g(x)$ and then plug those results where they belong into $f(x)$. If you plug each value of x into the functions, only one produces the final answer $f(x) = 9$. Rather than lead you through all of the choices, and all of the substitution behind the incorrect choices, it is time for a spoiler alert. The correct answer is (A). To see why, begin by substituting $x = 7$ into $g(x) = 2x + 1$.

$$\begin{aligned} g(x) &= 2x+1 \\ g(7) &= 2(7)+1 \\ g(7) &= 14+1 \\ g(7) &= 15 \end{aligned}$$

Now substitute $x = 7$ into $f(x)$.

$$\begin{aligned} f(x) &= \frac{[g(x)-6]^2}{g(x-3)} \\ f(7) &= \frac{[g(7)-6]^2}{g(7-3)} \\ f(7) &= \frac{[g(7)-6]^2}{g(4)} \end{aligned}$$

You have already determined that $g(7) = 15$, but now you need to calculate $g(4)$, the denominator of $f(7)$, as well: $g(4) = 2(4) + 1 = 9$. Substitute $g(7) = 15$ and $g(4)$ into $f(7)$.

If you need practice with functions, check out Problems 6.69–6.89.

$$\begin{aligned} f(7) &= \frac{[15-6]^2}{9} \\ &= \frac{[9]^2}{9} \\ &= \frac{81}{9} \\ &= 9 \end{aligned}$$

Because $f(x) = 9$ when $x = 7$, the correct answer is (A).

Variable-Heavy Questions

Do It Yourself (DIY) problems

2.6 If a question and its answer choices are full of variables and you are stuck, what simple step can you take to make the problem easier?

Sometimes SAT questions get very theoretical: "If a does b and c does d, but e is f times as good as $k(a - b)$, then *r u* ready to pull your hair out?" You want to scream, "Stop using letters and please, for the love of all that is good in the world, just use an actual number!" Well, if the SAT won't do that for you, just do it yourself (DIY)!

Make up values for the variables, and make sure you do so in a simple, straightforward way. After all, if you get to pick your own numbers, why would you pick ugly ones like fractions or eight-digit mega-numbers? Out of thin air, you can pull friendly numbers like 1, 2, and 3 and replace those aggravating variables with good, old-fashioned numbers that make sense. However, stay away from zero, because that may cause important parts of a formula to disappear.

2.7 Apply the DIY technique to solve the SAT question below.

If a certain machine produces x units every 15 minutes and another machine produces y units every 30 minutes, how many units will both machines working together produce in z hours?

(A) $z(15x+30y)$

(B) $2z(2x+y)$

(C) $60z(4x+2y)$

(D) $60z\left(\frac{x}{4}+\frac{y}{z}\right)$

(E) $\frac{z}{4x+2y}$

You can choose any friendly numbers that you like, but in this solution, the first machine will produce $x = 1$ unit every 15 minutes, the second machine will produce $y = 2$ units every 30 minutes, and you are looking for the total output of both machines in $z = 3$ hours.

If the first machine produces 1 unit every 15 minutes, then it produces 4 units per hour, which is 12 units in 3 hours. The second machine produces 2 units every 30 minutes, so it also produces 4 units every hour and 12 units in 3 hours. That means both machines account for a total of $12 + 12 = 24$ units in $z = 3$ hours.

To see which answer choice is correct, substitute $x = 1$, $y = 2$, and $z = 3$ into each until you get 24. Warning: More than one answer choice may produce the correct answer, depending upon the values you chose. If this happens, select new values for x, y, and z, and repeat the process, but only substitute them into the choices that produced matching correct results.

In this question, only one choice produces the value 24 when $x = 1$, $y = 2$, and $z = 3$. The correct answer is (B), as demonstrated below.

$$\begin{aligned} 2z(2x+y) &= 2(3)[2(1)+2] \\ &= 6[2+2] \\ &= 6[4] \\ &= 24 \end{aligned}$$

2.8 Apply the DIY technique to solve the SAT question below.

In a certain store, blue shirts are one-fifth the price of red shirts. If b blue shirts cost d dollars, what is the cost of r red shirts?

(A) $5dr$

(B) $5brd$

(C) $\dfrac{rd}{5b}$

(D) $\dfrac{5rd}{b}$

(E) $\dfrac{5b}{d}(r)$

Begin by assigning prices to the shirts. For example, if blue shirts cost $4 each, then red shirts would be five times as expensive, $20 each. Now select values for b and r. How many blue and red shirts will you purchase in the problem? To make things easy, let $b = 2$ blue shirts and $r = 3$ red shirts. The pair of blue shirts will cost $d = 8$ dollars and the three red shirts will cost $20 \cdot 3 = 60$ dollars.

Your goal is to determine which formula produces the value 60 (the cost of $r = 3$ red shirts) given $b = 2$, $r = 3$, and $d = 8$. The correct answer is choice (D).

$$\begin{aligned} \frac{5rd}{b} &= \frac{5(3)(8)}{2} \\ &= \frac{120}{2} \\ &= 60 \end{aligned}$$

Note: Problems 2.9–2.10 refer to the SAT question below.

At a certain zoo, p pounds of food will feed a group of w walruses for t weeks. How many pounds of food will only two of the walruses eat in three days?

(A) $\frac{2pt}{7w}$

(B) $\frac{3w}{7pt}$

(C) $\frac{7pt}{2w}$

(D) $\frac{6p}{7tw}$

(E) $\frac{42tw}{p}$

2.9 Apply the DIY technique to answer the question.

This questions requires a fair amount of setup. Take ownership of the problem—do not be afraid to charge right in and create your own walrus-based reality. It may be a difficult tusk, but someone has to do it.

There is no excuse for a pun this terrible. None.

If $w = 4$ walruses each eat 3 pounds of food per day, then you need 12 pounds of food for the whole group each day. This translates to $12(7) = 84$ pounds of food per week. In $t = 2$ weeks, the group of walruses will require twice as much food, $p = 168$ pounds.

If you are only feeding two walruses, then only 6 pounds of food per day is required—both eat 3 pounds each day. The question asks how much food is needed for two walruses over three days, so the answer should be $(6)(3) = 18$ pounds. Substitute $p = 168$, $w = 4$, and $t = 2$ into the formulas to determine which produces a value of 18. The correct answer is (D).

$$\begin{aligned}\frac{6p}{7wt} &= \frac{6(168)}{7(4)(2)}\\ &= \frac{1{,}008}{56}\\ &= 18\end{aligned}$$

Remember: You should ALWAYS apply the DIY technique in these situations when possible. Deriving the formulas takes a lot of time. However, it may help to understand where the formulas come from.

Note: Problems 2.9–2.10 refer to the SAT question in Problem 2.9.

2.10 Answer the question by deriving the formula mathematically instead of applying the DIY technique.

Although the solution to Problem 2.9 is riddled with real numbers instead of variables, you can use the same logic to work your way through the formula symbolically. The key step is, like in Problem 2.9, calculating how much each individual walrus eats per day, because the final goal is to calculate how much food a specific number of walruses consume in a specific number of days.

If p pounds of food are needed to feed w walruses for t weeks, p/w represents the pounds of food required per walrus for t weeks. Divide p/w by t to calculate how many pounds are needed per walrus for a single week.

$$\frac{p}{w} \div t = \frac{p}{w} \cdot \frac{1}{t} = \frac{p}{tw}$$

Dividing by t is the same thing as multiplying by 1/t.

You now know how much one walrus eats in a single week. Divide that by 7 to calculate how much a single walrus eats in one day.

$$\frac{p}{tw} \div 7 = \frac{p}{tw} \cdot \frac{1}{7} = \frac{p}{7tw}$$

Multiply by 3 to calculate the amount of food one walrus consumes in 3 days.

$$\frac{3p}{7tw}$$

The question asks you to calculate how much food 2 walruses consume in 3 days, so multiply by 2. The result is choice (D).

$$2 \cdot \frac{3p}{7tw} = \frac{6p}{7tw}$$

To Guess or Not to Guess?

The risk-reward conundrum

2.11 Calculate the raw score of a student who guesses randomly on each of the 44 multiple-choice math questions on the SAT.

According to Problem 1.10, you earn one point for every correct answer on the SAT and lose one-quarter of a point for every incorrect multiple-choice answer. Because there are five answer choices for each multiple-choice question, a student who guesses randomly will answer approximately one-fifth of the questions correctly. Multiply 44 by 1/5 to calculate the student's total number of correct answers.

$$\begin{aligned} 44 \cdot \frac{1}{5} &= \frac{44}{5} \\ &= 8.8 \\ &\approx 9 \end{aligned}$$

The student will earn 8.8, or approximately 9, points for correct answers. However, the student will answer $44 - 9 = 35$ answers incorrectly. Multiply 35 by 1/4 to calculate the total penalty incurred for incorrect guesses.

$$\begin{aligned} 35 \cdot \frac{1}{4} &= \frac{35}{4} \\ &= 8.75 \\ &\approx 9 \end{aligned}$$

Decimal penalties are rounded to the nearest whole number, so a total of 9 points should be subtracted from the student's score.

$$\begin{aligned}\text{raw score} &= \text{correct answers} - \text{penalty for incorrect answers}\\ &= 9 - 9\\ &= 0\end{aligned}$$

Random guessing results in a raw score of 0.

The penalties keep you from automatically getting 20% of the questions right, even if you know nothing about math. The system is rigged to give you no credit at all if you are utterly stumped and have to guess randomly.

2.12 Calculate the raw score of a student who can successfully eliminate exactly one incorrect answer choice on each of the 44 multiple-choice math questions on the SAT test but guesses randomly among the remaining four answer choices.

If a student can successfully eliminate one incorrect answer choice in each question, he or she is now operating with four answer choices instead of five. A student guessing randomly from those choices will select the correct answer one-fourth of the time.

$$44 \cdot \frac{1}{4} = \frac{44}{4} = 11$$

The student will answer 11 questions correctly and 44 – 11 = 33 questions incorrectly. Calculate the penalty for the incorrect multiple-choice answers.

$$\begin{aligned}33 \cdot \frac{1}{4} &= \frac{33}{4}\\ &= 8.25\\ &\approx 8\end{aligned}$$

Calculate the student's raw score.

$$\begin{aligned}\text{raw score} &= \text{correct answers} - \text{penalty for incorrect answers}\\ &= 11 - 8\\ &= 3\end{aligned}$$

2.13 Based on your answers to Problems 2.11 and 2.12, when should you guess on a multiple-choice SAT question?

In Problem 2.11, you demonstrate that students who guess randomly should earn a raw score of zero. However, Problem 2.12 demonstrates that eliminating even a single incorrect answer choice makes guessing worthwhile. Therefore, you should only answer a multiple-choice question if you can eliminate at least one incorrect answer choice. Otherwise, you should omit the question.

The more incorrect answer choices you eliminate, the more dramatically your score rises, even if you guess.

Multiple Multiple-Choice Questions

I, II, and III only

2.14 How do multiple-choice questions that contain statements with Roman numerals (typically I, II, and III) differ from multiple-choice questions that do not?

The answer choices do not actually contain answers; instead, they contain references to the statements. For example, "I only," "II and III only," and "I, II, and III" are common answer choices. These questions are no more difficult than traditional questions. They are, however, an extremely handy tool for test designers, because this format allows them to design questions with more than one correct answer! Because the questions contain multiple statements, and multiple statements can be true, this book refers to them as "multiple multiple-choice questions."

Well, there's still only one correct A, B, C, D, or E answer. However, up to all three of the statements labeled I, II, and III could be true.

2.15 Answer the following multiple multiple-choice SAT question.

Which of the following are solutions to the equation $x^2 + 3x = 0$?

I. –3
II. 0
III. 3

(A) I only
(B) II only
(C) III only
(D) I and II only
(E) II and III only

Apply the plug and chug technique, substituting –3, 0, and 3 into the equation to determine which values make the equation true. Notice that $x = -3$ and $x = 0$ both satisfy the equation, as demonstrated below, but $x = 3$ does not.

$$\boxed{x=-3}\qquad \begin{aligned} x^2+3x&=0\\ (-3)^2+3(-3)&=0\\ 9-9&=0\\ 0&=0 \text{ True}\end{aligned}$$

$$\boxed{x=0}\qquad \begin{aligned} x^2+3x&=0\\ 0^2+3(0)&=0\\ 0+0&=0\\ 0&=0 \text{ True}\end{aligned}$$

$$\boxed{x=3}\qquad \begin{aligned} x^2+3x&=0\\ 3^2+3(3)&=0\\ 9+9&=0\\ 18&\ne 0 \text{ False}\end{aligned}$$

Therefore, I is a solution and II is a solution, but III ($x = 3$) is not. Choice (D) identifies the correct solutions (I and II only), so it is the correct answer.

2.16 Answer the following multiple multiple-choice SAT question.

If x and y are even numbers and $x < y < 0$, then which of the following statements must be true?

I. xy is an even number
II. $xy > 0$
III. $y - x \geq 0$

(A) I only
(B) I and II only
(C) II and III only
(D) I and III only
(E) I, II, and III

You may find it helpful to apply the DIY technique and supply your own values for x and y. If you do, make sure that x and y are even, x and y are negative (because they are both less than 0), and x is more negative than (less than) y. For example, $x = -6$ and $y = -2$ satisfy these conditions.

Consider each of the statements independently:

- Statement I asks whether the product of x and y must be even. The product of any two even numbers, including $(-6)(-2) = 12$, is an even number, so statement I is true.
- Statement II asks whether the product xy must be greater than zero—in other words, whether the product is positive. The answer is yes; a negative number multiplied by a negative number always produces a positive result. In this case, $(-6)(-2) = 12$, which is positive.
- Statement III asks whether subtracting x from y gives you a positive (or zero) result. Substitute $x = -6$ and $y = -2$ into the equation.

$$\begin{aligned} y - x &\geq 0 \\ -2 - (-6) &\geq 0 \\ -2 + 6 &\geq 0 \\ 4 &\geq 0 \quad \text{True} \end{aligned}$$

All three statements are true, so the correct answer is (E).

You might think to yourself, "Sure, this time it's true, but what if I picked different values for x and y? Did I just get lucky with these numbers?" That's a great question to ask, and if you're doubtful, pick new values of x and y and test those until you're sure the statement MUST BE true.

2.17 Answer the following multiple multiple-choice SAT question.

If a is a positive integer and $\sqrt{a} = b\sqrt{c}$, which of the following statements MUST be true? Assume that $b\sqrt{c}$ is a real number.

I. $b > 0$
II. $a > c$
III. $a = b^2c$

(A) I only
(B) III only
(C) I and II only
(D) I and III only
(E) I, II, and III

The question contains the statement $\sqrt{a} = b\sqrt{c}$, which looks like a radical statement $\sqrt{a}$ and a simplified version of that radical statement, $b\sqrt{c}$. If you apply the DIY technique, you can select any simple radical expression and simplify it to get values for a, b, and c.

For example, because $\sqrt{50} = 5\sqrt{2}$, you can assign the values $a = 50$, $b = 5$, and $c = 2$. Using these values, you conclude that statement I is true $(5 > 0)$, statement II is true $(50 > 2)$, and statement III is true $(50 = 5^2[2] = 25[2] = 50)$. You may be tempted to select answer (E), but you would be incorrect. Just because the statements are true for *these* values of a, b, and c, they are not true for *all* possible values of a, b, and c.

If you select values for a, b, and c, be careful not to get too attached to them. Just because one set of values makes statement II true, you can't conclude that the statement is ALWAYS true. It only needs to be false once.

- Statement I: Is b always greater than 0? Yes. The square root of a positive number will always be positive. That means $b\sqrt{c}$ has to be positive, and if $b\sqrt{c}$ is also a real number (as stated in the problem), then c must be positive. (Taking the square root of a negative number produces imaginary numbers.) Thus, all of the values must be positive.
- Statement II: Is a always greater than c? Using the sample values $a = 50$, $b = 5$, and $c = 2$, it sure seems so. However, $\sqrt{a} = b\sqrt{c}$ does not always have to represent the simplified version of a radical expression. For example, consider the equation below, in which both sides are equal to 5.

$$\sqrt{25} = \frac{1}{2}\sqrt{100}$$

  In this case, $a = 25$ is less than $c = 100$. You might be crying foul: "But b is a fraction! I thought you had to stick to positive integers!" Re-read the question—only a has to be a positive integer. Statement II is not always true.

- Statement III: Are a and b^2c always equal? Yes. Squaring both sides of the equation $\sqrt{a} = b\sqrt{c}$ produces the equivalent, true statement $a = b^2c$.

Therefore, the correct answer is (D).

2.18 Answer the following multiple multiple-choice SAT question.

In the equation below, x, y, and z are integers and $z > 0$. Which of the following statements MUST be true?

$$z = \frac{1}{2}(x+y)$$

I. x and y are even
II. $x > 0$ and $y > 0$
III. z is the arithmetic mean of x and y

(A) I only
(B) II only
(C) III only
(D) I and II only
(E) I, II, and III

Examine the statements one at a time, applying the DIY technique by selecting values for x, y, and z. You need to decide if the statements *must* be true, so if you find only one case in which a statement *could be* false, then that statement should not be included in your final answer.

- Statement I: If x and y are even, their sum will be even and therefore divisible by 2. For example, substitute $x = 6$ and $y = 4$ into the equation to verify that z is a integer.

$$\begin{aligned} z &= \frac{1}{2}(6+4) \\ &= \frac{1}{2}(10) \\ &= 5 \end{aligned}$$

However, x and y do not *have* to be even. If both are odd, then their sum is still even and z is an integer. For example, z is an integer when $x = 3$ and $y = 5$.

$$\begin{aligned} z &= \frac{1}{2}(3+5) \\ &= \frac{1}{2}(8) \\ &= 4 \end{aligned}$$

Therefore, statement I is not necessarily true.

- Statement II: If x and y are positive, then z will be positive, as required by the question. However, do x and y both need to be positive to ensure z is positive? No. If you read the question carefully, you notice that there are no restrictions that require x or y to be positive. Calculate z when $x = -6$ and $y = 20$.

$$\begin{aligned} z &= \frac{1}{2}(-6+20) \\ &= \frac{1}{2}(14) \\ &= 7 \end{aligned}$$

  Because x, y, and z are all integers and $z > 0$, all the conditions described in the question are satisfied. You conclude that x and y do not have to be greater than zero.

- Statement III: The arithmetic mean (or average) of two numbers is equal to the sum of the numbers divided by 2. In this case, z is the arithmetic mean of x and y because z is equal to the sum of x and y multiplied by 1/2. (Dividing by 2 and multiplying by 1/2 are equivalent.) Therefore, statement III is always true, no matter what the signs of x, y, and z are. In fact, statement III is true whether or not x, y, and z are integers.

You conclude that only statement III is true, so the correct answer is (C).

Chapter 3

STUDENT-PRODUCED RESPONSE STRATEGIES

Grid-ins

Not all of the questions on the SAT are multiple-choice. One cluster of math questions, which always appears at the end of the 18-question mathematics section, contains 10 "student-produced response" questions, which are more commonly known as "grid-ins." To record your answer for a grid-in question, you do not select from choices (A), (B), (C), (D), or (E), as you do for multiple-choice questions. Instead, you use a grid to record the actual answer.

The grids are fairly straightforward but surprisingly powerful—they allow you to record whole number, fraction, and decimal answers. However, with great power comes great responsibility (or so I am led to believe by countless superhero movies). Every grid-in answer must meet very stringent requirements, or you run the risk of receiving no credit for a correct answer! In this chapter, you practice using the grid and learn how to record your answers to ensure you receive your maximum score for these 10 traditionally tricky questions.

As Chapter 1 warned, students tend to score lower on grid-in questions than they do on multiple-choice. Is it because they weren't prepared for the grid-ins ("Hey, where'd the answer choices go?") or they didn't understand the rules for gridding answers ("Do I have to simplify fractions?")? By the end of this chapter, you'll understand exactly what to do when faced with a grid-in question.

Get to Know the Grid

Bubble, bubble, toil, and trouble

3.1 The answer to a particular student-produced response SAT question is 25. Three student responses (labeled I, II, and III) are duplicated below. Which response is correct?

I.

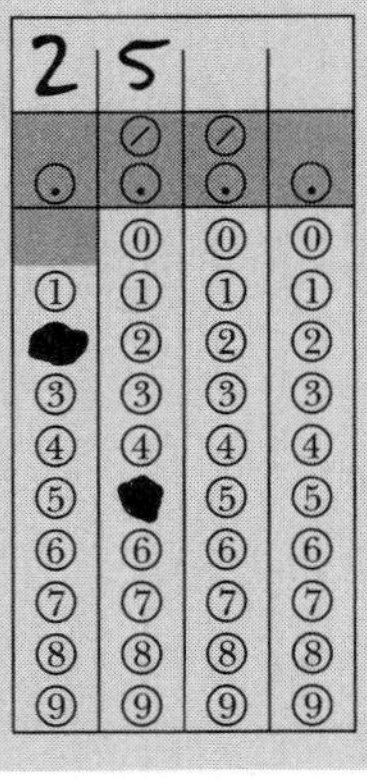

II.

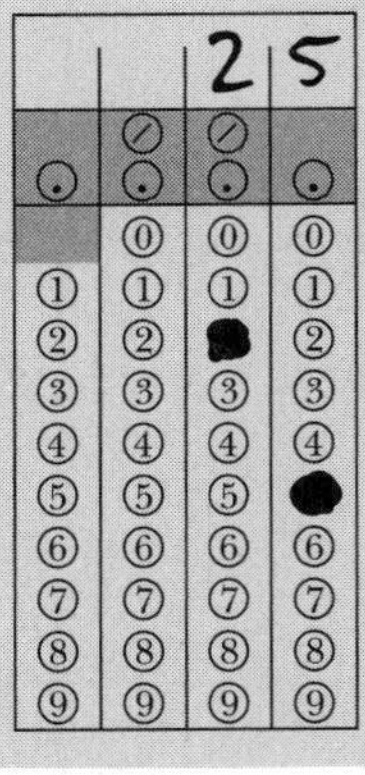

III.

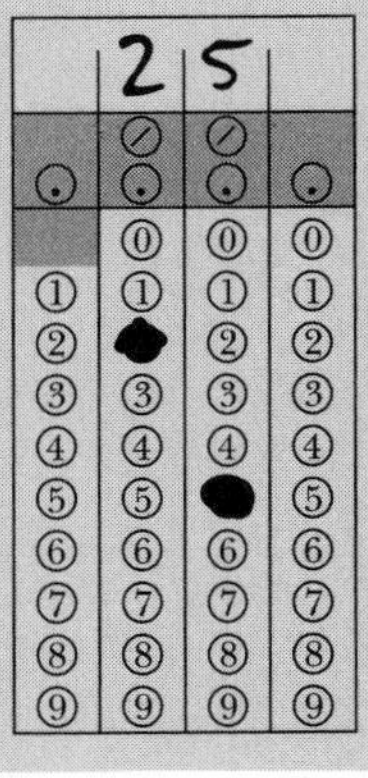

"Bubbling" means darkening in an oval, or bubble, on your answer sheet. For example, on a multiple-choice question, you bubble (C) on your answer sheet to select answer choice (C).

All three students calculated the correct answer, 25. However, they each bubbled their responses in a different position on the grid. Student I began the answer in the leftmost column, leaving the two rightmost columns blank. Student II ended the answer in the rightmost column, leaving the two leftmost columns blank. Student III centered the answer, leaving the far left and far right columns blank.

The computer that scores the SAT answer sheets would mark all three of these answers correct; the position of the answer does not matter.

However, you should follow the lead of Student I. Start in the left column, especially if the answer has multiple digits and/or fills up all four spaces.

3.2 The answer to a particular student-produced response SAT question is 173. A student response is duplicated below, in which the student calculated the correct answer but did not receive any credit. Identify the student's error.

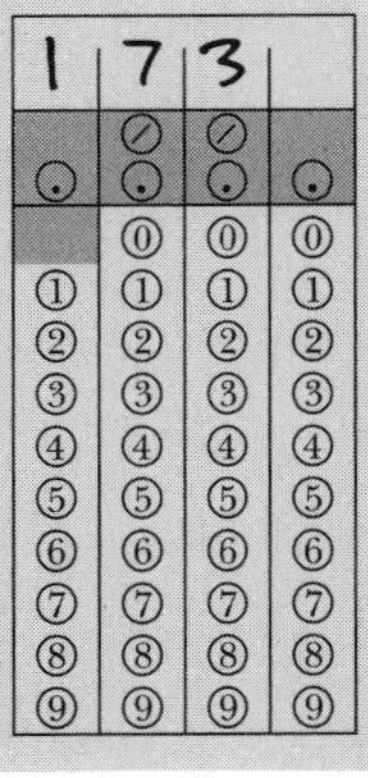

The student calculated the correct answer and wrote it, one digit at a time, above the columns of answer bubbles. However, the student neglected to actually bubble in the corresponding numbers in each column. The bubbles are the most important part of the grid-in answer! You do not even have to include the written answer "173" at the top of the grid, but you should because it makes checking your answers easier and faster. The computer that scores the SAT will not read your handwriting—it will only score the question based on the bubbles that you darkened in each column.

3.3 Is it possible for a grid-in question to have more than one correct answer? If so, how should you respond?

Grid-in questions may have multiple correct answers; in fact, such questions are quite common. When a question asks you to identify one possible solution, do exactly that. No matter how many solutions there may be, you need only identify one.

The mathematically curious part of your brain might ask questions such as "What are the boundaries for the correct answers?" and "How many possible answers are there?" While math teachers usually encourage that curiosity and the intellectual explorations that follow, the SAT is neither the time nor the place to answer such questions. Find a correct answer, bubble it on the answer sheet, and move on.

3.4 How do you record negative answers to a grid-in question?

The answer grid does not allow for negative answers, so any grid-in answer must be greater than or equal to zero. Remember that grid-in questions may have multiple answers (as Problem 3.3 explains). Some of those answers may be negative, but they are not considered valid answers on the SAT test, because there is no way to indicate negative values.

For example, imagine that an SAT question asks you to identify an odd number that is less than 3 but greater than –9. All of the following answers are correct: –7, –5, –3, –1, and 1. However, the only positive integer listed is 1, so that is the only answer that will receive credit as a correct grid-in response.

3.5 When should you guess on a grid-in question?

There is no penalty for an incorrect answer on a grid-in question, unlike multiple-choice questions, which incur a 0.25-point penalty for incorrect answers. If you cannot answer a grid-in question despite your best efforts, you should guess and move on. (You should also circle the question in your test booklet to make it easier to identify, in case you have time remaining to work on it at the end of the section.)

3.6 If the answer to a grid-in question is a fraction or a decimal, what is the maximum number of digits you can record for that answer on the grid?

The decimal point and fraction bar each occupy one of the four columns on the answer grid, leaving a maximum of three other columns in which you could record digits.

Avoiding Bubbling Blunders

Make sure to bubble correct answers correctly!

"Truncating" means chopping off the digits that don't fit into the grid. "Rounding" means you leave the digit in the rightmost column alone if the next digit in the decimal is 0, 1, 2, 3, or 4—otherwise you add 1 to the digit in the rightmost column.

3.7 If a decimal answer to a grid-in question contains more digits than the grid can hold, should you round or truncate your answer?

As long as you provide an accurate answer that uses all of the columns in the grid, you may either round or truncate your answers; both techniques are considered correct. For example, the answer 3.1682 can be truncated to 3.16 or rounded to 3.17 on your answer sheet, as illustrated below.

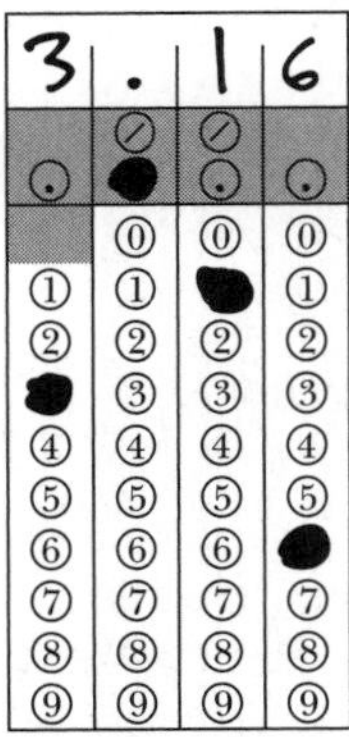

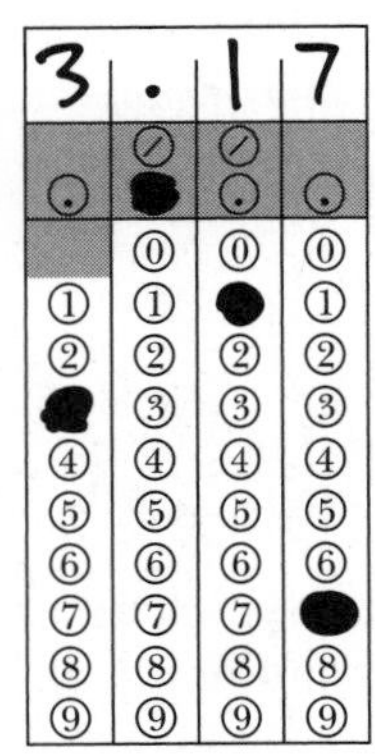

However, you can truncate and round too much. For example, the answer 3.1682 cannot be bubbled as 3.1 or 3.2. Here is the guideline you should follow: Record the most accurate answer you can in the space provided. In this example, the answer did not fit in the four columns of the grid, but you can use all four of those columns to record as accurate an answer as you can. Otherwise, your answer will be marked wrong.

3.8 The answer to a particular grid-in question is $\frac{1}{9}$. Which, if any, of the responses below represents a correct answer for that question on the SAT? Note that more than one of the answer choices may be true.

(A) 1/9 (B) 0.11 (C) .1 (D) .11 (E) .111

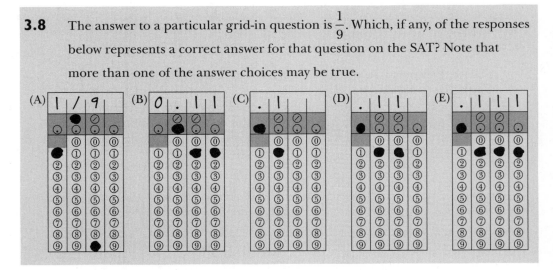

Consider the answer choices one at a time:

- (A) is correct. The student correctly bubbled the fraction, the numerator 1 divided by the denominator 9.

- (B) is incorrect. The fraction 1/9 is equivalent to the repeating decimal 0.11111111..., or $0.\overline{1}$. However, choice (B) places a 0 in the leftmost column. Notice that there is no bubble for 0 in that column! Why? The test is reminding you to record the most accurate answer that a four-column grid can hold. Do not start with a placeholder 0.
- (C) is incorrect. If choice (B) was not accurate enough, choice (C) is even worse. Mathematically, it is just plain wrong to state that 1/9 = 0.1. You read "0.1" as one tenth, so 0.1 = 1/10, not 1/9.
- (D) is incorrect. It represents the same answer as choice (B), 0.11. Like choice (B), it does not represent the most accurate answer possible in the space provided. You should not leave the last column empty.
- (E) is correct. While 1/9 does not technically equal 0.111, this is the most accurate decimal response that will fit in the grid. Truncating and rounding both produce this answer.

In Problems 3.9–3.10, assume that the answer to a particular SAT grid-in question is $\frac{6}{23}$.

3.9 Record the answer as a fraction in the grid below.

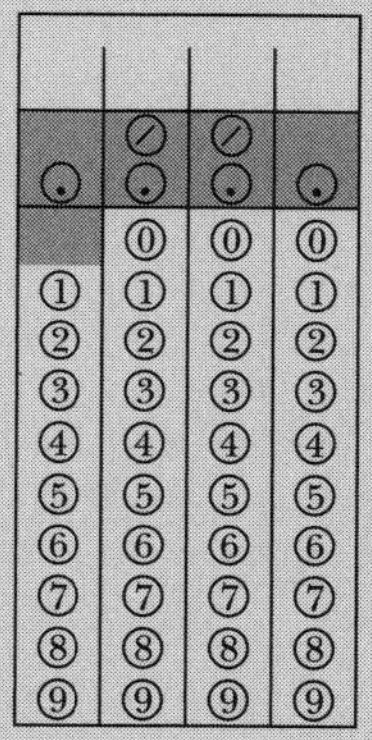

Bubble "6/23" to record the answer in the grid. Remember that the fraction bar occupies its own column in the answer—in this case it is the second column.

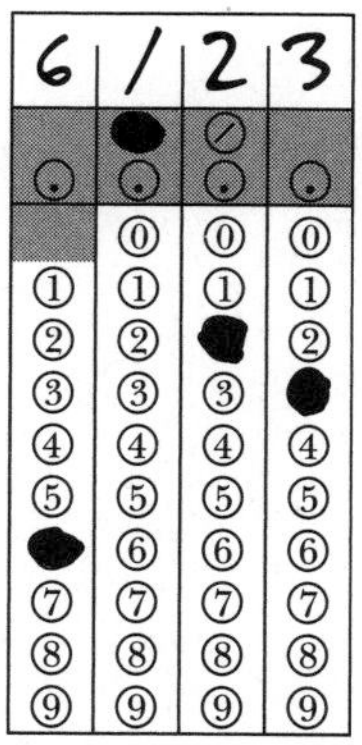

6 / 2 3

In Problems 3.9–3.10, assume that the answer to a particular SAT grid-in question is $\frac{6}{23}$.

3.10 Record the answer as a decimal in the grid below.

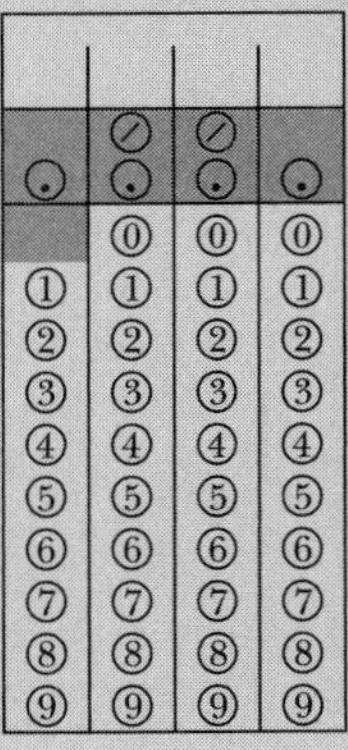

To compute the decimal equivalent of the fraction 6/23, use your calculator to divide 6 by 23.

$$6 \div 23 \approx 0.26086956522$$

You should include as many digits as possible and omit the placeholder 0 that appears left of the decimal point. The decimal point in the answer will occupy the leftmost column of the answer grid, leaving you space for three more digits in your answer—you can only bubble through the thousandths place.

If you choose to round the answer to the thousandths place, you would bubble .261. If, however, you choose to truncate the answer after those three digits, you should bubble .260. Do not omit that final 0 if you truncate, or you are suggesting that 6/23 is exactly equal to 0.26, which is not true. Both of the answers below are correct.

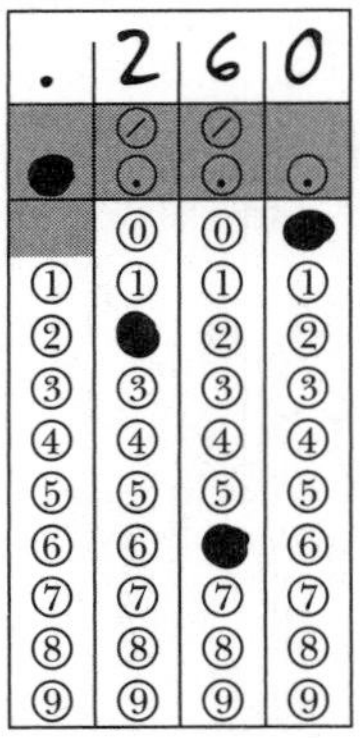

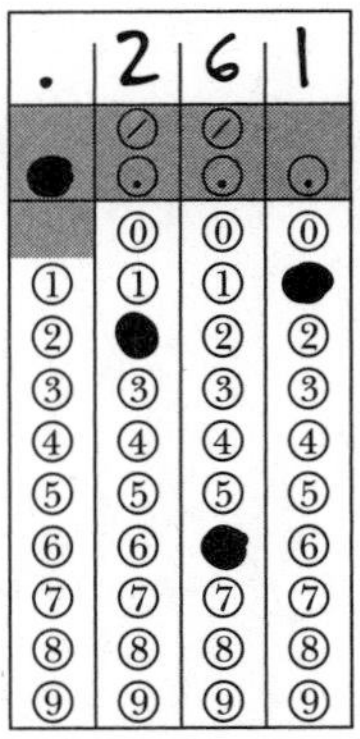

Remember, you want your answer to be as accurate as possible in the limited space provided. In this case, leaving the final zero out of your answer makes it less accurate.

Note: In Problems 3.11–3.12, assume the answer to a particular SAT grid-in question is $\frac{8}{10}$.

3.11 How many different responses in fraction form are possible?

You do not have to reduce fractions to lowest terms, so the answer 8/10 is correct. However, you can reduce the fraction if you prefer, so 4/5 is also correct. Therefore, there are two possible correct answers in fraction form. Remember that the step of reducing, while relatively simple when compared to the other mathematical tasks required of you during the SAT, is another place for errors to creep into your work. Why take the risk of simplifying incorrectly and losing credit for a correct answer? For this reason, some students always record their answers as decimals, which is a great idea.

If you absolutely insist on reducing fractions, did you know that most graphing calculators are equipped to reduce fractions for you? For example, on the TI-82, TI-83, and TI-84, you can type 8/10, press the "MATH" button, and then select "▶FRAC"; the calculator will reduce 8/10 to lowest terms: 4/5.

Note: In Problems 3.11–3.12, assume the answer to a particular SAT grid-in question is $\frac{8}{10}$.

3.12 How many different correct decimal responses are possible?

Note that $8 \div 10 = 0.8$, so there is only one decimal answer you can bubble to get this question correct: .8. Only two columns are needed, one for the decimal point and one for the digit 8. As Problem 3.1 explains, you can choose to left-justify, right-justify, or center your answer—that does not matter.

3.13 If a particular SAT grid-in question has an answer of $\frac{161}{56}$, how many different correct responses are possible?

The answer 161/56 will not fit into the four columns you are allotted, so in order to record your answer as a fraction, you must reduce to lowest terms. Divide the numerator 161 and denominator 56 by the greatest common factor, 7.

Or use your calculator to reduce the fraction for you, as explained in Problem 3.11.

$$\frac{161}{56} = \frac{161 \div 7}{56 \div 7} = \frac{23}{8}$$

The reduced fraction 23/8 will fit into the answer grid. To calculate the equivalent decimal, divide 23 by 8 (or divide 161 by 56).

$$\frac{161}{56} = \frac{23}{8} = 2.875$$

The answer grid does not have space for all four digits and the decimal point, so you can either truncate the decimal or round it; 2.87 and 2.88 are both correct answers. Therefore, there are three possible correct responses: 23/8, 2.87, and 2.88.

3.14 If a particular SAT grid-in question has a mixed number answer of $1\frac{2}{7}$, how many different responses in fraction form are possible?

The answer is a mixed number because it contains a whole number part (1) and a fractional part (2/7). The SAT does not understand mixed number answers, so the following answer is ***not correct***.

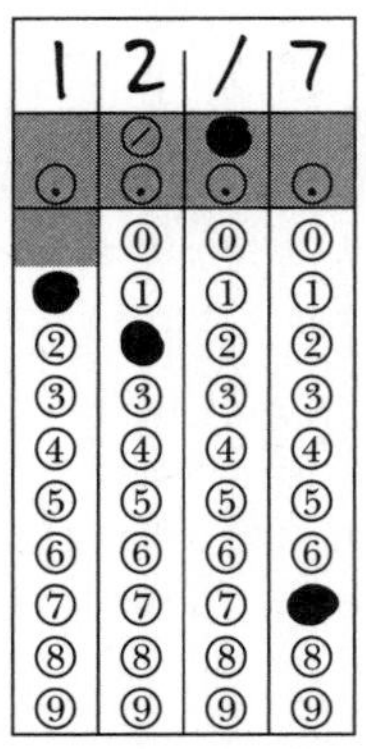

Incorrect response

The SAT interprets the answer above as the improper fraction $\frac{12}{7}$, not $1\frac{2}{7}$. If you wish to record your answer in fraction form, you need to convert the mixed number to an improper fraction by multiplying the denominator (7) by the whole number (1), adding the numerator (2), and then dividing that result by the denominator (7), as demonstrated below.

$$1\frac{2}{7}=\frac{7\cdot 1+2}{7}=\frac{9}{7}$$

There is only one correct answer in fractional form: 9/7.

Although this problem specifically asks you for an answer in fraction form, you could always convert the mixed number to a decimal. Note that $9 \div 7 \approx 1.28571429$, so the truncated answer 1.28 and the rounded answer 1.29 are also correct.

Practice Problems

Gird yourself for grids

3.15 Answer the following SAT grid-in question and record your response on the grid below.

If $3(4x-1)-5x=19$, what is the value of x?

Solve the equation for x by distributing 3 through the quantity $(4x-1)$, combining like terms, and isolating x on the left side of the equation.

$$\begin{aligned} 3(4x-1)-5x&=19\\ 3(4x)+3(-1)-5x&=19\\ 12x-3-5x&=19\\ 7x-3&=19\\ 7x&=22\\ x&=\frac{22}{7} \end{aligned}$$

Note that $22/7\approx 3.\overline{142857}$, so both of the following answers are correct.

3.16 Answer the following SAT grid-in question and record your response on the grid below.

If $x = 5\frac{2}{3}$, $y = 3\frac{1}{2}$, and $z = 4y + 5x$, what is the value of z?

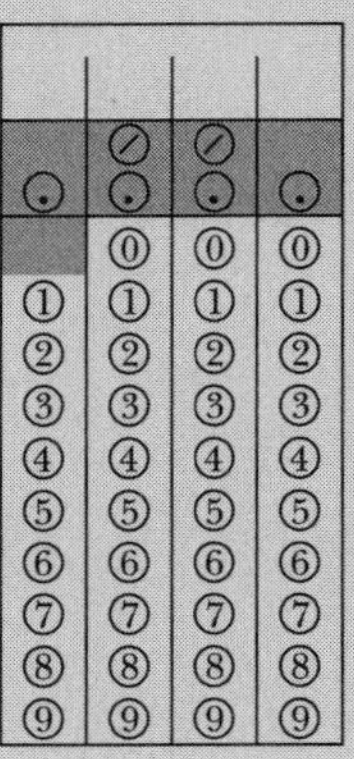

In a traditional math classroom, you would rewrite the mixed numbers as improper fractions, plug x and y into the expression $4y + 5x$, generate common denominators, add the fractions, and then reduce. This takes a lot of time—time you may not have to spare on the SAT. Embrace the time-saving power of your calculator!

Your graphing calculator may differ slightly, but to evaluate the expression $4y + 5x$ for the given values of x and y, you type the following expression on a Texas Instruments calculator:

$$4(3 + 1/2) + 5(5 + 2/3)$$

The result is 127/3, or $42.\bar{3}$. The fraction will not fit into the grid, so you have to bubble the answer in decimal form, as demonstrated below.

3.17 Answer the following SAT grid-in question and record your response on the grid below.

If $(2x + 3)(3x - 1)(8x - 11) = 0$, what is one possible value of x?

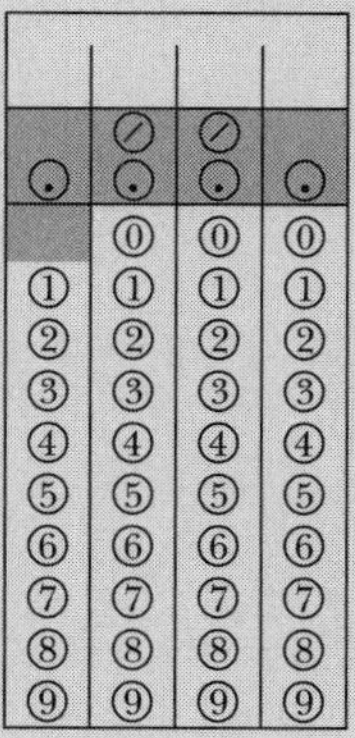

The three quantities in the equation $(2x+3)$, $(3x-1)$, and $(8x-11)$ have a product of 0, so the equation is true if any of those quantities is equal to 0. Set each equal to 0 individually and solve the equations.

$$\begin{array}{lll} 2x+3=0 & 3x-1=0 & 8x-11=0 \\ 2x=-3 & 3x=1 & 8x=11 \\ x=-\frac{3}{2} & x=\frac{1}{3} & x=\frac{11}{8} \end{array}$$

There are three values of x that satisfy the equation, but one of those values is negative ($x = -3/2$), so it cannot be recorded in the answer grid (as Problem 3.4 explains). To answer this question correctly, you should bubble either of the remaining solutions ($x = 1/3$ or $x = 11/8$) in fraction or decimal form. All five of the following responses earn full credit.

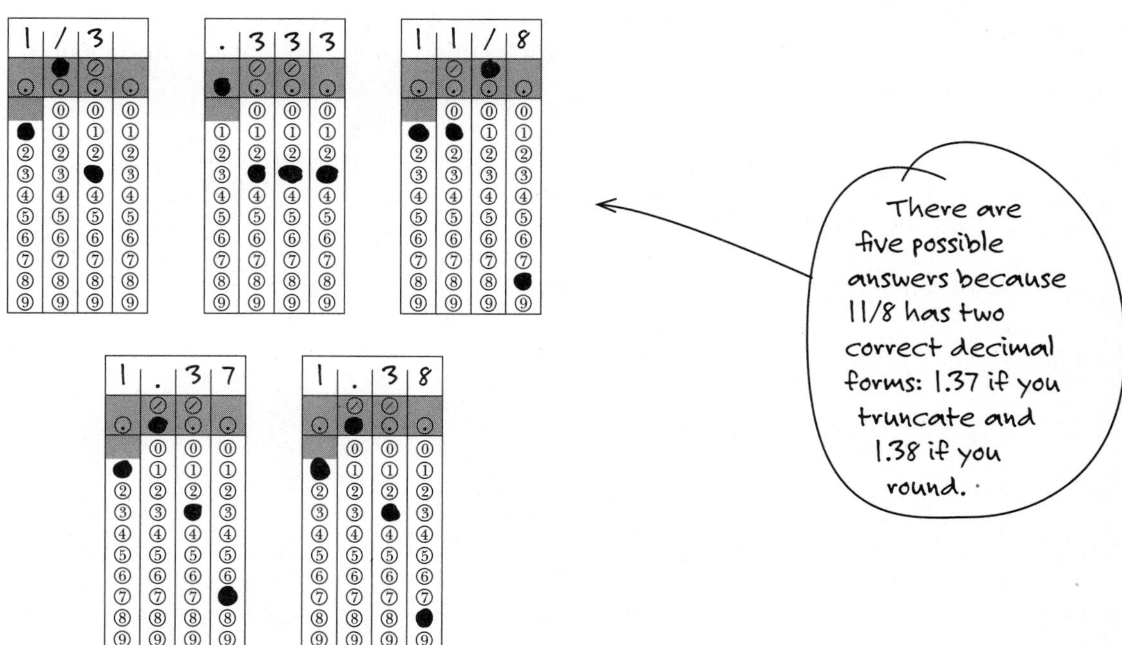

1/3
.333
11/8
1.37
1.38
There are five possible answers because 11/8 has two correct decimal forms: 1.37 if you truncate and 1.38 if you round.

3.18 Answer the following SAT grid-in question and record your response on the grid below.

$$\begin{cases} 8x+27y=-8 \\ 4x-9y=1 \end{cases}$$

If (x, y) is the solution to the above system of equations, what is the value of $\frac{y}{x}$?

	⊘	⊘	
⊙	⊙	⊙	⊙
	0	0	0
1	1	1	1
2	2	2	2
3	3	3	3
4	4	4	4
5	5	5	5
6	6	6	6
7	7	7	7
8	8	8	8
9	9	9	9

This question requires a little work, because you need to solve the system of equations. The quickest approach is to multiply the bottom equation by –2 and add the equations together to calculate y. This is called the elimination method, and it is explained in Problems 6.61–6.64.

$$\begin{cases} 8x+27y=-8 \\ -2(4x-9y=1) \end{cases} \rightarrow \begin{cases} 8x+27y=-8 \\ -8x+18y=-2 \end{cases} \rightarrow \begin{aligned} 45y&=-10 \\ y&=-\frac{10}{45} \\ y&=-\frac{2}{9} \end{aligned}$$

To calculate x, substitute $y = -2/9$ into one of the original equations of the system. In the solution below, $y = -2/9$ is substituted into $4x - 9y = 1$.

$$\begin{aligned} 4x-9\left(-\frac{2}{9}\right)&=1 \\ 4x+\frac{18}{9}&=1 \\ 4x+2&=1 \\ 4x&=-1 \\ x&=-\frac{1}{4} \end{aligned}$$

The question asks you to compute y/x. Use your calculator to evaluate the expression or apply the technique described in Problem 4.14 to simplify the complex fraction.

$$\frac{y}{x} = \frac{-2/9}{-1/4}$$
$$= \frac{8}{9}$$
$$= 0.\overline{8}$$

All three of the responses below are correct.

8/9

.888

.889

Part Two

MATH REVIEW AND PRACTICE

Dust off the rust and sharpen your skills

In Part I, you get to know the SAT a little better, and you explore some strategies for the different kinds of questions you will face on test day. Now, in Part 2 of the book, it is time to dig down into the math to make sure none of the mathematics takes you by surprise. Over the course of the next eight chapters, you will step your way through every topic the College Board identifies as "fair game" for SAT questions.

Work through the even chapters in this part of the book as though you were taking private lessons with a math tutor. Do not be afraid if the problems confuse or perplex you—the answer explanations will teach you not only how to solve them but also how to understand the underlying concepts better. Some of these questions will feel like SAT questions, and others will feel like problems you would get for homework in an algebra or geometry class. Go with the flow—they are all designed to maximize your understanding of the broadest possible range of topics without wasting your time.

The odd chapters are your chance to apply the skills you've just reviewed to SAT-style questions, both multiple-choice and grid-ins. Those chapters are organized a little differently—the answer explanations are located at the end of the chapter instead of immediately following each question. This will help you stay disciplined and curb any temptation to peek at the answers. In addition, the explanations will focus on question-specific tips and strategies to help you get the answer as cleverly and efficiently as possible.

If you need to focus only on specific SAT topics, feel free to jump right to them. Otherwise, work your way through the book in order, one question at a time, one chapter at a time. You may not know what you need to review until you try problems and get stuck, so don't skip any! If you feel a little nervous or overwhelmed, that's okay! This book doesn't assume you eat, sleep, and breathe math. It will explain everything in detail, so all you need to bring to the table is a willingness to work hard—the rest comes naturally. Take a deep breath, turn the page, and take the first step toward a better math score!

Chapter 4

MATH REVIEW: NUMBERS AND OPERATIONS

22% of the SAT

A full quarter of the SAT is dedicated to a set of skills and concepts you could loosely define as "pre-algebra, arithmetic, and logical thinking." Although nothing in this chapter requires an engineering degree to complete, the SAT is very good at making easy and medium questions sound difficult. Why? For one thing, test writers do not shy away from precise mathematical vocabulary. Once you understand that a complex-looking phrase like "consecutive integers" translates into "numbers in a row that do not contain fractions," the problems are a little less intimidating.

Think of this chapter as a social event, and think of the various content areas as old friends you remember fondly or acquaintances you barely remember. Who knows? You may even bump into a few strangers you have never met. Take your time and mingle with all of the problems, getting to know them better, so that when you run into each other again on test day, you are ready.

Do you ever find yourself lamenting "I don't know what this problem is asking, let alone how to even start it"? Don't let that panic and stress overwhelm you. You don't need to be a math genius to do well on the SAT—you just need to develop your critical thinking skills, and now's the perfect time to do so!

Don't skip right to the answers if you get stuck on these problems. Allow yourself to squirm, scratch your chin in deep thought, or go take a walk for inspiration. You're not under any sort of time limit, and the more time you spend wrestling with these problems now, the more confident you will be on test day.

Integers, Squares, Square Roots, and Fractions

Get started with the basics

4.1 In the chart below, *even* refers to an even integer and *odd* refers to an odd integer. Complete the right column of each row, classifying the sum or product indicated. For example, the first row states that *even* + *even* = *even*, because the sum of two even integers is always even.

even + *even* =	*even*
odd + *odd* =	
odd + *even* =	
even · *even* =	
odd · *odd* =	
odd · *even* =	

First of all, an integer is a positive number than includes no fractional part. In other words, the integers include the group of numbers called the counting numbers or the natural numbers (1, 2, 3, 4, 5, etc.), the opposites of the counting numbers (–1, –2, –3, –4, –5, etc.), and 0. Even integers are divisible by 2, but odd numbers are not:

even integers: ..., −10, −8, −6, −4, −2, 0, 2, 4, 6, 8, 10, ...

odd integers: ..., −9, −7, −5, −3, −1, 1, 3, 5, 7, 9, ...

You should select actual even and odd numbers to plug into the formulas, turning "odd + even" into something like "3 + 4." This is the DIY technique described in Problem 2.6.

Odd and even integers follow predictable patterns when you add and multiply them. You do not need to memorize these patterns, because you can figure them out at any time by selecting even and odd integers and testing them. For example, if you add the odd integers 3 and 7, the result is the even integer 10. Therefore, *odd* + *odd* = *even*. Multiplying the odd integer 1 by the odd integer 5 produces the odd integer 5, so *odd* · *odd* = *odd*. Repeat this process for each row to generate the solution table below.

even + *even* =	*even*
odd + *odd* =	*even*
odd + *even* =	*odd*
even · *even* =	*even*
odd · *odd* =	*odd*
odd · *even* =	*even*

4.2 Assuming x is an even integer, answer the following questions:

a. What is the next consecutive integer, and is that integer even or odd?

b. What are the next two consecutive even integers larger than x?

c. What are the two largest odd integers that are less than x?

Like in Problem 4.1, assigning a value to the variable is helpful. For example, $x = 4$ is an even integer. Use this value to answer each of the questions.

If you want to jump forward or backward one consecutive integer, add or subtract 1. If you want to jump forward or backward to the next consecutive integer with the same odd/even classification, add or subtract 2.

a. The next consecutive integer is $x + 1$, exactly one greater than x. If $x = 4$, then the next consecutive integer is $4 + 1 = 5$, an odd number. Consecutive integers alternate between even and odd.

b. If x is an even integer, the next consecutive integer is $x + 2$, exactly two greater than x. To generate the second consecutive even integer, add 2 once again: $(x + 2) + 2 = x + 4$. If $x = 4$, then $x + 2 = 6$ and $x + 4 = 8$ are the next two consecutive even integers.

c. If you are identifying integers less than x, you will have to subtract from x. The largest odd integer less than x is the integer that immediately precedes it: $x - 1$. If $x = 4$, then the largest odd integer that is less than x is $x - 1 = 3$. To find the next odd integer, subtract 2 more: $(x - 1) - 2 = x - 3$. Once again, if $x = 4$, then $x - 3 = 1$ is the second largest odd integer that is less than x.

4.3 If the sum of three consecutive odd integers is 81, what is the product of the largest and smallest of those three integers?

Assume x represents the smallest of the three consecutive odd integers. As Problem 4.2 explains, the integers alternate even and odd. If x is an odd integer, then you have to skip ahead two integers to get to the next odd integer $(x + 2)$ and then skip ahead two more to get to the odd integer after that: $(x + 4)$. Therefore, the three integers are x, $x + 2$, and $x + 4$. The problem states that the sum of those numbers is 81. In other words, if you add x, $x + 2$, and $x + 4$, the answer is 81.

$$\begin{array}{ccccccc} \text{smallest integer} & + & \text{next odd integer} & + & \text{next odd integer} & = 81 \\ x & + & x+2 & + & x+4 & = 81 \end{array}$$

Combine like terms on the left side of the equation: $x + x + x = 3x$ and $2 + 4 = 6$.

$$3x + 6 = 81$$

Solve the equation by subtracting 6 from both sides and then dividing both sides by 3.

$$\begin{aligned} 3x &= 81 - 6 \\ 3x &= 75 \\ x &= \frac{75}{3} \\ x &= 25 \end{aligned}$$

Remember, x represents the smallest of the three consecutive odd integers, so the full list is 25, 27, and 29. The problem asks you to calculate the product of the largest and smallest of those integers: $25 \cdot 29 = 725$.

You could apply the DIY approach and just select values for x, y, and z from the beginning, such as x = 3, y = 4, and z = 6. However, in this chapter, you focus on understanding the math happening behind the scenes. In other words, this is a "put on your math thinking cap" chapter, not a "find the shortcut" chapter.

4.4 If x and y are consecutive integers and z is an even number, is $z(x + y)$ even or odd?

The problem states that z is an even number, so replace z in the expression with "even number."

$$z(x + y) = (\text{even number}) \cdot (x + y)$$

Consecutive integers alternate between odd and even, so if x and y are consecutive integers, one of those integers is odd and one is even. According to the chart in Problem 4.1, odd + even = odd, so $x + y$ is an odd number.

$$\begin{aligned} z(x+y) &= (\text{even number})(x+y) \\ &= (\text{even number})(\text{odd number}) \end{aligned}$$

Once again refer to the chart in Problem 4.1: An even number multiplied by an odd number produces an even number. Therefore, $z(x + y)$ is even.

4.5 If n is a positive integer and $n \div 5$ has a remainder of 3, what is the remainder of $(n + 8) \div 3$?

You are not given a value for integer n, but do not get hung up on trying to figure out what n may or may not be equal to. Remainder problems like this can be solved using a purely visual approach that not only gives you the correct answer, but helps you understand *why* the answer is correct.

Imagine you have one big pile of n pieces of candy. You are asked to divide that candy into groups of 5 pieces, but when you do so, you have 3 pieces left over. This is illustrated in the following diagram; pieces of candy are represented by small squares, and full groups are circled. In this drawing, you were able to make 4 piles of candy, each consisting of 5 pieces, but the number of piles is not important. Because you do not know how many pieces n you started with, focus not on the number of complete piles but what the incomplete pile represents—the remaining, 3 lonely, leftover pieces.

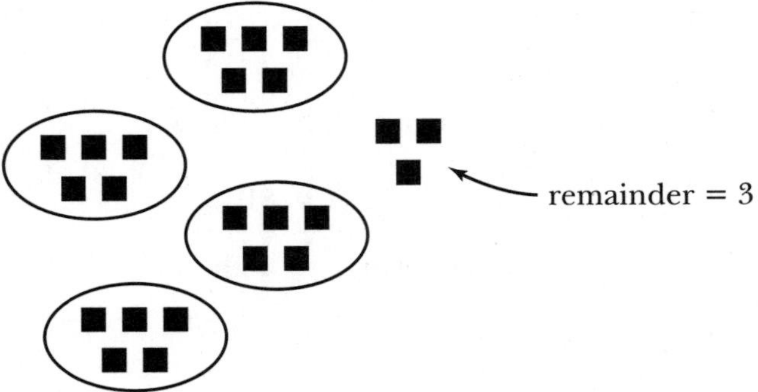

If you add 8 pieces to your original stash of candy, for a total of $n + 8$ pieces, how does this affect the remainder? The next diagram illustrates those additional 8 pieces as white squares.

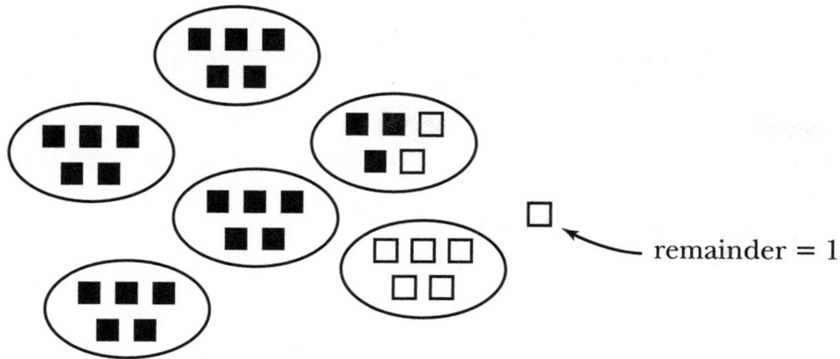

Notice that two of the squares complete the group of 3 squares that once was the remainder. Another 5 squares form a complete, 5-piece pile of candy, and only 1 piece of candy is left out in the cold. The new remainder is 1.

4.6 Simplify the expression: $\sqrt{144}+2\sqrt{36}$.

The numbers 144 and 36 are called "perfect squares" because they can be produced by multiplying an integer times itself: $12 \cdot 12 = 144$ and $6 \cdot 6 = 36$. You can also write these using exponents: $12^2 = 144$ and $6^2 = 36$. The small floating numbers—the exponents—explain how many times a number should be multiplied times itself. Therefore, $12^2 = 12 \cdot 12$, and $12^5 = 12 \cdot 12 \cdot 12 \cdot 12 \cdot 12$.

Make sure to memorize the first 15 perfect squares; they are listed below. You will see squared values throughout the SAT, and the presence of these numbers is usually a clue that you need to use a formula that contains a squared value in it.

$1^2=1$	$2^2=4$	$3^2=9$	$4^2=16$	$5^2=25$
$6^2=36$	$7^2=49$	$8^2=64$	$9^2=81$	$10^2=100$
$11^2=121$	$12^2=144$	$13^2=169$	$14^2=196$	$15^2=225$

The symbols that look like checkmarks in this problem are square roots, and they are the opposites of squares. If $6^2 = 36$ (which is read "six squared equals 36"), then $\sqrt{36}=6$ (which is read "the square root of 36 equals 6"). Similarly, $\sqrt{144}=12$.

$$\begin{aligned}\sqrt{144}+2\sqrt{36} &= 12+2(6)\\ &= 12+12\\ &= 24\end{aligned}$$

You have to multiply before you add. In this problem, you have to multiply 2(6) to get 12 (instead of adding 12 + 2 and then multiplying by 6). Have you heard of the phrase "Please Excuse My Dear Aunt Sally," which helps you remember the correct order of operations? It stands for (P)arentheses first, then (E)xponents, followed by (M)ultiplication and (D)ivision left-to-right in the same step, and then finally (A)ddition and (S)ubtraction left-to-right in the same step.

4.7 Rewrite the expression as an integer: $64^{3/2}$.

If an integer is raised to a rational (fractional) exponent, you can rewrite it as a radical expression. The denominator tells you what kind of radical will be used. In this case, the denominator is 2, so you should use a square root. The numerator (in this case 3) tells you the power to which you should raise the square root. Therefore, $64^{3/2}$ is equal to the square root of 64 raised to the third power.

If the denominator were 3, you'd use a cube root: $\sqrt[3]{}$. Cube roots undo perfect cubes, in the same way that square roots undo perfect squares.

$$64^{3/2} = \left(\sqrt{64}\right)^3$$

Notice that 64 is a perfect square: $8^2 = 64$. Therefore, $\sqrt{64} = 8$. Substitute this value into the expression and simplify.

$$\begin{aligned} 64^{3/2} &= \left(\sqrt{64}\right)^3 \\ &= (8)^3 \\ &= 8 \cdot 8 \cdot 8 \\ &= 512 \end{aligned}$$

4.8 In the expression $\sqrt{x^3}$, x represents a positive integer less than 10. For how many values of x is $\sqrt{x^3}$ also a positive integer?

If x is a positive integer less than 10, then $x = 1, 2, 3, 4, 5, 6, 7, 8$, or 9. The question is referencing the cubed values of each x, the value when x is raised to the third power. List those nine cubed values.

$$1^3 = 1,\ 2^3 = 8,\ 3^3 = 27,\ 4^3 = 64,\ 5^3 = 125,\ 6^3 = 216,\ 7^3 = 343,\ 8^3 = 512,\ 9^3 = 729$$

The problem is asking you to determine how many of those nine values are perfect squares. Take the square root of each (using your calculator when the numbers get large); if that square root is an integer (instead of a lengthy decimal), it represents a value of x for which $\sqrt{x^3}$ is also a positive integer.

The correct answer is three. For three values of x (1, 4, and 9), the square root of x^3 is also an integer: $\sqrt{1} = 1$, $\sqrt{64} = 8$, and $\sqrt{729} = 27$.

4.9 Simplify the expression: $\left(x^2\right)^3 + \left(x^2 \cdot x^3\right) + \dfrac{x^4}{x^{-2}}$.

This problem contains three separate laws of exponents, all wrapped up in one expression. Simplify one term at a time:

- $\left(x^2\right)^3$: If something raised to a power (in this case x is raised to a power of 2) is then raised to another power (in this case 3), you should multiply the powers together: $\left(x^2\right)^3 = x^{2 \cdot 3} = x^6$.
- $(x^2 \cdot x^3)$: If two exponential expressions with the same base (in this case x) are multiplied, the product is equal to that common base raised to the sum of the powers: $x^2 \cdot x^3 = x^{2+3} = x^5$.

- $\frac{x^4}{x^{-2}}$: If two exponential expressions with the same base (in this case x) are divided, the quotient is equal to the common base raised to the difference of the powers. In other words, subtract the exponent in the denominator from the exponent in the numerator.

$$\frac{x^4}{x^{-2}} = x^{4-(-2)} = x^{4+2} = x^6$$

Replace each term in the original expression with the simplified values above.

$$\begin{aligned}(x^2)^3 + (x^2 \cdot x^3) + \frac{x^4}{x^{-2}} &= x^6 + x^5 + x^6 \\ &= 2x^6 + x^5\end{aligned}$$

Notice that you can add like terms x^6 and x^6 to get $2x^6$. You cannot combine $2x^6$ and x^5, because they are not like terms.

Like terms contain the same variable raised to the same exponent.

4.10 If $8^5 = 4^{3n}$, what is the value of n?

Solving exponential equations is much easier when the terms contain the same base. Unfortunately, the bases of the exponential expressions in this equation are 8 and 4. Happily, most SAT equations involving exponents *can* be written in terms of the same base, and that is true here. Notice that both bases, 8 and 4, are equal to 2 raised to a power: $8 = 2^3$ and $4 = 2^2$.

$$\begin{aligned}8^5 &= 4^{3n} \\ (2^3)^5 &= (2^2)^{3n}\end{aligned}$$

Now apply the first exponential rule listed in Problem 4.9: If something is raised to a power (such as 2 raised to the power of 3) and that expression is raised to another power (like 5), multiply the powers together. Do this on both sides of the equation.

$$\begin{aligned}2^{3\cdot5} &= 2^{2\cdot3n} \\ 2^{15} &= 2^{6n}\end{aligned}$$

When $2^x = 2^y$, then $x = y$. Think about it: 2 can only equal 2 if it is raised to the same power.

On the left side of the equation, 2 is raised to the fifteenth power. That means 2 must be raised to the fifteenth power on the right side of the equation as well. In other words, $6n = 15$. Solve for n.

$$\begin{aligned}6n &= 15 \\ n &= \frac{15}{6} \\ n &= \frac{5}{2}\end{aligned}$$

4.11 Reduce the fraction to lowest terms and identify the greatest common factor: $\frac{105}{147}$.

To reduce a fraction, you need to identify any factors shared by the numerator and denominator. What does that mean? If a number divides evenly into 105 and 147, then it is considered a common factor. You can divide both 105 and 147 by that common factor, and the resulting fraction (even though it looks different) is equivalent to the original fraction.

If you do not know any of the divisibility tricks (such as how to tell at a glance if 105 is divisible by 3), you should not waste time memorizing them. Use your calculator to divide the numerator and denominator by small integers until you identify something that divides evenly into both. For example, both are divisible by 3.

$$\frac{105}{147} = \frac{105 \div 3}{147 \div 3} = \frac{35}{49}$$

Now look for common factors of 35 and 49. Both are divisible by 7.

$$\frac{35 \div 7}{49 \div 7} = \frac{5}{7}$$

Because 5 and 7 do not share any common factors (except for the factor 1, which divides evenly into everything so it doesn't really count), you have expressed the original fraction in lowest terms.

The greatest common factor of 105 and 147 is equal to the product of the numbers you divided out of the fraction. Remember, you first divided the numerator and denominator by 3 and then by 7, so the greatest common factor is $3 \cdot 7 = 21$.

4.12 Simplify the expression: $\frac{1}{4} + \frac{5}{2} - \frac{2}{3}$.

Although you can compute this with your calculator, you should review how to calculate it by hand. You can only add or subtract fractions that have a common (equal) denominator, but these fractions do not. To identify the least common denominator, work with the largest denominator, in this case 4. Do the other denominators divide into 4 evenly? No. $4 \div 2$ does divide evenly, but $4 \div 3$ does not. Therefore, 4 is not the least common denominator.

Your next step is to multiply the largest denominator by 2. In this problem, $4 \cdot 2 = 8$. Do the other denominators in the problem (2 and 3) divide evenly into 8? Unfortunately, once again the answer is no. That troublesome 3 still leaves a remainder when you divide it into 8.

Keep pressing on. Now multiply the largest denominator by 3. You get $4 \cdot 3 = 12$. Finally, the other denominators divide in evenly: $12 \div 2 = 6$ and $12 \div 3 = 4$. That means 12 is the least common denominator. To write each fraction in terms of the least common denominator, figure out what number each denominator should be multiplied by in order to get a product of 12. Then multiply not only the denominator by that number but the numerator as well.

$$\frac{1}{4}+\frac{5}{2}-\frac{2}{3}=\frac{1}{4}\left(\frac{3}{3}\right)+\frac{5}{2}\left(\frac{6}{6}\right)-\frac{2}{3}\left(\frac{4}{4}\right)$$
$$=\frac{3}{12}+\frac{30}{12}-\frac{8}{12}$$
$$=\frac{3+30-8}{12}$$
$$=\frac{25}{12}$$

4.13 Express the product as an improper fraction reduced to lowest terms: $4\frac{2}{3}\cdot 5\frac{1}{2}$.

The problem asks you to write your answer as an improper fraction, so convert the mixed numbers to improper fractions before you multiply. To convert a mixed number into an improper fraction, apply the formula $A\frac{b}{c}=\frac{cA+b}{c}$. In other words, multiply the denominator c by the whole number A and add the numerator b to generate the numerator of the improper fraction. The denominator of the improper fraction is c; it matches the denominator of the mixed number.

"Improper fractions" have numerators that are larger than their denominators, but there's nothing mathematically improper about them. In fact, most math teachers prefer them to mixed numbers.

$$4\frac{2}{3}\cdot 5\frac{1}{2}=\frac{3\cdot 4+2}{3}\cdot\frac{2\cdot 5+1}{2}$$
$$=\frac{12+2}{3}\cdot\frac{10+1}{2}$$
$$=\frac{14}{3}\cdot\frac{11}{2}$$

To multiply two fractions, multiply the numerators and denominators separately.

$$=\frac{14\cdot 11}{3\cdot 2}$$
$$=\frac{154}{6}$$

Divide the numerator and denominator by 2, the greatest common factor, to reduce the improper fraction to lowest terms.

$$=\frac{154\div 2}{6\div 2}$$
$$=\frac{77}{3}$$

4.14 Simplify the complex fraction: $\dfrac{\frac{2}{9}}{3\frac{1}{4}}$.

A complex fraction is a fraction that, itself, contains a fraction in the numerator and/or the denominator. In this case, both the numerator and denominator are fractions. Before you simplify, use the technique described in Problem 4.13 to rewrite the mixed number in the denominator as an improper fraction.

$$\frac{\frac{2}{9}}{3\frac{1}{4}} = \frac{\frac{2}{9}}{\frac{4\cdot 3+1}{4}}$$
$$= \frac{\frac{2}{9}}{\frac{13}{4}}$$

A fraction is both a real number and a division problem—a quotient. For example, the fraction 1/2 is equivalent to the quotient 1 ÷ 2. By extension, this complex fraction is the quotient of two fractions: 2/9 ÷ 13/4. How do you divide by a fraction? Multiply by its reciprocal. In other words, dividing by 13/4 is the same as multiplying by 4/13.

The reciprocal of a fraction reverses the numerator and denominator. For example, the reciprocal of A/B is B/A.

$$= \frac{2}{9} \div \frac{13}{4}$$
$$= \frac{2}{9} \cdot \frac{4}{13}$$
$$= \frac{8}{117}$$

4.15 Express the product in scientific notation: $(2.9 \times 10^{-13})\ (8.45 \times 10^{7})$.

Scientific notation allows you to express large or small numbers without using tons and tons of zeros. This problem looks a lot uglier without scientific notation. To generate the decimal equivalent of 2.9×10^{-13}, you need to move the decimal point left 13 places (and insert place-holder zeros as needed. The number 8.45×10^{7} has a positive exponent, so to generate its decimal equivalent, you need to move the decimal 7 places to the right, again adding zeros as needed. Therefore, negative powers of 10 represent small numbers and positive powers of 10 represent large numbers.

$$2.9 \times 10^{-13} = 0.00000000000029$$
$$8.45 \times 10^{7} = 84{,}500{,}000$$

Happily, you can multiply and divide numbers while they are still in scientific notation. Multiply the decimal portions and the powers of 10 separately: (2.9)(8.45) = 24.505 and $(10^{-13})(10^{7}) = 10^{-13+7} = 10^{-6}$.

You may be tempted to conclude that the product is 24.505×10^{-6}, but a number in scientific notation has exactly one non-zero digit left of the decimal point. Therefore the final answer should have a decimal portion of 2.4505 instead of 24.505. You cannot just move that decimal point indiscriminately; the movement of the decimal should be reflected in the power of 10. In this case, the number you are representing (24.505) is larger than the number in your answer (2.4505), so the power of 10 should be larger as well—one larger because you move the decimal one place.

$$\begin{aligned} 24.505 \times 10^{-6} &= 2.4505 \times 10^{-6+1} \\ &= 2.4505 \times 10^{-5} \end{aligned}$$

Ratios and Proportions

A is to B as C is to D

4.16 A local specialty bookstore sells only novels and biographies. If $\frac{3}{8}$ of its 1,560 books are biographies, how many of its books are novels?

To calculate 3/8 of 1,560, multiply the values. There is no need to compute the answer by hand; use your calculator.

$$\frac{3}{8}(1{,}560) = 585$$

The bookstore carries 585 biographies. Subtract this number from the total books sold (1,560) to calculate the number of novels stocked at the bookstore.

$$1{,}560 - 585 = 975 \text{ novels}$$

You can also solve this problem by noticing that 5/8 of the books in the store must be novels if 3/8 are biographies, because $3/8 + 5/8 = 1$. Multiplying 1,560 by 5/8 also produces the correct answer of 975.

4.17 If the ratio of x to y is $3:5$, what is the ratio of $2x$ to $3y$?

The ratio $x:y$ can be written as the fraction x/y.

$$\frac{x}{y} = \frac{3}{5}$$

This equation, called a proportion, states that the ratio of x to y is equal to the ratio of 3 to 5. Your goal is to transform the left side of the proportion (the ratio $x:y$) into the ratio specified by the problem, $(2x)/(3y)$. Multiply the numerator x by 2, and multiply the denominator y by 3. In order to ensure the equation remains true, you must do the same thing on the right side of the equation—multiply that fraction's numerator by 2 and its denominator by 3.

$$\begin{aligned} \frac{x \cdot 2}{y \cdot 3} &= \frac{3 \cdot 2}{5 \cdot 3} \\ \frac{2x}{3y} &= \frac{6}{15} \end{aligned}$$

Reduce the fraction 6/15 to lowest terms.

$$\frac{2x}{3y}=\frac{2}{5}$$

The ratio of $2x$ to $3y$ is 2 to 5.

4.18 In a certain college course, all 20 students earned a semester grade of A, B, or C. If 5 students earned Cs and the ratio of As to Bs among the remaining student grades was 1 to 4, how many students earned a B?

If 5 of the 20 students earned a C, then $20-5=15$ students earned either an A or a B semester grade. The ratio of As to Bs is 1 to 4, so for every 1 student that earned an A, 4 students earned a B. If you let x equal the number of As, then $4x$ represents the number of Bs. This allows you to construct the equation below.

$$\begin{array}{ccccc}\text{total number of As} & + & \text{total number of Bs} & = & \text{total number of As and Bs}\\ x & + & 4x & = & 15\end{array}$$

Solve the equation $x+4x=15$ for x.

$$\begin{aligned}x+4x&=15\\5x&=15\\x&=\frac{15}{5}\\x&=3\end{aligned}$$

Remember that $x=3$ represents the number of As and $4x$ represents the number of Bs. Therefore, $4(3)=12$ students earned Bs.

4.19 If 160 hot dogs are required to feed 90 campers, what is the minimum number of hot dogs needed to feed 115 campers? Assume the rate of consumption is constant.

In other words, dividing the number of hot dogs by the number of campers gives you the same number of hot dogs per camper, regardless of how many campers there are. This is a clue that you will use a proportion to solve the problem, because a proportion sets two rates (or ratios) equal to each other.

Use a proportion to organize the information in the problem. In other words, create an equation that sets two ratios equal, as demonstrated below.

$$\frac{\text{number of hot dogs required}}{\text{number of campers}}=\frac{\text{number of hot dogs required}}{\text{number of campers}}$$

On the left side of the equation, state that 160 hot dogs feed 90 campers. On the right side of the equation, state that some number of hot dogs x is needed to feed 115 campers.

$$\frac{160}{90}=\frac{x}{115}$$

You solve a proportion by cross-multiplying. In other words, multiply the numerator of one fraction by the denominator of the other fraction and set those products equal.

$$\begin{aligned}160\cdot 115&=90\cdot x\\18{,}400&=90x\end{aligned}$$

Solve the equation for x.

$$\frac{18,400}{90} = x$$
$$204.\overline{4} = x$$

You need slightly more than 204 hot dogs to feed 115 campers, so 205 hot dogs are needed.

You can't buy 0.4 of a hot dog, so the answer is either 204 or 205. You need 204.44444444 hot dogs, so 204 hot dogs are not enough.

4.20 The ratio of Erin's current age to Sara's current age is 2:3. In three years, the sum of their ages will be 36. What is Sara's current age?

If e is Erin's current age and s is Sara's current age, then you can create the following proportion.

$$\frac{e}{s} = \frac{2}{3}$$

If you cross-multiply, you create an equation that compares their ages.

$$3e = 2s$$

For the moment, this is as far as you can go with this information. Now turn your attention to their future ages. In three years, Erin is $e + 3$ years old and Sara is $s + 3$ years old. The sum of their ages will be 36.

$$(e+3)+(s+3) = 36$$
$$e+s+3+3 = 36$$
$$e+s+6 = 36$$

Subtract 6 from both sides of that equation to simplify it.

$$e + s = 30$$

Solve this equation for e by subtracting s from both sides.

$$e = 30 - s$$

Now that you have calculated Erin's current age e in terms of Sara's current age s, substitute $e = 30 - s$ into the equation $3e = 2s$ that you generated earlier. Solve for s.

You are technically solving a system of equations, $3e = 2s$ and $e + s = 30$, using the substitution method. This might be a ratio/proportion problem, but you need to know some algebra to actually figure out the answer. If you need to review systems of equations, check out Problems 6.58–6.65.

$$3e = 2s$$
$$3(30-s) = 2s$$
$$90 - 3s = 2s$$
$$90 = 5s$$
$$\frac{90}{5} = s$$
$$18 = s$$

Sara's current age is $s = 18$.

4.21 At a certain birthday party, the ratio of boys to girls is 2:5. If there are 15 more girls than boys in attendance, what is the total number of boys and girls at the party?

If b represents the number of boys, then $b + 15$ represents the number of girls. Create a proportion stating that the ratio of boys to girls is 2 to 5.

$$\frac{\text{number of boys}}{\text{number of girls}} = \frac{2}{5}$$
$$\frac{b}{b+15} = \frac{2}{5}$$

Cross-multiply and solve for b.

$$\begin{aligned} b \cdot 5 &= (b+15)(2) \\ 5b &= 2b + 30 \\ 3b &= 30 \\ b &= \frac{30}{3} \\ b &= 10 \end{aligned}$$

There are $b = 10$ boys at the party and $b + 15 = 25$ girls at the party. A total of $10 + 25 = 35$ boys and girls are at the party.

4.22 13 is 40% of what number?

Most percentage problems can be solved with the following percentage formula:

$$\frac{\text{part}}{\text{whole}} = \frac{\text{percentage}}{100}$$

Write the percentage as a whole number, not a decimal. In this problem, the percentage is 40, not 0.4.

In this problem, you are told that 13 represents 40 percent of some larger number; 13 is some *part* of a larger *whole*, and 40 is the percentage.

$$\frac{13}{\text{whole}} = \frac{40}{100}$$

Cross-multiply and solve for the unknown value.

$$\begin{aligned} 13 \cdot 100 &= \text{whole}(40) \\ 1{,}300 &= \text{whole}(40) \\ \frac{1{,}300}{40} &= \text{whole} \\ 32.5 &= \text{whole} \end{aligned}$$

You conclude that 13 is 40% of 32.5. Note that 65/2, the rational (or fractional) form of 32.5, is also correct.

4.23 In a recent television talent show competition, viewers selected a winner by voting either for a country singer or for an opera singer. The ratio of the votes for the country singer to the total of 4.5 million votes cast was 8:13. What percentage of the votes (rounded to the nearest tenth) was cast for the opera singer?

If p represents the number of votes cast for the opera singer (in millions), then $4.5 - p$ represents the number of votes cast for the country singer (in millions). Create a proportion using the information given in the problem.

$$\frac{\text{number of votes cast for country singer}}{\text{total votes cast}} = \frac{8}{13}$$
$$\frac{4.5 - p}{4.5} = \frac{8}{13}$$

Cross-multiply and solve for p.

$$\begin{aligned}
13(4.5 - p) &= 8(4.5) \\
58.5 - 13p &= 36 \\
-13p &= -22.5 \\
p &= \frac{-22.5}{-13} \\
p &\approx 1.7307692 \text{ million votes} \\
p &\approx 1{,}730{,}769 \text{ votes}
\end{aligned}$$

The problem asks you to calculate the percentage of the total votes that were cast for the opera singer. Divide p by 4.5 million.

$$\begin{aligned}
\text{perecntage of votes for opera singer} &= \frac{\text{number of votes for opera singer}}{\text{total votes cast}} \\
&= \frac{1{,}730{,}769}{4{,}500{,}000} \\
&\approx 0.3846 \\
&\approx 38.5\%
\end{aligned}$$

4.24 The sales tax in a particular state is t percent. An electronics store in that state is discounting all of its items s percent off of ticketed prices. If you buy an item from that store with a ticketed price of p dollars, what is the price you will pay, including sales tax? Assume sales tax is collected on the discounted, not ticketed, price.

Before you begin thinking about the prices of the items, you should restate all of the percentages in this problem as decimals. Usually this is accomplished by moving the decimal point two places to the left; for example, you can calculate 45% of a value by multiplying that value by 0.45. However, you are not given a numeric value for t or for s. To express them as decimals, you need to divide by 100 (which effectively moves the decimal point two places to the left; for example, 45/100 = 0.45).

$$s\% = \frac{s}{100} \qquad t\% = \frac{t}{100}$$

If the ticketed price of the item is p, then the store's discount is $s/100$, or $s\%$, times that price. Calculate the new price by subtracting the discount from the original price.

$$\begin{aligned}\text{discounted price} &= \text{original price} - \text{discount}\\ &= p - \frac{s}{100}\cdot p\\ &= p - \frac{ps}{100}\end{aligned}$$

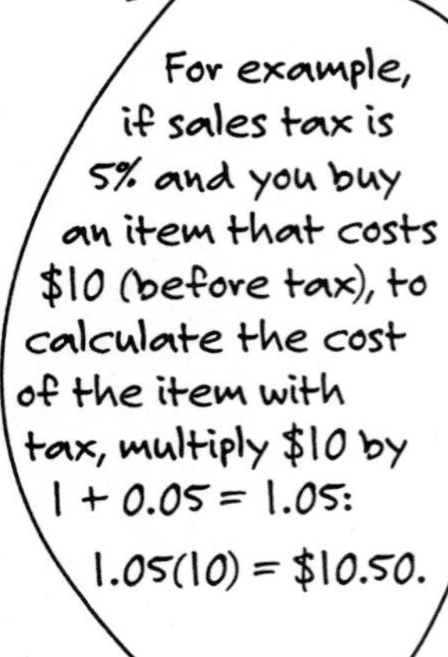

For example, if sales tax is 5% and you buy an item that costs $10 (before tax), to calculate the cost of the item with tax, multiply $10 by 1 + 0.05 = 1.05: 1.05(10) = $10.50.

The tax on the item is $t\%$ of the discounted price, calculated above. To calculate the price with tax, add $t/100$ (which is $t\%$ expressed as a decimal) to 1 and multiply by the discounted price.

$$\begin{aligned}\text{total cost with tax} &= (\text{discounted price})(1 + \text{sales tax percentage in decimal form})\\ &= \left(p - \frac{ps}{100}\right)\left(1 + \frac{t}{100}\right)\end{aligned}$$

If you use a least common denominator to express the fractions, it produces the following equivalent answer, which is equally correct.

$$= \left(\frac{100p - ps}{100}\right)\left(\frac{100 + t}{100}\right)$$

4.25 On a certain 50-question test, a student answers 80 percent of the first 35 questions correctly but only answers 60 percent of the last 15 questions correctly. What percentage of the 50 questions did the student answer correctly?

Calculate the number of correct answers in each part independently; in other words, calculate 80% of 35 and 60% of 15.

$$\begin{aligned}\text{total correct answers} &= \text{correct answers in first part} + \text{correct answers in second part}\\ &= (.80)(35) + (.60)(15)\\ &= 28 + 9\\ &= 37\end{aligned}$$

Thus, the student answered 37 of the 50 questions correctly, $37/50 = 0.74 = 74\%$ of the 50 test questions.

4.26 On the number line below, what is the coordinate of x?

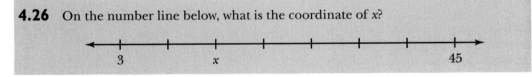

Given a number line, you can assume that the marks are evenly spaced—the distance from one mark to the next is a fixed distance. However, you cannot assume that the distance between each mark is equal to 1. Usually, your first goal in a number line problem is to identify the fixed distance between marks.

Let n represent the distance from one mark to the next. In this diagram, the coordinate of the first mark is 3, so the second mark has a coordinate of $3 + n$, the next mark has a coordinate of $3 + 2n$, etc. The final mark is seven units away from the first mark, so its coordinate is $3 + 7n$, as illustrated below.

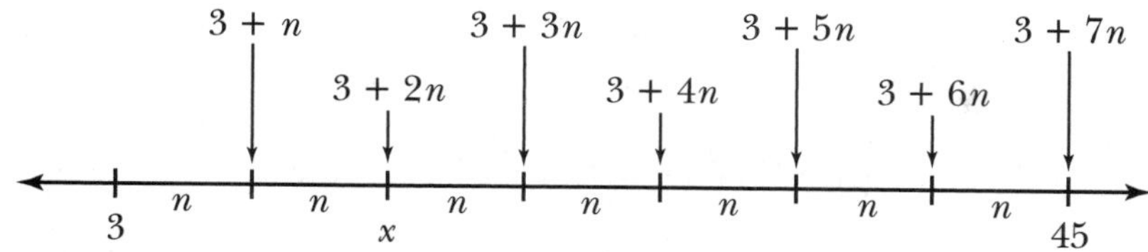

The problem states that the final coordinate is equal to 45, so $3 + 7n = 45$. Solve this equation for n.

$$\begin{aligned}3+7n&=45\\7n&=45-3\\7n&=42\\n&=\frac{42}{7}\\n&=6\end{aligned}$$

Each mark on the number line represents a distance of 6. Notice that x in the original problem is equal to $3 + 2n$.

$$\begin{aligned}x&=3+2n\\x&=3+2(6)\\x&=3+12\\x&=15\end{aligned}$$

The coordinate of x is 15; it is two marks right of 3 on the number line, and each of those marks represents a distance of 6.

Note: Problems 4.27–4.28 refer to the number line below.

4.27 What is the ratio of AF to EG?

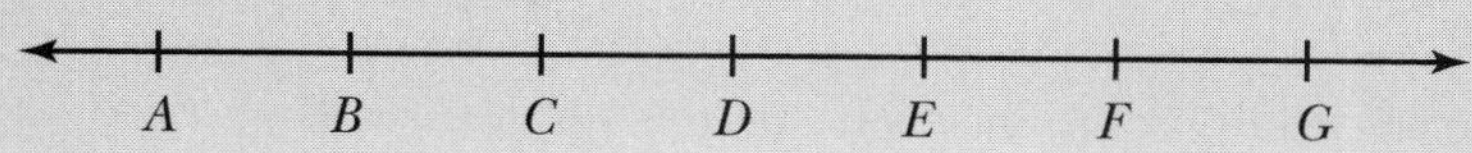

As Problem 4.26 explains, the marks on a number line are evenly spaced, and the same distance lies between any two adjacent marks. If you let n represent that unknown, fixed distance, you can express AF (the length of the segment beginning at A and ending at F) and EG in terms of n, as illustrated below.

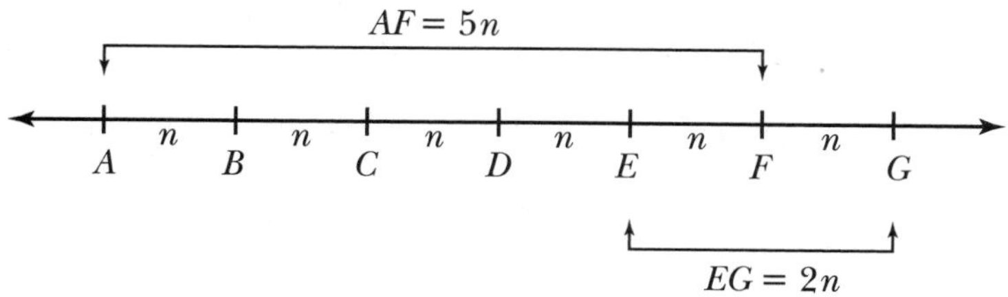

You do not know what n is, but you know that AF is equal to five of those unknown lengths. Similarly, the length EG is equal to two of those lengths. The ratio of AF to EG is equal to the ratio of $5n$ to $2n$. Express the ratio as a fraction and simplify to lowest terms, dividing the numerator and denominator by n.

$$\frac{AF}{EG} = \frac{5n}{2n} = \frac{5\cancel{n}}{2\cancel{n}} = \frac{5}{2}$$

The ratio of AF to EG is 5 to 2.

Note: Problems 4.27–4.28 refer to the number line in Problem 4.27.

4.28 If $BE + AF = 24$, what is the value of AG?

Let n represent the fixed distance between adjacent marks on the number line. The length BE is three units long, so its length is $3n$; AF is 5 units long, so its length is $5n$.

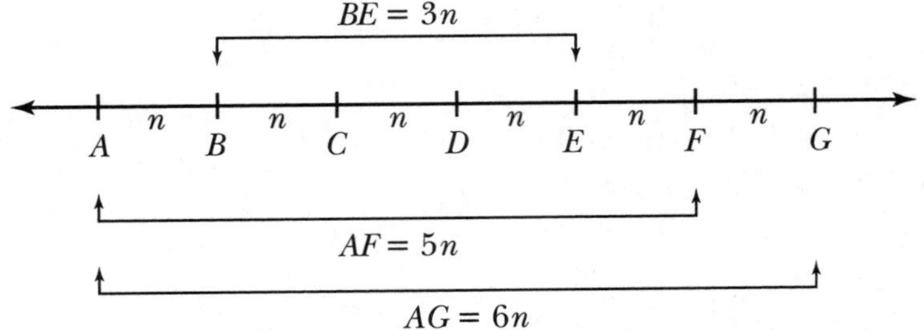

The problem states that $BE + AF = 24$. Substitute the values of BE and AF identified above into the equation and solve for n.

$$\begin{aligned} BE + AF &= 24 \\ 3n + 5n &= 24 \\ 8n &= 24 \\ n &= \frac{24}{8} \\ n &= 3 \end{aligned}$$

The length AG is 6 units, each with a length of $n = 3$.

$$AG = 6n = 6(3) = 18$$

Basic Number Theory

Factors, multiples, primes

4.29 List the positive prime numbers less than 30 and explain how to determine whether or not a number is prime.

A prime number is not divisible by any number other than itself and 1. Said another way, prime numbers have no factors other than 1 and the numbers themselves. Therefore, 5 is a prime number because only 1 and 5 divide evenly into 5, whereas 10 is not a prime number because four numbers divide into 10 evenly: 1, 2, 5, and 10.

X divides evenly into Y when Y ÷ X does not have a remainder.

Note that 1 is not considered a prime number, so the positive prime numbers less than 30 are 2, 3, 5, 7, 11, 13, 17, 19, 23, and 29.

4.30 What is the sum of the common factors of 36 and 54?

A factor is a number that divides evenly into another number. For example, 4 is a factor of 8 because 4 divides evenly into 8; the quotient $8 \div 4$ has no remainder. Common factors are factors that are shared by two or more numbers. In order to identify the common factors of 36 and 54, list all the factors of each number.

factors of 36: 1, 2, 3, 4, 6, 9, 12, 18, 36
factors of 54: 1, 2, 3, 6, 9, 18, 27, 54

The common factors are 1, 2, 3, 6, 9, and 18. Calculate the sum of the common factors to complete the problem.

$$1 + 2 + 3 + 6 + 9 + 18 = 39$$

4.31 If a is a multiple of x and b is a factor of x, can you conclude that a is a multiple of b? Why or why not?

If a is a multiple of x, then x times some number is equal to a. For example, 20 is a multiple of 10 because 10 times some number (in this case, 2) is equal to 20. Let n represent the unknown number, and you can deduce that $x \cdot n = a$.

If b is a factor of x, then b times some number is equal to x. For example, 2 is a factor of 10 because 2 times some number (in this case, 5) is equal to 10. Let p represent the unknown number: $b \cdot p = x$.

Substitute $x = b \cdot p$ into the equation $x \cdot n = a$.

$$x \cdot n = a$$
$$(b \cdot p) \cdot n = a$$
$$b \cdot p \cdot n = a$$

You can conclude that a is a multiple of b, because b times some number (in this case $p \cdot n$) is equal to a.

If you need further convincing that this is true, use the real values introduced above. If $a = 20$ is a multiple of $x = 10$ and $b = 2$ is a factor of $x = 10$, can you conclude that $a = 20$ is a multiple of $b = 2$? Yes, because 2 times some number (in this case, 10) is equal to 20.

Note: In Problems 4.32–4.34, you use the prime factorizations of two integers to calculate their greatest common factor and least common multiple.

4.32 Use a factor tree to identify the prime factorizations of 36 and 90.

A factor tree is a visual tool that helps you identify the prime factorization of a number. To construct a factor tree for 36, identify two numbers that have a product of 36. For example, $6 \cdot 6 = 36$. Those two factors of 6 are the first branches of the tree. From those numbers, additional branches grow. Because $2 \cdot 3 = 6$, branches of 2 and 3 extend from each 6.

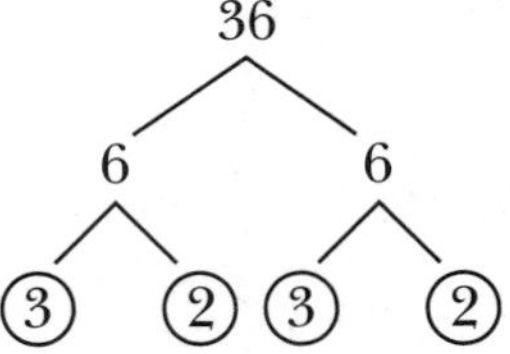

Each branch ends at a prime number. In the preceding diagram, each prime number is circled to indicate that no branches will extend below it. You conclude that the prime factorization of 36 is $2 \cdot 2 \cdot 3 \cdot 3$, or $2^2 \cdot 3^2$. Often, there are different ways you can start a factor tree. For example, rather than $6 \cdot 6$, you can begin with $9 \cdot 4$—a pair of numbers that also has a product of 36. Although the numbers will sprout from different branches of the tree, the final result is the same, $2^2 \cdot 3^2$, as illustrated below.

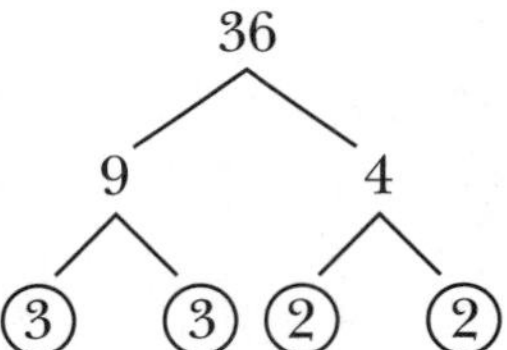

Now construct a factor tree for 90.

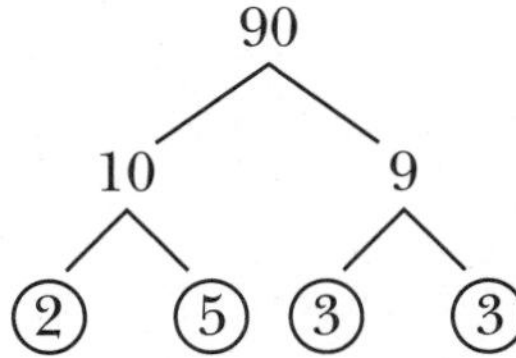

The prime factorization for 90 is $2 \cdot 3^2 \cdot 5$.

Note: In Problems 4.32–4.34, you use the prime factorizations of two integers to calculate their greatest common factor and least common multiple.

4.33 Use the prime factorizations you generated in Problem 4.32 to calculate the greatest common factor of 36 and 90.

According to Problem 4.32, $36 = 2^2 \cdot 3^2$ and $90 = 2 \cdot 3^2 \cdot 5$. To generate the greatest common factor (or GCF) of those numbers, first identify the prime factors that appear in both factorizations, in this case 2 and 3. Although 5 appears in the prime factorization of 90, it does not appear in the prime factorization of 36, so it is not a common factor of both.

Once you have identified the common prime factors, list each one raised to the lowest power from the factorizations. For example, 2^2 appears in the prime factorization of 36, and 2^1 (or simply 2) appears in the prime factorization of 90. The greatest common factor will contain 2 raised to the lower power: 2^1. The other common prime factor, 3, is squared in both factorizations. Thus, you should square 3 in the greatest common factor: 3^2.

Once you raise the common prime factors to the lowest powers, you can multiply those numbers together to calculate the greatest common factor.

$$\begin{aligned} \text{GCF} &= 2^1 \cdot 3^2 \\ &= 2 \cdot 9 \\ &= 18 \end{aligned}$$

Note: In Problems 4.32–4.34, you use the prime factorizations of two integers to calculate their greatest common factor and least common multiple.

4.34 Use the prime factorizations you generated in Problem 4.32 to calculate the least common multiple of 36 and 90.

Fun fact: The least common denominator for a group of denominators is their least common multiple.

This technique is similar to the technique presented in Problem 4.33, with a slight twist. Instead of selecting the common factors, you select *all* of the prime factors that appear in either factorization. In this case, that includes 2, 3, and 5. Raise each of these numbers to the highest powers associated with them in either factorization.

In this problem, prime factor 2 should be raised to the second power (because 2^2 is a factor of 36), 3 should be raised to the second power (because 3^2 is a factor of 36 and 90), and 5 should be raised to the first power (because 5^1 is a factor of 90). Multiply those values to calculate the least common multiple (LCM).

$$\begin{aligned} \text{LCM} &= 2^2 \cdot 3^2 \cdot 5^1 \\ &= 4 \cdot 9 \cdot 5 \\ &= 180 \end{aligned}$$

Note: Problems 4.35–4.36 refer to the integers 200 and 280.

4.35 Calculate the greatest common factor of the integers.

Use the technique described in Problem 4.32 to create factor trees and generate the prime factorizations of 200 and 280.

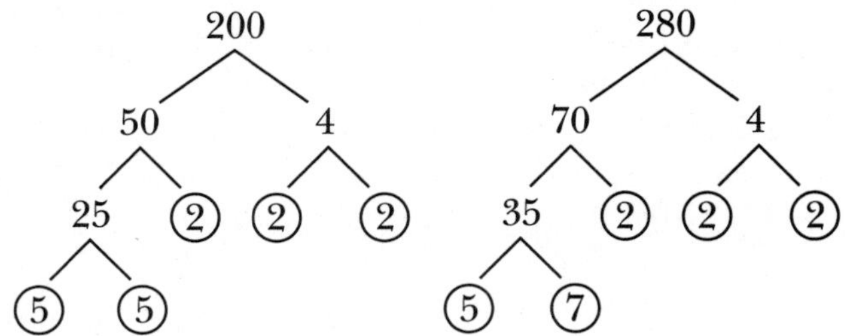

The prime factorizations are $200 = 2^3 \cdot 5^2$ and $280 = 2^3 \cdot 5 \cdot 7$. According to Problem 4.33, the greatest common factor is the product of the shared prime factors when each is raised to the lowest power from the factorizations. The integers 200 and 280 share two common prime factors, 2 and 5. The lowest power of prime factor 2 is 3, and the lowest power of prime factor 5 is 1.

$$\text{greatest common factor} = 2^3 \cdot 5 = 8 \cdot 5 = 40$$

Note: Problems 4.35–4.36 refer to the integers 200 and 280.

4.36 Calculate the difference of the least common multiple and the greatest common factor of the integers.

According to Problem 4.35, the prime factorizations of the integers are $200 = 2^3 \cdot 5^2$ and $280 = 2^3 \cdot 5 \cdot 7$. To calculate the least common multiple of 200 and 280, raise any prime factor that appears in either factorization to its highest power.

$$\text{least common multiple} = 2^3 \cdot 5^2 \cdot 7 = 8 \cdot 25 \cdot 7 = 1{,}400$$

The problem asks you to calculate the difference of the least common multiple and the greatest common factor of 200 and 280. According to Problem 4.35, the greatest common factor of 200 and 280 is 40.

$$\text{difference of LCM and GCF} = \text{LCM} - \text{GCF} = 1{,}400 - 40 = 1{,}360$$

Sets and Logical Reasoning

Union, intersection, Venn diagrams

Note: In Problems 4.37–4.39, A is the set of integers that are factors of 18 and B is the set of integers that are factors of 64.

4.37 List the elements of the set $A \cup B$.

A set is a collection of items called elements. This problem refers to two different sets, with names *A* and *B*. Set *A* contains all of the integers that are factors of 18; any integer that divides evenly into 18 belongs to set *A*. Set *B* contains all of the integers that divide evenly into 64. List the elements of both sets.

$$A = \{1, 2, 3, 6, 9, 18\} \qquad B = \{1, 2, 4, 8, 16, 32, 64\}$$

The problem asks you to identify the elements of $A \cup B$, which is read "*A* union *B*" or "the union of sets *A* and *B*." The union of two sets is the collection of all elements appearing in those sets. Any element that appears either in set *A* or set *B* also appears in the union. One thing to keep in mind: If an element appears in more than one set—for example the element 2 appears in both sets *A* and *B*—you only list that element once in the union.

$$A \cup B = \{1, 2, 3, 4, 6, 8, 9, 16, 18, 32, 64\}$$

Note: In Problems 4.37–4.39, A is the set of integers that are factors of 18 and B is the set of integers that are factors of 64.

4.38 List the elements of the set $A \cap B$.

In Problem 4.37, you began by listing the elements of *A* and *B*.

$$A = \{1, 2, 3, 6, 9, 18\} \qquad B = \{1, 2, 4, 8, 16, 32, 64\}$$

This problem asks you to identify $A \cap B$, read "*A* intersect *B*" or "the intersection of sets *A* and *B*." The intersection of two sets is comprised of the elements that appear in both sets. Only elements 1 and 2 appear in sets *A* and *B*.

Here's a visual way to remember what the set symbols mean: ∪nion i∩tersection

$$A \cap B = \{1, 2\}$$

Note: In Problems 4.37–4.39, A is the set of integers that are factors of 18 and B is the set of integers that are factors of 64.

4.39 If set *C* is defined as the positive multiples of 3 that are less than 36, can you conclude that $B \cup (A \cap C) = A \cup B$?

To evaluate the expression B + (A – C), you would follow the order of operations, subtracting inside parentheses first. The same rule holds true for set operations. Simplify inside parentheses first.

List the elements of all three sets.

$$A = \{1, 2, 3, 6, 9, 18\}$$
$$B = \{1, 2, 4, 8, 16, 32, 64\}$$
$$C = \{3, 6, 9, 12, 15, 18, 21, 24, 27, 30, 33\}$$

This problem asks you to perform two set operations, a union and an intersection. Begin with the operation within parentheses, $A \cap C$. List the elements that are shared by sets *A* and *C*.

$$A \cap C = \{3, 6, 9, 18\}$$

Now rewrite the problem, replacing $A \cap C$ with the list of elements you identified.

$$B \cup (A \cap C) = B \cup \{3, 6, 9, 18\}$$

The union of B and $\{3, 6, 9, 18\}$ is the collection of all elements appearing in either set.

$$B \cup \{3, 6, 9, 18\} = \{1, 2, 3, 4, 6, 8, 9, 16, 18, 32, 64\}$$

Note that $B \cup (A \cap C)$ contains precisely the same elements as $A \cup B$, as identified in Problem 4.37. Therefore, $B \cup (A \cap C) = A \cup B$.

Note: Problems 4.40–4.41 refer to the Venn diagram below, the results of a taste test between two popular sodas. X is the set of people who liked soda X, and Y is the set of people who liked soda Y.

4.40 What percentage of the participants liked only one of the sodas?

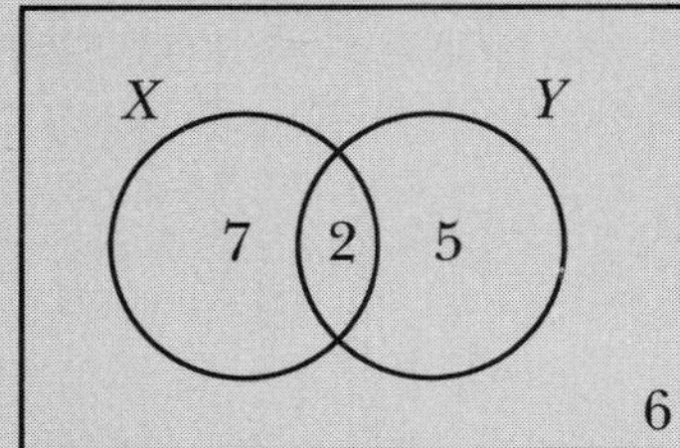

A Venn diagram is a simple visual tool that identifies the number of elements within at least two sets, including the number of elements that are shared between sets. Begin this problem by interpreting the numbers in the diagram:

7 = the number of people who liked only soda X
5 = the number of people who liked only soda Y
2 = the number of people who liked both sodas
6 = the number of people who liked neither soda

A total of $7 + 5 + 2 + 6 = 20$ people participated in the taste test. Of that group, 7 people liked only soda X and 5 people liked only soda Y. Therefore, $7 + 5 = 12$ people liked only one of the sodas, which represents $12/20 = 0.6 = 60\%$ of the total population.

Note: Problems 4.40–4.41 refer to the Venn diagram in Problem 4.40, the results of a taste test between two popular sodas. X is the set of people who liked soda X, and Y is the set of people who liked soda Y.

4.41 What is the ratio of people who liked soda X to the people who liked neither soda?

The number of elements in set X represents the number of people who liked soda X, the shaded region in the following diagram.

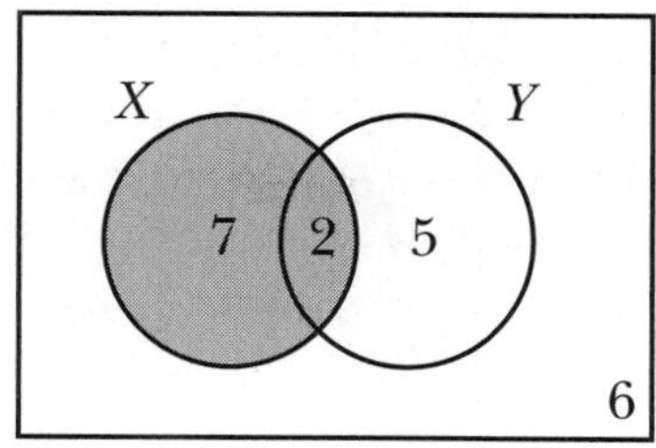

A total of 9 people liked soda X; 7 of those people only liked soda X, and 2 of those people liked both of the sodas.

As Problem 4.40 explains, the number 6 in the lower-right corner of the diagram represents the number of people who are in neither set X nor in set Y, the group that did not like either of the sodas.

The problem asks you to calculate the ratio of the people who liked soda X to the people who liked neither soda. Create a fraction to represent the ratio and express it in lowest terms.

$$\frac{\text{people who liked soda } X}{\text{people who liked neither soda}} = \frac{9}{6} = \frac{3}{2}$$

Note: Problems 4.42–4.45 refer to the Venn diagram below, the results of a survey of teenagers who were asked to identify their hobbies. In the diagram, T is the set of students who listed watching television as a hobby, G is the set of students who listed playing video games as a hobby, and S is the set of students who listed social networking as a hobby.

4.42 How many students participated in this survey?

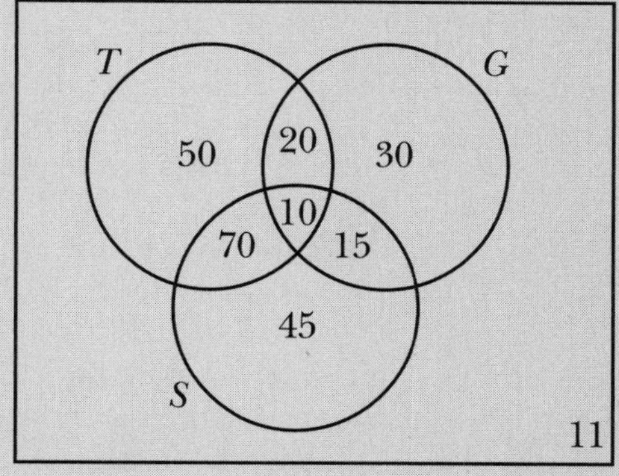

Add all of the numbers in the diagram to calculate the total number of participants.

$$50 + 20 + 30 + 70 + 10 + 15 + 45 + 11 = 251 \text{ participants}$$

Don't forget the 11 that's hiding down in the lower-right corner.

Note: Problems 4.42–4.45 refer to the Venn diagram in Problem 4.42, the results of a survey of teenagers who were asked to identify their hobbies. In the diagram, T is the set of students who listed watching television as a hobby, G is the set of students who listed playing video games as a hobby, and S is the set of students who listed social networking as a hobby.

4.43 How many students listed playing video games as a hobby?

Set G represents the group of students whose hobby is video games. Note that the answer is not 30; that is the number of students who *only* listed video games as a hobby. You also need to include the 20 students who like television and video games, the 15 students who like social media and video games, and the 10 students who listed all three hobbies. In other words, you need to add all of the numbers in the circle representing set G, as illustrated below.

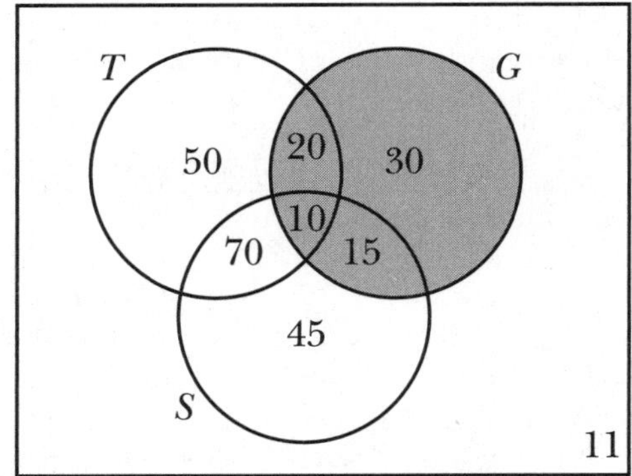

You conclude that $30 + 20 + 15 + 10 = 75$ students listed video games as a hobby.

Note: Problems 4.42–4.45 refer to the Venn diagram in Problem 4.42, the results of a survey of teenagers who were asked to identify their hobbies. In the diagram, T is the set of students who listed watching television as a hobby, G is the set of students who listed playing video games as a hobby, and S is the set of students who listed social networking as a hobby.

The word "OR" indicates the union of sets, whereas the word "AND" indicates intersection. In other words, this problem is asking you to identify $(T \cup G) - S$.

4.44 How many students listed television or video games as hobbies but did not list social media?

The solution to this problem is the sum of the values in sets T and G that do not overlap set S, the numbers in the shaded region of the following diagram.

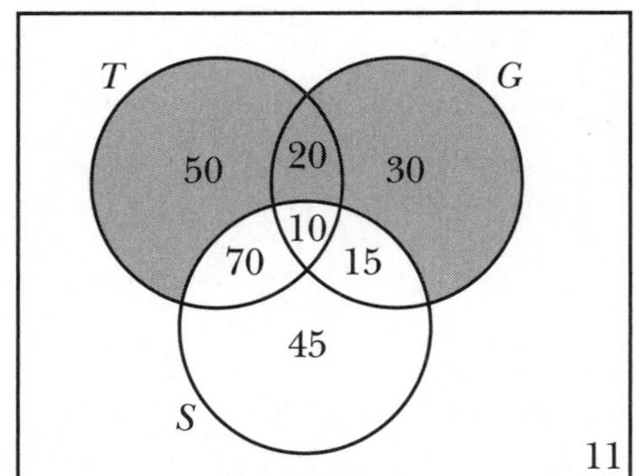

You conclude that $50 + 20 + 30 = 100$ students listed television or video games as hobbies but not social media.

Note: Problems 4.42–4.45 refer to the Venn diagram in Problem 4.42, the results of a survey of teenagers who were asked to identify their hobbies. In the diagram, T is the set of students who listed watching television as a hobby, G is the set of students who listed playing video games as a hobby, and S is the set of students who listed social networking as a hobby.

4.45 Given the sets A and B defined below, which set contains more elements?

$A = T - (S \cap T)$
$B = (G \cap T) + (G \cap S) + [(G \cap S) \cap T]$

Use the diagram to calculate how many elements belong to each term in the expressions. Note that $(G \cap S) \cap T$ is the region of the diagram in which all three circles overlap.

set	number of elements
T	$50+20+10+70=150$
$S \cap T$	$70+10=80$
$G \cap T$	$20+10=30$
$G \cap S$	$10+15=25$
$(G \cap S) \cap T$	10

Use these values to calculate the sizes of sets A and B.

$$\begin{aligned} A &= T-(S \cap T) \\ &= 150-80 \\ &= 70 \end{aligned} \qquad \begin{aligned} B &= (G \cap T)+(G \cap S)+[(G \cap S) \cap T] \\ &= 30+25+10 \\ &= 65 \end{aligned}$$

Set A contains five more elements than set B.

4.46 A private high school is studying the math courses taken by its freshman class during the current academic year. In the Venn diagram below, A is the set of students currently enrolled in an algebra course; B is the set of students currently enrolled in a geometry course. If the ratio of A to B is 3 to 4, how many of the freshmen at that school are enrolled in neither an algebra course nor a geometry course?

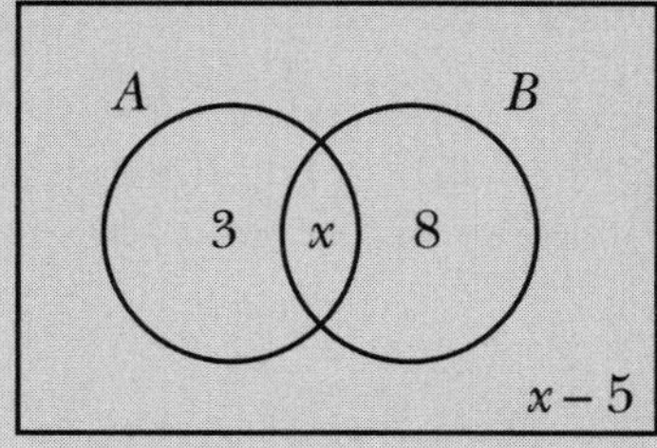

The value $x - 5$ in the lower-right corner of the diagram represents the number of students enrolled neither in an algebra course nor in a geometry course. Once you calculate x, you can calculate $x - 5$ to solve the problem. At the moment, however, you only know that the ratio of A to B is 3 to 4, so construct a proportion using this information.

$$\frac{A}{B} = \frac{3}{4}$$

Note that x students are enrolled in both classes, so the total number of students enrolled in algebra is $x + 3$ and the total number of students enrolled in geometry is $x + 8$. Substitute these values into the proportion and solve for x.

Cross-multiply to solve the proportion, multiplying the numerator (x+3) of one fraction by the denominator (4) of the other fraction and vice versa.

$$\frac{x+3}{x+8} = \frac{3}{4}$$
$$4(x+3) = 3(x+8)$$
$$4x+12 = 3x+24$$
$$4x-3x = 24-12$$
$$x = 12$$

You conclude that $x - 5 = 12 - 5 = 7$ students are enrolled neither in algebra nor in geometry.

4.47 Adrienne and Mark are two students in the same class in which a quiz was recently administered. Adrienne's score was the seventh lowest in the class, and Mark's score was the sixth highest. If Adrienne's score was greater than Mark's and two student scores separated them, how many students in the class took the quiz?

This is a logical reasoning problem. To solve it, you do not need to memorize any formulas or apply any complex algebraic properties—you need only think in an organized and logical manner. To solve this (and many logical reasoning problems), you should draw a diagram to help you visualize the information given. For example, you know that Adrienne's score is higher than Mark's and two other student scores are wedged between them. The following diagram captures this information; A represents Adrienne's score and M represents Mark's score.

$\underline{A}$ ___ ___ $\underline{M}$

If Adrienne's score is the seventh lowest, six students had lower scores than hers. Use blanks to represent the unknown scores, as illustrated below.

To make sure you draw the correct number of blanks, double check by naming them one by one as you move from Adrienne's score to the right: "Sixth lowest, fifth lowest, fourth lowest, third lowest, second lowest, lowest," a total of six lower scores.

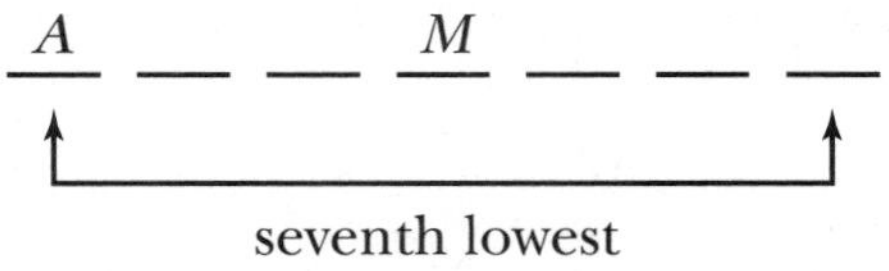

$\underline{A}$ ___ ___ $\underline{M}$ ___ ___ ___

seventh lowest

If Mark's score is the sixth highest, five students had scores higher than his (including Adrienne's score and the two scores separating him from Adrienne's score).

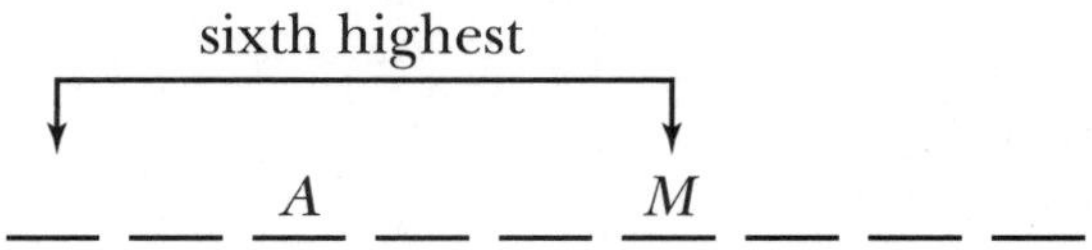

sixth highest

___ ___ $\underline{A}$ ___ ___ $\underline{M}$ ___ ___ ___

Your diagram contains nine blanks, so there are nine students in the class.

4.48 In the diagram below, the numbered boxes represent student desks in a classroom seating arrangement. Six students occupy those desks: three boys (named Arthur, Frank, and John) and three girls (named Anne, Ellen, and Jessica). The students are seated according to the following rules:

- Seats 1 and 6 are occupied by boys.
- Students whose names start with the same letter are always seated next to each other in the same row.
- Arthur sits at desk 2.

At what numbered desk does each student sit?

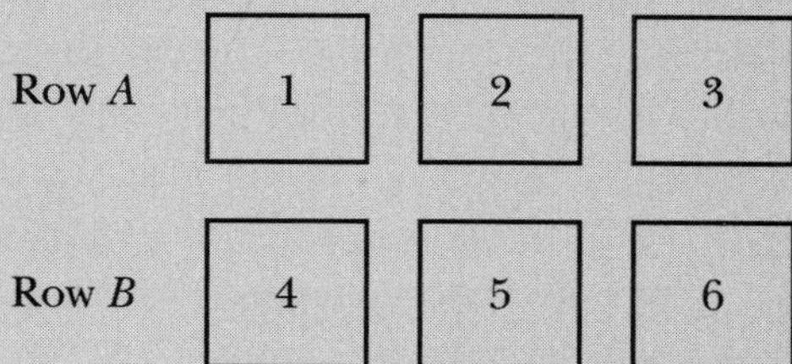

Use the seating rules to add information to the diagram. You know that Arthur sits at desk two. Desks 1 and 6 are occupied by the other two boys, so either Frank or John sits at those desks.

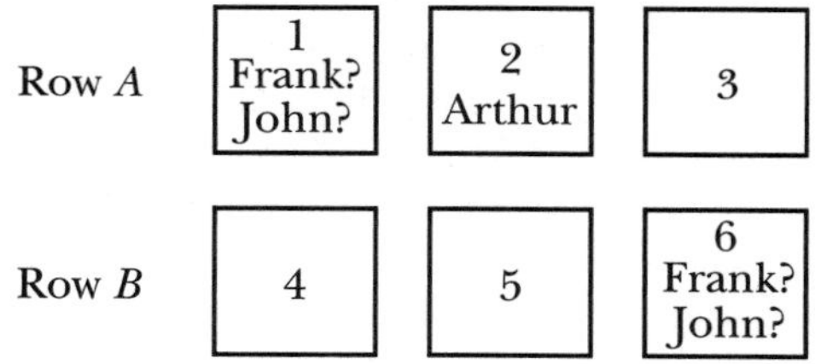

Students whose names start with the same first letter sit next to each other in the same row. That means Anne must sit at desk 3, next to Arthur. The only other two names that begin with the same letter are Jessica and John. Jessica cannot sit next to John if he is at desk 1, so John must be at desk 6, and Jessica must be at desk 5.

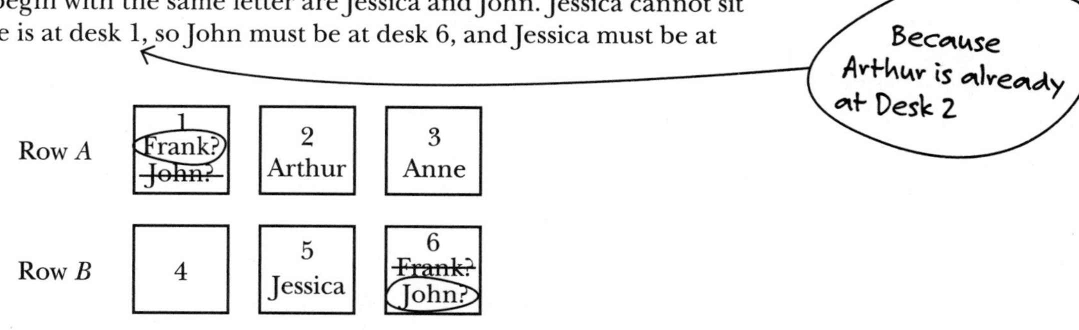

Only Ellen is left, so she must sit at desk 4. You conclude that Frank sits at desk 1, Arthur sits at desk 2, Anne sits at desk 3, Ellen sits at desk 4, Jessica sits at desk 5, and John sits at desk 6.

Sequences and Counting Problems

Lists and combinations

Note: Problems 4.49–4.51 refer to a sequence of integers in which each term is 4 more than the term that precedes it. The first term of the sequence is 5.

4.49 Write the first 10 terms of the sequence.

A sequence in which you generate subsequent terms by adding or subtracting the same value is called an arithmetic sequence. The number that is repeatedly added or subtracted to generate the sequence is called the common difference and is usually represented by the variable d. The terms of a sequence are represented by the same variable with a subscript that explains where that term falls in the sequence. The first term might be labeled x_1, the second term x_2, etc.

In this sequence, $x_1 = 5$, so the second term is 4 larger: $x_2 = 5 + 4 = 9$. Add 4 more to generate the third term: $x_3 = 9 + 4 = 13$. Repeat this process to generate the first 10 terms of the sequence.

$$x_1 = 5,\ x_2 = 9,\ x_3 = 13,\ x_4 = 17,\ x_5 = 21,$$
$$x_6 = 25,\ x_7 = 29,\ x_8 = 33,\ x_9 = 37, \text{ and } x_{10} = 41$$

Note: Problems 4.49–4.51 refer to a sequence of integers in which each term is 4 more than the term that precedes it. The first term of the sequence is 5.

4.50 Calculate the 58th term of the sequence.

Writing all 57 terms of the sequence just to identify the 58th term is a tedious process, but it is certainly possible. To save time, you can apply the formula that identifies the nth term in any arithmetic sequence: $a_n = a_1 + (n - 1)d$, where a_1 is the first term of the sequence and d is the common difference between successive terms. Substitute $n = 58$, $a_1 = 5$, and $d = 4$ to calculate a_{58}, the 58th term in the sequence.

$$\begin{aligned} a_n &= a_1 + (n-1)d \\ a_{58} &= 5 + (58-1)4 \\ a_{58} &= 5 + (57)4 \\ a_{58} &= 5 + 228 \\ a_{58} &= 233 \end{aligned}$$

Note: Problems 4.49–4.51 refer to a sequence of integers in which each term is 4 more than the term that precedes it. The first term of the sequence is 5.

4.51 Calculate the sum of the first 25 terms of the sequence.

You can complete this problem by listing the first 25 terms and then adding them, or you can apply the formula below to calculate the sum. In this formula, S_n is the sum of the first n terms of an arithmetic sequence, a_1 is the first term, and a_n is the nth term.

$$S_n = \frac{n}{2}(a_1 + a_n)$$

In this problem $n = 25$, but you are not given the value of $a_n = a_{25}$ to plug into the formula above. Use the formula introduced in Problem 4.50 to calculate a_{25}, the 25th term in the sequence.

$$\begin{aligned}a_n &= a_1 + (n-1)d\\ a_{25} &= 5 + (25-1)4\\ a_{25} &= 5 + (24)4\\ a_{25} &= 101\end{aligned}$$

Substitute $n = 25$, $a_1 = 5$, and $a_{25} = 101$ into the formula to calculate the sum of the first n terms.

$$\begin{aligned}S_n &= \frac{n}{2}(a_1 + a_n)\\ S_{25} &= \frac{25}{2}(5+101)\\ S_{25} &= \frac{25}{2}(106)\\ S_{25} &= 25\left(\frac{106}{2}\right)\\ S_{25} &= 25(53)\\ S_{25} &= 1,325\end{aligned}$$

Half of 25 times 106 is equal to 25 times half of 106. Basically, you're multiplying 25, 1/2, and 106. The order in which you multiply those numbers doesn't matter.

4.52 In a certain sequence, each term is n less than the term that precedes it. If the second term of the sequence is 12 and the sixth term is –16, what is the first term?

Draw a diagram that illustrates the information you are given. In the diagram below, each of the blanks represents one term of the sequence. You know that $a_2 = 12$, so add d to that term to calculate the next term: $a_3 = a_2 + d = 12 + d$. Add d again to calculate the fourth term: $a_4 = 12 + d + d = 12 + 2d$. Fill in all of the blanks using this technique, noting that a_1 is d less than a_2—you subtract d from 12 to generate the first term.

$12 - d$	12	$12 + d$	$12 + 2d$	$12 + 3d$	-16 ($\leftarrow 12 + 4d$)
a_1	a_2	a_3	a_4	a_5	a_6

You already know that the sixth term is $a_6 = -16$, but now you also know that the sixth term must be equal to $12 + 4d$. Set these values equal to calculate d.

$$\begin{aligned}12 + 4d &= -16\\ 4d &= -16 - 12\\ 4d &= -28\\ d &= \frac{-28}{4}\\ d &= -7\end{aligned}$$

Each term in this sequence is less than the term before it, so d is negative. You still generate terms by adding the same common difference over and over again, whether that common difference is positive or negative.

Each term is 7 less than the term that precedes it. Calculate the first term, $a_1 = 12 - d$.

$$\begin{aligned} a_1 &= 12 - d \\ &= 12 - (-7) \\ &= 12 + 7 \\ &= 19 \end{aligned}$$

Note: Problems 4.53–4.55 refer to a sequence of real numbers in which each term is two times as large as the term preceding it. The first term of the sequence is 3.

4.53 List the first 10 terms of the sequence.

Unlike the arithmetic sequences of Problems 4.49–4.52, each term of a geometric sequence is r times as large as the term that precedes it, where r is called the common ratio. In this geometric sequence, the first term is $x_1 = 3$ and the second term is $r = 2$ times as large: $x_2 = 3(2) = 6$. Multiply x_2 by 2 to calculate the third term of the sequence: $x_3 = 6(2) = 12$. Repeat this process to generate the first 10 terms of the geometric sequence.

$$x_1 = 3,\ x_2 = 6,\ x_3 = 12,\ x_4 = 24,\ x_5 = 48,$$
$$x_6 = 96,\ x_7 = 192,\ x_8 = 384,\ x_9 = 768, \text{ and } x_{10} = 1{,}536$$

Note: Problems 4.53–4.55 refer to a sequence of real numbers in which each term is two times as large as the term preceding it. The first term of the sequence is 3.

4.54 Calculate the 25th term in the sequence.

Observe the pattern of the terms in a geometric sequence:

$$\begin{aligned} &a_1 \\ &a_2 = a_1 \cdot r \\ &a_3 = a_1 \cdot r \cdot r = a_1 \cdot r^2 \\ &a_4 = a_1 \cdot r \cdot r \cdot r = a_1 \cdot r^3 \\ &a_5 = a_1 \cdot r \cdot r \cdot r \cdot r = a_1 \cdot r^4 \end{aligned}$$

This is similar to the arithmetic sequence formula $a_1 + (n - 1)d$. In that formula, you add one less d to a_1 than the number of the term.

Each term is equal to the first term (a_1) multiplied by a few r's. How many r's? One less than the number of the term. For example, to generate the fifth term, a_5, multiply the first term by $5 - 1 = 4$ r's. This rule is captured in the formula $a_n = a_1 \cdot r^{n-1}$, which can be used to calculate the nth term of a geometric sequence. This problem asks you to calculate a_{25}, so substitute $a_1 = 3$, $r = 2$, and $n = 25$ into the formula.

$$\begin{aligned} a_n &= a_1 \cdot r^{n-1} \\ a_{25} &= 3 \cdot (2)^{25-1} \\ a_{25} &= 3 \cdot 2^{24} \\ a_{25} &= 3 \cdot 16{,}777{,}216 \\ a_{25} &= 50{,}331{,}648 \end{aligned}$$

Note: Problems 4.53–4.55 refer to a sequence of real numbers in which each term is two times as large as the term preceding it. The first term of the sequence is 3.

4.55 Calculate the sum of the first 20 terms of the sequence.

You could calculate each of the first 20 terms and then add them to solve this problem, but as you can see in Problem 4.54, the numbers get fairly large. Instead, apply the formula below, which calculates the sum S_n of the first n terms of a geometric sequence.

$$S_n = a_1\left(\frac{r^n - 1}{r - 1}\right)$$
$$S_{20} = 3\left(\frac{2^{20} - 1}{2 - 1}\right)$$
$$S_{20} = 3\left(\frac{1{,}048{,}576 - 1}{1}\right)$$
$$S_{20} = 3(1{,}048{,}575)$$
$$S_{20} = 3{,}145{,}725$$

4.56 In a certain sequence of positive terms, each term is n times the previous term. If the third term is 90 and the fifth term is 10, what is the seventh term?

Remember, each term in a geometric series is r times as large as the preceding term. If the third term is 90, then the fourth term is $90r$ and the fifth term is $90 \cdot r \cdot r = 90r^2$, as illustrated below.

a_1	a_2	a_3	a_4	a_5	a_6	a_7
		90	$90 \cdot r$	10 ($\leftarrow 90 \cdot r^2$)	$10 \cdot r$	$10 \cdot r^2$

Because $a_5 = 90r^2$ and the problem states that $a_5 = 10$, you know that $90r^2$ and 10 are equal. Write this as an equation and solve it for r.

$$90r^2 = 10$$
$$r^2 = \frac{10}{90}$$
$$r^2 = \frac{1}{9}$$
$$\sqrt{r^2} = \sqrt{\frac{1}{9}}$$
$$r = \frac{1}{3}$$

Usually, this answer would be ±1/3. You need to place a "plus or minus" sign on the right side of an equation after square rooting both sides. Why? Because $(1/3)^2$ and $(-1/3)^2$ both equal 1/9. However, this problem specifically states that the sequence consists of all positive numbers, and if r were equal to −1/3, the sequence would have negative numbers in it.

Every term in the sequence is one-third the term that precedes it. That means $a_4 = 90(1/3) = 30$, and one-third of $a_4 = 30$ is equal to 10, the fifth term of the sequence. To generate the sixth term, multiply the fifth term by r. Then, multiply it by r once again to generate the seventh term, a_7.

$$a_6 = a_5 \cdot r \qquad a_7 = a_6 \cdot r$$
$$= 10 \cdot \frac{1}{3} \qquad = \frac{10}{3} \cdot \frac{1}{3}$$
$$= \frac{10}{3} \qquad = \frac{10}{9}$$

You conclude that $a_7 = 10/9$.

4.57 When 5 is divided by 13, what digit of the quotient is located 203 places right of the decimal point?

This decimal is a sequence problem in disguise!

Use your calculator to compute the quotient 5 ÷ 13. The result is a repeating decimal, a pattern of 6 digits that cycles infinitely.

$$5 \div 13 = 0.384615384615384615384615\ldots$$
$$= 0.\overline{384615}$$

You can interpret the repeating digits as a sequence of numbers, in which $a_1 = 3$, $a_2 = 8$, $a_3 = 4$, $a_4 = 6$, $a_5 = 1$, and $a_6 = 5$. The seventh digit begins the pattern again: $a_7 = 3$, $a_8 = 8$, $a_9 = 4$, $a_{10} = 6$, $a_{11} = 1$, and $a_{12} = 5$. After every sixth term, the pattern repeats, so divide 203 by 6 to determine how many full groups of the six-digit sequence occur within the first 203 decimal places.

$$203 \div 6 = 33.8\overline{3}$$

There are 33 full groups of the six-digit pattern within the first 203 digits of the decimal. In other words, the first $33 \cdot 6 = 198$ terms are occupied by the pattern repeated over and over. The 199th term, the digit that is 199 places right of the decimal point, starts the pattern again: $a_{199} = 3$, $a_{200} = 8$, $a_{201} = 4$, $a_{202} = 6$, $a_{203} = 1$, $a_{204} = 5$. You conclude that the digit of the quotient 203 places right of the decimal point is $a_{203} = 1$.

4.58 If the nth term of a certain sequence is $a_n = \dfrac{(-1)^n(n+1)}{n-1}$, what is the value of $a_3 - a_4$?

Substitute $n = 3$ and $n = 4$ into the formula independently to calculate a_3 and a_4, the third and fourth terms of the sequence.

$$a_3 = \frac{(-1)^3(3+1)}{3-1} = \frac{(-1)(4)}{2} = -\frac{4}{2} = -2$$

$$a_4 = \frac{(-1)^4(4+1)}{4-1} = \frac{(1)(5)}{3} = \frac{5}{3}$$

The problem asks you to calculate the difference of a_3 and a_4.

$$a_3 - a_4 = -2 - \frac{5}{3} = -\frac{6}{3} - \frac{5}{3} = -\frac{11}{3}$$

4.59 In the sequence below, each term is *p* less than 4 times the previous term. The first term in the sequence is 3. What is the fifth term?

3, 7, 23, ...

This is neither an arithmetic nor a geometric sequence; it is a mashup of both. The first term of the sequence is $a_1 = 3$, so the second term is *p* less than 4 times as large: $a_2 = 4(a_1) - p$. The problem lists the first three terms of the sequence, including $a_2 = 7$. You have now stated a_2 in two different ways, but the values must be equal. Set them equal and solve for *p*.

$$\begin{aligned} 4(a_1) - p &= 7 \\ 4(3) - p &= 7 \\ 12 - p &= 7 \\ -p &= 7 - 12 \\ -p &= -5 \\ p &= 5 \end{aligned}$$

Each term is 5 less than 4 times the previous term. To verify the value of *p*, multiply $a_2 = 7$ by 4 and subtract 5; the result is $a_3 = 23$, the third term listed in the problem.

$$\begin{aligned} a_3 &= 4(a_2) - p \\ &= 4(7) - 5 \\ &= 28 - 5 \\ &= 23 \end{aligned}$$

Repeat this pattern to calculate a_4 and then a_5.

$$\begin{aligned} a_4 &= 4(a_3) - p \\ &= 4(23) - 5 \\ &= 92 - 5 \\ &= 87 \end{aligned} \qquad \begin{aligned} a_5 &= 4(a_4) - p \\ &= 4(87) - 5 \\ &= 348 - 5 \\ &= 343 \end{aligned}$$

The fifth term of the sequence is 343.

Note: In Problems 4.60–4.62, you have to choose a four-digit combination using the integers 0–9 to secure a hotel safe. The rules for selecting the integers vary in each problem.

4.60 How many combinations can be generated for the safe if none of the digits may repeat?

The fundamental counting principle states that if there are *x* ways to do one thing and *y* ways to do another thing, then there are $x \cdot y$ ways to do both things. This problem is called a counting problem, as it is named for the fundamental counting principle. Choosing a combination for the safe is, essentially, a series of four tasks—each task consists of selecting a single digit for the safe's combination. Your goal is to determine how many ways you can accomplish each task, based on the rules set by the problem, and then multiply those values together.

Begin by drawing a diagram, in which each task is represented by a box.

Digit 1 | Digit 2 | Digit 3 | Digit 4

How many ways can you select the first digit in the combination? There are ten digits from which to choose (0, 1, 2, 3, 4, 5, 6, 7, 8, and 9), so there are 10 ways to complete the task. Update the diagram with this value.

10 | Digit 2 | Digit 3 | Digit 4

How many ways can you select the second digit in the combination? The rules state that you cannot repeat digits, so now you have only 9 values from which to select. That leaves a pool of 8 values for the third number in the combination and then 7 values for the final number in the combination.

10 | 9 | 8 | 7

According to the fundamental counting principle, if there are 10 ways to select the first digit, 9 ways to select the second digit, 8 ways to select the third digit, and 7 ways to select the fourth digit, then there are $10 \cdot 9 \cdot 8 \cdot 7 = 5{,}040$ possible combinations if none of the digits may repeat.

Note: In Problems 4.60–4.62, you have to choose a four-digit combination using the integers 0–9 to secure a hotel safe. The rules for selecting the integers vary in each problem.

4.61 How many combinations can be generated for the safe if digits may be repeated and the combination must start and end with the same number?

Once again begin with a drawing that represents the tasks.

Digit 1 | Digit 2 | Digit 3 | Digit 4

You can select any of the 10 digits, 0–9, as the first digit of the combination, but that number must repeat as the final digit in the combination. Therefore, there are 10 ways to accomplish the first task but there is only 1 way to accomplish the final task.

10 | Digit 2 | Digit 3 | 1

Unlike Problem 4.60, this problem allows digits of the combination to repeat, so there are 10 ways to choose the second and third digits in the combination.

10 | 10 | 10 | 1

According to the fundamental counting principle, there are $10 \cdot 10 \cdot 10 \cdot 1 = 1{,}000$ possible combinations for the safe under the restrictions set by this problem.

Note: In Problems 4.60–4.62, you have to choose a four-digit combination using the integers 0–9 to secure a hotel safe. The rules for selecting the integers vary in each problem.

4.62 How many combinations can be generated for the safe if the first digit must be odd, consecutive digits alternate between even and odd, and digits cannot repeat?

There are 5 odd values from which to choose for the first digit of the combination: 1, 3, 5, 7, or 9. There are 5 ways to select the second digit (0, 2, 4, 6, or 8).

[5] [5] [Digit 3] [Digit 4]

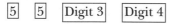

To complete the combination, you have to choose one more odd and one more even digit. Because digits cannot repeat, that leaves you with 4 even and 4 odd digits from which to choose.

[5] [5] [4] [4]

According to the fundamental counting principle, there are $5 \cdot 5 \cdot 4 \cdot 4 = 400$ possible combinations.

4.63 In a particular state, automobile license plates contain seven characters, four letters of the alphabet (A–Z) followed by three digits (0–9). Recently, the state decided to exclude the letter O and the digit 0 from future license plates because they were too similar. Assuming letters may repeat on a license plate but digits may not, how many unique license plates can the state create?

Draw a diagram representing the seven tasks that must be completed to create a license plate. Four letters must be chosen, followed by three digits.

[Letter 1] [Letter 2] [Letter 3] [Letter 4] [Digit 1] [Digit 2] [Digit 3]

There are 26 letters in the alphabet, but the letter O may not be chosen. Therefore, there are 25 ways to select each letter. (Letters may repeat, so each time you select a letter, you can choose from all 25 available candidates.)

[25] [25] [25] [25] [Digit 1] [Digit 2] [Digit 3]

There are only 9 digits from which to select for the first digit in the license plate, because 0 has been excluded. Digits may not repeat, so there are only 8 candidates for the second digit and then 7 digits for the third number.

[25] [25] [25] [25] [9] [8] [7]

According to the fundamental counting principle, the state can create $25 \cdot 25 \cdot 25 \cdot 25 \cdot 9 \cdot 8 \cdot 7 = 196{,}875{,}000$ unique license plates.

4.64 A certain ice cream shop offers five flavors: vanilla, chocolate, strawberry, chocolate chip, and butter pecan. How many ways can a customer order a two-scoop cone, if she may select the same or different flavors for each scoop?

This problem is tricky, because unlike Problems 4.60–4.63, the order in which the customer selects the favors does not matter. In other words, if the customer selects vanilla for scoop 1 and chocolate for scoop 2, she has designed the same cone as a customer who selects chocolate for scoop 1 and vanilla for scoop 2.

The best way to combat problems like these (without memorizing additional formulas) is to create a diagram that illustrates all of the possibilities. In the chart below, for example, choices for scoop 1 are listed horizontally and choices for scoop 2 are listed vertically. The remaining cells represent combinations of those flavors, and each unique flavor combination is numbered.

	vanilla	chocolate	strawberry	chocolate chip	butter pecan
vanilla	1	2	3	4	5
chocolate	2	6	7	8	9
strawberry	3	7	10	11	12
chocolate chip	4	8	11	13	14
butter pecan	5	9	12	14	15

Flavor combination 1 is vanilla-vanilla, flavor combination 2 is vanilla-chocolate, flavor combination 3 is vanilla-strawberry, etc. There are 15 unique flavor combinations for a two-scoop cone.

4.65 Four friends enter a tug of war contest as part of a school outdoor field day competition. The rules state that two of the four friends must be randomly selected to compete in the first round. How many different ways can you select a two-player team from the group of four friends?

Like Problem 4.64, the order in which team members are selected does not matter. In other words, a team that contains Jody and Rachel is the same as a team containing Rachel and Jody. Let *A*, *B*, *C*, and *D* represent the four friends, and list all possible two-person teams.

AB *BA* *CA* *DA*
AC *BC* *CB* *DB*
AD *BD* *CD* *DC*

Now cross out any repeated teams. For example, team *AB* appears in the upper-left corner, and team *BA* represents the same two players, so *BA* is a repeated team. Remember, the problem is asking you to determine how many *different* teams you can create.

AB ~~*BA*~~ ~~*CA*~~ ~~*DA*~~
AC *BC* ~~*CB*~~ ~~*DB*~~
AD *BD* *CD* ~~*DC*~~

There are six different ways to select a two-person team from a group of four people.

Arithmetic Word Problems

Where math and the "real world" collide

Note: Problems 4.66–4.67 refer to the sign below, the posted ticket prices at a certain movie theater.

4.66 What is the least expensive way to purchase enough tickets for 12 people?

Ticket Prices	
10-pack of tickets	\$85
4-pack of tickets	\$40
Single ticket	\$15

The more tickets you purchase in one bundle, the less the per-ticket price. Divide each posted price by the number of tickets included at that price to calculate the price per ticket.

10-pack price per ticket: $\$85 \div 10 = \$8.50/\text{ticket}$
4-pack price per ticket: $\$40 \div 4 = \$10/\text{ticket}$
Single ticket: $\$15/\text{ticket}$

The 10-packs include the least expensive tickets, which leaves 2 more tickets that need to be purchased for a total of 12. If you buy those 2 tickets individually, it costs $2 \cdot \$15 = \$30$, which is less expensive than the 4-pack of tickets that costs \$40.

Therefore, the least expensive way to purchase the tickets is via one 10-pack and two individual tickets, for a total of $\$85 + \$30 = \$115$.

Note: Problems 4.66–4.67 refer to the sign in Problem 4.66, the posted ticket prices at a certain movie theater.

4.67 What is the least expensive way to purchase enough tickets for 7 people?

There are four different ways to purchase enough tickets for 7 people. You could purchase 7 individual tickets, a 4-pack and 3 individual tickets, two 4-packs of tickets, or a 10-pack of tickets.

The last two options leave you with extra tickets, but you still have "enough tickets for 7 people," as the problem states.

The cost of each option is calculated below.

$$\begin{aligned}
\text{7 individual tickets:} &\quad 7(\$15) = \$105 \\
\text{4-pack} + \text{3 individual tickets:} &\quad \$40 + 3(\$15) = \$85 \\
\text{Two 4-packs of tickets:} &\quad 2(\$40) = \$80 \\
\text{One 10-pack of tickets:} &\quad = \$85
\end{aligned}$$

The least expensive way to purchase enough tickets for 7 people is to purchase two 4-packs of tickets.

4.68 A school district owns 25 buses. Three of the buses seat 16 children each. Half of the remaining buses seat 42 children each. The rest of the buses seat 54 children each. How many children can the school district transport simultaneously using all 25 buses?

Three of the buses seat 16 children each. This accounts for 3 of the buses, leaving 25 – 3 = 22 buses to investigate. Half of those 22 buses, in other words 11 buses, seat 42 children each. The rest of the buses (the remaining 11 buses, the second half of the group of 22 buses) seat 54 children each. Multiply the occupancy of each kind of bus by the number of buses and add the results to calculate the total number of children that can be transported at once.

$$\begin{aligned}&(16\text{ children})(3\text{ buses})+(42\text{ children})(11\text{ buses})+(54\text{ children})(11\text{ buses})\\&=48+462+594\\&=1{,}104\text{ children}\end{aligned}$$

4.69 The total cost for a band to hold a concert at a local venue is \$2,500. If the band wishes to earn a \$7,000 profit from concert ticket sales and the venue seats 900 people, how much should the band charge per seat, assuming that 90% of the tickets will sell? Round your answer to the nearest dollar.

In order to cover costs and earn the desired profit, the band will need to raise \$2,500 + \$7,000 = \$9,500 in ticket sales. If 90% of the 900 available tickets sell, the band will sell 0.90(900) = 810 tickets. Divide the total needed ticket sales by the number of tickets to calculate how much the band must charge per seat.

$$\begin{aligned}\text{price per seat}&=\frac{\text{total ticket sales needed}}{\text{total number of tickets}}\\&=\frac{\$9{,}500}{810}\\&\approx \$11.7283951\\&\approx \$12\end{aligned}$$

4.70 Torrie purchases a hat, a shirt, and a scarf during an event in which the store does not charge sales tax. Assuming the following information is true, how much does each of the three items cost?

- The combined price of the hat and shirt is \$23
- The combined price of the shirt and scarf is \$32
- The combined price of the scarf and hat is \$27

Compare the first two bulleted items. The hat and the shirt together cost \$23, but if you swap out the hat for the scarf, the combined price climbs 32 – 23 = 9 dollars. That means the scarf is \$9 more expensive than the hat. If h represents the price of the hat, then $h + 9$ is the price of the scarf. According to the last bullet, if you add the prices of the scarf and hat, the sum is \$27. Substitute h and $h + 9$ into that equation.

$$\begin{aligned}\text{scarf price}+\text{hat price}&=27\\(h+9)+h&=27\\2h+9&=27\end{aligned}$$

Solve the equation for h.

$$\begin{aligned} 2h &= 27 - 9 \\ 2h &= 18 \\ h &= \frac{18}{2} \\ h &= 9 \end{aligned}$$

The price of the hat is \$9, and the price of the scarf is $h + 9$ = \$18. According to the first bullet, the combined price of the hat and shirt is \$23, so the price of the shirt is \$23 – \$9 = \$14.

4.71 The chart below contains information about the total gold, silver, and bronze medals won by China and Russia in the 2012 Summer Olympics. If the total number of bronze medals earned by China is one less than the total number of gold medals earned by Russia, what is the total number of medals both countries earned?

	Gold	Silver	Bronze	*Total*
China	38	27		
Russia			32	82
Total		53		

To solve this problem, you need to calculate values missing from the chart. At the start, there is only one value you can calculate. If a total of 53 silver medals were earned and China earned 27 of them, then Russia earned 53 – 27 = 26 silver medals. Now you can calculate Russia's gold medals by subtracting its silver and bronze medal counts from the total count: 82 – 32 – 26 = 24 gold medals. At this point, you can calculate the total number of gold medals won by both countries: 38 + 24 = 62.

	Gold	Silver	Bronze	*Total*
China	38	27		
Russia	24	26	32	82
Total	62	53		

You calculate this number first, this number second, & this number third.

You would be stuck at this point, unable to continue, if it were not for the information given in the problem: "the total number of bronze medals earned by China is one less than the total number of gold medals earned by Russia." Russia earned 24 gold medals, so China earned 23 bronze medals; the total number of bronze medals earned by both countries is 23 + 32 = 55.

	Gold	Silver	Bronze	*Total*
China	38	27	23	
Russia	24	26	32	82
Total	62	53	55	

Complete the chart by calculating China's total medal count and verifying that the sums of the bottom row and right column match. That matching sum, 170, is the total number of medals earned by both countries.

	Gold	Silver	Bronze	***Total***
China	38	27	23	88
Russia	24	26	32	82
Total	62	53	55	170

4.72 A certain movie theater seats 400 people. Fifteen minutes before a movie begins, the theater is half full, and 60 percent of the audience is children. During the 15 minutes before the movie begins, new audience members arrive at a ratio of 1 parent for every 2 children. When the movie begins, the theater is $\frac{7}{8}$ full. What percentage of the seats, rounded to the nearest whole number, is occupied by children when the movie starts?

Fifteen minutes before the movie, half of the 400 seats are full; in other words, 200 seats are full. Sixty percent of that group is children, so 15 minutes before the movie begins, 0.6(200) = 120 children are seated. For the next 15 minutes, the theater continues to fill until it is 7/8 full; when the movie starts, 400(7/8) = 350 seats are occupied. That means 350 – 200 = 150 members of the total audience arrive during the 15-minute period before the movie.

To calculate 60% of 200, convert 60% to a decimal by moving the decimal point two places to the left (60% = 0.60) and multiply 200 by that decimal.

For every 1 parent who arrives during that 15-minute period, 2 children arrive. In other words, two out of every three, or 2/3, of the 150 people who enter the theater during that 15-minute period are children: 150(2/3) = 100 children.

If 120 children are seated 15 minutes before the movie and 100 more children arrive before the movie starts, then 220 of the final 350 audience members are children, which represents 220/350 ≈ 0.62857143 ≈ 63% of the audience.

4.73 Approximately 8,000 people pass through a security checkpoint at a constant rate each day at a certain airport. Each person must pass through a metal detector. If the security checkpoint is open from 9 a.m. to 8 p.m. each day and it takes an average of 30 seconds for each person to pass through a metal detector, how many metal detectors does the airport need? Report your answer as a whole number.

There are a number of ways to solve this problem; in this solution, you calculate how many people can pass through a single metal detector in one day and then divide that number into the 8,000 total people to calculate the number of required metal detectors.

Each metal detector will scan passengers between 9 a.m. and 8 p.m., a total of 11 hours per day. Multiply by 60 to calculate the number of minutes each metal detector is scanning passengers: $11 \cdot 60 = 660$ minutes. It takes an average of 30 seconds to scan one passenger, so the metal detector can scan two people each minute. Therefore, each metal detector can scan $660 \cdot 2 = 1{,}320$ people each day.

Calculate $8{,}000 \div 1{,}320$ to divide the total number of people who visit the airport each day into groups that can be serviced by a single metal detector: $8{,}000 \div 1{,}320 \approx 6.06$. The airport needs a minimum of 7 metal detectors.

Don't round to the nearest whole number. Six is the wrong answer! The airport needs 6.06 metal detectors, which is MORE than six. If the airport only buys six metal detectors, it can't process all of the passengers before closing time.

4.74 A watch at a certain department store was purchased for $28 (not including sales tax) on October 5. The price had been discounted each day for three days. The purchase price represented a 15% discount of the October 4 price, which was a 25% discount of the October 3 price, which was a 10% discount of the October 2 price. What was the price of the watch on October 2?

This watch was priced to move! In what seems a desperate attempt to liquidate inventory, the store discounted the watch each day for three consecutive days, with a final price of $28. It is helpful to create an illustration to sort through the information you are given, especially because the problem lists the discounts in reverse chronological order.

Let p represent the price of the watch on October 2. On October 3, price p was discounted by 10 percent. Therefore, the price on October 3 was $100 - 10 = 90$ percent of p. Rewrite 90 percent as a decimal ($90\% = 0.90 = 0.9$) and multiply by p to generate the price on October 3.

p	$0.9p$		28
October 2	October 3	October 4	October 5

On October 4, the watch was discounted by an additional 25 percent, which means the price on October 4 was $100 - 25 = 75$ percent of the price on October 3. Multiply $0.9p$ (the price on October 3) by 0.75 to calculate the price on October 4: $0.9p(0.75) = 0.675p$.

p	$0.9p$	$0.675p$	28
October 2	October 3	October 4	October 5

The price on October 5 reflects another discount, this time of 15 percent. Thus, the October 5 price is $100 - 15 = 85$ percent of the October 4 price: $0.675p(.85) = 0.57375p$.

p	$0.9p$	$0.675p$	$0.57375p = 28$
October 2	October 3	October 4	October 5

To solve the equation $0.57375p = 28$ for p, divide both sides by the coefficient of p.

$$0.57375p = 28$$
$$p = \frac{28}{0.57375}$$
$$p \approx 48.801743$$

The price of the watch on October 2 was $p = \$48.80$.

You can check your answer by applying the discounts in order:
Oct. 2: $48.80
Oct. 3: 10% off = $43.92
Oct. 4: 25% off = $32.94
Oct. 5: 15% off = $28.00

4.75 Tim and Alicia sell cars at a local automobile dealership. On Monday of this week, Tim sold three more cars than Alicia. On Tuesday, Alicia sold four fewer cars than Tim. On Wednesday, Tim sold two fewer cars than Alicia. If Tim sells a total of nine cars on Thursday, how many will Alicia have to sell on Thursday to match Tim's total sales for the week?

On Monday, Tim sold three more cars than Alicia. On Tuesday, Tim sells four more than Alicia, so Tim is now seven cars in the lead. Alicia sells two more cars than Tim on Wednesday, cutting slightly into Tim's lead, which now stands at five cars. Tim sells nine cars on Thursday, which puts him a total of $5 + 9 = 14$ cars ahead of Alicia. She will need to sell 14 cars on Thursday to match his total sales.

4.76 Chanelle rode a bicycle from her house to the beach at an average speed of 15 miles per hour, and it took her 35 minutes to complete the trip. When she bicycled home later that day, she took the same route and traveled at an average speed of 17.5 miles per hour. What was Chanelle's average speed, in miles per hour, for both parts of the trip? Round your answer to the nearest hundredth.

Travel problems require you to use the formula $d = rt$, which states that distance d is equal to the rate r at which you travel multiplied by the time t that you travel. Chanelle's trip to the beach took $t = 35$ minutes at a rate of $r = 15$ miles per hour (mph). Given this information, you can apply the $d = rt$ formula to calculate the distance d between her house and the beach. However, all of your values must be written in matching units.

Her trip took 35 *minutes* but her speed is expressed in miles per *hour*. The problem asks you to give the final answer in terms of miles per *hour*, so you need to convert 35 minutes into hours. To do so, divide 35 by 60.

$$t = \frac{35 \text{ minutes}}{60 \text{ minutes}} = \frac{7}{12} \text{ hours}$$

Now substitute $r = 15$ mph and the value of t you calculated into $d = rt$ to compute the distance Chanelle biked to the beach.

$$\begin{aligned} d &= rt \\ &= (15)\left(\frac{7}{12}\right) \\ &= 8.75 \text{ miles} \end{aligned}$$

Now calculate the time it took for Chanelle to travel home from the beach. She took the same route, so the distances both to and from the beach are $d = 8.75$ miles. Her rate on the return trip was $r = 17.5$ mph.

$$\begin{aligned} d &= rt \\ 8.75 &= 17.5t \\ \frac{8.75}{17.5} &= t \\ 0.5 &= t \end{aligned}$$

Chanelle's return trip took 0.5, or 1/2, hours. To calculate her average speed, once again apply the $d = rt$ formula. This time, d represents the total distance traveled: 8.75 miles to the beach and 8.75 miles back home, which is a total of 17.5 miles. Similarly, t is the total time she traveled, 7/12 hours to the beach and 1/2 hours home, which is a total of 13/12 hours. Substitute these values into the equation and solve for r, her average speed in miles per hour.

$$d = rt$$

$$17.5 = r\left(\frac{13}{12}\right)$$

$$\left(\frac{12}{13}\right)(17.5) = r$$

$$16.1538462 \approx r$$

Chanelle's average speed was 16.15 mph.

4.77 Mitch's car has a 12-gallon gas tank. On a recent trip, Mitch used the full tank of gas to travel 450 miles. If Mitch's fuel efficiency averages 35 miles per gallon for the first third of the tank, what is his average fuel efficiency for the remaining two-thirds of the tank?

The first third of the tank is 12 gallons ÷ 3 = 4 gallons. On those 4 gallons of gas, Mitch averages 35 miles per gallon, which means he travels 4(35) = 140 miles. The remainder of the trip is 450 – 140 = 310 miles and consumes 12 – 4 = 8 gallons of gas, the remaining two-thirds of the tank. His fuel efficiency is equal to the total miles traveled divided by the gas used.

$$\begin{aligned}\text{fuel efficiency} &= \frac{\text{distance traveled}}{\text{fuel consumed}}\\ &= \frac{310 \text{ miles}}{8 \text{ gallons}}\\ &= 38.75 \text{ miles per gallon}\end{aligned}$$

4.78 Eliza is trying to decide between two cell phone payment plans. Plan A charges a flat fee of 15 cents per call and an additional 10 cents for each minute of the call. Plan B charges a flat fee of $80 per month for unlimited calls of unlimited length. If Eliza makes an average of 50 calls per month, what is the longest average time she can spend on those calls (in whole minutes) in order to ensure that plan A is less expensive than plan B?

To solve this problem, calculate the call length that causes plans A and B to cost the same. In other words, when does plan A cost $80, the price of plan B?

If Eliza makes an average of 50 calls, she is charged 0.15 dollars per call, for a total of $7.50. That leaves $80 – $7.50 = $72.50 for per-minute charges. If minutes cost ten cents (or 0.1 dollars) each, she can afford 72.5 ÷ 0.1 = 725 minutes. Remember that she makes an average of 50 calls, so each call could last approximately 725 ÷ 50 = 14.5 minutes.

Based on these calculations, if Eliza's 50 monthly calls last an average of 14.5 minutes, plans A and B will cost the same, $80. The longest average call time, in whole minutes, for which plan A would be less expensive is 14 minutes, the answer 14.5 rounded down to the nearest whole number.

If she makes 50 calls that average 15 minutes long, she spends 750 minutes on the phone, which costs 750(0.1) = $75.00. When you add the $7.50 in total connection charges to place the calls, she spends $82.50 per month, which is more than $80.

4.79 An inheritance is divided evenly between four sisters: Carol, Eileen, Kathy, and Sharon. Eileen gives one-third of her share to Kathy, and Carol gives one-fifth of her share to Kathy. If Kathy owes Sharon an amount of money equal to half of her original share, what percentage of the total inheritance is Kathy's once she repays her debt to Sharon?

In this problem, you work with fractions that represent portions of the original inheritance, which is divided evenly into four shares. Therefore, at the outset, each sister has 1/4 of the total inheritance. Eileen gives 1/3 of her 1/4 share, or (1/3)(1/4) = 1/12 of the total inheritance, to Kathy. Carol gives 1/5 of her 1/4 share, or (1/5)(1/4) = 1/20 of the total inheritance to Kathy. That means Kathy now owns 1/4 + 1/12 + 1/20 of the total inheritance. Add the fractions.

$$\begin{aligned}\frac{1}{4}+\frac{1}{12}+\frac{1}{20}&=\frac{1}{4}\left(\frac{15}{15}\right)+\frac{1}{12}\left(\frac{5}{5}\right)+\frac{1}{20}\left(\frac{3}{3}\right)\\&=\frac{15}{60}+\frac{5}{60}+\frac{3}{60}\\&=\frac{23}{60}\end{aligned}$$

Once Kathy has received money from two of her sisters, she owns 23/60 of the inheritance. Kathy owes Sharon money, 1/2 of her original 1/4 share. Therefore, Kathy owes Sharon (1/2)(1/4) = 1/8 of the inheritance. Subtract 1/8 from 23/60 to calculate how much Kathy owns after reconciling her debt with Sharon.

$$\begin{aligned}\frac{23}{60}-\frac{1}{8}&=\frac{23}{60}\left(\frac{2}{2}\right)-\frac{1}{8}\left(\frac{15}{15}\right)\\&=\frac{46}{120}-\frac{15}{120}\\&=\frac{31}{120}\end{aligned}$$

Kathy's final share of the inheritance is $31/120 \approx 25.8\overline{3}\%$, slightly more than her original share of 1/4 = 25%.

4.80 A chemist creates a mixture by combining Solutions A, B, C, and D in the ratio 3 : 5 : 2 : 1, respectively. If the chemist needs to create 90 ounces of the mixture, how many ounces of Solution C are required?

In order to mix the solution, 3 parts of A, 5 parts of B, 2 parts of C, and 1 part of D must be combined. In other words, 3 + 5 + 2 + 1 = 11 parts are required to complete the mixture. Of those 11 parts, 2 represent Solution C, so C is 2/11 of the total mixture.

The chemist creates 90 ounces of the mixture. Multiply 90 by 2/11 to calculate the amount of Solution C in the mixture.

$$\begin{aligned}\frac{2}{11}(90)&=\frac{180}{11}\text{ ounces}\\&=16.\overline{36}\text{ ounces}\end{aligned}$$

4.81 Kaley was paid an advance of \$4,000 to write a book about digital artistry. Copies of the book sell for \$25, 40% of which is profit for the publisher. Kaley earns 8% of that profit for each book sold, but her first royalty payment will not be awarded until her earnings have exceeded the initial advance payment. If her first royalty payment was \$249.60, how many copies of her book were sold?

Each book sells for \$25, and 40% of that price is profit: (0.40)(25) = \$10. Kaley earns 8% of the \$10 profit, or (0.08)(\$10) = \$0.80 per book. Unfortunately, Kaley will see no money beyond the initial advance of \$4,000 until she repays the publisher that advance in \$0.80 increments for each book sold. She will need to sell \$4,000 ÷ \$0.80 = 5,000 books just to repay the advance.

She has received a royalty check, which means at least 5,000 books sold and the advance has been recouped by the publisher. Her royalty payment was \$249.60, so she sold 249.60 ÷ 0.80 = 312 books above the break-even point for the advance. Kaley's total book sales were 5,000 + 312 = 5,312.

4.82 A rental car company charges \$600 for a weekly (7-day) rental plan that includes unlimited mileage. They also offer a daily rental plan that costs \$80 that includes 10 miles. If you select the daily rental plan, for every mile (or portion thereof) over 10 miles, you are charged \$0.75. What is the highest distance (in whole miles) you can drive in a 7-day period for which the daily rental plan is less expensive than the weekly rental plan?

If you select the daily rental plan each day for 7 days, it costs \$80(7) = \$560, which is \$40 less than the \$600 week-long rental. For each of the 7 days, the daily rental plan allows you to drive 10 miles for free, so you can drive 7(10) = 70 miles without incurring any additional fees. Once you drive farther than 70 miles, you have to pay \$0.75 per mile.

Remember, you are saving \$40 by selecting the daily rental plan. Divide \$40 by \$0.75 to calculate how many additional miles you can travel with the money you saved.

$$\$40 \div \$0.75 = 53.\overline{3}$$

Driving exactly $70 + 53.\overline{3} = 123.\overline{3}$ miles under the daily rental plan is exactly the same price as the weekly rental plan. Thus, 123 is the highest whole number of miles you can drive so that the daily rental plan is less expensive; driving 124 miles would move the total cost of the daily rental plan above \$600.

According to the terms of the daily plan, driving 53.33333 miles counts as driving 54 miles. You are charged for every mile OR PORTION THEREOF. The fee for 54 miles over the 70 free miles is 54(\$0.75) = \$40.50, which would bring the total cost of the daily plan to \$560 + \$40.50 = \$600.50.

Chapter 5

SAT PRACTICE: NUMBERS AND OPERATIONS

Test the Skills You Practiced in Chapter 4

This chapter contains 25 SAT-style questions to help you practice the skills and concepts you explored in Chapter 4. These questions are clustered by difficulty—early questions are easier than later questions. A portion of a student test booklet is reproduced on the following page; use it to record your answers. Full answer explanations are located at the end of the chapter.

Try not to peek at the answers until you have worked through all 25 questions. If you want to practice SAT pacing, try to finish all of the questions in no more than 30-35 minutes.

IMPORTANT NOTE: On the real SAT, if a section contains multiple-choice and grid-in questions, the difficulty resets when you hit the grid-ins. (See Problem 1.4 for more information.) When you hit the practice SATs in Chapter 12, that's exactly what will happen.

However, in these 25-question practice sets (Chapters 5, 7, 9, and 11), the difficulty does not reset. Questions 1–9 will be fairly easy, questions 10–17 will be of medium difficulty, and questions 18–25 are more difficult.

Practice Questions: Numbers and Operations

5.1 ⒶⒷⒸⒹⒺ
5.2 ⒶⒷⒸⒹⒺ
5.3 ⒶⒷⒸⒹⒺ
5.4 ⒶⒷⒸⒹⒺ
5.5 ⒶⒷⒸⒹⒺ
5.6 ⒶⒷⒸⒹⒺ
5.7 ⒶⒷⒸⒹⒺ
5.8 ⒶⒷⒸⒹⒺ
5.9 ⒶⒷⒸⒹⒺ
5.10 ⒶⒷⒸⒹⒺ
5.11 ⒶⒷⒸⒹⒺ
5.12 ⒶⒷⒸⒹⒺ
5.13 ⒶⒷⒸⒹⒺ
5.14 ⒶⒷⒸⒹⒺ
5.15 ⒶⒷⒸⒹⒺ

5.16 **5.17** **5.18** **5.19** **5.20**

5.21 **5.22**

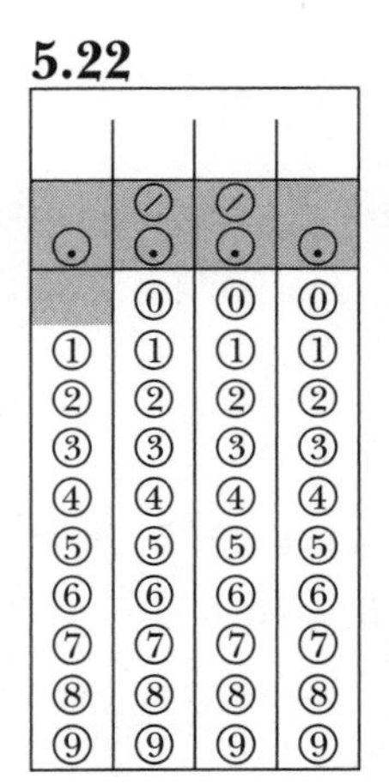

5.23 **5.24** **5.25**

5.1 What percent of 70 is 14?

(A) 5
(B) 15
(C) 17
(D) 20
(E) 30

5.2 If four consecutive integers have a sum of 90, what is the average of those four integers?

(A) 21
(B) 22.5
(C) 24.5
(D) 29
(E) 30

5.3 What is the value of $\left(\sqrt{16}\right)^3$?

(A) 4
(B) $4\sqrt[3]{4}$
(C) 8
(D) 48
(E) 64

5.4 Working at a constant rate, Jill can mow her lawn in 1.5 hours. What percentage of the lawn does she mow in 15 minutes?

(A) 6
(B) 10
(C) $12\frac{2}{3}$
(D) $16\frac{2}{3}$
(E) 60

5.5 If n is the number of unique factors of 18 and p is the number of unique prime factors of 24, what is the value of p^n?

(A) 16
(B) 32
(C) 48
(D) 64
(E) 512

5.6 At a certain pet store, the price of a hermit crab is $8 less than the price of a mouse, and the price of a ferret is $3 more than three times the price of a mouse. Each of the prices includes sales tax. Troy pays a total of $70 to buy one mouse, one ferret, and one hermit crab. How much less would he have paid if he only purchased the ferret?

(A) 15
(B) 22
(C) 42
(D) 48
(E) 55

5.7 Megan is three years older than Elena. Sam is one year younger than Elena. Joey is five years older than Sam. Which of the following statements must be true?

I. Joey is one year older than Megan
II. Joey and Sam's average age is equal to Megan and Elena's average age
III. Joey's age is twice Sam's age

(A) I only
(B) III only
(C) I and II only
(D) II and III only
(E) I, II, and III

5.8 If a number is a multiple of 20, then it must also be divisible by each of the following EXCEPT

(A) 2
(B) 5
(C) 10
(D) 15
(E) 20

5.9 If A is the set of prime numbers and B is the set of odd numbers less than 25, how many elements belong to the set $A \cap B$?

(A) 6
(B) 7
(C) 8
(D) 14
(E) 15

5.10 If the sum of three consecutive integers is positive, which of the following must be true?

I. At least one of the integers must be odd
II. All three integers must be positive
III. At most, one of the integers can be negative

(A) I only
(B) II only
(C) III only
(D) I and II only
(E) II and III only

5.11 In a certain sequence, each term is five less than the square of the previous term. The first term of the sequence is 1. What is the sum of the first three terms of the sequence?

(A) –15
(B) 8
(C) 12
(D) 16
(E) 19

5.12 The ratio of adults to children at a certain sporting event is 5:1. All of the following numbers represent possible attendance totals at the event EXCEPT

(A) 2,938
(B) 2,988
(C) 3,018
(D) 3,078
(E) 3,168

5.13 A certain car's gas tank holds g gallons of gas. The price of gas at a local fueling station is d dollars per gallon. If only a quarter of the car's gas tank is full, how much does it cost to fill the tank?

(A) $\frac{0.75d}{g}$

(B) $\frac{d}{0.75g}$

(C) $\frac{g}{4d}$

(D) $\frac{3dg}{4}$

(E) $\frac{75d}{100g}$

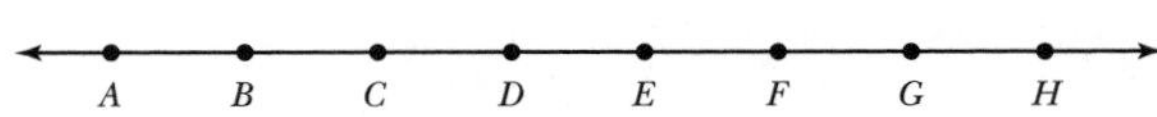

5.14 In the diagram above, eight evenly spaced points are labeled A through H. What is the ratio of $\frac{CH}{BF}$ to $\frac{AD}{FG}$?

(A) $\frac{5}{12}$

(B) $\frac{3}{5}$

(C) $\frac{4}{5}$

(D) $\frac{5}{3}$

(E) $\frac{12}{5}$

Questions 5.15 and 5.16 refer to the following information.

Alex created a computer program that calculates the prime factorization of any positive integer. The program displays the factorization by listing pairs of numbers. The first number in each pair is a prime factor, and the second number is the power to which that prime factor is raised in the factorization. Pairs are separated by periods. For example, if Alex directs the program to calculate the prime factorization of 20, the computer displays "2(2).5(1)," which is equal to $2^2 \cdot 5^1$.

5.15 Which of the following is the prime factorization of 360, as generated by the program?

(A) 2(1).5(1).6(2)
(B) 3(3).2(2).5(2)
(C) 3(2).8(5)
(D) 6(3).12(2)
(E) 2(3).3(2).5(1)

5.16 According to the program, the prime factorizations of two integers are 2(3).7(1) and 2(2).7(2). What is the least common multiple of the two integers?

5.17 If x is $\frac{3}{8}$ of y and y is $\frac{2}{3}$ of z, then z is what percentage of x?

5.18 A store allows children to design and build their own stuffed animals by selecting an animal, the type of stuffing, and a colored jewel that is sewn onto the animal when it is complete. If children can select between 40 different animals, 6 kinds of stuffing, and 9 different colored jewels, how many unique products can the children create?

5.19 Joan is playing on the swings at the playground. After a few minutes, she reaches her maximum height of 12 feet above ground while swinging forward. Each of the next six times she swings forward, her maximum height is 10 percent less than her previous maximum height. How much did her height decrease, in inches, from her maximum height of 12 feet to her height at the end of those six swings?

5.20 Maggie, running at a constant rate, can complete a 200-meter race in 45 seconds. If Sandy runs at a constant rate that is $\frac{2}{5}$ Maggie's rate, how many seconds would it take Sandy to run a 150-meter race?

Club	*Total Enrollment*
Kickball	11
Art	12
Choir	13

5.21 A certain elementary school class contains 25 students, and each student is required to enroll in at least one afterschool club. The students can select one of three clubs: kickball, art, or chorus. The total enrollments for all three clubs are listed in the chart above. If five students in each club enrolled only in that club and only one student enrolled in all three clubs, how many total students are enrolled in the kickball and choir clubs?

5.22 A certain candy is produced in three different flavors: lime, strawberry, and mint. The total number of mint candies produced each day is two times the number of lime candies produced each day, and the total number of lime and strawberry candies produced each day represents 70 percent of total candy production. If 3,000 mint candies are produced each day, how many strawberry candies are produced each day?

5.23 A certain country contains four adjacent time zones A, B, C, and D. The time in A is one hour less than the time in B, the time in B is one hour less than the time in C, and the time in C is one hour less than the time in D. Katie, a flight attendant, is scheduled to work on four consecutive flights, as described in the table above. If the first flight departs from time zone C at 6:00 a.m. local time, at what local time does the fourth flight land? (Omit the colon and "a.m." or "p.m." in your answer. For example, bubble the answer "10:36 p.m." as "1036.")

5.24 A group of 50 adults was asked if they listened to music programming or talk shows on the radio as they commuted to work. Eight adults reported that they listened to both music and talk radio; 12 percent of the adults did not listen to either. The ratio of adults that listened only to music to the adults that listened to talk radio was 1:3. How many adults reported that they listened to music?

5.25 A certain card game uses a standard deck of 52 cards from which the jokers have been removed. To play the game, you draw four cards from the full deck of cards. A winning hand satisfies the following conditions:

1. Your hand must include the ace of hearts.
2. It contains one card of each suit—one heart, one spade, one diamond, and one club.
3. None of the ranks of the cards may match. For example, a winning hand cannot contain a 2 of clubs and a 2 of diamonds.

How many winning hands are possible?

Solutions

5.1 **D**. Divide 14 by 70 to determine what percentage 14 is of 70.

$$14 \div 70 = 0.2 = 20\%$$

5.2 **B**. You might be tempted to represent the four consecutive integers as x, $x + 1$, $x + 2$, and $x + 3$ and then create an equation you can solve for x—the technique demonstrated in Problem 4.3. However, this question can be solved more quickly and without solving any equations.

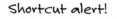

Shortcut alert!

To take the average of four numbers, you add those numbers together and divide by 4. The question already states that the sum of the four consecutive integers is 90, so all you need to do is divide 90 by 4 to calculate the average: $90 \div 4 = 22.5$.

5.3 **E**. Simplify the square root first, as it is contained within parentheses: $\left(\sqrt{16}\right)^3 = (4)^3$. Now raise 4 to the third power: $4^3 = 4 \cdot 4 \cdot 4 = 64$.

5.4 **D**. Convert both times to the same unit of measure, either minutes or hours. In the solution below, 1.5 hours is expressed as 90 minutes. Apply the formula from Problem 4.22 to calculate the percentage, setting part = 15, whole = 90, and percentage = x, the value for which you are solving.

$$\begin{aligned}\frac{\text{part}}{\text{whole}} &= \frac{\text{percentage}}{100}\\ \frac{15}{90} &= \frac{x}{100}\\ 90x &= 1{,}500\\ x &= \frac{1{,}500}{90}\\ x &= 16.\overline{6}\\ x &= 16\frac{2}{3}\end{aligned}$$

5.5 **D**. There are $n = 6$ factors of 18: 1, 2, 3, 6, 9, and 18. The prime factorization of 24 is $2^3 \cdot 3$, so there are $p = 2$ unique prime factors of 24. You conclude that $p^n = 2^6 = 64$.

Use a factor tree to identify a prime factorization. If you need to review factor trees, look at Problem 4.32.

5.6 **B**. Let m represent the price of the mouse. The price h of the hermit crab is \$8 less than the mouse, so $h = m - 8$. The price f of the ferret is \$3 more than three times the price of a mouse, so $f = 3m + 3$. The sum of the three pet prices is \$70. Add the prices (expressed in terms of m) and solve the resulting equation for m.

$$\begin{aligned}m + h + f &= 70\\ m + (m-8) + (3m+3) &= 70\\ (m + m + 3m) + (-8 + 3) &= 70\\ 5m - 5 &= 70\\ 5m &= 75\\ m &= \frac{75}{5}\\ m &= 15\end{aligned}$$

The price of the mouse is m = \$15, the price of the hermit crab is $h = 15 - 8 = \$7$, and the price of the ferret is $f = 3m + 3 = \$48$. If Troy purchased only the ferret, he would have spent \$70 – \$48 = \$22 less than purchasing all three pets.

5.7 C. Create a diagram that illustrates the relative ages of the people named in the question in chronological order. The higher a name appears in the diagram, the younger they are. The diagram below indicates that Megan is three years older than Elena by placing Megan three lines below Elena.

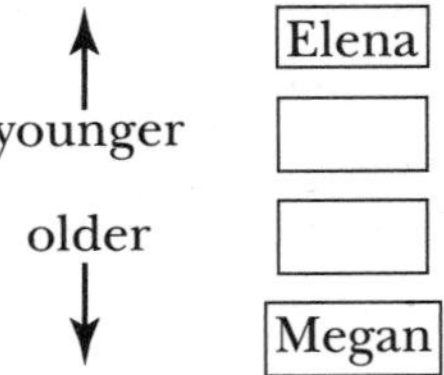

Sam is one year younger than Elena, so she is one line above Elena. Joey is five years older than Sam, so he is listed five lines below Sam.

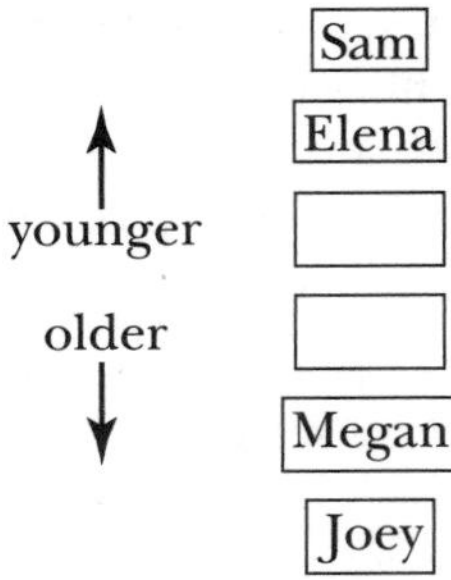

Examine each of the statements I–III, one at a time. Statements I and III are the easiest to verify. Statement I is true because Joey is one line below Megan. You cannot determine the ages of the people, so statement III is false. If Sam were 5 years old then statement III would be true, because Joey would be 10. However, this information is not provided by the question.

You will explore averages in more detail in Chapter 10.

Statement II is a little trickier. It is true, even though you cannot actually determine the age of each person. Visually, the average of two numbers is the balance point between those numbers when listed on a number line. Imagine that the illustration you created is a see-saw. Assuming each person weighed the same, the balance point for Sam and Joey is the same as the balance point for Elena and Megan, the dotted line in the diagram below. Therefore, their average ages are the same.

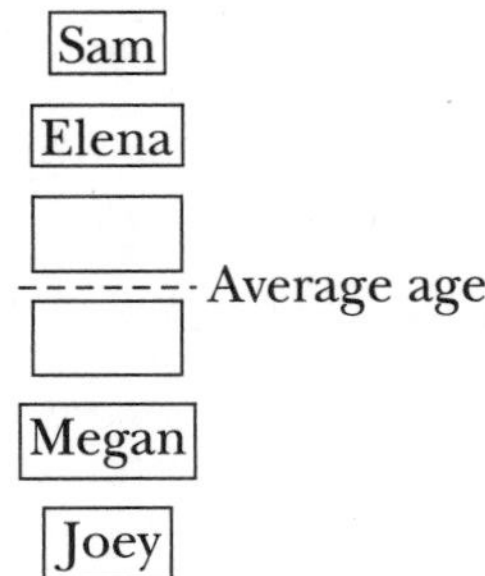

Only statements I and II are true, so the correct answer is C.

5.8 **D**. Apply the DIY technique to answer this question by selecting a multiple of 20, such as 40, 60, 80, or 100. If, for example, you select 100, you note that 100 is divisible by 2, 5, 10, and 20, but not 15; dividing 100 by 15 produces a remainder.

See Problem 2.6.

Note that all multiples of 20 would not have yielded the correct answer. For example, 60 is a multiple of 20 and divisible by all of the choices, including 15. If this happens to you on the SAT, don't panic. Select a different multiple and try again.

5.9 **C**. Your goal is to identify elements that belong to both sets, so think to yourself, "What odd numbers less than 25 are also prime numbers?" The answer is the following set: {3, 5, 7, 11, 13, 17, 19, 23}. There are 8 numbers in that set.

5.10 **A**. Select a few sets of three consecutive positive integers to get a feel for their sums. Focus on integers whose sum is near zero, because this helps you determine which integers will have a positive and negative sum.

integers	sum
$-2, -1, 0$	-3
$-1, 0, 1$	0
$0, 1, 2$	3
$1, 2, 3$	6
$2, 3, 4$	9

Only the last three rows of the chart are applicable—those are the only rows that have a positive sum. Now examine the statements one by one. Statement I is true; each of the rows contains at least one odd number. Three consecutive integers are either "even, odd, even" or "odd, even, odd"; in both cases, the integers contain an odd number.

Statement II may seem true at first glance, but it is actually false. In the third row of the chart, consecutive integers 0, 1, and 2 have a positive sum of 3. Note that 0 is neither a positive nor a negative number, so in this case, only two of the three numbers are positive. Only one counterexample like this is enough to state that all three do not *have* to be positive.

Statement III is false. As the chart demonstrates, a sum that contains any negative integer will be less than or equal to 0, which is not a positive sum.

5.11 **B**. The first term of the sequence is $a_1 = 1$. To generate the second term, square a_1 and subtract 5.

In other words, raise a_1 to the second power.

$$\begin{aligned} a_2 &= (a_1)^2 - 5 \\ &= 1^2 - 5 \\ &= 1 - 5 \\ &= -4 \end{aligned}$$

Repeat the process to generate the third term.

$$\begin{aligned} a_3 &= (a_2)^2 - 5 \\ &= (-4)^2 - 5 \\ &= 16 - 5 \\ &= 11 \end{aligned}$$

The sum of the first three terms is $1 + (-4) + 11 = 8$.

5.12 **A**. If the ratio of adults to children is 5:1, then 5 times some unknown number added to 1 times that same unknown number equals the total attendance. Let x represent that number: total attendance = $5x + 1x = 6x$. This means that the total attendance is a multiple of 6, so the total attendance must be divisible by 6. Only 2,938 has a remainder when you divide it by 6.

5.13 **D**. Apply the DIY technique, selecting values for d and g. For example, if the tank holds $g = 8$ gallons of gas, it contains 2 gallons when it is one-quarter full. If gas costs $d = 3$ dollars per gallon, then the $8 - 2 = 6$ gallons of gas needed to fill the tank would cost $6 \cdot 3 = 18$ dollars.

If you substitute $g = 8$ and $d = 3$ into the answer choices, only (D) generates the correct cost.

$$\frac{3dg}{4} = \frac{3(3)(8)}{4} = \frac{72}{4} = 18$$

5.14 **A**. This question asks you to calculate two ratios and then to calculate the ratio between the ratios. The distance between consecutive points is fixed; although you do not know what that fixed distance is, you do not need it to answer the question. You only need to count the number of fixed distances between the endpoints of each segment and compare them. For example, there are 5 fixed distances between C and H; there are 4 between B and F. Therefore, $CH/BF = 5/4$. Similarly, $AD = 3$ and $FG = 1$, so $AD/FG = 3/1$.

Because the ratio of x to y (or x:y) is x ÷ y.

Now calculate the ratio of the ratios by dividing CH/BF by AD/FG.

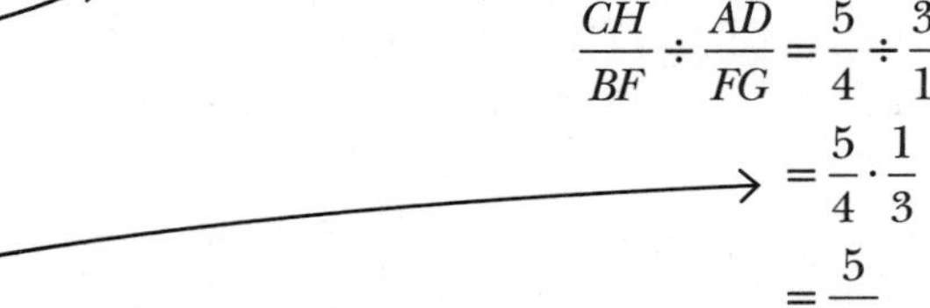

$$\frac{CH}{BF} \div \frac{AD}{FG} = \frac{5}{4} \div \frac{3}{1} = \frac{5}{4} \cdot \frac{1}{3} = \frac{5}{12}$$

To divide a fraction by a fraction, change division to multiplication and take the reciprocal of the fraction by which you are dividing.

5.15 **E**. Problem 4.32 explains how to draw a factor tree in order to identify the prime factorizations of integers. Use this technique to generate the prime factorization of 360.

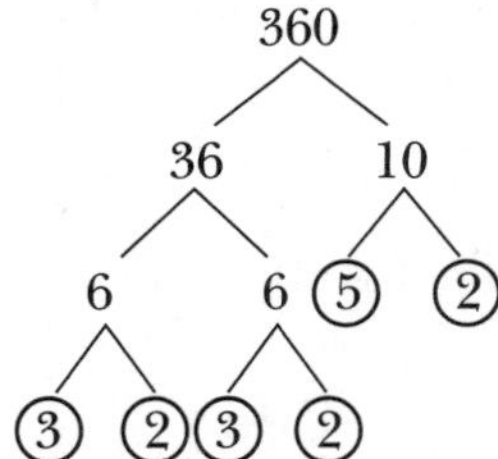

The prime factorization of 360 is $2^3 \cdot 3^2 \cdot 5^1$, which the computer program would display as 2(3).3(2).5(1).

5.16 **392.** The prime factorizations of the integers are $2^3 \cdot 7^1$ and $2^2 \cdot 7^2$. As Problem 4.34 explains, you generate the least common multiple (LCM) by multiplying all of the prime factors, raising each to the highest power from either factorization. These two prime factorizations contain only two prime factors, 2 and 7; the highest exponent of 2 is 3, and the highest exponent of 7 is 2.

$$\begin{aligned} \text{LCM} &= 2^3 \cdot 7^2 \\ &= 8 \cdot 49 \\ &= 392 \end{aligned}$$

5.17 **400.** If x is $\frac{3}{8}$ of y and y is $\frac{2}{3}$ of z, then you can create the following two equations.

$$x = \frac{3}{8}y \qquad y = \frac{2}{3}z$$

Substitute $y = \frac{2}{3}z$ into the first equation and reduce the fraction to lowest terms.

$$\begin{aligned} x &= \frac{3}{8}y \\ x &= \frac{3}{8}\left(\frac{2}{3}z\right) \\ x &= \frac{6}{24}z \\ x &= \frac{1}{4}z \end{aligned}$$

This equation reads "x is one-fourth the size of z." In other words, z is four times as large as x, so z is 400% of x.

5.18 **2160.** According to the fundamental counting principle, if there are 40 ways to select an animal, 6 ways to select a type of stuffing, and 9 ways to select a type of colored jewel, then there are $40 \cdot 6 \cdot 9 = 2{,}160$ unique ways to create a stuffed animal.

5.19 **67.4** or **67.5.** Joan's maximum swing height decreases six times. Each time, her maximum height is 10 percent less than, or $100 - 10 = 90$ percent of, her previous maximum height. The question specifically tells you to calculate the answer in inches, so multiply by 12 to convert 12 feet into inches.

Because there are 12 inches in a foot. If your answer was 6.37 or 6.38, you did not convert from feet into inches.

Multiply the maximum height of 144 inches by 90 percent (or 0.9) six times to calculate the height she reaches in the following six swings.

Initial maximum height:	144 inches
Height of swing 1:	$144(0.90) = 129.6$ inches
Height of swing 2:	$129.6(0.90) = 116.64$ inches
Height of swing 3:	$116.64(0.90) = 104.976$ inches
Height of swing 4:	$104.976(0.90) = 94.4784$ inches
Height of swing 5:	$94.4784(0.90) = 85.03056$ inches
Height of swing 6:	$85.03056(0.90) = 76.527504$ inches

Joan's maximum height decreased $144 - 76.527504 = 67.472496$ inches.

5.20 **84.3** or **84.4.** This question is similar to Problem 4.76; both require you to apply the formula $d = r \cdot t$. Maggie can run $d = 200$ meters in $t = 45$ seconds. Substitute these values into the formula to calculate her constant rate r.

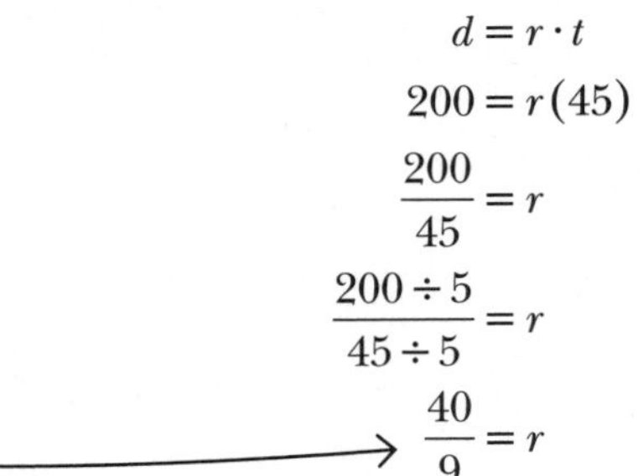

$$\begin{aligned} d &= r \cdot t \\ 200 &= r(45) \\ \frac{200}{45} &= r \\ \frac{200 \div 5}{45 \div 5} &= r \\ \frac{40}{9} &= r \end{aligned}$$

You could also use the decimal form of her rate: $R = 4.\overline{4}$.

Maggie's constant rate is 40/9 meters per second. Sandy's rate is 2/5 times Maggie's rate.

$$\begin{aligned} \text{Sandy's rate} &= \frac{2}{5}\left(\frac{40}{9}\right) \\ &= \frac{80}{45} \\ &= \frac{80 \div 5}{45 \div 5} \\ &= \frac{16}{9} \end{aligned}$$

Substitute $d = 150$ and $r = 16/9$ into the $d = r \cdot t$ formula to calculate the time t, in seconds, it takes Sandy to run a 150-meter race.

$$\begin{aligned} d &= r \cdot t \\ 150 &= \frac{16}{9}t \\ \frac{150}{1}\left(\frac{9}{16}\right) &= t \\ \frac{1{,}350}{16} &= t \\ 84.375 &= t \end{aligned}$$

5.21 **4.** Create a Venn diagram that illustrates the information given in the question. In the following diagram, K represents the set of students in the kickball club, A represents the set of students in the art club, and C represents the set of students in the chorus club. Each set contains 5 students that are unique to that set, 5 students who are not enrolled in any other clubs. The intersection of all three sets contains 1 student who participated in all three clubs. All students are required to participate, so the number in the lower-right corner of the diagram, representing the number of students who do not belong to any of the three sets, is 0. Three values, labeled x, y, and z, cannot be immediately determined based on the given information, but you do know that the sum of x, y, and z is $25 - 16 = 9$.

There are 25 students, so the total of all the regions in the diagram must be 25. At the moment, you have values for 5 of the regions: 5, 5, 5, 1, and 0. The three remaining regions (x, y, and z) account for the remaining 9 students.

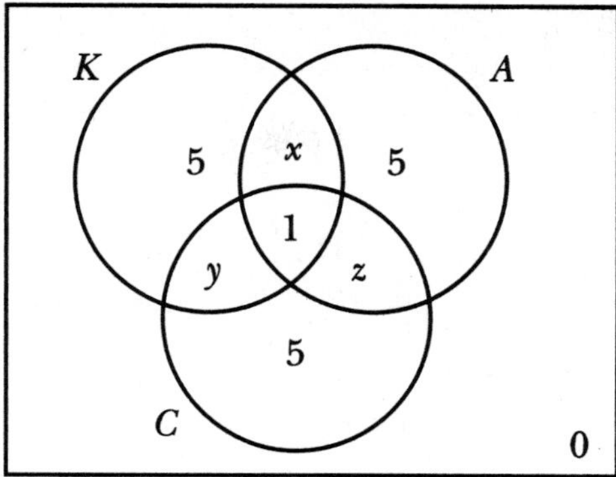

Add the numbers in each set and set it equal to the enrollment totals provided in the chart. For example, according to the Venn diagram, $5 + x + y + 1$ students participate in kickball. The chart indicates that 11 students are enrolled in kickball. Set those values equal and simplify.

$$\begin{aligned} 5+x+y+1&=11 \\ x+y+6&=11 \\ x+y&=11-6 \\ x+y&=5 \end{aligned}$$

Repeat this process for art and choir.

Art Enrollment	Choir Enrollment
$5+x+1+z=12$	$5+y+1+z=13$
$6+x+z=12$	$6+y+z=13$
$x+z=12-6$	$y+z=13-6$
$x+z=6$	$y+z=7$

Notice that $x + y = 5$ and $x + z = 6$, so z must be exactly 1 greater than y. Similarly, $x + y = 5$ and $y + z = 7$, so z must be 2 greater than x. If z is 1 greater than y and 2 greater than x, then x, y, and z are consecutive integers: x, $y = x + 1$, and $z = x + 2$. Recall that the sum of x, y, and z must be 9, as explained earlier.

$$\begin{aligned} x+y+z&=9 \\ x+(x+1)+(x+2)&=9 \\ 3x+3&=9 \\ 3x&=6 \\ x&=\frac{6}{3} \\ x&=2 \end{aligned}$$

Thus, $x = 2$, $y = 3$, and $z = 4$. Complete the Venn diagram with the new information.

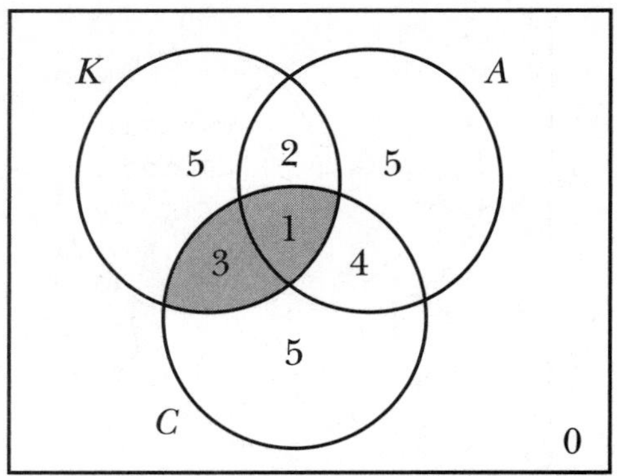

The question asks you to calculate the number of students in the intersection $K \cap C$, the shaded region in the diagram. Three students are enrolled in kickball and choir clubs; one student is enrolled in kickball, choir, and art clubs. Therefore, $3 + 1 = 4$ students are enrolled in kickball and choir clubs.

5.22 **5500**. As you read the question, translate the important information into algebraic equations. Let l represent the number of lime candies, s represent the number of strawberry candies, and m represent the number of mint candies produced each day. According to the question, $m = 2l$ and $l + s = 0.7(l + s + m)$.

You are given $m = 3{,}000$. Substitute that into the equation $m = 2l$ to calculate l.

$$\begin{aligned} m &= 2l \\ 3{,}000 &= 2l \\ \frac{3{,}000}{2} &= l \\ 1{,}500 &= l \end{aligned}$$

Substitute $m = 3{,}000$ and $l = 1{,}500$ into the equation $l + s = 0.7(l + s + m)$ to calculate s.

$$\begin{aligned} l + s &= 0.7(l + s + m) \\ 1{,}500 + s &= 0.7(1{,}500 + s + 3{,}000) \\ 1{,}500 + s &= 0.7(4{,}500 + s) \\ 1{,}500 + s &= 0.7(4{,}500) + 0.7s \\ 1{,}500 + s &= 3{,}150 + 0.7s \\ s - 0.7s &= 3{,}150 - 1{,}500 \\ 0.3s &= 1{,}650 \\ s &= \frac{1{,}650}{0.3} \\ s &= 5{,}500 \end{aligned}$$

This solution requires a bit of algebra, so it could also have appeared in the Algebra and Functions practice test. It is located here because percentages are a key component of the question. Difficult questions often require you to apply skills from different content areas.

5.23 **250**. Investigate each of Katie's flights independently, focusing on calculating the time her next flight leaves.

If you are used to U.S. time zones, D is Eastern Time, C is Central Time, B is Mountain Time, and A is Pacific Time.

- Flight 1 leaves at 6:00 a.m. from time zone *C*. The flight lasts 2.5 hours, so it lands at 8:30 a.m. (in *C* time). However, the flight lands in *A*, which is two hours less than *C*. Thus, the flight lands at 6:30 a.m. in *A* time. She waits 45 minutes, or until 7:15 a.m. *A* time, until the next flight.
- Flight 2 leaves at 7:15 a.m. from time zone *A*. This flight lasts 4 hours, so it lands at 11:15 a.m. (in *A* time). Because the flight lands in time zone *B*, you need to add one hour to convert to *B* time, 12:15 p.m. She waits 30 minutes until the next flight, so it leaves at 12:45 p.m. *B* time.
- Flight 3 leaves at 12:45 p.m. *B* time. The flight lasts 5.75 hours and lands at 6:30 p.m. (in *B* time). The flight lands in time zone *D*, which is two hours ahead of *B*, so add 2 hours; the arrival time is 8:30 p.m. *D* time. She waits 80 minutes, or until 9:50 p.m. *D* time for the final flight to depart.
- Flight 4 leaves at 9:50 p.m. *D* time. The final flight of Katie's air odyssey is 8 hours, so she lands at 5:50 a.m. *D* time. However, she flew from *D* to *A*, so three hours must be subtracted to convert to local *A* time.

You conclude that Katie lands at 2:50 a.m., in local *A* time.

5.24 **17.** This question verbally describes a situation that is more succinctly illustrated in a Venn diagram. In the diagram below, *M* is the set of adults that listen to music on the radio and *T* is the set of adults that listen to talk radio. According to the question, 8 adults listen to music and talk radio, so $M \cap T = 8$. The question also states that 12 percent of 50 adults (or $0.12 \cdot 50 = 6$ adults) belong neither to set *T* nor to set *M*.

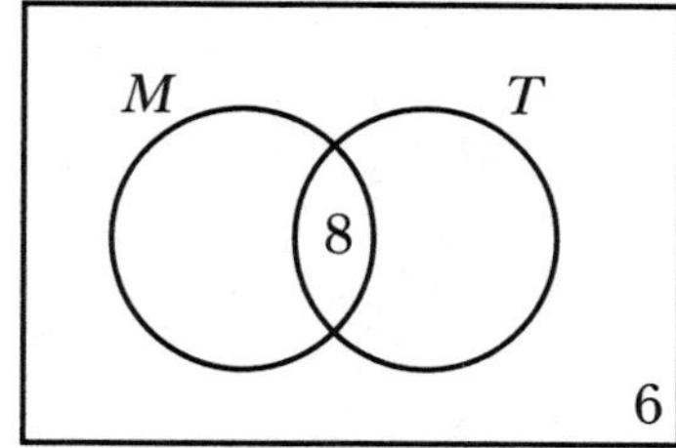

Subtract 8 and 6 from 50 to determine how many more adults must be accounted for in the diagram: $50 - 8 - 6 = 36$. There are two spaces left to fill, the non-intersecting portions of sets *M* and *T*, so the numbers in those positions must have a sum of 36.

If x is the number of people who listen only to music, then $3x$ is the number of people who listen only to talk radio, because the question states the values are in the ratio 1:3. Recall that the sum of x and $3x$ must be 36.

$$\begin{aligned} x + 3x &= 36 \\ 4x &= 36 \\ x &= \frac{36}{4} \\ x &= 9 \end{aligned}$$

Therefore, $x = 9$ adults listen only to music and $3x = 27$ adults listen only to talk radio.

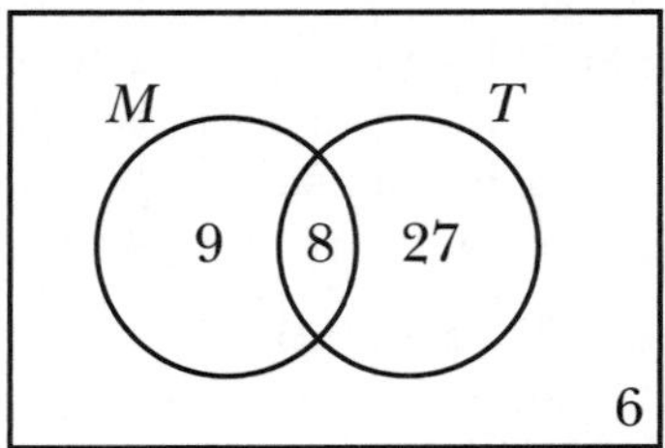

To answer the question, you need to calculate the number of adults who listen to music, the number of elements in set M. This includes the adults that listen only to music and the adults that listen both to music and talk radio. The answer is $9 + 8 = 17$.

5.25 **7920.** Apply the fundamental counting principle, counting the number of ways you can select each card. One of the cards must be the ace of hearts, but it does not matter which, so for the sake of simplicity, you can assume that it is the first card.

Number of ways to select ace of hearts	Number of ways to select Card 2	Number of ways to select Card 3	Number of ways to select Card 4

The cards are: Ace, 2, 3, 4, 5, 6, 7, 8, 9, 10, Jack, Queen, and King.

Only 1 of the 52 cards satisfies the first condition—there is only 1 ace of hearts. Each of the remaining cards needs to be a different suit. A deck is separated into four different suits, so the deck contains $52 \div 4 = 13$ cards of each suit. How many ways can you select the second card? Consider the diagram below, which lists all of the cards of each rank. When you select the ace of hearts as Card 1, that eliminates all of the heart cards and all of the aces as candidates for Card 2.

♥	~~A~~	~~2~~	~~3~~	~~4~~	~~5~~	~~6~~	~~7~~	~~8~~	~~9~~	~~10~~	~~J~~	~~Q~~	~~K~~
♦	~~A~~	2	3	4	5	6	7	8	9	10	J	Q	K
♣	~~A~~	2	3	4	5	6	7	8	9	10	J	Q	K
♠	~~A~~	2	3	4	5	6	7	8	9	10	J	Q	K

There are 36 cards left from which to select Card 2.

1	36	Number of ways to select card 3	Number of ways to select card 4

It doesn't matter which card you choose here. You will eliminate the same number of cards from the pool, 2 cards of the same rank and 11 cards of the same suit.

Now imagine that you draw the 2 of diamonds as Card 2. All of the remaining diamonds and 2s are eliminated as possible choices for Card 3.

♥	~~A~~	~~2~~	~~3~~	~~4~~	~~5~~	~~6~~	~~7~~	~~8~~	~~9~~	~~10~~	~~J~~	~~Q~~	~~K~~
♦	~~A~~	~~2~~	~~3~~	~~4~~	~~5~~	~~6~~	~~7~~	~~8~~	~~9~~	~~10~~	~~J~~	~~Q~~	~~K~~
♣	~~A~~	~~2~~	3	4	5	6	7	8	9	10	J	Q	K
♠	~~A~~	~~2~~	3	4	5	6	7	8	9	10	J	Q	K

There are 22 ways to select Card 3.

Imagine that you draw the 3 of clubs as Card 3. Update the chart to reflect that Card 4 cannot be any of the remaining clubs or the 3 of spades.

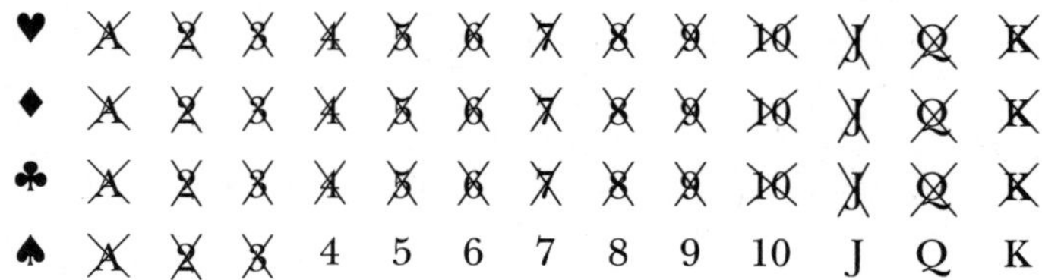

There are 10 ways to select Card 4.

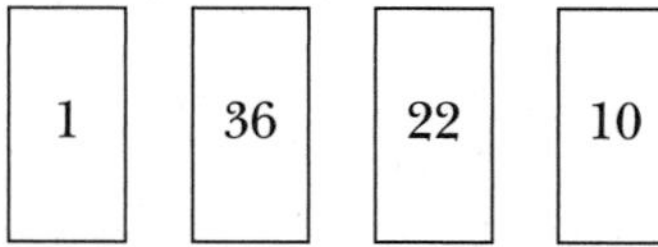

The final answer is $1 \cdot 36 \cdot 22 \cdot 10 = 7{,}920$ possible winning hands.

Chapter 6

MATH REVIEW: ALGEBRA AND FUNCTIONS

37% of the SAT

Almost two out of every five questions on the SAT fall into the "algebra and functions" category. Why? There are two reasons. First, algebraic reasoning is a fundamental skill required by most colleges. Second, algebra is a collection of many different skills and concepts, so it can only be assessed accurately with a large collection of questions. In this chapter, you practice all of the skills and concepts listed below:

- Absolute values
- Direct and indirect variation
- Domain and range of a function
- Evaluating expressions
- Factoring expressions
- Fractional exponents
- Graphs of equations, functions, and inequalities
- Linear and quadratic equations
- Linear and quadratic inequalities
- Mathematical modeling with functions
- Rational equations and inequalities
- Systems of equations and inequalities
- Transformations of functions
- Translating expressions
- Word problems that require equations to solve

While the SAT does not include complex numbers, logarithms, or trigonometry, there are still a lot of concepts to master. As you use the problems in this chapter to review your algebra skills, check off the topics above once you feel comfortable with them. After your review is complete, move on to Chapter 7 to practice SAT-style algebra and function questions.

This is the longest chapter in the book, with over 100 questions, because it reviews the largest category of questions on the SAT. You will also find that content in this chapter sneaks its way into other categories of the test, meaning that algebra and functions questions actually represent MORE than 37% of the test. If you need more in-depth practice with algebra, pick up The Humongous Book of Algebra Problems and work your way through that book as well.

Algebraic Expressions

Working with variables

6.1 If $x = 2$ and $y = -6$, what is the value of $\frac{x+y}{y-x}$?

Substitute $x = 2$ and $y = -6$ into the expression and simplify the fraction to lowest terms.

$$\begin{aligned}\frac{x+y}{y-x} &= \frac{2+(-6)}{-6-(2)} \\ &= \frac{2-6}{-6-2} \\ &= \frac{-4}{-8} \\ &= \frac{-4\div(-4)}{-8\div(-4)} \\ &= \frac{1}{2}\end{aligned}$$

6.2 What is the value of $3m^2 - 7m + 1$ when $m = -1$?

Substitute $m = -1$ into the expression and simplify.

$$\begin{aligned}3m^2 - 7m + 1 &= 3(-1)^2 - 7(-1) + 1 \\ &= 3(1) - 7(-1) + 1 \\ &= 3 + 7 + 1 \\ &= 11\end{aligned}$$

6.3 If $x = 3y$ and $y = 4z$, what is the value of z in terms of x?

The question asks you to express z in terms of x, which means your final answer should begin with "$z =$" and contain x's and real numbers after the equal sign. Therefore, you need to eliminate the variable y from the equations. Substitute $y = 4z$ into the equation $x = 3y$.

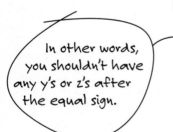

In other words, you shouldn't have any y's or z's after the equal sign.

$$\begin{aligned}x &= 3y \\ x &= 3(4z) \\ x &= 12z\end{aligned}$$

To express z in terms of x, solve the equation for z.

$$\begin{aligned}\frac{x}{12} &= \frac{12z}{12} \\ \frac{x}{12} &= z \\ z &= \frac{x}{12}\end{aligned}$$

6.4 On a recent algebra quiz, David had d correct answers. Michael's correct answers numbered 12 more than one-third of David's correct answers. If David answered 24 questions correctly, how many questions did Michael answer correctly?

If d is the number of David's correct answers, Michael has $(1/3)d + 12$ correct answers. Substitute $d = 24$ into the expression.

$$\begin{aligned} \text{Michael's correct answers} &= \frac{1}{3}d + 12 \\ &= \frac{1}{3}(24) + 12 \\ &= \frac{24}{3} + 12 \\ &= 8 + 12 \\ &= 20 \end{aligned}$$

6.5 If y is nine less than three times the value of x, what is the value of $x - 3$ in terms of y?

The problem states that y is nine less than three times the value of x, so $y = 3x - 9$. You are asked to draw conclusions about the value of $x - 3$. Notice that dividing $3x - 9$ by 3 produces the desired expression, $x - 3$. If you divide one side of the equation by 3, in order to preserve equality, you also need to divide the other side of the equation by 3.

$$\begin{aligned} y &= 3x - 9 \\ \frac{y}{3} &= \frac{3x}{3} - \frac{9}{3} \\ \frac{y}{3} &= x - 3 \end{aligned}$$

You conclude that the value of $x - 3$ in terms of y is $y/3$.

6.6 If $\frac{2x}{y} = \frac{4}{3}$, what is the value of $\frac{x}{9y}$?

The given proportion contains the fraction $2x/y$ on the left side of the equal sign. You are asked to draw conclusions about the value of $x/9y$. Multiplying the left side of the equation by 1/18 transforms $2x/y$ into the desired fraction $x/9y$, but if you multiply one side of an equation by 1/18, you also need to multiply the other side by 1/18.

Here's how you get 1/18. Multiplying $2x/y$ by some unknown value equals $x/9y$. To solve for ?, multiply both sides by the reciprocal of $2x/y$.

$$\left(\frac{y}{2x}\right)\left(\frac{2x}{y}\right) \cdot ? = \left(\frac{y}{2x}\right)\left(\frac{x}{9y}\right)$$

$$? = \frac{\cancel{x}\cancel{y}}{18\cancel{x}\cancel{y}}$$

$$? = \frac{1}{18}$$

$$\frac{2x}{y}=\frac{4}{3}$$

$$\left(\frac{1}{18}\right)\left(\frac{2x}{y}\right)=\left(\frac{1}{18}\right)\left(\frac{4}{3}\right)$$

$$\frac{2x}{18y}=\frac{4}{54}$$

Reduce the fractions to lowest terms.

$$\frac{x}{9y}=\frac{2}{27}$$

You conclude that the value of $x/9y$ is $2/27$.

6.7 Which is larger when $x = -2$: (A) $x - x^2$ or (B) $x\sqrt{x^2}$?

Substitute $x = -2$ into both of the expressions.

If you plot −4 and −6 on a number line, −4 is right of −6, which means −4 is greater than −6.

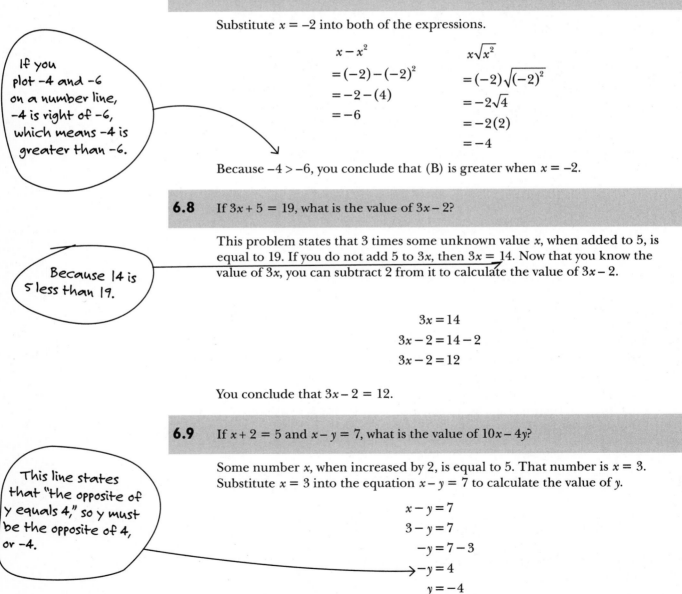

$$\begin{aligned} &x-x^2 \\ &=(-2)-(-2)^2 \\ &=-2-(4) \\ &=-6 \end{aligned} \qquad \begin{aligned} &x\sqrt{x^2} \\ &=(-2)\sqrt{(-2)^2} \\ &=-2\sqrt{4} \\ &=-2(2) \\ &=-4 \end{aligned}$$

Because $-4 > -6$, you conclude that (B) is greater when $x = -2$.

6.8 If $3x + 5 = 19$, what is the value of $3x - 2$?

This problem states that 3 times some unknown value x, when added to 5, is equal to 19. If you do not add 5 to $3x$, then $3x = 14$. Now that you know the value of $3x$, you can subtract 2 from it to calculate the value of $3x - 2$.

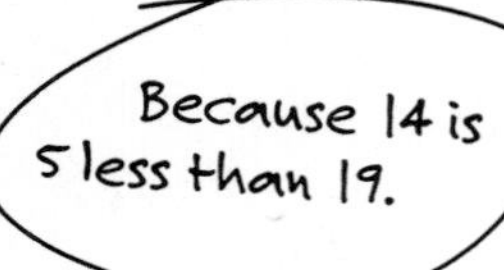
Because 14 is
5 less than 19.

$$\begin{aligned} 3x&=14 \\ 3x-2&=14-2 \\ 3x-2&=12 \end{aligned}$$

You conclude that $3x - 2 = 12$.

6.9 If $x + 2 = 5$ and $x - y = 7$, what is the value of $10x - 4y$?

Some number x, when increased by 2, is equal to 5. That number is $x = 3$. Substitute $x = 3$ into the equation $x - y = 7$ to calculate the value of y.

This line states that "the opposite of y equals 4," so y must be the opposite of 4, or −4.

$$\begin{aligned} x-y&=7 \\ 3-y&=7 \\ -y&=7-3 \\ -y&=4 \\ y&=-4 \end{aligned}$$

Evaluate the expression $10x - 4y$ for $x = 3$ and $y = -4$.

$$\begin{aligned}10x - 4y &= 10(3) - 4(-4)\\ &= 30 + 16\\ &= 46\end{aligned}$$

Factoring Expressions

Pulling out the greatest common factor

Note: Problems 6.10–6.11 explain how to factor the expression $18x^3 + 24x^2$.

6.10 Identify the greatest common factor of the coefficients in the expression.

You can generate the prime factorizations of 18 and 24 by constructing factor trees, as demonstrated in Problem 4.32. You conclude that $18 = 2 \cdot 3^2$ and $24 = 2^3 \cdot 3$. Both factorizations contain prime factors 2 and 3, so the greatest common factor (GCF) is the product of those shared factors when each is raised to the lowest power from either factorization: GCF $= 2^1 \cdot 3^1 = 6$.

Factor trees are not required to calculate the GCF, especially when the numbers involved are not very large. You may be able to identify the GCF simply by asking yourself, "What is the largest number that divides into 18 and 24 evenly?" If you guess too low, you can still derive the correct answer.

For example, If you hypothesized that 2 was the GCF of 18 and 24, divide both of those numbers by 2 to see if the resulting quotients have common factors: $18 \div 2 = 9$ and $24 \div 2 = 12$. Because 9 and 12 are both divisible by 3, you should once again divide the numbers by the common factor: $9 \div 3 = 3$ and $12 \div 3 = 4$. The quotients 3 and 4 share no common factors, so the GCF of 18 and 24 is the product of the two common factors you identified: $2 \cdot 3 = 6$.

Note: Problems 6.10–6.11 explain how to factor the expression $18x^3 + 24x^2$.

6.11 Factor the expression.

When you factor an expression, you "unmultiply," identifying the GCF of the terms and, essentially, dividing it out of the terms. This is a useful tool when simplifying expressions, especially fractions.

According to Problem 6.10, the GCF of 18 and 24 is 6. Factor it out of the terms, writing it outside a set of parentheses and dividing each of the terms within by the GCF, as demonstrated below.

$$\begin{aligned}18x^3 + 24x^2 &= 6\left(\frac{18}{6}x^3 + \frac{24}{6}x^2\right)\\ &= 6\left(3x^3 + 4x^2\right)\end{aligned}$$

You are not yet finished factoring the expression. Notice that both terms contain a factor of x raised to an exponential power. Select the lowest power of x and factor it out of the terms, once again dividing the terms by the common factor, x^2.

When you divide a variable to a power by the same variable to another power, the answer is that variable raised to the top power minus the bottom power. In this case, $x^3 \div x^2 = x^{3-2} = x^1$, or x. Note that $x^0 = 1$, because anything raised to the 0 power is 1 (except for 0, but don't worry about why).

$$= 6x^2\left(\frac{3x^3}{x^2} + \frac{4x^2}{x^2}\right)$$
$$= 6x^2\left(3x^{3-2} + 4x^{2-2}\right)$$
$$= 6x^2\left(3x^1 + 4x^0\right)$$
$$= 6x^2\left(3x + 4\cdot 1\right)$$
$$= 6x^2\left(3x + 4\right)$$

Notice that the result is the GCF of the original terms, $6x^2$, multiplied by two terms in parentheses. If you distribute the GCF to the terms in the parentheses, you get the terms from the original problem: $6x^2(3x) = 18x^3$ and $6x^2(4) = 24x^2$.

6.12 Factor the expression: $4x^3y^2 - 20x^2y^4 + 36xy^5$.

The GCF of the coefficients is 4. All of the terms contain an x- and a y-term, so the GCF will contain x and y raised to their lowest powers in any of the terms: x^1 and y^2. Therefore, the GCF of the three terms in this expression is $4xy^2$. Write this term outside a set of parentheses, and within the parentheses, divide the GCF into each of the terms.

$$4x^3y^2 - 20x^2y^4 + 36xy^5 = 4xy^2\left(\frac{4x^3y^2}{4xy^2} - \frac{20x^2y^4}{4xy^2} + \frac{36xy^5}{4xy^2}\right)$$
$$= 4xy^2\left(\frac{4}{4}x^{3-1}y^{2-2} - \frac{20}{4}x^{2-1}y^{4-2} + \frac{36}{4}x^{1-1}y^{5-2}\right)$$
$$= 4xy^2\left(1x^2y^0 - 5x^1y^2 + 9x^0y^3\right)$$
$$= 4xy^2\left(x^2 - 5xy^2 + 9y^3\right)$$

Remember that you can distribute the GCF to the terms in the parentheses to check your work. If you have calculated correctly, the result should be the three terms from the original expression.

6.13 Explain how to factor the quadratic expression $x^2 + Ax + B$, based on the values of A and B.

The highest power of the variable x in this expression is 2; it occurs in the first term, x^2. That highest power is called the *degree* of the expression, so this expression has degree 2. Expressions with degree 2 are called *quadratic expressions.* The coefficient of the term containing the variable raised to that highest power is called the *leading coefficient.* The coefficient of the term x^2—the term containing the variable raised to the highest power, as noted earlier—is 1, so the leading coefficient of this expression is 1. In conclusion, the three-term expression $x^2 + Ax + B$ is a quadratic expression with leading coefficient 1.

To factor a quadratic expression with a leading coefficient of 1, you need to identify two numbers that satisfy two specific requirements. First, the sum of the numbers must equal the coefficient of the x-term. In the expression $x^2 + Ax + B$, the coefficient of the x-term is A. Second, the product of the two numbers must be equal to the constant, in this case B.

The "constant" in an expression is the term that does not contain any variables. Think of it this way: It is not variable because it is constant. In the expression $x^2 + Ax + B$, the only term that does not contain the variable x is B, so B is the constant.

Let r and s represent the two numbers as described above. They have a sum of A, which means $r + s = A$. Their product is B, so $r \cdot s = B$. Once you have identified r and s, you conclude that the factored form of the quadratic expression is $(x + r)(x + s)$. Although this process may sound complicated, in practice, it is actually very straightforward. In fact, it basically boils down to a puzzle—a number game.

Note: Problems 6.14–6.15 refer to the expression $x^2 - 8x + 15$.

6.14 Factor the expression.

You are given a quadratic expression with leading coefficient 1, so you can apply the technique described in Problem 6.13. Your goal is to find two numbers that have a sum of –8 and a product of 15. In Problem 6.13, those numbers are described as r and s.

Notice that the numbers have a negative sum but a positive product. Two numbers only have a positive product when they have the same sign—a positive times a positive equals a positive and a negative times a negative equals a positive. Therefore, both of the numbers you are trying to identify must be negative, or they could not have a negative sum! The two negative numbers with a sum of –8 and a product of 15 are $r = -3$ and $s = -5$ (or $s = -3$ and $r = -5$, because the order in which you name the numbers does not matter). Factor the expression, as explained in Problem 6.13.

$$\begin{aligned} x^2 - 8x + 15 &= (x + r)(x + s) \\ &= (x + (-3))(x + (-5)) \\ &= (x - 3)(x - 5) \end{aligned}$$

Note: Problems 6.14–6.15 refer to the expression $x^2 - 8x + 15$.

6.15 Verify your solution to Problem 6.14 by multiplying the factors and verifying that the product is the original expression.

According to Problem 6.14, the factored form of $x^2 - 8x + 15$ is $(x - 3)(x - 5)$. To multiply the factors $(x - 3)$ and $(x - 5)$, multiply one of the factors by each term of the other factor and add the results. For example, multiply the factor $(x - 5)$ by the two terms of the other factor, x and –3.

$$\begin{aligned} (x - 3)(x - 5) &= x(x - 5) + (-3)(x - 5) \\ &= x^2 - 5x + (-3x) + (-3)(-5) \\ &= x^2 - 5x - 3x + 15 \\ &= x^2 - 8x + 15 \end{aligned}$$

You might have used something called the FOIL technique to multiply $(x-3)(x-5)$. The letters in FOIL explain which pairs of terms you should multiply:

F = First. Multiply the first terms in each factor: $x \cdot x = x^2$.

O = Outside. Multiply the terms at the outside, far ends of the product, x at the far left and -5 at the far right: $x(-5) = -5x$.

I = Inside. Multiply the inner terms of the product together: $(-3)(x) = -3x$.

L = Last. Multiply the last terms in each factor: $(-3)(-5) = 15$.

The final result is the same: $x^2 - 3x - 5x + 15 = x^2 - 8x + 15$.

6.16 Factor the expression $x^2 + 6x + 9$ and verify the solution by multiplying the factors.

You are given a quadratic expression with leading coefficient 1, so apply the technique described in Problem 6.13. Your goal is to identify two numbers that have a sum of 6 and a product of 9. The "numbers" are actually the same! Only the numbers 3 and 3 have a sum of 6 and a product of 9. Therefore, $x^2 + 6x + 9 = (x+3)(x+3)$.

You could also state that $x^2 + 6x + 9 = (x+3)^2$.

Verify your solution by multiplying the factors.

$$\begin{aligned}(x+3)(x+3) &= x(x+3)+3(x+3)\\ &= x^2+3x+3x+9\\ &= x^2+6x+9\end{aligned}$$

6.17 Factor the expression: $m^2 + 9m - 22$.

To factor this quadratic, which has leading coefficient 1, identify the two numbers with a sum of 9 and a product of –22. Because the product is negative, the two numbers must have different signs—multiplying two numbers with the same sign always results in a positive product, as explained in Problem 6.14. Only the numbers 11 and –2 fit these conditions: $11 + (-2) = 9$ and $11(-2) = -22$. You conclude that $m^2 + 9m - 22 = (m-2)(m+11)$.

6.18 Factor the expression $x^2 - 49$ and verify your answer by multiplying the factors.

You could apply the method from Problem 6.13, but this expression is a special case—a two-term quadratic expression in which one perfect square is subtracted from another perfect square. This special case, called a *difference of perfect squares*, is written $a^2 - b^2$ and is factored according to this pattern: $(a+b)(a-b)$.

Remember, a perfect square is the result of something multiplied times itself, like $100 = 10 \cdot 10$ or $9y^2 = 3y \cdot 3y$.

In this problem, you are given $x^2 - 49$. If $a^2 - b^2 = x^2 - 49$, then $a = x$ and $b = 7$, as demonstrated below.

$$\begin{aligned}a^2 - b^2 &= (x)^2 - (7)^2\\ &= x^2 - 49\end{aligned}$$

Any difference of perfect squares follows the factor pattern $a^2 - b^2 = (a+b)(a-b)$, so you conclude that $x^2 - 49 = (x+7)(x-7)$.

6.19 Factor the expression: $p^4 - 81$.

This expression is a difference of perfect squares. Set $a = p^2$ and $b = 9$, and apply the difference of perfect squares factoring pattern.

$$a^2 - b^2 = (a+b)(a-b)$$
$$(p^2)^2 - (9)^2 = (p^2+9)(p^2-9)$$
$$p^4 - 81 = (p^2+9)(p^2-9)$$

You are not yet finished factoring the expression! One of the factors also is a difference of perfect squares: $p^2 - 9 = (p+3)(p-3)$.

$$p^4 - 81 = (p^2+9)(p+3)(p-3)$$

6.20 Factor the expression: $2x^2 + 11x + 5$.

When the leading coefficient of a quadratic expression is not 1, you need to modify the factoring process. Your final goal remains the same: find two numbers with a specific sum and a specific difference, but now you must take the leading coefficient into account.

In order to factor $Ax^2 + Bx + C$ (if $A \neq 0$ and $A \neq 1$), you need to identify two numbers whose sum is B and whose product is $A \cdot C$. In this problem, you have to find two numbers whose sum is 11 and whose product is $2 \cdot 5 = 10$. Those two numbers are 1 and 10, because $1 + 10 = 11$ and $1(10) = 10$.

This is the same, whether the leading coefficient is 1 or not. The sum of the mystery numbers always equals the x-coefficient. The only thing that changes is the product of the mystery numbers, which equals A · C instead of C.

You will now apply a process called *factoring by decomposition*. Rewrite the original expression, replacing the middle term, $11x$, with the sum of the mystery numbers you identified: $11x = 1x + 10x$.

$$2x^2 + 11x + 5 = 2x^2 + 1x + 10x + 5$$

Break the four-term expression into two mini-expressions, each containing two terms: $2x^2 + 1x$ and $10x + 5$.

$$= (2x^2 + x) + (10x + 5)$$

Each of those mini-expressions contains a greatest common factor: x and 5, respectively. Factor the mini-expressions.

$$= x(2x+1) + 5(2x+1)$$

Now each term contains the same quantity, $(2x + 1)$. Factor that entire quantity out of both terms, leaving behind x in the first term and 5 in the second term.

$$= (2x+1)(x+5)$$

You conclude that $2x^2 + 11x + 5 = (2x+1)(x+5)$. Some students do not apply the factoring by decomposition process to factor expressions such as $2x^2 + 11x + 5$. Instead, they experiment until they stumble across the correct answer. Either method is acceptable, although factoring by decomposition is far more mathematically rigorous.

If you can figure out the answer without going through all of these steps, by all means, do it! Factoring by decomposition is best used as a last resort.

6.21 Factor the expression: $6x^3 - x^2 - 12x$.

The terms contain greatest common factor x. Begin by factoring it out of the expression.

$$6x^3 - x^2 - 12x = x[6x^2 - x - 12]$$

The quantity in brackets is a quadratic expression with leading coefficient 6. Apply the technique described in Problem 6.20 to factor it. Your goal is to find two numbers that have a difference of –1 and a product of 6(–12) = –72. The two numbers are 8 and –9.

$$\begin{aligned} &= x\left[6x^2 + 8x - 9x - 12\right] \\ &= x\left[\left(6x^2 + 8x\right) + \left(-9x - 12\right)\right] \\ &= x\left[2x(3x+4) - 3(3x+4)\right] \\ &= x(3x+4)(2x-3) \end{aligned}$$

Linear and Rational Equations

Solve simple and fractional equations

6.22 Solve the equation for y: $8y + 10 = -3y - 1$.

The highest power of the variable in this equation is 1, so this is a linear equation. To solve a linear equation in one variable—in this case the variable is y—isolate the variable on one side of the equation. To move the y-terms to the left side of the equation, add $3y$ to both sides.

$$\begin{aligned} 8y + 3y + 10 &= -3y + 3y - 1 \\ 11y + 10 &= -1 \end{aligned}$$

Move the constants to the right side of the equation by subtracting 10 from both sides.

$$\begin{aligned} 11y + 10 - 10 &= -1 - 10 \\ 11y &= -11 \end{aligned}$$

Finally, solve for y by eliminating its coefficient; divide both sides of the equation by 11.

$$\begin{aligned} \frac{11y}{11} &= \frac{-11}{11} \\ y &= -1 \end{aligned}$$

The solution to the equation is $y = -1$.

6.23 Solve the equation for x: $3(x + 2) - 7x = 4(9 - 5x) + 2$.

Distribute 3 to the terms in the quantity $(x + 2)$, distribute 4 to the terms in the quantity $(9 - 5x)$, and simplify the expressions on both sides of the equal sign.

$$3(x)+3(2)-7x=4(9)+4(-5x)+2$$
$$3x+6-7x=36-20x+2$$
$$(3x-7x)+6=-20x+(36+2)$$
$$-4x+6=-20x+38$$

Isolate x-terms on the left side of the equation and constants on the right side of the equation.

$$-4x+20x+6=-20x+20x+38$$
$$16x+6=38$$
$$16x+6-6=38-6$$
$$16x=32$$

Solve for x by eliminating its coefficient.

$$\frac{16x}{16}=\frac{32}{16}$$
$$x=2$$

6.24 Four times a number is equal to 15 less that number. What is the number?

Let x represent the unknown number. The problem, as stated, translates into the following equation.

$$4x = 15 - x$$

"15 less x" translates into 15 – x, but "15 LESS THAN x" translates into x – 15.

Solve the equation for x.

$$4x+x=15-x+x$$
$$5x=15$$
$$\frac{5x}{5}=\frac{15}{5}$$
$$x=3$$

6.25 Solve the equation for s: $\frac{3s}{4}+\frac{5s}{3}=10$.

In order to solve this equation, you need to work around the fractions. There are two ways to accomplish this—either by rewriting the fractions in terms of a common denominator (in which case $3s/4 = 9s/12$ and $5s/3 = 20s/12$) or by eliminating the fractions altogether, multiplying every term in the equation by the least common denominator, 12. The latter approach is preferred, unless you really love fractions.

The "multiply to get rid of fractions" technique only works well when the denominator you're getting rid of is a real number. If the denominators contain variables, applying this technique may cause solutions to vanish or create false solutions!

$$12\left[\frac{3s}{4}+\frac{5s}{3}\right]=12[10]$$
$$\frac{12\cdot 3s}{4}+\frac{12\cdot 5s}{3}=12\cdot 10$$
$$\frac{12}{4}\cdot 3s+\frac{12}{3}\cdot 5s=120$$
$$3(3s)+4(5s)=120$$

Solve the now-fractionless equation for *x*.

$$9s+20s=120$$
$$29s=120$$
$$\frac{29s}{29}=\frac{120}{29}$$
$$s=\frac{120}{29}$$

You might have been able to eliminate fractions from the equation, but they made a triumphant return in the solution, $s = 120/29$.

6.26 Solve the rational equation and identify any value of *x* for which the equation is undefined: $\frac{x-3}{x+2}=0$.

When the numerator of a fraction equals zero, the entire fraction equals zero. However, when the denominator of a fraction equals zero, the fraction is undefined. Why? You are never allowed to divide by zero.

To solve this equation, set the numerator equal to zero and solve for *x*. This identifies *potential* solutions to the equation.

$$x-3=0$$
$$x=3$$

Now set the denominator equal to zero to identify values of *x* for which the equation is undefined.

$$x+2=0$$
$$x=-2$$

As long as a potential solution is not a value for which the equation is undefined, the potential solution becomes a valid solution. You conclude that the solution to this rational equation is $x = 3$; the equation is undefined when $x = -2$.

6.27 Solve the rational equation and verify your answer: $\frac{x^2-5x+4}{x^2+2x-24}=0$.

In Problems 6.28–6.29, the rational equations are not single fractions set equal to zero, so you have to apply slightly different techniques.

Like the equation in Problem 6.26, this rational equation consists of a single fraction equal to zero. This is important, because the solutions to such equations are relatively simple to identify; they are the values of *x* for which the numerator equals zero but the denominator does not. Begin by factoring the expressions in the numerator and denominator.

$$\frac{(x-4)(x-1)}{(x+6)(x-4)}=0$$

Set each factor of the numerator equal to 0 and solve to identify potential solutions of the rational equation.

$$x-4=0 \qquad x-1=0$$
$$x=4 \qquad x=1$$

Now set the factors of the denominator equal to 0 to identify the values of x for which the rational equation is undefined.

$$x+6=0 \qquad x-4=0$$
$$x=-6 \qquad x=4$$

Only one of the potential solutions to the rational equation is actually a solution. Although $x = 4$ is a potential solution, it is also a value for which the rational equation is undefined. Therefore, $x = 4$ is not a solution.

You conclude that only $x = 1$ is a solution to the rational equation. Verify your answer by substituting $x = 1$ into the original rational equation; the result should be a true statement.

$$\frac{x^2-5x+4}{x^2+2x-24}=0$$

$$\frac{1^2-5(1)+4}{1^2+2(1)-24}=0$$

$$\frac{1-5+4}{1+2-24}=0$$

$$\frac{0}{-21}=0$$

$$0=0 \quad \textbf{True}$$

6.28 Solve the rational equation: $\frac{1}{x+1}=\frac{5}{2x-3}$.

This equation is a proportion—two fractions that are equal to each other. Solve by cross-multiplying the fractions. In other words, multiply the numerator of one fraction by the denominator of the other and vice versa.

$$1(2x-3)=5(x+1)$$
$$2x-3=5x+5$$
$$-3-5=5x-2x$$
$$-8=3x$$
$$-\frac{8}{3}=x$$

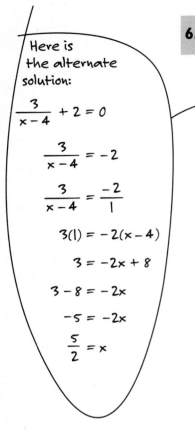

Here is the alternate solution:

$$\frac{3}{x-4}+2=0$$

$$\frac{3}{x-4}=-2$$

$$\frac{3}{x-4}=\frac{-2}{1}$$

$$3(1)=-2(x-4)$$

$$3=-2x+8$$

$$3-8=-2x$$

$$-5=-2x$$

$$\frac{5}{2}=x$$

6.29 Solve the rational equation: $\frac{3}{x-4}+2=0$.

You can apply two different techniques to solve this equation. You could subtract 2 from both sides of the equation, express –2 as the fraction –2/1, and then cross-multiply. The other technique is demonstrated below, in which 2 is expressed as a fraction with common denominator $x-4$.

$$\frac{3}{x-4}+\frac{2}{1}\left(\frac{x-4}{x-4}\right)=0$$

$$\frac{3}{x-4}+\frac{2(x-4)}{x-4}=0$$

$$\frac{3}{x-4}+\frac{2x-8}{x-4}=0$$

Because the fractions contain a common denominator, you can add their numerators.

$$\frac{3+2x-8}{x-4}=0$$

$$\frac{2x-5}{x-4}=0$$

The fraction is equal to 0 when $2x-5=0$; it is undefined when $x-4=0$.

$$2x-5=0$$

$$2x=5$$

$$x=\frac{5}{2}$$

The solution to the rational equation is $x = 5/2$. The equation is undefined when $x = 4$.

Linear and Rational Inequalities

Solving simple and fractional inequalities

6.30 Solve the linear inequality: $4x+3>2x-5$.

A linear inequality is very similar to a linear equation. Your goal is to isolate x on one side of the inequality sign. Begin by subtracting $2x$ and 3 from both sides of the inequality to group all of the x-terms on the left and all of the constants on the right.

$$4x-2x>-5-3$$

$$2x>-8$$

Now solve for x, dividing both sides of the inequality by its coefficient of 2. Be careful; if you multiply or divide both sides of an inequality by a negative value, you have to reverse the inequality sign—in other words, $>$ in this equation would become $<$. However, in this case you are dividing by a positive value, so the inequality sign is unchanged.

$$x > -\frac{8}{2}$$
$$x > -4$$

You conclude that all x values greater than –4 satisfy the inequality.

6.31 Solve the linear inequality: $2(1 - 5x) - 6x + 7 \geq 0$.

Simplify the left side of the inequality and solve for x.

$$\begin{aligned} 2(1-5x)-6x+7 &\geq 0 \\ 2(1)+2(-5x)-6x+7 &\geq 0 \\ 2-10x-6x+7 &\geq 0 \\ -16x+9 &\geq 0 \\ -16x &\geq -9 \end{aligned}$$

To solve for x, you must divide both sides of the inequality by –16. As Problem 6.30 explains, this requires you to reverse the inequality sign, changing it from $\geq$ to $\leq$.

$$\frac{-16x}{-16} \leq \frac{-9}{-16}$$
$$x \leq \frac{9}{16}$$

Note: Problems 6.32–6.33 explain how to solve the rational inequality $\frac{2x-3}{x+1} \leq 0$.

6.32 Identify the critical numbers for the solution.

Like linear inequalities, the solution to a rational inequality is a segment, or interval, of the number line. All of the x-values within that interval make the inequality true. The trick is determining how to slice a number line into the intervals that represent possible solutions.

Make sure you are working with a single fraction set equal to 0, or this technique does not work. If the inequality looks like anything else, your first goal is to get a fraction on the left side and zero on the right side. See Problem 6.34 for more information.

In Problems 6.30–6.31, the linear inequalities took care of the intervals for you; all you needed to do was solve the inequality as you would any linear equation, being careful to reverse the inequality symbol at the end when necessary.
In order to solve a rational inequality, you need to apply a slightly different strategy, one that requires critical numbers.

Critical numbers are the x-values for which an expression either equals zero or is undefined. In a rational expression, critical numbers are x-values that cause the numerator or denominator to equal 0. Set the numerator and denominator equal to 0 to identify the critical numbers.

$$\begin{aligned} 2x-3 &= 0 \\ 2x &= 3 \\ x &= \frac{3}{2} \end{aligned} \qquad \begin{aligned} x+1 &= 0 \\ x &= -1 \end{aligned}$$

There are two critical numbers for the rational expression, $x = -1$ and $x = 3/2$. Note that the fraction equals zero at $x = 3/2$ and is undefined at $x = -1$.

Note: Problems 6.32–6.33 explain how to solve the rational inequality $\frac{2x-3}{x+1} \le 0$.

6.33 Solve the inequality with a wiggle graph containing the critical numbers you identified in Problem 6.32.

A wiggle graph illustrates how critical numbers divide a number line into possible solution intervals. You use two kinds of points when you create a wiggle graph—solid dots for solutions and open dots (hollow circles) for non-solutions.

This inequality contains the symbol "≤," which means that the statement is true when the fraction is less than zero or when it is equal to zero. According to Problem 6.32, the fraction equals zero when $x = 3/2$, so $x = 3/2$ is a solution to the inequality. Plot it on the number line using a solid dot. The fraction is undefined when $x = -1$, so plot that critical number using an open dot. (If the fraction is undefined when $x = -1$, then -1 could not possibly be a solution.)

Wiggle graphs are used to determine where things fundamentally change for an equation or function. They are often used in calculus to determine where functions change direction, "wiggling" from one direction to the other, hence the name. In this problem, you are figuring out where a fraction's values wiggle from positive to negative or vice versa.

The two critical numbers divide the number line into three segments, as illustrated below: $x < -1$, $-1 < x \le 3/2$, and $x \ge 3/2$. All of the values in each segment have the same effect on the fraction. In other words, every x-value in a segment makes the fraction either positive or negative.

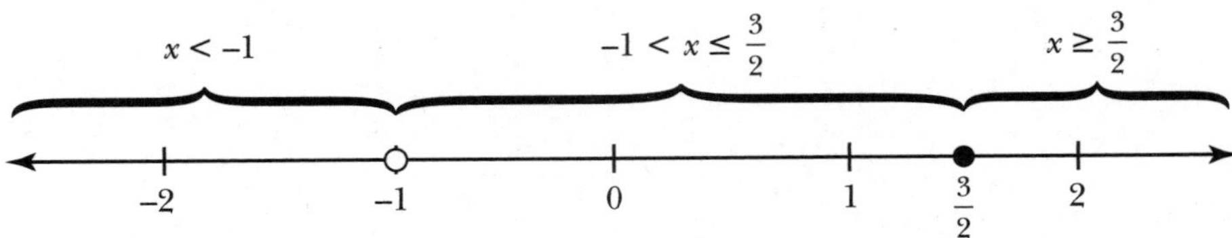

To check each interval, select one value and substitute it into the inequality. You should select simple integer values whenever possible. For example, you could choose $x = -2$ to represent the left interval in the diagram, $x = 0$ to represent the middle interval, and $x = 2$ to represent the right interval. Substitute these three values into the inequality to determine which produce true statements.

Test $x = -2$	Test $x = 0$	Test $x = 2$
$\frac{2(-2)-3}{-2+1} \le 0$	$\frac{2(0)-3}{0+1} \le 0$	$\frac{2(2)-3}{2+1} \le 0$
$\frac{-4-3}{-1} \le 0$	$\frac{-3}{1} \le 0$	$\frac{4-3}{3} \le 0$
$\frac{-7}{-1} \le 0$	$-3 \le 0$ **True**	$\frac{1}{3} \le 0$ **False**
$7 \le 0$ **False**		

The inequality is only true for the middle interval, the portion of the number line between $x = -1$ and $x = 3/2$. Remember that the solid dot at $x = 3/2$ needs to be included as part of the solution, but $x = -1$ must be excluded. You conclude that the solution to the inequality is $-1 < x \le 3/2$.

Note: Problems 6.34–6.35 explain how to solve the rational inequality $\frac{3}{x-2}+3>0$.

6.34 Identify the critical numbers for the solution.

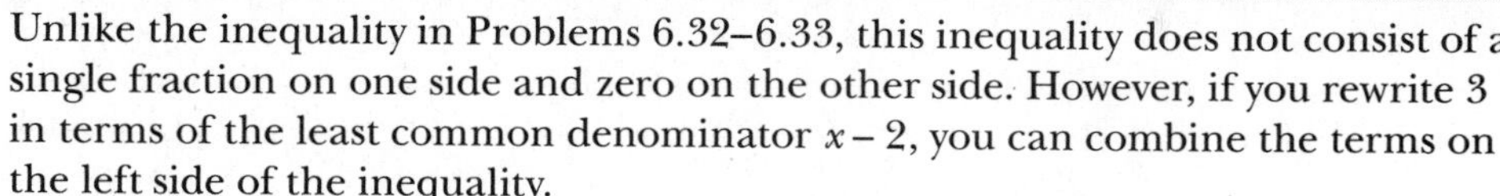

Unlike the inequality in Problems 6.32–6.33, this inequality does not consist of a single fraction on one side and zero on the other side. However, if you rewrite 3 in terms of the least common denominator $x-2$, you can combine the terms on the left side of the inequality.

$$\frac{3}{x-2}+3>0$$

$$\frac{3}{x-2}+\frac{3}{1}\left(\frac{x-2}{x-2}\right)>0$$

$$\frac{3}{x-2}+\frac{3(x)+3(-2)}{x-2}>0$$

$$\frac{3}{x-2}+\frac{3x-6}{x-2}>0$$

$$\frac{3+3x-6}{x-2}>0$$

$$\frac{3x-3}{x-2}>0$$

You now have the correct form you need to proceed with the solution—a fraction on the left side of the inequality symbol and a zero on the other side. The critical numbers are the values for which the numerator or denominator are equal to zero.

$$\begin{aligned} 3x-3&=0 \\ 3x&=3 \\ x&=\frac{3}{3} \\ x&=1 \end{aligned} \qquad \begin{aligned} x-2&=0 \\ x&=2 \end{aligned}$$

The critical numbers are $x = 1$ and $x = 2$. Neither of these values represents a solution to the inequality, because the rational expression equals 0 when $x = 1$ and is undefined when $x = 2$.

You want to know when the fraction is GREATER than zero, so an x-value that causes the fraction to EQUAL zero is not a solution. It is, however, a critical number. Just plot it using an open dot, like you plot critical numbers that make the fraction undefined.

Note: Problems 6.34–6.35 explain how to solve the rational inequality $\frac{3}{x-2}+3>0$.

6.35 Solve the inequality with a wiggle graph containing the critical numbers you identified in Problem 6.34.

According to Problem 6.34, the critical numbers are $x = 1$ and $x = 2$. They divide the number line into possible solution intervals, as illustrated below: $x < 1$, $1 < x < 2$, and $x > 2$. Notice that open dots are used, because neither critical number is a solution.

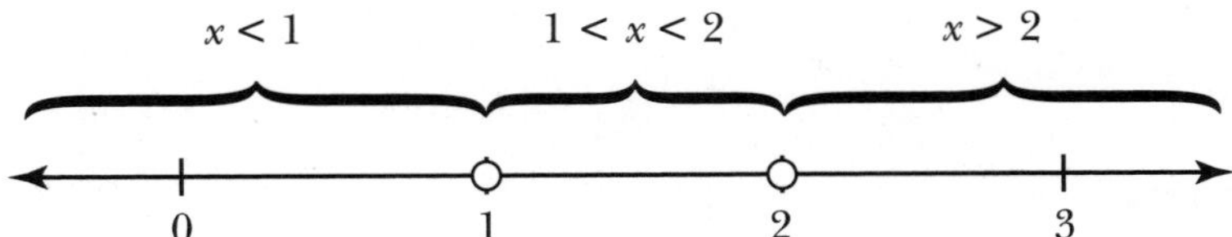

Select a value from each of the intervals. For example, you could select $x = 0$ to represent $x < 1$, you could select $x = 3/2$ to represent $1 < x < 2$, and you could select $x = 3$ to represent $x > 2$. You may wish to use a calculator to simplify some of the complex fractions.

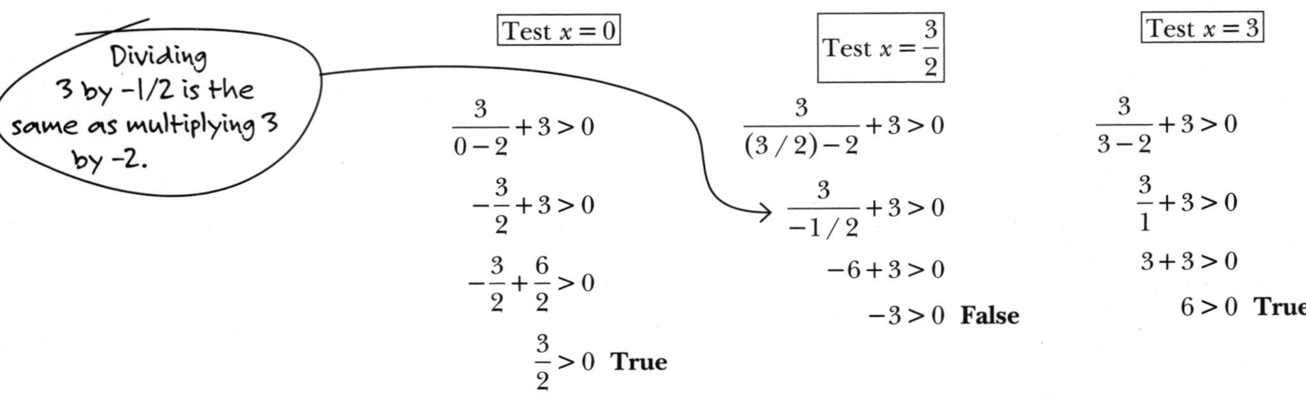

Dividing 3 by −1/2 is the same as multiplying 3 by −2.
Test $x = 0$
$\frac{3}{0-2}+3>0$
$-\frac{3}{2}+3>0$
$-\frac{3}{2}+\frac{6}{2}>0$
$\frac{3}{2}>0$ **True**
Test $x = \frac{3}{2}$
$\frac{3}{(3/2)-2}+3>0$
$\frac{3}{-1/2}+3>0$
$-6+3>0$
$-3>0$ **False**
Test $x = 3$
$\frac{3}{3-2}+3>0$
$\frac{3}{1}+3>0$
$3+3>0$
$6>0$ **True**

The inequality is true for the x-values in the right and left intervals of the wiggle graph, but it is false for the middle interval. You conclude that the solution to the inequality is any x-value such that $x < 1$ or $x > 2$.

Graphing Linear Equations and Inequalities

Including slope-intercept and point-slope formulas

6.36 Use the x- and y-intercepts of $2x - 3y = 6$ to graph the linear equation.

The x-intercept of a graph is the x-value at which the graph crosses the x-axis. That point of intersection always has a y-value of 0. Substitute $y = 0$ into the linear equation to calculate the x-intercept.

$$\begin{aligned} 2x - 3y &= 6 \\ 2x - 3(0) &= 6 \\ 2x - 0 &= 6 \\ 2x &= 6 \\ x &= \frac{6}{2} \\ x &= 3 \end{aligned}$$

Similarly, the y-intercept of a line is the y-value at which a graph intercepts the y-axis. Substitute $x = 0$ into the equation to calculate the y-intercept.

$$\begin{aligned} 2x - 3y &= 6 \\ 2(0) - 3y &= 6 \\ 0 - 3y &= 6 \\ -3y &= 6 \\ y &= \frac{6}{-3} \\ y &= -2 \end{aligned}$$

The graph of $2x - 3y = 6$ crosses the x-axis at point (3,0) and crosses the y-axis at point (0,–2), as illustrated in the following graph.

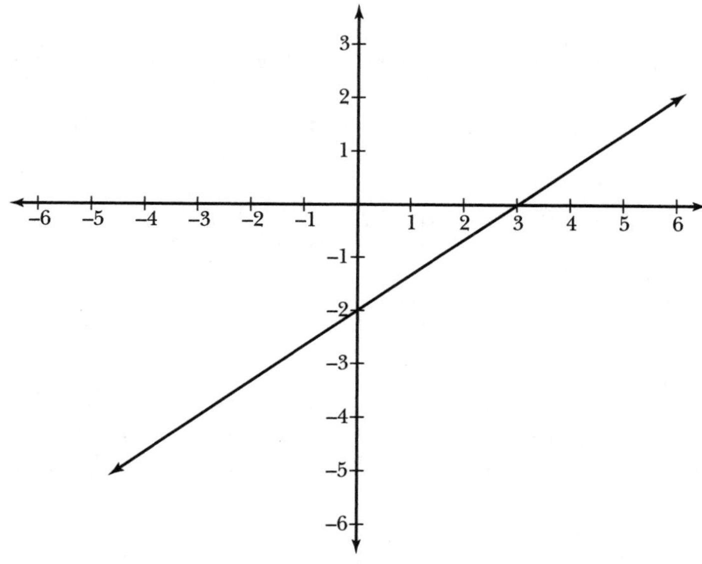

6.37 Calculate the slope of the linear equation by expressing it in slope-intercept form: $4x - 7y = 15$.

The slope-intercept form of a linear equation is $y = mx + b$, where m is the slope of the line and b is the y-intercept. To express a line in slope-intercept form, solve it for y.

$$\begin{aligned} 4x - 7y &= 15 \\ -7y &= -4x + 15 \\ \frac{-7y}{-7} &= \frac{-4}{-7}x + \frac{15}{-7} \\ y &= \frac{4}{7}x - \frac{15}{7} \end{aligned}$$

The coefficient of the x-term is also the slope of the line: $m = 4/7$.

6.38 Express the linear equation in slope-intercept form and graph it: $4(x - 3) + 2(y - 1) = 10x + y - 21$.

As Problem 6.37 explains, to express a linear equation in slope-intercept form, you need to solve it for y. Begin by simplifying the left side of the equation.

$$\begin{aligned} 4(x) + 4(-3) + 2(y) + 2(-1) &= 10x + y - 21 \\ 4x - 12 + 2y - 2 &= 10x + y - 21 \\ 4x + 2y - 14 &= 10x + y - 21 \end{aligned}$$

Now subtract $4x$ from, add 14 to, and subtract y from both sides of the equation to isolate the y-terms on the left side of the equal sign.

$$\begin{aligned} 2y - y &= 10x - 4x - 21 + 14 \\ y &= 6x - 7 \end{aligned}$$

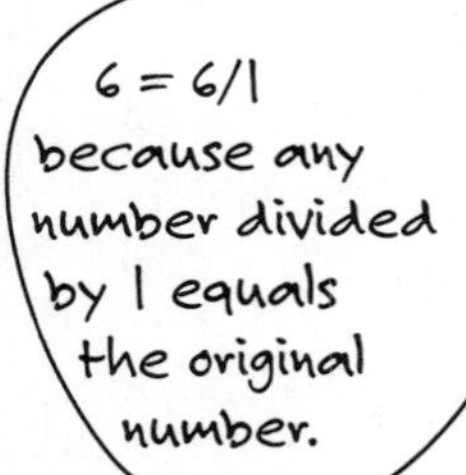

6 = 6/1
because any
number divided
by 1 equals
the original
number.

The slope of the line is 6, the coefficient of x. It is helpful to express the slope as the equivalent fraction 6/1. The y-intercept of the linear equation is –7, the constant. Slope represents the vertical and horizontal change between points on the line.

$$\text{slope} = \frac{\text{vertical change}}{\text{horizontal change}}$$

$$m = \frac{6}{1}$$

Plot the y-intercept, point (0,–7), and then use the slope to identify another point on the graph. Because the vertical change is 6 and the horizontal change is 1, you should count up 6 units and right 1 unit from the y-intercept to plot another point on the line, as illustrated in the following graph.

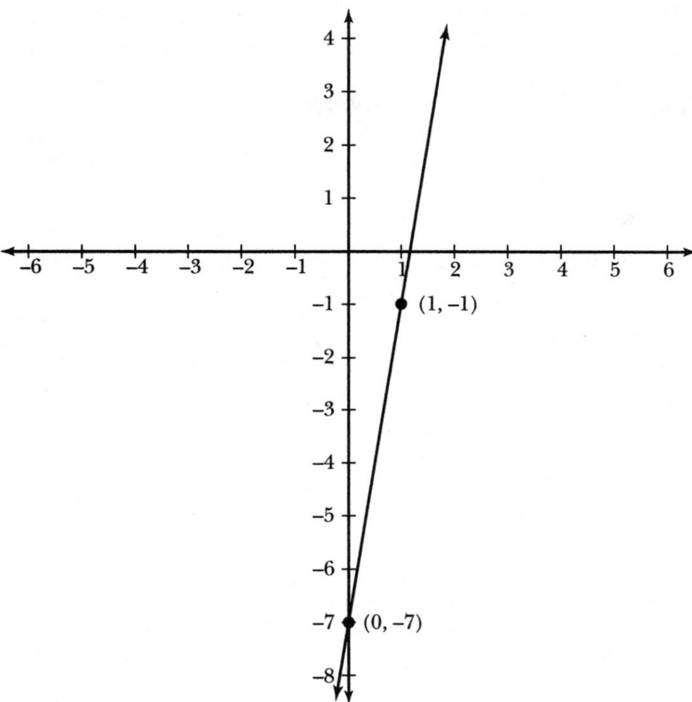

6.39 Line l has slope $m = -\frac{2}{3}$ and passes through point $(-8, 2)$. Write the equation of line l in standard form.

Given the slope m of a line and a point (x_1, y_1) on that line, you can apply the point-slope formula (below) to generate the equation of that line.

$$y - y_1 = m(x - x_1)$$

Substitute $m = -2/3$, $x_1 = -8$, and $y_1 = 2$ into the point-slope formula.

$$y - 2 = -\frac{2}{3}(x - (-8))$$
$$y - 2 = -\frac{2}{3}(x + 8)$$

The standard form of a line is $Ax + By = C$, where A, B, and C are integers and $A > 0$. Because standard form does not contain fractions, you should multiply both sides of the point-slope equation by the denominator of the sole fraction, 3. This eliminates the fractions and brings you one step closer to standard form.

$$3(y - 2) = \frac{3}{1}\left(-\frac{2}{3}\right)(x + 8)$$
$$3y - 6 = -\frac{6}{3}(x + 8)$$
$$3y - 6 = -2(x + 8)$$
$$3y - 6 = -2x - 16$$

Move the x- and y-terms to the left side of the equation (by adding $2x$ to both sides) and move the constants to the right side of the equation (by adding 6 to both sides).

$$2x+3y=-16+6$$
$$2x+3y=-10$$

Notice that each of the coefficients is an integer and the x-coefficient (2) is positive. Therefore, $2x + 3y = -10$ is the standard form of the linear equation.

6.40 Line k passes through points $(-4,-1)$ and $(3,5)$. Write the equation of k in slope-intercept form.

The slope of the line passing through points (x_1, y_1) and (x_2, y_2) is calculated according to the following formula.

$$m=\frac{y_2-y_1}{x_2-x_1}$$

Substitute $x_1 = -4$, $y_1 = -1$, $x_2 = 3$, and $y_2 = 5$ into the formula to calculate m.

$$m=\frac{5-(-1)}{3-(-4)}$$
$$=\frac{5+1}{3+4}$$
$$=\frac{6}{7}$$

You now have the slope of line k and a pair of points through which the line passes. Substitute $m = 6/7$, $x_1 = -4$, and $y_1 = -1$ into the point-slope formula to generate the equation of the line.

$$y-(-1)=\frac{6}{7}(x-(-4))$$
$$y+1=\frac{6}{7}(x+4)$$
$$y+1=\frac{6}{7}x+\frac{24}{7}$$

This solution uses point $(-4,-1)$, but line k also passes through $(3,5)$. If you substitute $x_1 = 3$ and $y_1 = 5$ into the point-slope formula, the final answer is the same:

$$y-y_1=m(x-x_1)$$
$$y-5=\frac{6}{7}(x-3)$$
$$y-5=\frac{6}{7}x-\frac{18}{7}$$
$$y=\frac{6}{7}x-\frac{18}{7}+5$$
$$y=\frac{6}{7}x-\frac{18}{7}+\frac{35}{7}$$
$$y=\frac{6}{7}x+\frac{35-18}{7}$$
$$y=\frac{6}{7}x+\frac{17}{7}$$

To express the equation in slope-intercept form, solve for y.

$$y=\frac{6}{7}x+\frac{24}{7}-1$$
$$y=\frac{6}{7}x+\frac{24}{7}-\frac{7}{7}$$
$$y=\frac{6}{7}x+\frac{17}{7}$$

6.41 Graph the inequality: $y \geq \frac{1}{2}x + 3$.

Temporarily disregard the inequality symbol, and imagine that it is an equal sign: $y = (1/2)x + 3$. This is a linear equation in slope intercept form—the slope of the line is $1/2$ and the y-intercept is 3. Draw a line in the coordinate plane that passes through the y-axis at $(0, 3)$ and has slope $1/2$; in other words, another point on the line is 1 unit above and 2 units to the right of the y-intercept.

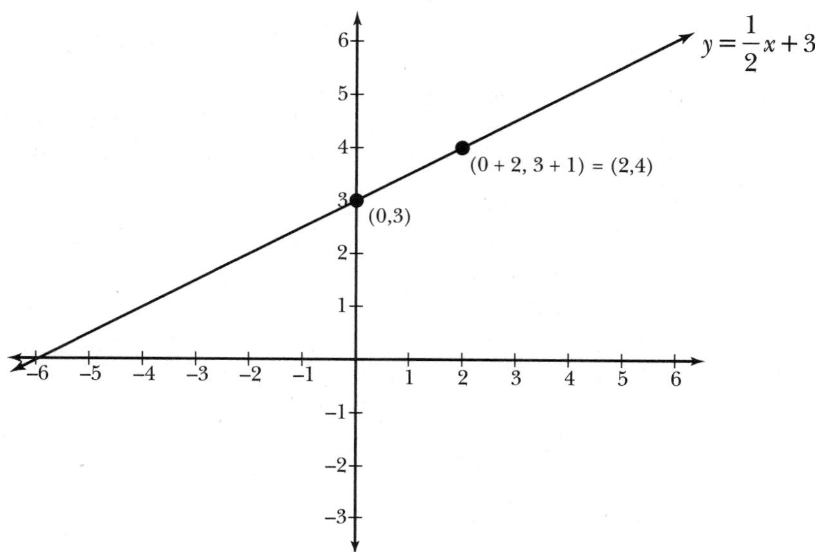

The graph of the inequality $y \geq (1/2)x + 3$ is very similar to the graph of the line $y = (1/2)x + 3$. Now that you have generated the graph of the line, consider the inequality symbol in the original problem. If it allows for the possibility of equality, then the line is part of the solution and must be part of the graph. This is illustrated by drawing the graph as a solid line. If the inequality symbol is < or >, then the linear graph should be dotted.

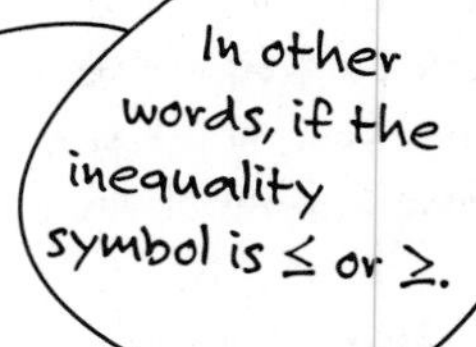

The inequality symbol in this problem is ≥, so the line in the graph is solid. Finally, the direction of the inequality symbol describes which region of the graph to shade. The solution to $y \geq (1/2)x + 3$ is the region *greater than* or equal to the line, so you should shade the region *above* the line, as illustrated in the following graph.

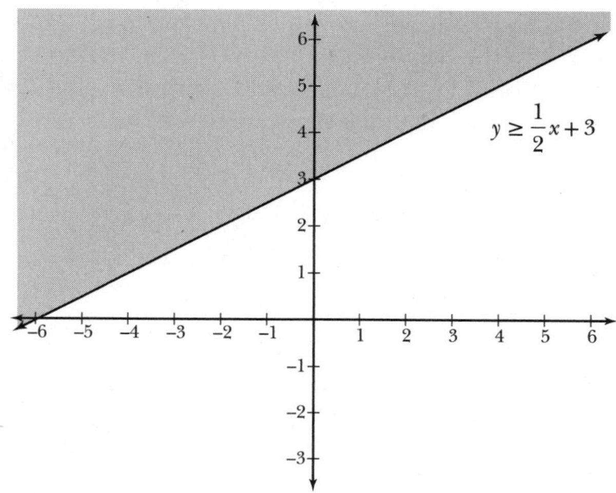

6.42 Graph the inequality: $-5x - 2y > -8$.

Express the equation in slope-intercept form by solving for y. Do not forget to reverse the inequality symbol when you divide by the negative coefficient of y.

$$-2y > 5x - 8$$
$$\frac{-2y}{-2} < \frac{5x}{-2} - \frac{8}{-2}$$
$$y < -\frac{5}{2}x + 4$$

That means you should travel down 5 units and right 2 units from the y-intercept to plot another point on the line.

For the moment, ignore the inequality symbol and graph the linear equation $y = -(5/2)x + 4$, which passes through the y-axis at (0,4) and has slope $m = -5/2$. Because the inequality symbol is <, the line should be dotted and you should shade the region beneath the line.

This is why you have to express the equation in slope-intercept form before you graph it. If you apply the intercepts method of Problem 6.36 to graph the line, you might see the > symbol and incorrectly shade above the line.

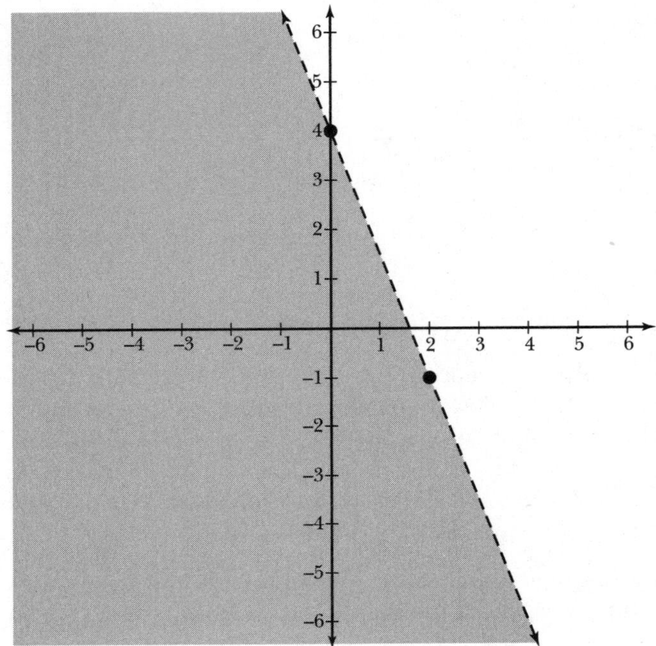

If you are having difficulty deciding which region of the graph to shade, select one point from the region you believe is correct and substitute its x- and y-coordinates into the original inequality. For example, the origin (0, 0) lies in the shaded solution region of this graph, so substituting $x = 0$ and $y = 0$ into the original inequality should make the statement true.

$$-5x - 2y > -8$$
$$-5(0) - 2(0) > -8$$
$$0 > -8 \quad \textbf{True}$$

6.43 Write the equation of the linear inequality graphed below.

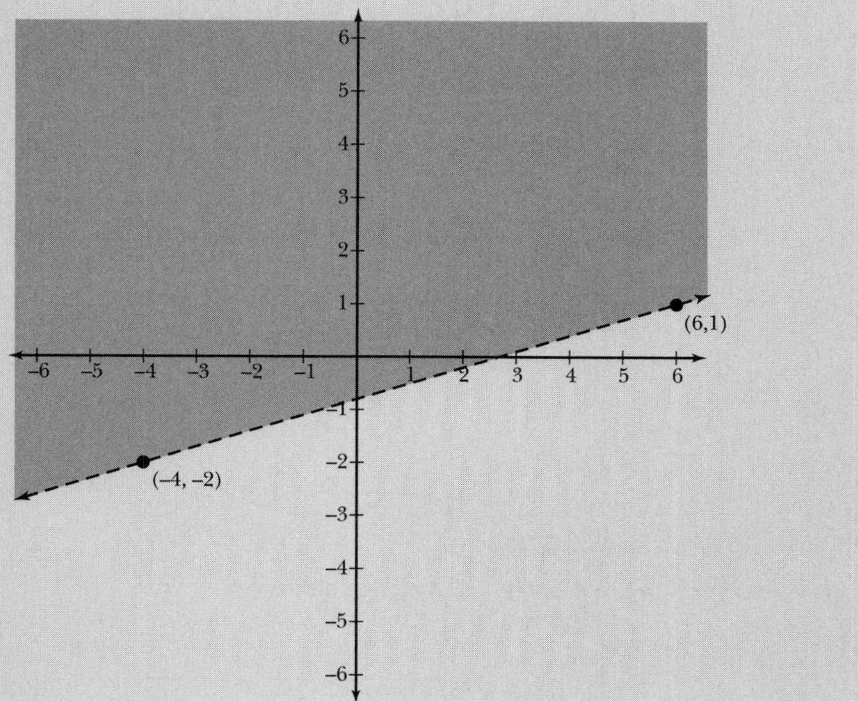

The dotted line passes through (–4,–2) and (6,1). Use the technique demonstrated in Problem 6.40 to identify the equation of that line. Begin by calculating its slope.

$$\begin{aligned} m &= \frac{y_2 - y_1}{x_2 - x_1} \\ &= \frac{1-(-2)}{6-(-4)} \\ &= \frac{1+2}{6+4} \\ &= \frac{3}{10} \end{aligned}$$

Now substitute $m = 3/10$, $x_1 = -4$ and $y_1 = -2$ into the point-slope formula to generate the equation of the dotted line.

$$\begin{aligned} y - y_1 &= m(x - x_1) \\ y-(-2) &= \frac{3}{10}(x-(-4)) \\ y+2 &= \frac{3}{10}(x)+\frac{3}{10}(+4) \\ y+2 &= \frac{3}{10}x+\frac{12}{10} \\ y+2 &= \frac{3}{10}x+\frac{6}{5} \end{aligned}$$

Solve the equation for y.

$$y = \frac{3}{10}x + \frac{6}{5} - 2$$
$$y = \frac{3}{10}x + \frac{6}{5} - \frac{10}{5}$$
$$y = \frac{3}{10}x - \frac{4}{5}$$

This is the equation of the dotted line but not the equation of the inequality. The region above the line is shaded (which means the inequality symbol must be > or ≥) and the line is dotted (which means that the inequality symbol must be < or >). The only symbol that fits both conditions is >, so use it to replace the equal sign in the equation.

$$y > \frac{3}{10}x - \frac{4}{5}$$

Nonlinear Equations

Quadratics, roots, and absolute values

6.44 Solve the equation: $\sqrt{x} + 3 = 19$.

Isolate the expression containing x on the left side of the equal sign.

$$\sqrt{x} = 19 - 3$$
$$\sqrt{x} = 16$$

Square both sides of the equation to solve for x.

$$\left(\sqrt{x}\right)^2 = 16^2$$
$$x = 256$$

Note that squaring a value and taking the square root of a value are inverse operations, so they cancel themselves out. Squaring the square root of x leaves only x.

Just like adding 4 to x and then subtracting 4 from x would cancel out: $x + 4 - 4 = x$ Addition and subtraction are also inverse operations.

6.45 Solve the equation: $\sqrt{3y - 1} = \sqrt{y - 13}$.

If these two radical expressions are equal, their radicands—the expressions within the radical symbols—must be equal. In other words, two square roots are equal only if you are taking the square roots of the same value. Set $3y - 1$ equal to $y - 13$ and solve for y.

$$3y - 1 = y - 13$$
$$3y - y = -13 + 1$$
$$2y = -12$$
$$y = \frac{-12}{2}$$
$$y = -6$$

The two square root expressions are equal when $y = -6$, but you *cannot* conclude that $y = -6$ is the solution. The reason becomes clear when you substitute it into the original equation.

$$\begin{aligned}\sqrt{3y-1} &= \sqrt{y-13}\\ \sqrt{3(-6)-1} &= \sqrt{-6-13}\\ \sqrt{-18-1} &= \sqrt{-19}\\ \sqrt{-19} &= \sqrt{-19}\end{aligned}$$

The square root of a negative number is not a real number—no real number multiplied times itself equals a negative number. Because the only potential solution did not produce real number results, you conclude that there is no solution to the equation.

Tip: When solving equations that contain roots (radical symbols), plug your answers back into the original equation to make sure they work.

6.46 Solve the equation: $5(2x-1)^2 = 45$.

Isolate the squared variable expression on the left side of the equation.

$$\begin{aligned}\frac{5(2x-1)^2}{5} &= \frac{45}{5}\\ (2x-1)^2 &= 9\end{aligned}$$

To solve for x, you need to eliminate the power to which the x-expression is raised. In Problem 6.44, you eliminate a square root by squaring both sides of an equation; here, you do the opposite. Take the square root of both sides of the equation to eliminate the square. One word of caution: Whenever you take an even root (like a square root) of both sides of an equation, you need to insert a "±" symbol right of the equal sign.

Here's why: A squared expression might actually be positive or negative. You don't know, because squaring it erases its sign. For example, the solution to $x^2 = 25$ is $x = \pm 5$. Both $(+5)^2$ and $(-5)^2$ equal 25.

$$\begin{aligned}\sqrt{(2x-1)^2} &= \pm\sqrt{9}\\ 2x-1 &= \pm 3\\ 2x &= 1 \pm 3\end{aligned}$$

You now have two equations: $2x = 1 + 3$ and $2x = 1 - 3$. Solve them both for x.

$$\begin{aligned}2x &= 1+3 & \qquad 2x &= 1-3\\ 2x &= 4 & 2x &= -2\\ x &= \frac{4}{2} & x &= -\frac{2}{2}\\ x &= 2 & x &= -1\end{aligned}$$

The solutions of $5(2x-1)^2 = 45$ are $x = -1$ and $x = 2$.

6.47 Solve the equation by factoring: $x^2 + 7x = 60$.

This is a quadratic equation, because the highest power of the variable is 2. Begin by subtracting 60 from both sides of the equation, which sets the quadratic expression equal to zero.

$$x^2 + 7x - 60 = 0$$

Now apply the technique described in Problems 6.13–6.17 to factor the quadratic expression.

$$(x+12)(x-5) = 0$$

Set each of the factors equal to zero and solve those equations for x.

$$\begin{aligned} x+12&=0 & x-5&=0 \\ x&=-12 & x&=5 \end{aligned}$$

The equation $x^2 + 7x = 60$ has two solutions: $x = -12$ and $x = 5$.

6.48 Apply the quadratic formula to solve the equation: $(x+4)^2 = x + 11$.

You cannot factor all quadratic expressions, but you can apply the quadratic formula to solve any quadratic equation. Specifically, the quadratic equation $ax^2 + bx + c = 0$ has the following solution:

$$x = \frac{-b \pm \sqrt{b^2 - 4ac}}{2a}$$

Before you can apply the quadratic formula, you must first ensure it has form $ax^2 + bx + c = 0$. To solve the equation in this problem, first expand the left side of the equation by applying the technique described in Problem 6.16; then subtract x and 11 from both sides of the equation, leaving only 0 to the right of the equal sign.

$$\begin{aligned} (x+4)^2 &= x+11 \\ (x+4)(x+4) &= x+11 \\ x(x)+x(4)+4(x)+4(4) &= x+11 \\ x^2+4x+4x+16 &= x+11 \\ x^2+8x+16 &= x+11 \\ x^2+(8x-x)+(16-11) &= 0 \\ x^2+7x+5 &= 0 \end{aligned}$$

Apply the quadratic formula to solve the equation, setting $a = 1$, $b = 7$, and $c = 5$.

$$\begin{aligned} x &= \frac{-b \pm \sqrt{b^2-4ac}}{2a} \\ &= \frac{-7 \pm \sqrt{7^2-4(1)(5)}}{2(1)} \\ &= \frac{-7 \pm \sqrt{49-20}}{2} \\ &= \frac{-7 \pm \sqrt{29}}{2} \\ &= -\frac{7}{2} \pm \frac{\sqrt{29}}{2} \end{aligned}$$

The equation has two solutions: $x = -\frac{7}{2} - \frac{\sqrt{29}}{2}$ and $x = -\frac{7}{2} + \frac{\sqrt{29}}{2}$.

6.49 Solve the equation: $|x|+2=5$.

This is a linear absolute value equation. To solve it, isolate the absolute value expression on the left side of the equal sign.

$$\begin{aligned}|x|+2&=5\\|x|&=5-2\\|x|&=3\end{aligned}$$

If the absolute value of x is 3, then $x = -3$ or $x = 3$. Remove the absolute value bars from the variable expression and add a "±" sign to the other side of the equation.

Remember, absolute values remove negative signs but do not affect positive signs. In other words, $|7| = 7$ and $|-2| = 2$.

$$x = \pm 3$$

The equation $|x|+2=5$ has two solutions: $x = -3$ and $x = 3$.

6.50 Solve the equation: $2|3x-4|=22$.

Follow the procedure outlined in Problem 6.49, beginning by dividing both sides of the equation by 2 to isolate the absolute value expression.

$$\begin{aligned}\frac{2|3x-4|}{2}&=\frac{22}{2}\\|3x-4|&=11\end{aligned}$$

Eliminate the absolute value bars from the left side of the equation and add a "±" sign to the right side.

$$3x-4 = \pm 11$$

You now have two equations: $3x-4 = -11$ and $3x-4 = +11$. Solve both for x.

$$\begin{aligned}3x-4&=-11\\3x&=-11+4\\3x&=-7\\x&=-\frac{7}{3}\end{aligned}\qquad\begin{aligned}3x-4&=11\\3x&=11+4\\3x&=15\\x&=\frac{15}{3}\\x&=5\end{aligned}$$

The equation $2|3x-4|=22$ has two solutions: $x=-\frac{7}{3}$ and $x=5$.

Note: Problems 6.51–6.52 explain how to solve the quadratic inequality $x^2 - 12x + 35 \geq 0$.

6.51 Identify the critical numbers for the quadratic expression.

Review Problems 6.32–6.33, in which you solve a rational inequality by using critical numbers to divide a number line, or wiggle graph, into intervals that represent possible solutions. Quadratic (and other polynomial inequalities with degree greater than or equal to 2) require the same technique. The critical numbers of this quadratic expression are the values of x for which the expression equals zero.

$$x^2 - 12x + 35 = 0$$

To solve this equation, factor the quadratic expression and set the factors equal to zero, using the technique modeled in Problem 6.47.

$$(x-5)(x-7)=0$$
$$x=5, \quad x=7$$

The critical numbers for the quadratic expression are $x = 5$ and $x = 7$. Because the original expression contains the ≥ symbol, values for which the expression is greater than *or equal to* zero satisfy the inequality. The expression is equal to zero when $x = 5$ and when $x = 7$, so those values that bound the intervals of the number line must be included in the solutions. You indicate this on the number line by using solid dots for the critical numbers, as illustrated below.

If the symbol in the inequality is < or >, use open dots on the number line.

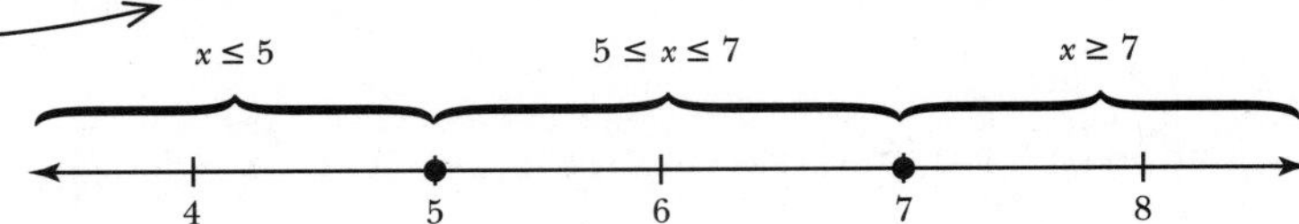

Note: Problems 6.51–6.52 explain how to solve the quadratic inequality $x^2 - 12x + 35 \ge 0$.

6.52 Use the critical numbers you identified in Problem 6.51 to calculate the solution. *Optional:* Verify your answer by graphing the quadratic expression on a calculator.

According to Problem 6.51, there are three sections of the number line—three intervals—that represent possible solutions: $x \le 5$, $5 \le x \le 7$, and $x \ge 7$. Select one value from each interval to substitute into the quadratic inequality to determine which values make the inequality true. In the solution below, the values $x = 0$, $x = 6$, and $x = 8$ are selected, respectively.

Test $x=0$	Test $x=6$	Test $x=8$
$0^2-12(0)+35\ge 0$	$6^2-12(6)+35\ge 0$	$8^2-12(8)+35\ge 0$
$0-0+35\ge 0$	$36-72+35\ge 0$	$64-96+35\ge 0$
$35\ge 0$	$-1\ge 0$	$3\ge 0$
True	**False**	**True**

Two intervals of the number line satisfy the inequality. The solution is $x \le 5$ or $x \ge 7$. If you graph $y = x^2 - 12x + 35$ on a graphing calculator, you will see that the entire graph lies above the x-axis except for a small portion that dips below the x-axis between $x = 5$ and $x = 7$. Therefore, the graph is either greater than or equal to zero when $x \le 5$ or when $x \ge 7$.

6.53 Solve the inequality: $x^2 - 3x < 54$.

Apply the technique demonstrated in Problems 6.51–6.52. Begin by subtracting 54 from both sides of the inequality; the right side of the inequality must be equal to 0.

$$x^2 - 3x - 54 < 0$$

Factor the quadratic expression and set the factors equal to 0 to identify the critical numbers.

$$(x-9)(x+6)=0$$
$$x=9, \quad x=-6$$

Because the inequality asks you to find values of x for which $x^2 - 3x - 54$ is *less than* zero, these critical numbers are *not* solutions. They are the numbers for which the expression *equals* zero, so use open dots to represent them on the number line, as illustrated below.

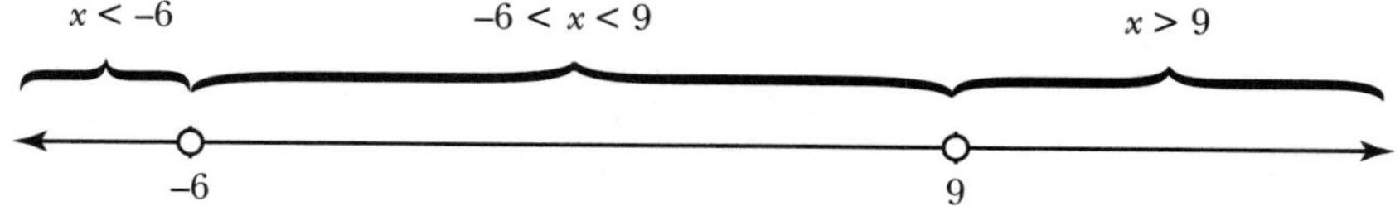

If you would rather, you can plug the test values into the original inequality instead: $x^2 - 3x < 54$.

Choose one value from each of the possible solution intervals, such as $x = -10$ from the interval $x < -6$, $x = 0$ from the interval $-6 < x < 9$, and $x = 10$ from the interval $x > 9$.

Test $x=-10$	Test $x=0$	Test $x=10$
$(-10)^2-3(-10)-54<0$	$0^2-3(0)-54<0$	$10^2-3(10)-54<0$
$100+30-54<0$	$0-0-54<0$	$100-30-54<0$
$76<0$	$-54<0$	$100-84<0$
False	**True**	$16<0$
		False

The solution is the interval $-6 < x < 9$.

Note: Problems 6.54–6.55 explain how to solve the absolute value inequality $|2x-3| \le 9$.

6.54 Explain how to rewrite a linear absolute value inequality containing a $\le$ or $<$ symbol as a compound, non-absolute value inequality.

This technique also works when the inequality symbol is $<$.

You can rewrite an absolute value inequality of the form $|ax+b| \le c$ as the compound inequality $-c \le ax + b \le c$. In other words, drop the absolute value bars, write the opposite of the constant c to the left of the expression (the opposite of the constant 9 in this problem is -9), and then separate $-c$ from the variable expression using a copy of the inequality symbol (in this case $\le$), as demonstrated below.

$$-9 \le 2x-3 \le 9$$

This is a compound inequality, because it is a shorthand way of representing two inequality statements: $-9 \le 2x - 3$ and $2x - 3 \le 9$. A compound inequality represents a single interval—a piece of the number line—bounded by two numbers. You calculate those two numbers in Problem 6.55.

Note: Problems 6.54–6.55 explain how to solve the absolute value inequality $|2x - 3| \le 9$.

6.55 Solve the complex inequality you identified in Problem 6.54 to solve the absolute value inequality.

To solve the compound inequality $-9 \le 2x - 3 \le 9$, solve for x in the middle of the statement. In other words, add 3 to *all three* expressions (-9, $2x - 3$, and 9) and then divide *all three* statements by 2, the coefficient of x.

$$\begin{aligned} -9+3 \le 2x&-3+3 \le 9+3 \\ -6 \quad \le \quad &2x \quad \le \; 12 \\ -\frac{6}{2} \quad \le \quad &\frac{2}{2}x \quad \le \; \frac{12}{2} \\ -3 \quad \le \quad &x \quad \le \; 6 \end{aligned}$$

All of the values between, and including, $x = -3$ and $x = 6$ satisfy the inequality $|2x - 3| \le 9$. The solution is the interval $-3 \le x \le 6$.

Note: Problems 6.56–6.57 explain how to solve the absolute value inequality $3|x + 1| - 8 > 7$.

6.56 Explain how to rewrite a linear absolute value inequality containing a $>$ or $\ge$ symbol as two non-absolute value inequalities.

Before you can rewrite the absolute value inequality, you need to isolate the absolute value expression on the left side of the inequality symbol and the constant on the right side. In other words, add 8 to both sides and then divide both sides by 3.

$$\begin{aligned} 3|x+1| - 8 &> 7 \\ 3|x+1| &> 7+8 \\ 3|x+1| &> 15 \\ \frac{3|x+1|}{3} &> \frac{15}{3} \\ |x+1| &> 5 \end{aligned}$$

You use the word "or" because one value of x can't satisfy both inequalities—it can only satisfy one OR the other.

You can rewrite the absolute value inequality $|ax + b| > c$ as two non-absolute value inequalities connected by the word "or."

$$ax + b > c \quad \text{or} \quad ax + b < -c$$

This technique also works when the inequality symbol is $\ge$.

In other words, the left inequality is simply a copy of the original inequality without absolute value bars. To generate the right inequality, reverse the inequality symbol and take the opposite of the constant on the right side of the inequality symbol.

$$x + 1 > 5 \quad \text{or} \quad x + 1 < -5$$

Note: Problems 6.56–6.57 explain how to solve the absolute value inequality $3|x+1|-8>7$.

6.57 Solve the inequalities you identified in Problem 6.56 to solve the absolute value inequality.

According to Problem 6.56, you can rewrite the absolute value inequality $3|x+1|-8>7$ as $x+1>5$ or $x+1<-5$. Solve the pair of inequality statements.

$$\begin{aligned} x+1&>5 & & & x+1&<-5 \\ x&>5-1 & &\text{or} & x&<-5-1 \\ x&>4 & & & x&<-6 \end{aligned}$$

All x-values that are either less than -6 or greater than 4 will satisfy the inequality $3|x+1|-8>7$. The solution is $x<-6$ or $x>4$.

Systems of Equations and Inequalities

Substitution, elimination, and shading

Note: In Problems 6.58–6.59, you solve the system of linear equations below using two different techniques.

$$\begin{cases} y=-x+1 \\ y=-\dfrac{1}{3}x+3 \end{cases}$$

6.58 Solve the system of equations by graphing.

Both of the equations in this system are expressed in slope-intercept form, which makes them easy to graph. Start with $y=-x+1$. It has a y-intercept of 1, so the line crosses the graph at (0,1). From that point, use the slope $m=-1/1$ to locate another point; move down one unit and right one unit and plot a point there. Draw a line through those two points to graph the line $y=-x+1$.

If a line has a negative slope, the graph decreases from left to right (it goes down). A line with a positive slope increases from left to right (it goes up). Both of these lines decrease from left to right.

To graph $y=-1/3\,x+3$, plot the y-intercept at (0,3) and use the slope $m=-1/3$ to plot another point on the line by traveling down one unit and right three units. Connect those points to graph the line.

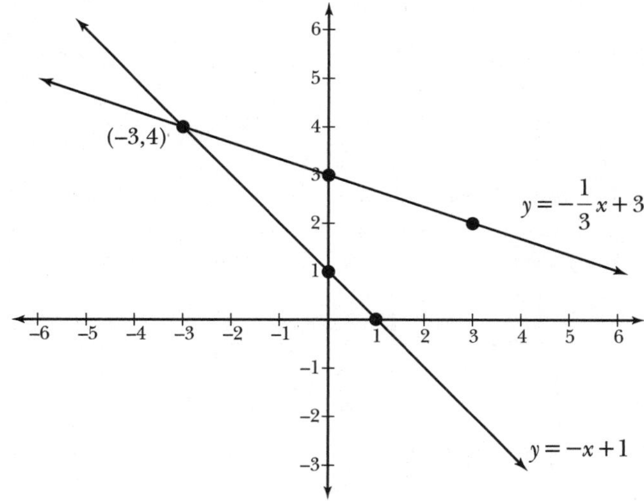

The two lines intersect at point (–3,4), so the only x- and y-coordinates that satisfy both equations in the system are $x=-3$ and $y=4$.

Note: In Problems 6.58–6.59, you solve the system of linear equations below using two different techniques.

$$\begin{cases} y = -x + 1 \\ y = -\frac{1}{3}x + 3 \end{cases}$$

6.59 Solve the system of equations using the substitution method and verify your solution to Problem 6.58.

In the substitution method, you solve one equation for either x or y and then substitute the resulting expression into the other equation. Both of these equations are solved for y, so you can substitute $y = -x + 1$ into $y = -1/3\,x + 3$. In other words, replace y in the second equation with $-x + 1$.

$$y = -\frac{1}{3}x + 3$$
$$-x + 1 = -\frac{1}{3}x + 3$$

Multiply all of the terms by 3 to eliminate the fraction and then solve for x.

$$3(-x) + 3(1) = 3\left(-\frac{1}{3}x\right) + 3(3)$$
$$-3x + 3 = -\frac{3}{3}x + 9$$
$$-3x + 3 = -1x + 9$$
$$-3x + 1x = 9 - 3$$
$$-2x = 6$$
$$x = -\frac{6}{2}$$
$$x = -3$$

The x-value of the solution is $x = -3$. Substitute this into either of the equations in the original system to calculate the corresponding y-value. In the solution below, $x = -3$ is substituted into $y = -x + 1$.

$$y = -(-3) + 1$$
$$y = 3 + 1$$
$$y = 4$$

To check your answer, substitute $x = -3$ and $y = 4$ into the equation of the system. They should make both equations true.

The solution to the system of equations is $x = -3$, $y = 4$. You can also report solutions as a coordinate pair: $(x, y) = (-3,4)$. This is the same solution you generated in Problem 6.58 using the graphing template.

6.60 Use the substitution method to solve the system of equations below.

$$\begin{cases} 4x+3y=5 \\ 6x+y=11 \end{cases}$$

Solve one of the equations for either x or y. Because the coefficient of y in the second equation is 1, it is easiest to solve that equation for y—you do not need to divide by the coefficient of y when you solve, which means fewer fractions to deal with.

$$\begin{aligned} 6x+y&=11 \\ y&=-6x+11 \end{aligned}$$

Replace y in the other equation with the equivalent expression $-6x + 11$.

$$\begin{aligned} 4x+3y&=5 \\ 4x+3(-6x+11)&=5 \\ 4x+3(-6x)+3(11)&=5 \\ 4x-18x+33&=5 \end{aligned}$$

Solve for x.

$$\begin{aligned} -14x&=5-33 \\ -14x&=-28 \\ x&=\frac{-28}{-14} \\ x&=2 \end{aligned}$$

The x-value of the solution is 2. Substitute it into either equation of the system to calculate the corresponding y-value. To save time, you can substitute it into $y = -6x + 11$, an equation of the system that you already solved for y.

$$\begin{aligned} y&=-6x+11 \\ &=-6(2)+11 \\ &=-12+11 \\ &=-1 \end{aligned}$$

The solution to the system of equations is $(x, y) = (2,-1)$.

Note: In Problems 6.61–6.62 you apply the elimination method to solve the system of equations below.

$$\begin{cases} 3x - 8y = -10 \\ x + 2y = -1 \end{cases}$$

6.61 Eliminate the x-variables to solve the system of equations.

In the elimination technique, you multiply all terms of one (or both) equations by a constant so that corresponding coefficients in the equations are opposites. Then, you add the equations of the system together to eliminate those terms from the system. The goal is the same as the substitution method; you can only solve a linear equation if it contains one variable, so the other variable needs to be eliminated whether you choose the substitution method or the elimination method to do so.

This problem asks you to eliminate the x-variables. The coefficients of the x-variables are 3 and 1. If you multiply the x-coefficient 1 by –3, you will get –3, which is the opposite of the other x-coefficient. Therefore, you should multiply *each* term of the second equation by –3.

$$\begin{cases} 3x - 8y = -10 \\ -3(x + 2y = -1) \end{cases} \rightarrow \begin{cases} 3x - 8y = -10 \\ -3x - 6y = 3 \end{cases}$$

Now add the equations together, combining like terms.

$$\begin{array}{rcrcr} 3x & - & 8y & = & -10 \\ -3x & - & 6y & = & 3 \\ \hline 0 & - & 14y & = & -7 \end{array}$$

The result is a linear equation in one variable: $-14y = -7$. Solve for y.

$$\begin{aligned} y &= \frac{-7}{-14} \\ &= \frac{1}{2} \end{aligned}$$

Substitute $y = 1/2$ into one of the equations in the original system to calculate the corresponding value of x.

$$\begin{aligned} 3x - 8y &= -10 \\ 3x - 8\left(\frac{1}{2}\right) &= -10 \\ 3x - 4 &= -10 \\ 3x &= -10 + 4 \\ 3x &= -6 \\ x &= -\frac{6}{3} \\ x &= -2 \end{aligned}$$

The solution to the system is $(x, y) = (-2, 1/2)$.

Note: In Problems 6.61–6.62 you apply the elimination method to solve the system of equations below.

$$\begin{cases} 3x - 8y = -10 \\ x + 2y = -1 \end{cases}$$

6.62 Verify your solution to Problem 6.61 by solving the system of equations again, this time eliminating the y-variables.

Why are you working out the same problem twice? To show you that it doesn't matter which variable you initially eliminate. Either way, you get the same final answer.

If you multiply the second equation by 4, the y-coefficients will be opposites: –8 and 4(2) = 8. Adding opposite coefficients eliminates them from the system. Therefore, you should multiply the entire second equation by 4.

$$\begin{cases} 3x - 8y = -10 \\ 4(x + 2y = -1) \end{cases} \rightarrow \begin{cases} 3x - 8y = -10 \\ 4x + 8y = -4 \end{cases}$$

You don't always have to change the second equation—sometimes you might change the first equation instead. If you have a choice, select smaller coefficients to manipulate, such as the x-coefficient 1 (instead of the x-coefficient 3) in Problem 6.61 and the y-coefficient 2 (instead of the y-coefficient -8) in this problem.

Now add the equations, combining like terms.

$$\begin{array}{rcrcr} 3x & - & 8y & = & -10 \\ 4x & + & 8y & = & -4 \\ \hline 7x & + & 0 & = & -14 \end{array}$$

Solve the resulting equation for x.

$$\begin{aligned} 7x &= -14 \\ x &= -\frac{14}{7} \\ x &= -2 \end{aligned}$$

Substitute $x = -2$ into one of the equations from the original system to calculate the corresponding value of y.

$$\begin{aligned} x + 2y &= -1 \\ -2 + 2y &= -1 \\ 2y &= -1 + 2 \\ 2y &= 1 \\ y &= \frac{1}{2} \end{aligned}$$

The solution to the system of equations is $(x,y) = (-2, 1/2)$, which is the same solution you calculated in Problem 6.61. Therefore, it does not matter whether you choose to eliminate x or y from the system; the final solution is the same.

6.63 Apply the elimination method to solve the system of linear equations below.

$$\begin{cases} 5x + 3y = 26 \\ 7x - 9y = 32 \end{cases}$$

If you multiply the first equation of the system by 3, the y-coefficients of the equations are opposites and can be eliminated.

$$\begin{cases} 3(5x + 3y = 26) \\ 7x - 9y = 32 \end{cases} \rightarrow \begin{cases} 15x + 9y = 78 \\ 7x - 9y = 32 \end{cases}$$

Add the equations, combining like terms.

$$\begin{array}{rcrcr} 15x & + & 9y & = & 78 \\ 7x & - & 9y & = & 32 \\ \hline 22x & + & 0 & = & 110 \end{array}$$

Solve the equation for x.

$$22x = 110$$
$$x = \frac{110}{22}$$
$$x = 5$$

Substitute $x = 5$ into one of the original equations from the system to calculate the corresponding value of y.

$$5x + 3y = 26$$
$$5(5) + 3y = 26$$
$$25 + 3y = 26$$
$$3y = 26 - 25$$
$$3y = 1$$
$$y = \frac{1}{3}$$

The solution to the system is $(x, y) = (5, 1/3)$.

6.64 Apply the elimination method to solve the system of equations below.

$$\begin{cases} 2x - 7y = 9 \\ 3x + 4y = 28 \end{cases}$$

To apply the elimination method in this problem, you need to multiply *both* of the equations by a constant. For example, you could multiply the first equation by 3 and the second equation by –2 to eliminate the x-coefficients. Alternately, you could multiply the first equation by 4 and the second equation by 7 to eliminate the y-coefficients.

In other words, to eliminate the x-coefficients, multiply the top equation by the x-coefficient of the bottom equation and vice versa. Then, multiply one of the equations by –1 if you need those x-terms to be opposites.

$$\begin{cases} 3(2x - 7y = 9) \\ -2(3x + 4y = 28) \end{cases} \rightarrow \begin{cases} 6x - 21y = 27 \\ -6x - 8y = -56 \end{cases}$$

Add the equations, combining like terms.

$$\begin{array}{rcrcr} 6x & - & 21y & = & 27 \\ -6x & - & 8y & = & -56 \\ \hline 0 & - & 29y & = & -29 \end{array}$$

Solve the equation for y.

$$-29y = -29$$
$$y = \frac{-29}{-29}$$
$$y = 1$$

Calculate the corresponding value of x.

$$\begin{aligned} 2x-7y&=9\\ 2x-7(1)&=9\\ 2x-7&=9\\ 2x&=9+7\\ 2x&=16\\ x&=\frac{16}{2}\\ x&=8 \end{aligned}$$

The solution to the system of equations is $(x,y) = (8,1)$.

6.65 Apply any method to solve the system of equations below, and explain your solution.

$$\begin{cases} 2x-y=3\\ 8x-4y=1 \end{cases}$$

If you add y to, and subtract 3 from, both sides of the first equation, you can solve it for y.

$$\begin{aligned} 2x-y&=3\\ 2x-3&=y \end{aligned}$$

Apply the substitution method, replacing y in the equation $8x-4y=1$ with the equivalent value $2x-3$.

If you use the substitution method, you solve one equation for a variable and then plug the result into the OTHER equation.

$$\begin{aligned} 8x-4y&=1\\ 8x-4(2x-3)&=1\\ 8x-4(2x)-4(-3)&=1\\ 8x-8x+12&=1\\ 0+12&=1\\ 12&=1 \quad \textbf{\textit{False}} \end{aligned}$$

All of the variables are eliminated unexpectedly from the equation, leaving you with a false statement: $12 = 1$. Because this statement is false, you conclude that *no* values of x and y satisfy both equations, so there is no solution to the system.

Why is there no solution? Consider the slopes of the lines using this shortcut: The slope of the line $Ax + By = C$ is $-A/B$. In other words, if the line is written in standard form, the slope is the opposite of the x-coefficient divided by the y-coefficient.

$$\text{slope of } 2x-y=3: \frac{-2}{-1}=2 \qquad \text{slope of } 8x-4y=1: \frac{-8}{-4}=2$$

Both of the lines have slope 2, so the lines are parallel. Think back to Problem 6.58, which explains that the solution to a system of linear equations is the point at which the graphs of the lines intersect. Parallel lines, by definition, do not intersect. No point of intersection means there is no solution for this system of equations.

PERPENDICULAR lines have slopes that are opposite reciprocals, so if lines s and t are perpendicular and line s has slope 4/5, then line t has slope −5/4.

6.66 Graph the solution to the system of linear inequalities below.

$$\begin{cases} y < 4 \\ y \ge x \end{cases}$$

The solution to a system of linear inequalities is entirely graphical—it is a shaded region of a coordinate plane inside which all of the coordinate pairs are solutions to the system of the equation. To construct the solution, graph each of the inequalities. The graphs are each composed of a line (either dotted or solid, depending on the inequality symbol) and shading (either above or below the line, depending on the inequality symbol).

The solution to the system of inequalities may include the graphs of the lines, but it always includes the region where the shading for all of the inequalities overlaps. (See Problems 6.41–6.43 to review graphing linear inequalities.)

Only solid lines are technically part of the solution graphs. Dotted lines act as boundaries but are not technically solutions themselves.

To graph $y < 4$, replace the inequality symbol with an equal sign: $y = 4$. This is a horizontal line 4 units above the x-axis. Because the inequality symbol is $<$, the line should be dotted and you should shade beneath it.

To graph $y \ge x$, again replace the inequality symbol with an equal sign: $y = x$. This equation is in slope-intercept form with slope $m = 1$ and y-intercept $b = 0$. The line should be solid and you should shade above it.

The following graph is the solution to the system of inequalities.

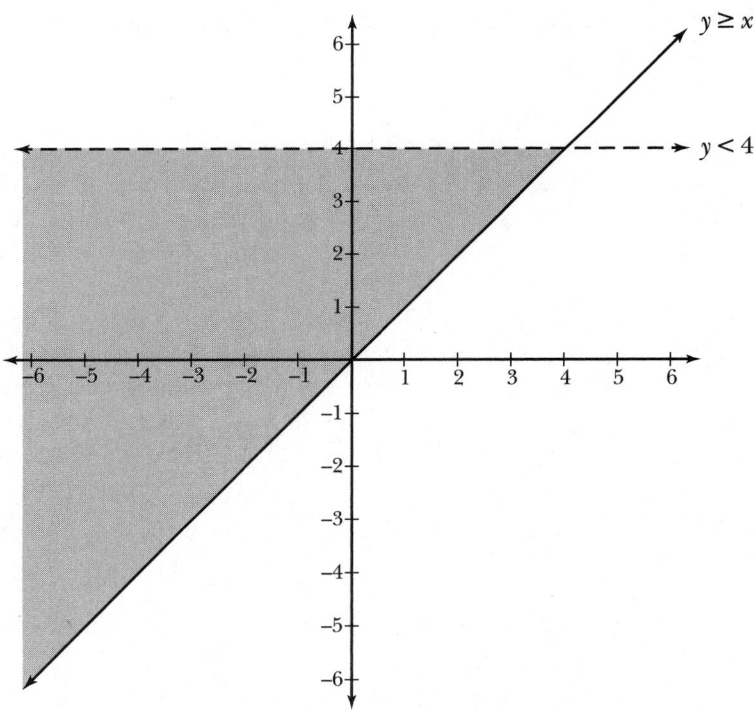

6.67 Graph the solution to the system of linear equations below.

$$\begin{cases} 2x + y < 6 \\ 3x - 4y > 12 \end{cases}$$

To ensure that you shade correctly, solve each of the inequalities for y to express them in slope-intercept form.

$$\begin{aligned} 2x + y &< 6 \\ y &< -2x + 6 \end{aligned}$$

$$\begin{aligned} 3x - 4y &> 12 \\ -4y &> -3x + 12 \\ \frac{-4}{-4}y &< \frac{-3}{-4}x + \frac{12}{-4} \\ y &< \frac{3}{4}x - 3 \end{aligned}$$

You have to reverse the inequality symbol because you're dividing by a negative number.

The solution to the system of inequalities is the region on which the shading for both inequalities overlaps.

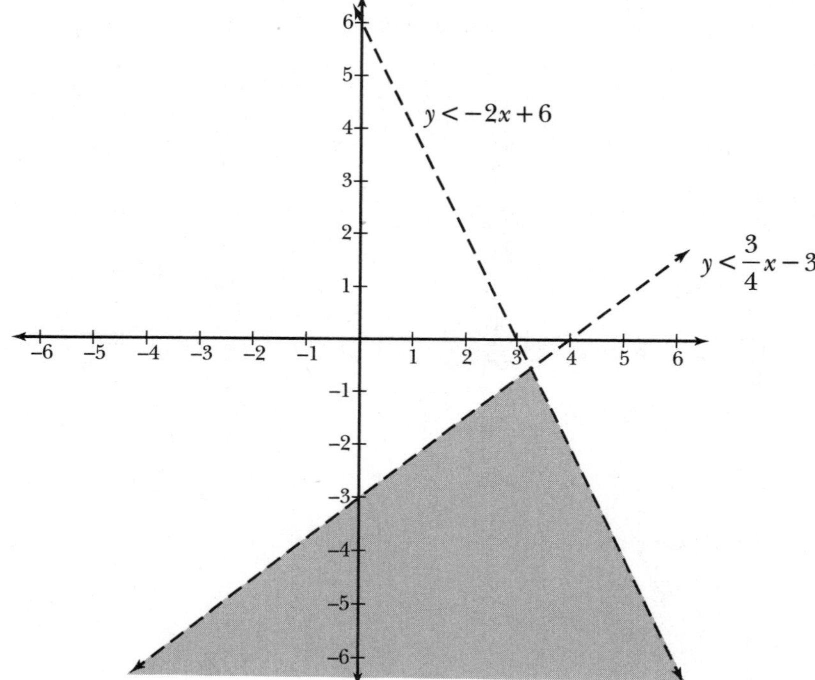

6.68 Graph the solution to the system of linear equations below.

$$\begin{cases} x \geq -3 \\ x \leq 5 \\ 2x + 4y > -8 \end{cases}$$

The graphs of the inequalities $x \geq -3$ and $x \leq 5$ are solid, vertical lines that form left and right boundaries of the region. Solve the remaining linear equation for y to express it in slope-intercept form.

$$\begin{aligned} 2x + 4y &> -8 \\ 4y &> -2x - 8 \\ \frac{4}{4}y &> -\frac{2}{4}x - \frac{8}{4} \\ y &> -\frac{1}{2}x - 2 \end{aligned}$$

The vertical lines are solid; the line with slope $m = -1/2$ and y-intercept -2 should be dotted, as illustrated in the following solution graph.

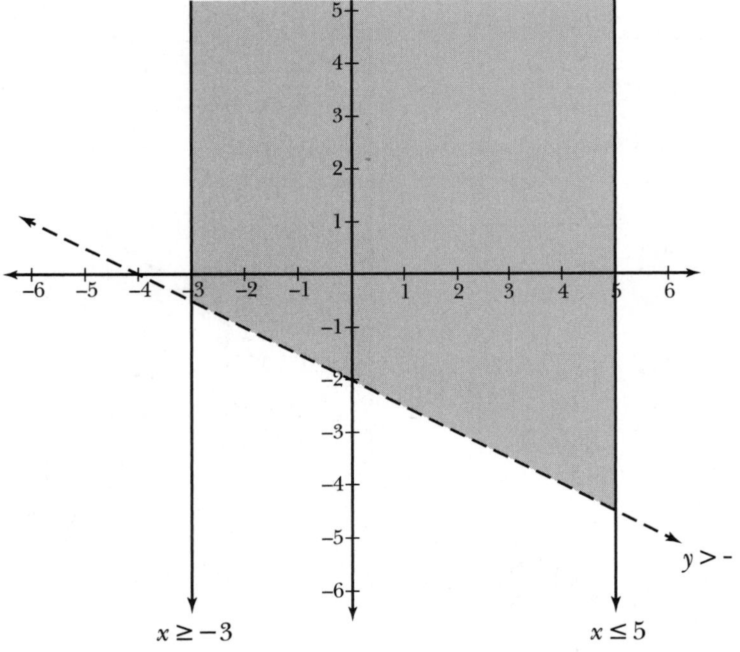

Functions

Evaluating, graphing, defining, translating, and modeling

6.69 Given $f(x)$, as defined below, calculate $f(3)$.

$$f(x)=\frac{x-x^2}{x+1}$$

The notation $f(3)$ is read "f of 3"; it represents the function $f(x)$ evaluated for $x = 3$. To calculate $f(3)$, substitute $x = 3$ into the function.

$$f(x)=\frac{x-x^2}{x+1}$$

$$f(3)=\frac{3-3^2}{3+1}$$

$$f(3)=\frac{3-9}{4}$$

$$f(3)=\frac{-6}{4}$$

$$f(3)=-\frac{3}{2}$$

Note: Problems 6.70–6.72 refer to the table below, which lists a few values of functions f(x) and g(x).

x	-1	0	1	2	3	4
$f(x)$	5	4	2	0	1	6
$g(x)$	7	6	4	3	0	-9

6.70 Evaluate $(f+g)(-1)$.

The function $(f+g)(x)$ is the sum of functions $f(x)$ and $g(x)$. In other words, $(f+g)(x) = f(x) + g(x)$. Therefore, $(f+g)(-1) = f(-1) + g(-1)$. According to the chart, $f(-1) = 5$ and $g(-1) = 7$. Therefore, $f(-1) + g(-1) = 5 + 7 = 12$.

Note: Problems 6.70–6.72 refer to the table in Problem 6.70, which lists a few values of functions f(x) and g(x).

6.71 Evaluate $g(f(3))$.

This problem asks you to compose functions, or substitute the value of one function into another function. Begin with the function inside parentheses, $f(3)$. The chart states that $f(3) = 1$. Replace $f(3)$ in the original expression with its value, 1.

$$g(f(3))=g(1)$$

Use the chart once again to evaluate $g(1)$.

$$g(1) = 4$$

You conclude that $g(f(3))=4$.

Note: Problems 6.70–6.72 refer to the table in Problem 6.70, which lists a few values of functions f(x) and g(x).

6.72 Evaluate $f\left(g^{-1}(3)\right)$.

The notation g^{-1} refers to the inverse of a function. Whereas $g(3)$ represents the value of $g(x)$ when you substitute $x = 3$ into it, $g^{-1}(3)$ asks the question "For what value of x does $g(x)$ *equal* 3?" The end result of $g(3)$ is a function output, something from the $g(x)$ row on the chart; the end result of $g^{-1}(3)$ is a function input, something from the x row on the chart.

Plugging $x = 2$ into $g(x)$ gives you 3. That means plugging $x = 3$ into the inverse function $g^{-1}(x)$ gives you 2.

Because $g(2) = 3$, you know that $g^{-1}(3) = 2$. Replace $g^{-1}(3)$ with the equivalent value, 2.

$$f\left(g^{-1}(3)\right) = f(2)$$

According to the chart, $f(2) = 0$. You conclude that $f\left(g^{-1}(3)\right) = 0$.

Note: Problems 6.73–6.74 refer to the following functions: $f(x) = x^2 + 4$, $g(x) = 2x - 5$, and $h(x) = 3x + 1$.

6.73 Calculate $f\left(g\left(h(-1)\right)\right)$.

Start with the innermost function value, $h(-1)$, and work your way outward.

$$\begin{aligned} h(-1) &= 3(-1)+1 \\ &= -3+1 \\ &= -2 \end{aligned}$$

Therefore, $f\left(g\left(h(-1)\right)\right) = f\left(g(-2)\right)$. Now evaluate $g(x)$ for $x = -2$.

$$\begin{aligned} g(-2) &= 2(-2)-5 \\ &= -4-5 \\ &= -9 \end{aligned}$$

Therefore, $f\left(g(-2)\right) = f(-9)$. Evaluate $f(x)$ for $x = -9$.

$$\begin{aligned} f(-9) &= (-9)^2+4 \\ &= 81+4 \\ &= 85 \end{aligned}$$

You conclude that $f\left(g\left(h(-1)\right)\right) = 85$.

Note: Problems 6.73–6.74 refer to the following functions: $f(x) = x^2 + 4$, $g(x) = 2x - 5$, and $h(x) = 3x + 1$.

6.74 Construct the graph of $h(g(x))$.

First, you must express $h(g(x))$ in terms of x. How? By substituting $g(x)$ into $h(x)$. If you were asked to calculate $h(2)$, you would replace the x's in $h(x)$ with 2. Therefore, to generate $h(g(x))$ you do the same thing. Replace the x's in $h(x)$ with $g(x)$.

$$h(x) = 3x + 1$$
$$h(g(x)) = 3(g(x)) + 1$$

Note that $g(x) = 2x - 5$.

$$\begin{aligned} h(g(x)) &= 3(2x-5)+1 \\ &= 3(2x)+3(-5)+1 \\ &= 6x-15+1 \\ &= 6x-14 \end{aligned}$$

This is a linear function in slope-intercept form. The y-intercept of $h(g(x))$ is –14 and the slope is 6, as graphed below.

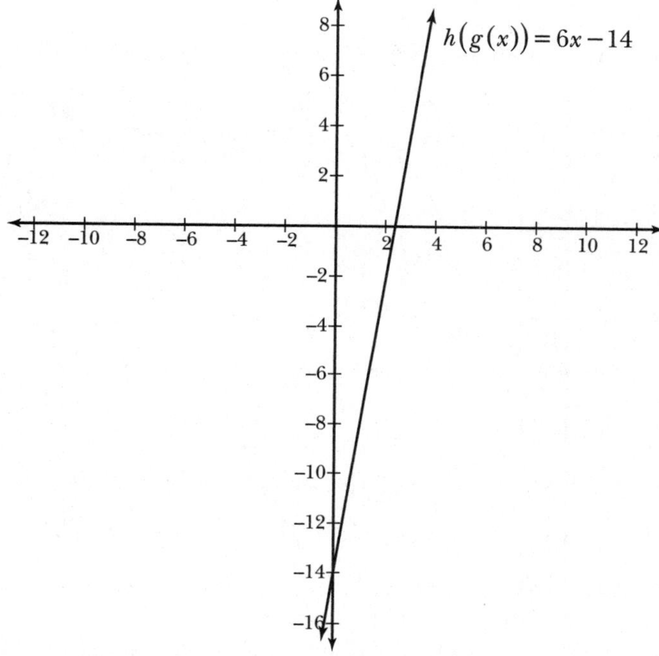

6.75 Explain why $f(x) = 6x^3 - 5x$ is an odd function

If $f(x)$ is an odd function, then $f(-x) = -f(x)$. In other words, if you replace x in the function with $-x$, the result is a function in which every term is the opposite of the corresponding term in the original function. Substitute $-x$ into $f(x)$ to see if the signs of the function's terms change.

$$f(x) = 6x^3 - 5x$$
$$f(-x) = 6(-x)^3 - 5(-x)$$
$$f(-x) = 6(-x^3) + 5x$$
$$f(-x) = -6x^3 + 5x$$

The terms of $f(-x)$ are $-6x^3$ and $5x$, which are the opposite of the terms in $f(x)$, $6x^3$ and $-5x$. Therefore, $f(x)$ is an odd function.

A function is odd when all of the terms contain a single variable raised to an odd power. That is why they are called odd functions.

6.76 Determine whether $g(x) = x^4 - 3x^2 + 5$ is an even function and explain your answer.

According to Problem 6.75, a function is odd if substituting $-x$ into it changes the signs of all the function's terms. The test for an even function also involves substituting $-x$ into a function, but in an even function, the signs of the terms will not change. Mathematically, function $g(x)$ is even if $g(-x) = g(x)$. The function in this problem is even because substituting $-x$ into x does not change *any* of the signs of the terms.

$$g(x) = x^4 - 3x^2 + 5$$
$$g(-x) = (-x)^4 - 3(-x)^2 + 5$$
$$g(-x) = x^4 - 3x^2 + 5$$

Because $g(x) = g(-x) = x^4 - 3x^2 + 5$, $g(x)$ is an even function.

6.77 Two functions, $j(x)$ and $k(x)$, are graphed below. Which of the functions is odd and which is even? Explain your answer.

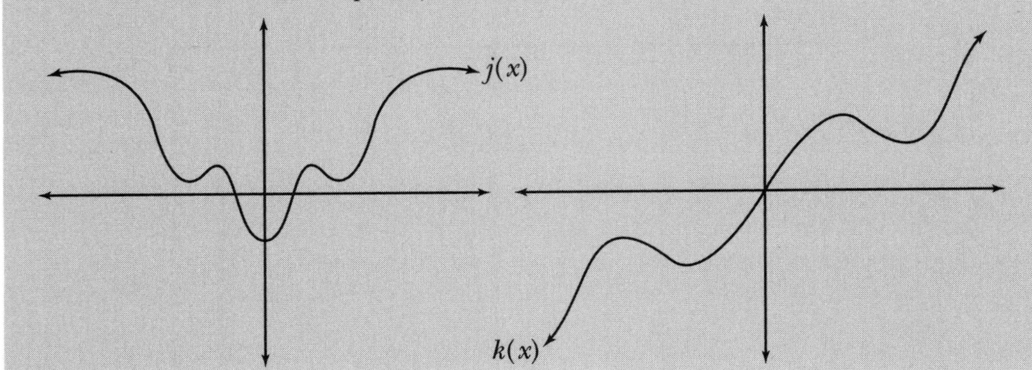

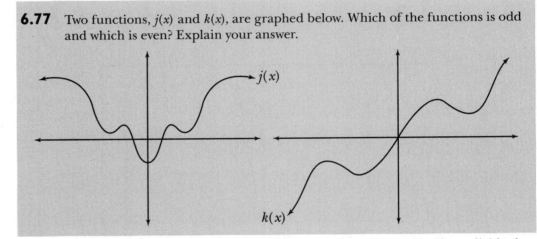

Even functions are symmetric about the y-axis. In other words, if you divide the graph vertically along the line $x = 0$, the two sides of the graphs are reflections of each other as though the y-axis were a mirror. The graph of $j(x)$ is symmetric about the y-axis, so $j(x)$ must be an even function.

Odd functions are origin-symmetric. In other words, the parts of the graph on opposite sides of the origin behave in opposite ways. The graph of $k(x)$ is origin-symmetric. In the first quadrant (the upper-right portion of the graph), $k(x)$ rises, falls slightly, and then rises again as you travel from the origin toward the right. As you travel left from the origin, the graph does the opposite: It falls, rises slightly, and then falls again.

The piece of the graph left of the origin is the same as the piece of the graph right of the origin, if you rotate either of the pieces 180°.

6.78 For real numbers x, y, and z, let $x \underset{z}{\blacktriangle} y$ be defined as follows: $x \underset{z}{\blacktriangle} y = xy^z$. Calculate $2 \underset{3}{\blacktriangle} 4$.

The problem defines a new function based on the values x, y, and z. To evaluate the function, substitute $x = 2$, $y = 4$, and $z = 3$ into xy^z.

$$x \underset{z}{\blacktriangle} y = xy^z$$

$$2 \underset{3}{\blacktriangle} 4 = (2)(4)^3$$

According to the order of operations, you need to evaluate exponents before you multiply. Therefore, you need to calculate 4^3 first.

$$2 \underset{3}{\blacktriangle} 4 = 2 \cdot 64$$

$$2 \underset{3}{\blacktriangle} 4 = 128$$

Note: Problems 6.79–6.80 refer to the graph of function p(x) below.

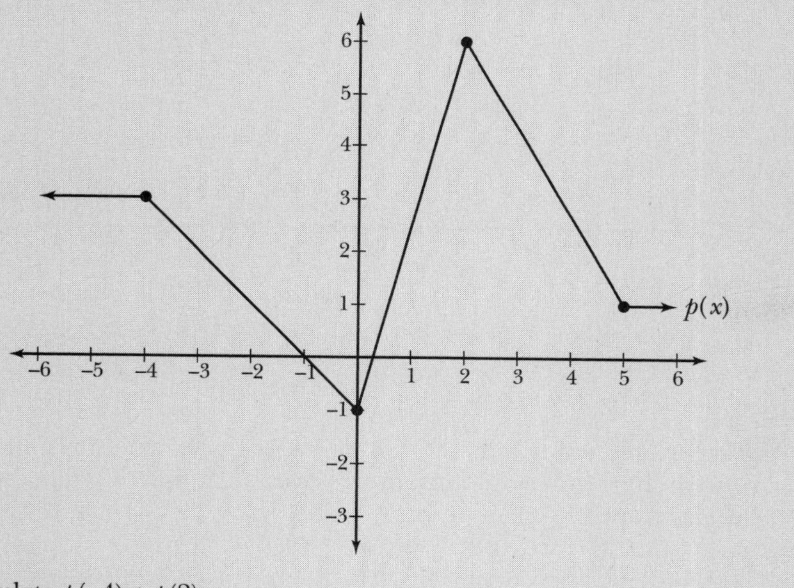

6.79 Calculate $p(-4) + p(2)$.

The value of $p(x)$ for a specific x-value is the height of the function at that x-value. For example, $p(-4) = 3$ because the function is 3 units above the x-axis at $x = -4$. Similarly, $p(2) = 6$. Therefore, $p(-4) + p(2) = 3 + 6 = 9$.

Note: Problems 6.79–6.80 refer to the graph of function p(x) in Problem 6.79.

6.80 For integers c and d, let $c \diamond d$ be defined in terms of the function $p(x)$, such that $c \diamond d = p(c) - d$. Evaluate $5 \diamond 6$.

Like Problem 6.78, this problem defines a new function. Substitute $c = 5$ and $d = 6$ into it.

$$c \diamond d = p(c) - d$$
$$5 \diamond 6 = p(5) - 6$$

The function $p(x)$ is one unit above the x-axis when $x = 5$, so $p(5) = 1$.

$$5 \diamond 6 = 1 - 6$$
$$5 \diamond 6 = -5$$

6.81 Given the graph of $f(x)$ below, construct the graphs of $f(x) - 3$ and $f(x + 2)$.

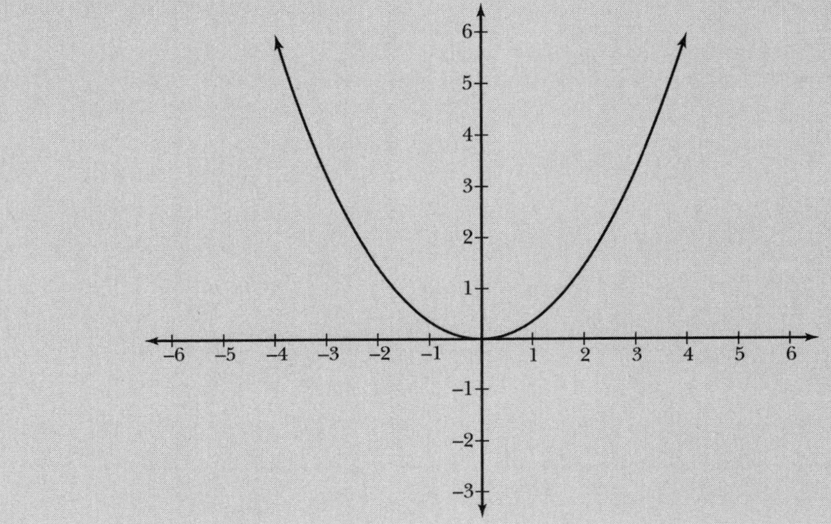

Adding or subtracting values to $f(x)$ moves its graph up or down the corresponding number of units. In this case, the graph of $f(x) - 3$ is 3 units below the graph of $f(x)$.

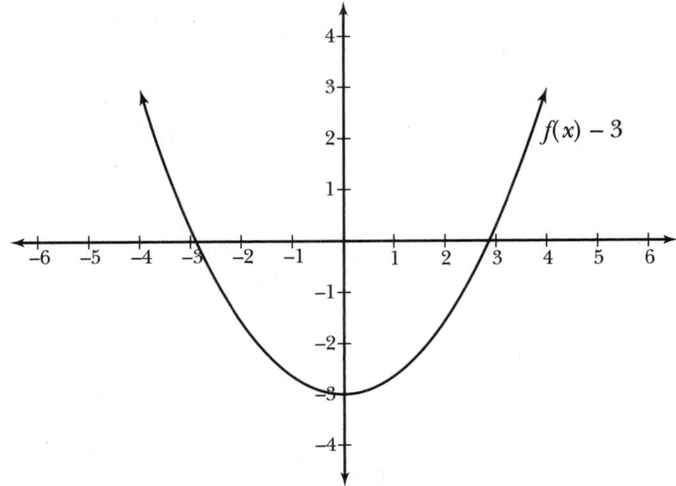

Adding values to, or subtracting them from, x within $f(x)$ moves the graph of a function right or left. If you subtract a value from x, that moves the graph to the right. Adding a value to x moves the graph to the left. In this case, the graph of $f(x + 2)$ is 2 units left of the graph of $f(x)$.

This is the opposite of what you might think. Adding moves the graph LEFT and subtracting moves it RIGHT.

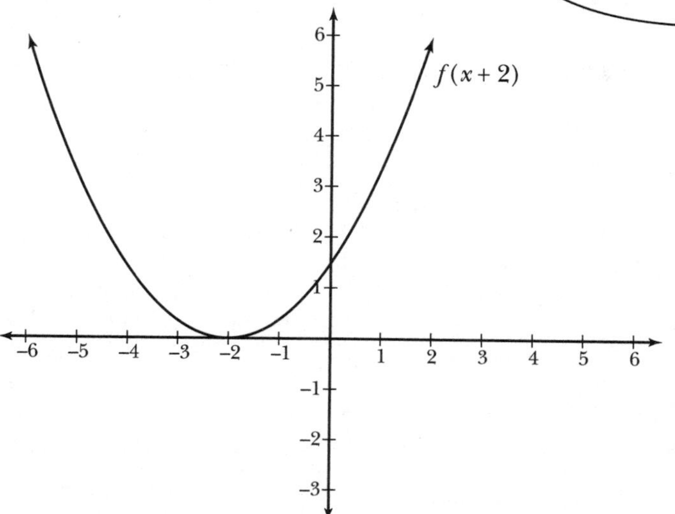

6.82 Given the graph of $g(x)$ below, construct the graph of $\frac{1}{2}g(x)$.

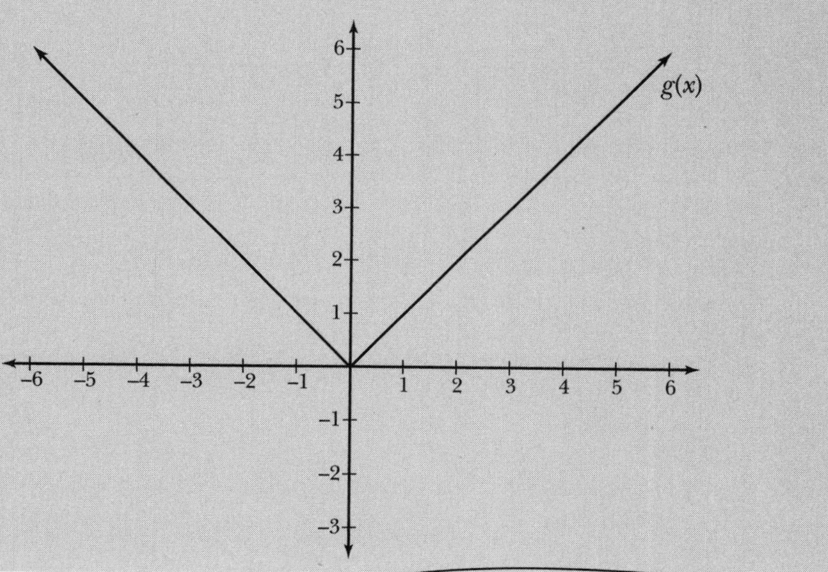

Multiplying a function by a constant either stretches or compresses the graph vertically. In this case, multiplying $g(x)$ by $1/2$ will compress the graph to one-half of its original height. At each value of x along their graphs, $(1/2)g(x)$ has function values that are one-half as far from the x-axis as the function values of $g(x)$.

In other words, a real number

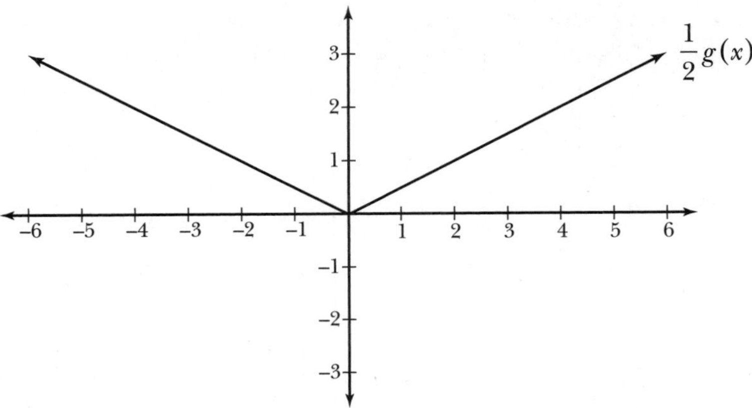

6.83 Given the graph of $h(x)$ below, construct the graphs of $-h(x)$ and $h(-x)$.

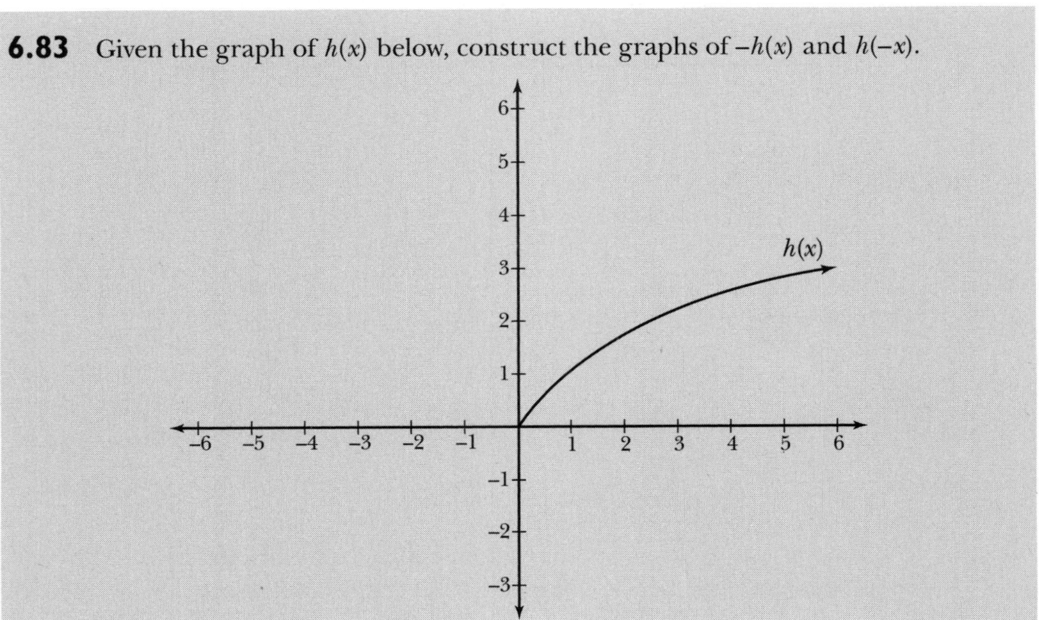

Multiplying a function by –1 reflects its graph across the x-axis, as illustrated below.

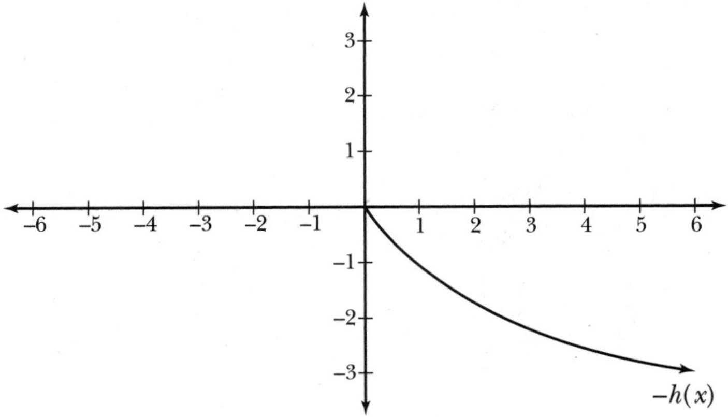

Multiplying x within a function by -1 reflects the graph across the y-axis.

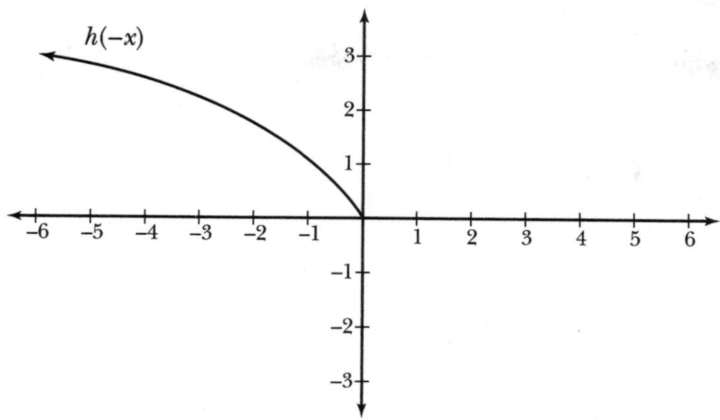

6.84 At a local sports event, promoters use an air-powered T-shirt cannon to propel souvenir shirts into crowds of fans during timeouts and delays in the game. The cannon is mounted on a raised platform at a fixed angle but can be swiveled to aim at different sections of the stadium. During tests, engineers determined that the function $s(t) = -4t^2 + 24t + 64$ represents the height of the shirt above the ground t seconds after the cannon is fired. According to the function, once a shirt is fired, how long will it stay in the air before it hits the ground?

The shirt is fired at time $t = 0$. It will hit the ground when its height above the ground, $s(t)$, equals 0. Set $s(t) = 0$ and solve the quadratic equation.

$$-4t^2 + 24t + 64 = 0$$

Divide each term of the equation by -4.

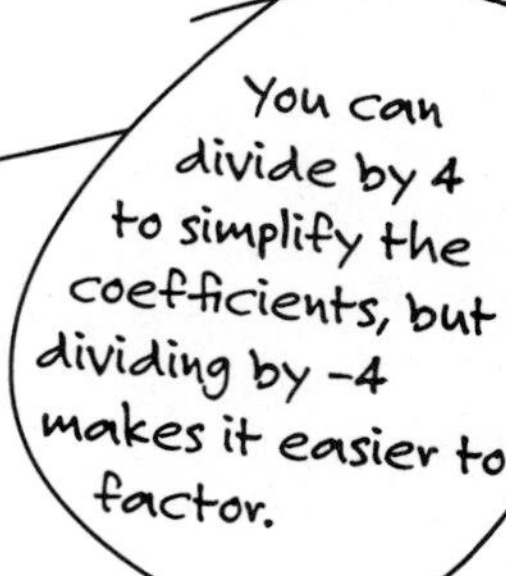

You can divide by 4 to simplify the coefficients, but dividing by -4 makes it easier to factor.

$$\frac{-4}{-4}t^2 + \frac{24}{-4}t + \frac{64}{-4} = \frac{0}{-4}$$

$$t^2 - 6t - 16 = 0$$

Factor the quadratic expression and set each factor equal to 0.

$$(t-8)(t+2) = 0$$

$$t = -2,\ t = 8$$

The solution $t = -2$ does not make sense in this problem—the shirt would have to hit the ground 2 seconds before it was fired! Therefore, you discard $t = -2$. The T-shirt will hit the ground $t = 8$ seconds after it is fired.

6.85 A business analyst determines that the demand for a certain product can be modeled by the linear function $d(x)$, where $d(x)$ represents the number of units in demand (in thousands of units) exactly x weeks after the product is launched. If the demand for the product is 300,000 units one week after launch and then dips to 220,000 units four weeks after launch, what is the projected demand for the product 12 weeks after launch? Round your answer to the nearest whole number.

The problem states that the function $d(x)$ is linear, but it does not tell you what that function is. However, it does give you enough information to construct the equation of the line. If each point on the graph of the function has coordinate pair $(x, d(x))$, the function passes through points $(1, 300)$ and $(4, 220)$. (Notice that demand is in thousands, so the $d(x)$ values in the coordinate pairs are 300 and 220 rather than 300,000 and 220,000.)

Apply the technique described in Problem 6.40 to calculate the equation of a line passing through two points. Begin by calculating the slope, m.

$$\begin{aligned} m &= \frac{y_2 - y_1}{x_2 - x_1} \\ &= \frac{220 - 300}{4 - 1} \\ &= -\frac{80}{3} \end{aligned}$$

Substitute the slope and one of the coordinate pairs into the point-slope formula: $m = -80/3$ and coordinate pair $x_1 = 1$, and $y_1 = 300$.

$$\begin{aligned} y - y_1 &= m(x - x_1) \\ y - 300 &= -\frac{80}{3}(x - 1) \end{aligned}$$

Solve for y.

$$\begin{aligned} y &= -\frac{80}{3}x + \frac{80}{3} + 300 \\ y &= -\frac{80}{3}x + \frac{80}{3} + \frac{900}{3} \\ y &= -\frac{80}{3}x + \frac{980}{3} \end{aligned}$$

The demand function is $d(x) = (-80/3)\,x + 980/3$. To calculate the demand after 12 weeks, substitute $x = 12$ into the function.

$$\begin{aligned} d(12) &= -\frac{80}{3}(12) + \frac{980}{3} \\ &= -\frac{960}{3} + \frac{980}{3} \\ &= \frac{20}{3} \\ &= 6.\overline{6} \end{aligned}$$

The demand after 12 weeks is approximately 6,666.666 units. You should round this value to the nearest whole number: 6,667 units.

Note: Problems 6.86–6.87 refer to function $f(x) = \left|3\sqrt{x} - 2x\right|$ and the graphs of functions g(x) and h(x) below. Assume that all segments of the graphs are linear.

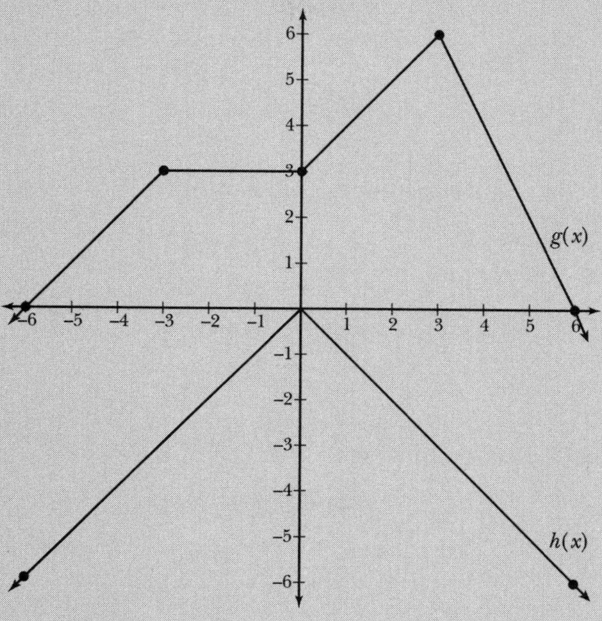

6.86 Calculate $g(h(-6))$.

Begin by evaluating the innermost function value within the parentheses, $h(-6)$. The function $h(x)$ consists of two line segments that intersect the origin. The left half of the graph has a slope of 1, and passes through the points $(-1,-1)$, $(-2,-2)$, $(-3,-3)$, etc. Because the graph passes through $(-6,-6)$, you can conclude that $h(-6) = -6$. Therefore, $g(h(-6)) = g(-6)$.

h(x) is actually the function $-|x|$, the absolute value function reflected over the x-axis.

Notice that $g(x)$ intersects the x-axis when $x = -6$, so $g(-6) = 0$. You conclude that $g(h(-6)) = 0$.

Note: Problems 6.86–6.87 refer to function $f(x) = \left|3\sqrt{x} - 2x\right|$ and the graphs of functions g(x) and h(x) below. Assume that all segments of the graphs are linear.

6.87 List the following values in order, from least to greatest: $f(4)$, $g(4)$, and $h(4)$.

The graph of $h(x)$ passes through point $(4,-4)$, so $h(4) = -4$. You can calculate $f(4)$ by substituting $x = 4$ into $f(x)$.

$$\begin{aligned} f(4) &= \left|3\sqrt{4} - 2(4)\right| \\ &= \left|3(2) - 2(4)\right| \\ &= |6 - 8| \\ &= |-2| \\ &= 2 \end{aligned}$$

To determine the value of $g(4)$, you need to identify the equation for the segment beginning at point $(3,6)$ and ending at point $(6,0)$. First, calculate the slope of the line passing through those points.

$$\begin{aligned} m &= \frac{y_2 - y_1}{x_2 - x_1} \\ &= \frac{0-6}{6-3} \\ &= \frac{-6}{3} \\ &= -2 \end{aligned}$$

Now apply the point-slope formula, substituting in $m = -2$, $x_1 = 3$, and $y_1 = 6$.

$$\begin{aligned} y - y_1 &= m(x - x_1) \\ y - 6 &= -2(x-3) \\ y - 6 &= -2x + 6 \\ y &= -2x + 12 \end{aligned}$$

Therefore, between $x = 3$ and $x = 6$, the graph of $g(x)$ has equation $g(x) = -2x + 12$. Now evaluate $g(4)$.

$$\begin{aligned} g(4) &= -2(4) + 12 \\ &= -8 + 12 \\ &= 4 \end{aligned}$$

You conclude that $h(4) < f(4) < g(4)$ because $-4 < 2 < 4$.

6.88 Use the graph of $j(x) = \frac{1}{x+1}$ to identify the domain and range of the function.

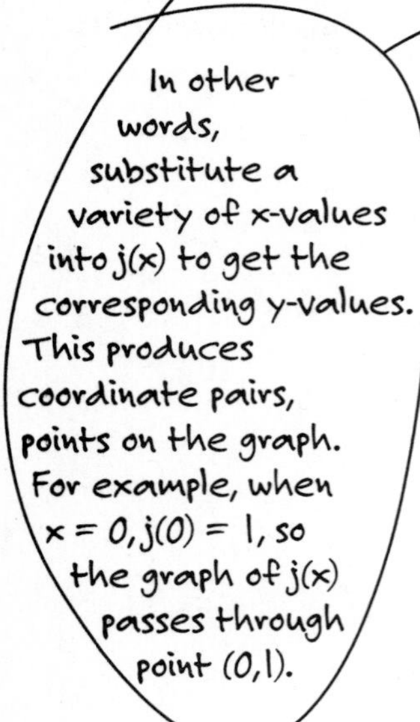

In other words, substitute a variety of x-values into j(x) to get the corresponding y-values. This produces coordinate pairs, points on the graph. For example, when x = 0, j(0) = 1, so the graph of j(x) passes through point (0,1).

Use a graphing calculator (or plot points) to graph $j(x)$. As illustrated below, the graph has a horizontal asymptote at $y = 0$ (the x-axis) and a vertical asymptote at $x = -1$. Vertical asymptotes are lines that a graph approaches, and gets infinitely close to, but does not cross at the ends of the x-axis.

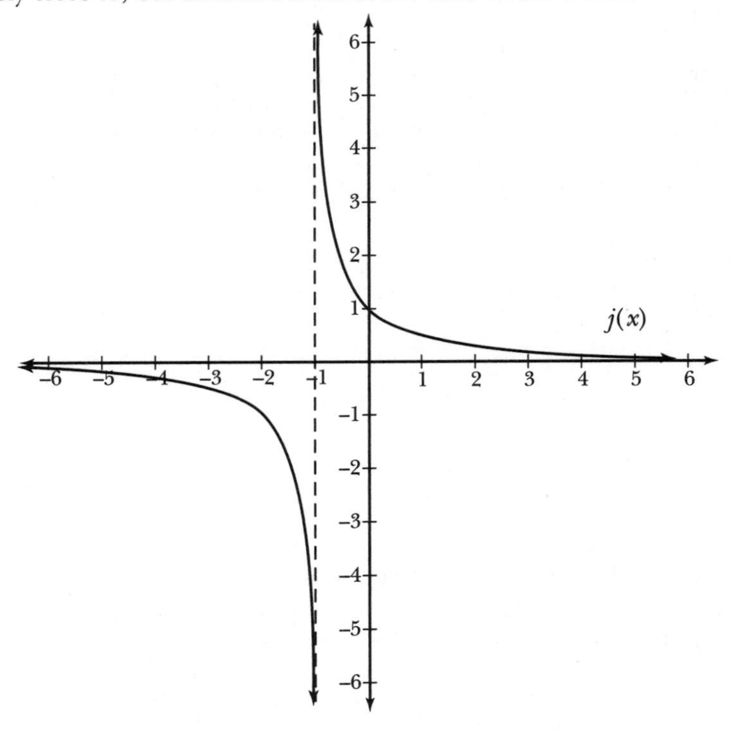

The domain of a function is the set of valid x-values that can be substituted into a function. In other words, they are the x-values that do not cause the function to be undefined. Graphically, you can visualize domain values as vertical lines that intersect the graph. If a vertical line intersects a function's graph, the corresponding x-value is part of that function's domain. If, however, a vertical line does not intersect the graph, then it is not part of the domain.

Only one vertical line does not intersect the graph of $j(x)$, the vertical asymptote at $x = -1$. Therefore, the domain of $j(x)$ is all real numbers except for $x = -1$. Notice that substituting $x = -1$ into the function results in division by zero, which is undefined.

$$j(-1) = \frac{1}{-1+1} = \frac{1}{0}$$

The domain is the set of valid *inputs* for a function. The range, on the other hand, is the set of valid *outputs* for a function, the set of values that result when you substitute members of the domain into a function. You can visualize range values as horizontal lines that intersect the graph. Any horizontal line you draw on the coordinate plane will intersect the graph of $j(x)$, except for the horizontal line $y = 0$. Therefore, the range of $j(x)$ is all real numbers except 0.

6.89 Use the graph of $k(x) = \sqrt{x-4}$ to identify the domain and range of the function.

Use a graphing calculator or plot points to construct the graph of $k(x)$.

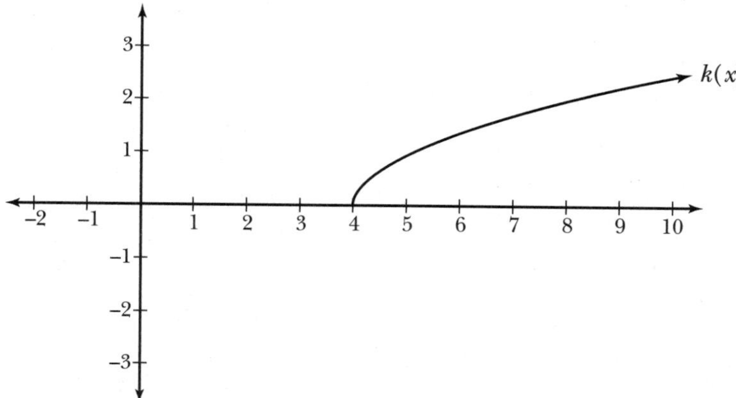

Apply the technique described in Problem 6.88 to identify the domain and range. The vertical line at $x = 4$ and all vertical lines to the right of $x = 4$ will intersect the graph, so the domain of $k(x)$ is all real numbers greater than or equal to 4.

If you plug numbers less than 4 into the function, you get a negative number inside the square root, which is not a real number.

The x-axis (horizontal line $y = 0$) intersects the graph of $k(x)$, as will any horizontal line above the x-axis, so the range of $k(x)$ is all real numbers greater than or equal to 0.

Direct and Inverse Variation

$y = kx$ and $y = k/x$

6.90 What formula defines the relationship between two variables x and y that vary directly?

If x and y vary directly, then one value is always equal to the other if multiplied by a specific constant k called the constant of proportionality. Mathematically, y varies directly (or "varies proportionally") with x when $y = kx$.

6.91 If x varies directly with y and $x = 3$ when $y = 5$, what is the value of y when $x = 18$?

If x and y vary directly, then $y = kx$, as Problem 6.90 explains. Substitute $x = 3$ and $y = 5$ into the equation to calculate k.

$$\begin{aligned} y &= kx \\ 5 &= k \cdot 3 \\ \frac{5}{3} &= k \end{aligned}$$

The constant of proportionality applies to all corresponding values of x and y, so when $x = 18$, k is still equal to 5/3. Substitute these values into the equation $y = kx$, and solve for y.

$$\begin{aligned} y &= kx \\ &= \frac{5}{3}(18) \\ &= \frac{90}{3} \\ &= 30 \end{aligned}$$

The problem states that $x = 3$ when $y = 5$. Because the values vary directly, when x grows 6 times as large (increasing from 3 to 18), y also grows 6 times as large (increasing from 5 to 30).

6.92 If x varies proportionally with y and $x = 4$ when $y = -7$, what is the value of x when $y = 12$?

Use the known pair of values ($x = 4$ when $y = -7$) to calculate the constant of proportionality, k.

$$\begin{aligned} y &= kx \\ -7 &= k \cdot 4 \\ -\frac{7}{4} &= k \end{aligned}$$

Substitute $k = -7/4$ and $y = 12$ into the direct variation equation to calculate the corresponding value of x.

$$y = kx$$
$$12 = -\frac{7}{4}x$$
$$\left(-\frac{4}{7}\right)12 = \left(-\frac{4}{7}\right)\left(-\frac{7}{4}x\right)$$
$$-\frac{48}{7} = x$$

6.93 What formula defines the relationship between two variables x and y that vary indirectly?

As Problem 6.91 explains, if x and y vary directly, then x is a fixed multiple of y—multiplying x by some fixed value k gives you y. If x and y vary indirectly, then the reciprocal of x is a fixed multiple of y. (Note that indirect variation may also be called "inverse variation" or "inverse proportionality.") Mathematically, the product of x and y is a constant value, k.

$$xy = k$$

If you divide both sides by x, you get the formula y = k/x that some textbooks list as the indirect variation formula.

When two values vary *directly* (proportionally), then they trend in the same direction. In other words, if one value increases so does the other; if one value decreases, the other decreases as well. However, when two values vary *indirectly*, they trend in opposite directions—when one value increases the other decreases, and vice versa.

6.94 If x varies indirectly with y and $x = -1$ when $y = 2$, what is the value of y when $x = 6$?

Substitute $x = -1$ and $y = 2$ into the indirect variation equation to calculate k.

$$xy = k$$
$$-1(2) = k$$
$$-2 = k$$

Now that you know the value of k, substitute it and $x = 6$ into the indirect variation equation once more to calculate the corresponding value of y.

$$xy = k$$
$$6y = -2$$
$$y = -\frac{2}{6}$$
$$y = -\frac{1}{3}$$

6.95 If x varies indirectly with y and $x = 10$ when $y = 1$, what is the value of x when $y = \frac{1}{5}$?

Substitute $x = 10$ and $y = 1$ into the indirect variation equation to calculate k.

$$xy = k$$
$$10(1) = k$$
$$10 = k$$

Now calculate the value of x that corresponds with $y = 1/5$.

$$xy = k$$
$$x\left(\frac{1}{5}\right) = 10$$
$$x\left(\frac{1}{5}\right)(5) = 10(5)$$
$$x = 50$$

6.96 In a certain restaurant famous for its hamburgers, research has shown that the number of customers waiting in line is directly proportional to the number of burgers that should be cooking to avoid delays in fulfilling those customers' orders. Specifically, when 3 people are in line, 4 burgers should be cooking. During the lunch rush, when 17 people are waiting in line, how many burgers should be cooking? Round your answer to the nearest whole number.

You could also use the equation $h = kl$, reversing the hamburger and people-in-line variables. You'll get the same final answer.

This is a direct variation word problem. It states that the number of people in line, l, is directly proportional to the number of hamburgers that should be cooking, h. Therefore, $l = kh$, such that k is the constant of proportionality. Calculate k.

$$l = kh$$
$$3 = k(4)$$
$$\frac{3}{4} = k$$

Now find the value of h that corresponds to $l = 17$.

$$l = kh$$
$$17 = \frac{3}{4}h$$
$$\left(\frac{4}{3}\right)17 = \frac{4}{3}\left(\frac{3}{4}h\right)$$
$$\frac{68}{3} = h$$
$$22.\overline{6} = h$$

If 17 people are in line, slightly more than 22 hamburgers are needed, so 23 hamburgers should be cooking.

6.97 Aaron believes that the number of hours he sleeps the night before a test is inversely proportional to the probability that he will fail that test. If he sleeps for 8 hours the night before a test, he is confident that there is only a 10% chance that he will fail. If he is correct, what is the probability that he will *pass* a test if Aaron sleeps for only 2 hours the night before it?

This problem states that s, the number of hours Aaron sleeps before a test, and f, the percentage probability that he will fail the test, are inversely proportional. Therefore, $fs = k$, such that k is the constant of proportionality. Calculate k.

$$\begin{aligned} fs &= k \\ 10(8) &= k \\ 80 &= k \end{aligned}$$

Now calculate the value of f that corresponds to $s = 2$.

$$\begin{aligned} fs &= k \\ f(2) &= 80 \\ f &= \frac{80}{2} \\ f &= 40 \end{aligned}$$

If Aaron sleeps only 2 hours the night before a test, he believes there is a 40 percent chance that he will fail the test. Therefore, there is a $100 - 40 = 60$ percent chance he will pass it.

The question asks for the probability that he will PASS, not the probability that he will FAIL.

Algebraic Word Problems

Word problems that require equations to solve

6.98 A parent preparing b gift bags for a child's birthday party has enough candy to place p pieces in each of the bags. There are c pieces of candy left over once the pieces have been evenly distributed to all of the bags. How many pieces of candy did the parent buy, in terms of b, c, and p?

The parent prepares b gift bags, and each contains p pieces of candy. This requires $b \cdot p$ pieces of candy. If c pieces of candy are not placed into bags, then the parent began with $bp + c$ pieces of candy.

To check your answer, apply the DIY technique. If the parent filled $b = 10$ bags with $p = 4$ pieces of candy and had $p = 2$ pieces left over, the parent purchased a total of $10(4) + 2 = 42$ pieces of candy.

See Problems 2.6–2.10 to review the DIY technique.

6.99 Max uses packages of disposable razors at a constant rate, p packages every w weeks. Each package contains r razors. If Max can use one razor for d days before replacing it, express d in terms of p, r, and w.

Max must be super organized when it comes to shaving if he feels compelled to buy razors at fixed intervals. Going to shave and discovering that he is out of razors must be his worst nightmare. This may speak to deep, dark psychological issues for Max, but lucky for you, none of those come into play in the solution.

To solve problems like this, do not jump right to the answer. Instead, reread the problem and list everything that you know to be true, based on the given information. For example, during each of his regularly spaced trips to the store, Max buys p packages of razors, and each of those packages contains r razors. In other words, each time he shops for razors, Max buys a total of $p \cdot r$ razors. These razors last for w weeks, or $7w$ days.

If $p \cdot r$ razors last $7w$ days, how long does each individual razor last? How many days does Max get out of each razor? Believe it or not, that question is its own answer. You are asking, "How many days per razor?"; in this case, "per" indicates a fraction: days per razor = days ÷ razors, or $(7w) \div (pr)$.

$$d = \frac{7w}{pr}$$

Verify this answer using the DIY method. If Max buys $p = 2$ packages of razors (each containing $r = 7$ razors) every $w = 4$ weeks, then he buys $pr = 14$ razors that last him $7w = 28$ days. Each razor only lasts for 2 days, because $28 \div 14 = 2$.

$$\begin{aligned} d &= \frac{7w}{pr} \\ &= \frac{7(4)}{2(7)} \\ &= \frac{28}{14} \\ &= 2 \end{aligned}$$

Max needs to stop buying such lousy razors.

6.100 A certain brand of canned mixed nuts contains peanuts, walnuts, and cashews. In each can, there are twice as many cashews as walnuts, and the number of peanuts is 25 less than twice the number of cashews. If a can contains 500 nuts, how many peanuts, walnuts, and cashews does it contain?

The problem describes the number of walnuts w and the number of peanuts p in terms of the number of cashews, c. There are half as many walnuts as cashews: $w = c/2$. The number of peanuts is 25 less than twice the number of cashews: $p = 2c - 25$. One can contains 500 nuts.

$$\text{number of cashews} + \text{number of walnuts} + \text{number of peanuts} = 500$$

Replace the phrases in the equation with the correct quantities, in terms of c.

$$c + w + p = 500$$

$$c + \left(\frac{c}{2}\right) + (2c - 25) = 500$$

Multiply all of the terms in the equation by 2 to eliminate the fraction. Then solve for c.

Don't forget to multiply 500 by 2.

$$2(c) + 2\left(\frac{c}{2}\right) + 2(2c - 25) = 2(500)$$

$$2c + c + 4c - 50 = 1{,}000$$

$$7c - 50 = 1{,}000$$

$$7c = 1{,}050$$

$$c = \frac{1{,}050}{7}$$

$$c = 150$$

There are $c = 150$ cashews in every can. Substitute this value into w and p, as defined above, to calculate how many walnuts and peanuts are in every can.

$$w = \frac{c}{2} = \frac{150}{2} = 75 \qquad p = 2c - 25 = 2(150) - 25 = 300 - 25 = 275$$

There are a total of 75 walnuts, 150 cashews, and 275 peanuts in every can of mixed nuts.

6.101 A certain store produces generic brands of medication based on best-selling, name brand medication. The prices for the generic medications are always one dollar less than k times the cost of the name brand equivalent, where k is a fixed real number. Name brand aspirin costs \$9.50 at the store, and the generic equivalent costs \$5.84. If the generic store brand antacid costs \$10.52, how much does the name brand antacid cost?

A generic price g is equal to \$1 less than k times the name brand price n. In other words, $g = kn - 1$. Use the information about aspirin prices ($g = 5.84$ when $n = 9.50$) to calculate the fixed value of k.

$$g = kn - 1$$

$$5.84 = k(9.50) - 1$$

$$5.84 + 1 = 9.5k$$

$$\frac{6.84}{9.5} = k$$

$$0.72 = k$$

Now that you know the value of k, substitute $k = 0.72$ and $g = 10.52$ into the formula $g = kn - 1$ to calculate the corresponding value of n.

$$g = kn - 1$$
$$10.52 = 0.72(n) - 1$$
$$10.52 + 1 = 0.72n$$
$$\frac{11.52}{0.72} = n$$
$$16 = n$$

If the generic brand of antacid costs \$10.52, the equivalent name brand antacid costs \$16.00.

6.102 Jill is 10 years younger than Brian. In 5 years, Jill will be two-thirds of Brian's age. How old will Jill be seven years from now?

Let j represent Jill's current age and let b represent Brian's current age. You know that Jill is 10 years younger than Brian, so $j = b - 10$. In five years, Jill will be $j + 5$ years old and Brian will be $b + 5$ years old. At that time, Jill's age will be two-thirds of Brian's age, so $j + 5 = 2/3\,(b + 5)$. Multiply the terms of that equation by 3 to eliminate the fraction. Then, combine like terms and simplify, as demonstrated below.

$$j + 5 = \frac{2}{3}(b + 5)$$
$$3(j) + 3(5) = 3\left(\frac{2}{3}\right)(b + 5)$$
$$3j + 15 = 2(b + 5)$$
$$3j + 15 = 2b + 10$$
$$3j - 2b = 10 - 15$$
$$3j - 2b = -5$$

You now have two equations that describe the relationship between variables j and b: $j = b - 10$ and $3j - 2b = -5$, a system of equations. Apply the substitution method, as described in Problem 6.59, to solve the system.

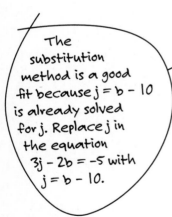

The substitution method is a good fit because j = b - 10 is already solved for j. Replace j in the equation 3j - 2b = -5 with j = b - 10.

$$3j - 2b = -5$$
$$3(b - 10) - 2b = -5$$
$$3b - 30 - 2b = -5$$
$$3b - 2b = -5 + 30$$
$$b = 25$$

Brian is currently 25 years old, so Jill is currently 25 – 10 = 15 years old. The problem asks you to calculate Jill's age seven years from now. In seven years, Jill will be 15 + 7 = 22 years old.

6.103 A machine at a manufacturing plant cuts large, rectangular sheets of aluminum into smaller squares. The length of each uncut, rectangular sheet is four times its width. The machine cuts the entire sheet into a total of 36 squares, each with area 9 square feet, with no aluminum left over. What is the length of the uncut rectangular sheet of aluminum?

The large rectangular sheet is cut into 36 squares, and each of those squares has area 9 square feet. Therefore, the area of the rectangular sheet is $36 \cdot 9 = 324$ square feet. The area of a rectangle is equal to its length l multiplied by its width w, so you know that $lw = 324$. According to the problem, the length of the rectangle is four times its width: $l = 4w$.

You now have two equations that describe the relationship between l and w. Apply the substitution method to solve the system.

$$\begin{aligned} lw &= 324 \\ 4w(w) &= 324 \\ 4w^2 &= 324 \\ w^2 &= \frac{324}{4} \\ w^2 &= 81 \\ w &= \sqrt{81} \\ w &= 9 \end{aligned}$$

The width of the uncut rectangular sheet is $w = 9$ feet, so the length of the sheet is $l = 4(9) = 36$ feet.

6.104 Two photocopying machines (one large and one small) in a certain office produce copies at a constant rate. Working simultaneously, the machines produce 13 copies per minute. Over a 15-minute time span, the large machine produces 45 more copies than the small machine. How many copies can the small machine produce in one hour?

Let s and l represent the number of copies the small and large copy machines, respectively, produce in one minute. According to the problem, $s + l = 13$. Over a 15-minute time span, the small machine produces $15s$ copies; the large machine produces $15l$ copies.

The problem states that the large copier produces 45 more copies than the small copier, so $15l = 15s + 45$. You can simplify this equation by dividing each of its terms by 15.

$$\begin{aligned} \frac{15l}{15} &= \frac{15s}{15} + \frac{45}{15} \\ l &= s + 3 \end{aligned}$$

You now have a system of two linear equations in terms of two unknowns.

$$\begin{cases} s + l = 13 \\ l = s + 3 \end{cases}$$

Because the second equation of the system is solved for l, you can apply the substitution method to solve the system.

$$
\begin{aligned}
s+l&=13\\
s+(s+3)&=13\\
2s+3&=13\\
2s&=13-3\\
2s&=10\\
s&=\frac{10}{2}\\
s&=5
\end{aligned}
$$

If $s = 5$, then $l = s + 3 = 8$. The small machine produces 5 copies per minute and the large machine produces 8. The problem asks you to calculate the number of copies the small machine can produce in one hour. You know that the machine produces $s = 5$ copies per minute, so it can produce $60(5) = 300$ copies in one hour.

Chapter 7

SAT PRACTICE: ALGEBRA AND FUNCTIONS

Test the skills you practiced in Chapter 6

This chapter contains 25 SAT-style questions to help you practice the skills and concepts you explored in Chapter 6. The questions are clustered by difficulty—early questions are easier than later questions. A portion of a student test booklet is reproduced on the following page; use it to record your answers. Full answer explanations are located at the end of the chapter.

Try not to peek at the answers until you have worked through all 25 questions. If you want to practice SAT pacing, try to finish all of the questions in no more than 30-35 minutes.

Practice Questions: Algebra and Functions

7.1 Ⓐ Ⓑ Ⓒ Ⓓ Ⓔ
7.2 Ⓐ Ⓑ Ⓒ Ⓓ Ⓔ
7.3 Ⓐ Ⓑ Ⓒ Ⓓ Ⓔ
7.4 Ⓐ Ⓑ Ⓒ Ⓓ Ⓔ
7.5 Ⓐ Ⓑ Ⓒ Ⓓ Ⓔ
7.6 Ⓐ Ⓑ Ⓒ Ⓓ Ⓔ
7.7 Ⓐ Ⓑ Ⓒ Ⓓ Ⓔ
7.8 Ⓐ Ⓑ Ⓒ Ⓓ Ⓔ
7.9 Ⓐ Ⓑ Ⓒ Ⓓ Ⓔ
7.10 Ⓐ Ⓑ Ⓒ Ⓓ Ⓔ
7.11 Ⓐ Ⓑ Ⓒ Ⓓ Ⓔ
7.12 Ⓐ Ⓑ Ⓒ Ⓓ Ⓔ
7.13 Ⓐ Ⓑ Ⓒ Ⓓ Ⓔ
7.14 Ⓐ Ⓑ Ⓒ Ⓓ Ⓔ
7.15 Ⓐ Ⓑ Ⓒ Ⓓ Ⓔ

7.16

7.17

7.18

7.19

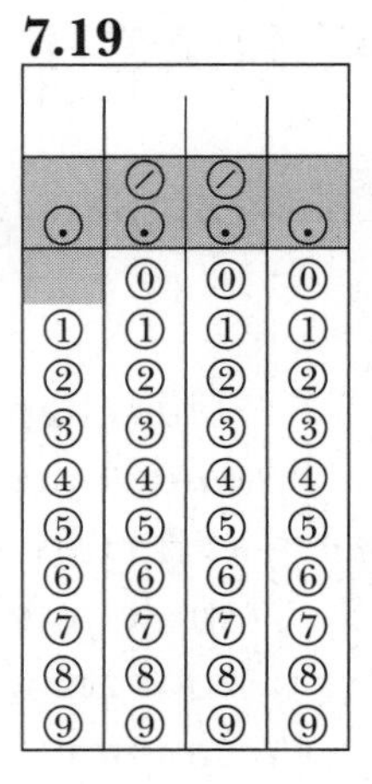

7.20

7.21

7.22

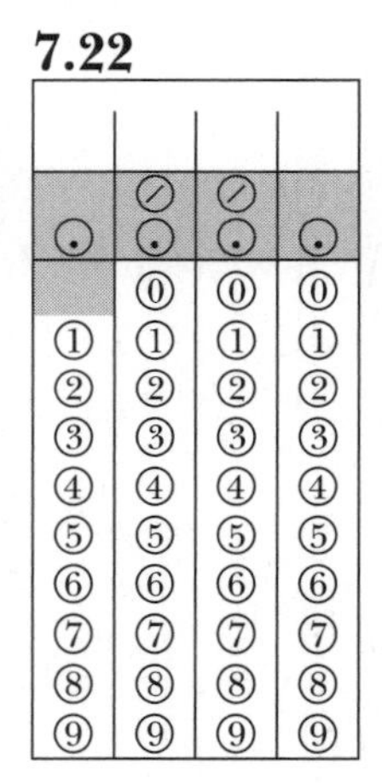

7.23

7.24

7.25

7.1 If two less than one-third of a number is equal to five, what is three more than that number?

(A) 7
(B) 10
(C) 13
(D) 21
(E) 24

7.2 If $x < 0$ and $|x+8| = 11$, what is the value of x?

(A) 19
(B) 3
(C) –3
(D) –16
(E) –19

7.3 Frank, Greg, and Hannah ate dinner together at a certain restaurant, and each was billed separately for his or her meal. (Tips were not included in the price of each meal.) Frank's meal cost twice as much as Greg's meal, and Hannah's meal cost $5 more than Greg's meal. If Hannah's meal cost $13.50, and each person tipped 20 percent of the cost of his or her meal, what was the total tip for all three people?

(A) $4.80
(B) $6.30
(C) $7.80
(D) $8.50
(E) $9.00

7.4 If $\sqrt{5y-4} = 9$, what is the value of y?

(A) $\frac{7}{5}$

(B) $\frac{13}{5}$

(C) 4

(D) $\frac{36}{5}$

(E) 17

$$f(x) = x^2 \quad g(x) = x^3 \quad h(x) = \sqrt{x^8}$$
$$j(x) = x^4 + x^3$$

7.5 Given the functions $f(x)$, $g(x)$, $h(x)$, and $j(x)$ defined above, which of the following lists the values of $f(-2)$, $g(-2)$, $h(-2)$, and $j(-2)$ in order, from least to greatest?

(A) $f(-2), g(-2), h(-2), j(-2)$
(B) $f(-2), g(-2), j(-2), h(-2)$
(C) $f(-2), h(-2), g(-2), j(-2)$
(D) $g(-2), f(-2), h(-2), j(-2)$
(E) $g(-2), f(-2), j(-2), h(-2)$

7.6 If x, y, and z are positive integers and $6 \cdot \frac{x^2 z^4}{y} = 3x^3 y^2 z$, what is x in terms of y and z?

(A) $3y^3 z^3$

(B) $\frac{2z^3}{y^3}$

(C) $\frac{18z^5}{y^3}$

(D) $\frac{yz^2}{2\sqrt{yz}}$

(E) $\sqrt[5]{\frac{y^2}{2z^4}}$

x	-2	0	1	4
y	-6	0	0	-12

7.7 The table above lists corresponding values of x and y for which of the following equations?

(A) $y = 3x$
(B) $y = 5x + 4$
(C) $y = x - x^2$
(D) $y = x^2 - 10$
(E) $y = x^3 - x$

Note: Questions 7.8 and 7.9 refer to the graph of $g(t)$ below, which consists of two curved sections and two linear sections.

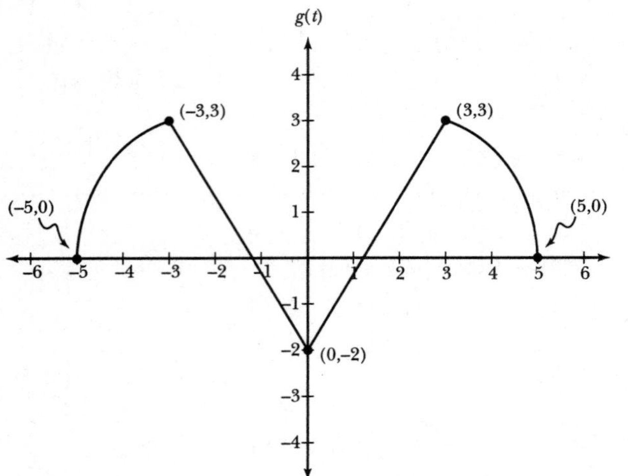

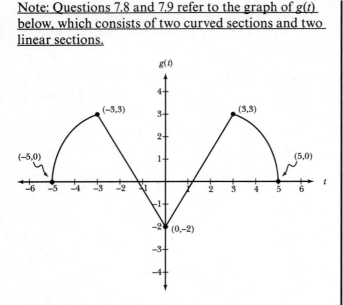

7.8 Based on the graph of $g(t)$, the function ___________.

(A) is even
(B) is odd
(C) is one-to-one
(D) is monotonic
(E) has an inverse

7.9 If a and b are integers such that $g(b) = 3$ and $g(0) = a$, what is the greatest value of $g(a + b)$?

(A) -5
(B) -2
(C) $-\frac{1}{3}$
(D) 0
(E) 3

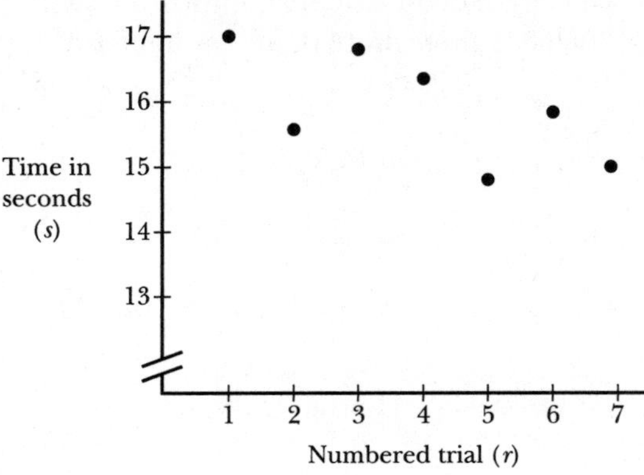

7.10 Pam is training for the 100-meter dash at an upcoming track meet. Today, she ran seven trial races and recorded her times on the graph above. Which of the following equations best expresses the relationship between her times s (in seconds) and the numbered trials r?

(A) $s(r) = 0.\bar{3}r$
(B) $s(r) = -3r$
(C) $s(r) = -0.\bar{3}r + 17.3$
(D) $s(r) = 0.\bar{3}r - 17.3$
(E) $s(r) = 3r + 17.3$

7.11 If a, b, and c are consecutive integers such that $a < b < c$, which of the following is equal to b^2?

(A) $a^2 - 2a + 1$
(B) $c^2 + 2c + 1$
(C) $ab + 2b + a + 3$
(D) $ac - a + c - 1$
(E) $bc - 2b - c + 1$

7.12 If p varies inversely with q and $q = 6$ when $p = 4$, what is the value of p when $q = 9$?

(A) $\frac{2}{27}$
(B) $\frac{1}{9}$
(C) $\frac{1}{6}$
(D) $\frac{8}{3}$
(E) 6

7.13 If $x \nabla y = |x - y|$, which of the following statements must be true?

I. $x \nabla y = y \nabla x$
II. $x \nabla (y \nabla z) = (x \nabla y) \nabla z$
III. $ax \nabla ay = a|x - y|$, such that a is a real number

(A) I only
(B) I and II only
(C) I and III only
(D) I, II, and III
(E) None of the statements are true

7.14 If $j - 4k + 3 = 6j + k + 2$, then what is the value of $j + k$?

(A) 1
(B) $\frac{1}{5}$
(C) -1
(D) $-\frac{3}{2}$
(E) -5

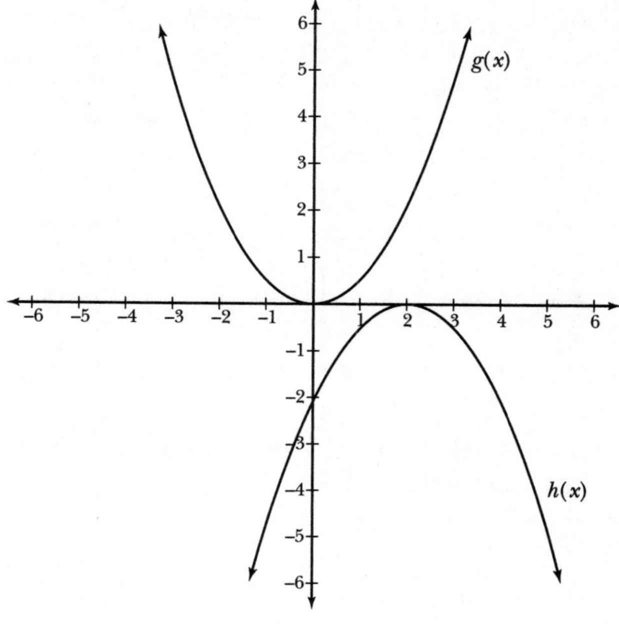

7.15 Given the graphs of $g(x)$ and $h(x)$ above, which of the following statements best describes the relationship between the functions?

(A) $h(x) = -g(x) + 2$
(B) $h(x) = -g(x) - 2$
(C) $h(x) = -g(x - 2)$
(D) $h(x) = -g(x + 2)$
(E) $h(x) = 2 - g(x + 2)$

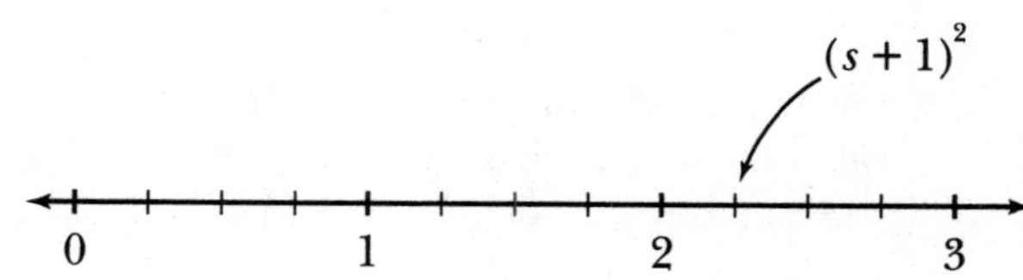

7.16 The number line above is divided into segments of equal length. What is one possible value of s?

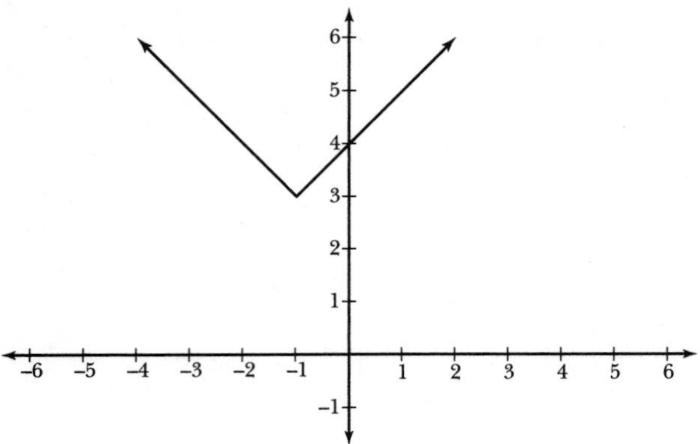

7.17 Given the graph of $f(x)=|x+a|+b$ above, what is the value of $b^3 - a^2$?

$$\begin{cases} 6y+1=x+3 \\ 4y=\frac{1}{2}z \\ yz=2x \end{cases}$$

7.18 Given the system of equations above, what is the least value of $x_1 \cdot y_1 \cdot z_1$, where (x_1, y_1, z_1) is a solution to the system of equations?

x	-2	-1	0	1	2
$g(x)$	-6	-3	-1	2	5

7.19 Given the table above, which lists values of a function $g(x)$ for five different values of x, what is the value of $g^{-1}(g^{-1}(-3))+g(g(1))$?

$$x^2+x-12>0 \qquad |3x-4|<11$$

7.20 Identify the integer value of x that satisfies both of the above inequalities.

7.21 If one perfect square less a second perfect square is equal to 12 and the sum of their square roots is 9, what is the absolute value of the difference of their square roots?

7.22 Let the function p be defined by $p(x) = x^2 - 3x$. If $p(1 + y) = p(y - 1)$, what is the value of y?

7.23 If $2x + 3y$ is equal to 80 percent of $8x$, what is the value of $\frac{5y}{8x}$?

7.24 A particle moves left and right along a horizontal wire such that its position at time t seconds is $p(t) = t^2 - at + b$ millimeters left or right of the center of the wire. (If $p(t) < 0$, the particle is $p(t)$ millimeters left of the center, and if $p(t) > 0$, the particle is $p(t)$ mm right of the center.) At $t = 1$ second, the particle is 1 millimeter left of the center, and at time $t = 10$ seconds, the particle is 3.5 centimeters right of the center. At time $t = 7$ seconds, how far (in millimeters) is the particle from the center of the wire?

$$\blacklozenge a \blacklozenge = \frac{\text{remainder of } a \div 3}{\text{remainder of } a \div 4}$$

7.25 Let the function $\blacklozenge a \blacklozenge$ be defined for positive integers greater than 4 according to the rule stated above. What is the least value for which $\blacklozenge a \blacklozenge = 1$?

Solutions

7.1 **E.** Translate "two less than one-third of a number is equal to five" into an equation in terms of x, and then solve the equation for x.

$$\frac{1}{3}x-2=5$$
$$\frac{1}{3}x=7$$
$$3\left(\frac{1}{3}x\right)=3(7)$$
$$x=21$$

The number is 21, so three more than the number is 21 + 3 = 24.

7.2 **E.** Apply the plug and chug method to answer this question. Note that answer choices A and B are not valid, because x must be less than 0, as specified by the question. The correct answer is choice E, $x = -19$, as demonstrated below.

$$|x+8|=11$$
$$|-19+8|=11$$
$$|-11|=11$$
$$11=11 \textit{ True}$$

7.3 **C.** Hannah's meal cost \$5 more than Greg's meal. The question states that Hannah's meal cost \$13.50, so Greg's meal cost \$13.50 – \$5 = \$8.50. Frank's meal was twice as expensive as Greg's meal, so Frank's meal cost 2(\$8.50) = \$17. Calculate the total cost of all three meals.

$$\$8.50 + \$13.50 + \$17 = \$39$$

Each person tipped 20 percent of the cost of his or her meal, so the total tip is 20 percent of the total bill. Multiply \$39 by 0.2 (20 percent expressed in decimal form) to calculate the total tip.

$$0.2\,(\$39) = \$7.80$$

7.4 **E.** Because three of the answer choices are fractions, it is easier to solve the equation than to apply the plug and chug method. Square both sides of the equation to eliminate the square root, and then solve for y.

$$\sqrt{5y-4}=9$$
$$\left(\sqrt{5y-4}\right)^2=9^2$$
$$5y-4=81$$
$$5y=81+4$$
$$5y=85$$
$$y=\frac{85}{5}$$
$$y=17$$

7.5 **E.** Evaluate each of the functions for $x = -2$.

$f(x)=x^2$ $f(-2)=(-2)^2$ $f(-2)=4$	$g(x)=x^3$ $g(-2)=(-2)^3$ $g(-2)=-8$
$h(x)=\sqrt{x^8}$ $h(-2)=\sqrt{(-2)^8}$ $h(-2)=\sqrt{256}$ $h(-2)=16$	$j(x)=x^4+x^3$ $j(-2)=(-2)^4+(-2)^3$ $j(-2)=16+(-8)$ $j(-2)=8$

The function values, from least to greatest, are $g(-2) = -8$, $f(-2) = 4$, $j(-2) = 8$, and $h(-2) = 16$.

7.6 **B.** To answer this question, you need to solve the equation for x. Multiply both sides of the equation by $1/x^2$ to consolidate the x-terms left of the equal sign.

$$6\cdot\frac{x^2z^4}{y}\left(\frac{1}{x^2}\right)=\left(3x^3y^2z\right)\left(\frac{1}{x^2}\right)$$

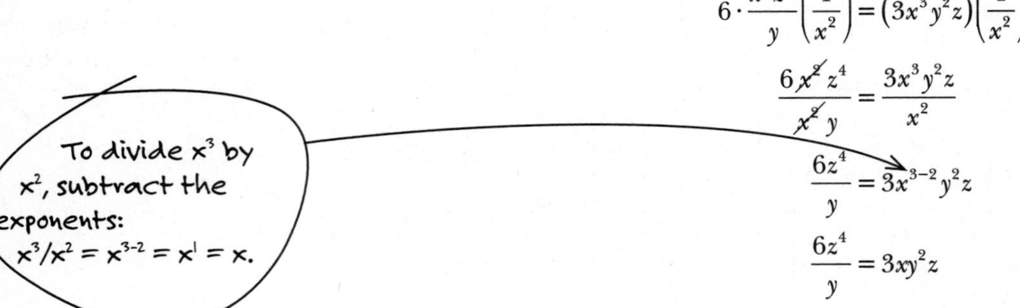

To divide x^3 by x^2, subtract the exponents:
$x^3/x^2 = x^{3-2} = x^1 = x$.

$$\frac{6\cancel{x^2}z^4}{\cancel{x^2}y}=\frac{3x^3y^2z}{x^2}$$

$$\frac{6z^4}{y}=3x^{3-2}y^2z$$

$$\frac{6z^4}{y}=3xy^2z$$

Now multiply both sides of the equation by $1/3$, $1/y^2$, and $1/z$ to isolate x right of the equal sign.

$$\frac{6z^4}{y}\left(\frac{1}{3}\cdot\frac{1}{y^2}\cdot\frac{1}{z}\right)=\left(3xy^2z\right)\left(\frac{1}{3}\cdot\frac{1}{y^2}\cdot\frac{1}{z}\right)$$

$$\left(\frac{6}{3}\right)\left(\frac{1}{y\cdot y^2}\right)\left(\frac{z^4}{z}\right)=\left(\frac{3}{3}\right)x\left(\frac{y^2}{y^2}\right)\left(\frac{z}{z}\right)$$

$$\frac{2z^3}{y^3}=x$$

7.7 **C.** If you substitute each value of x into the correct equation among the answer choices, the result will be the corresponding value of y. Beware! Do not select your answer too soon using the plug and chug method. All five answer choices return a y-value of -6 when you substitute $x = -2$ into the equations. In fact, three of the answer choices also return the correct value $y = 0$ when $x = 0$. Only one of the equations ($y = x - x^2$) returns the correct y-values for all four given values of x, as demonstrated below.

$x=-2$	$x=0$	$x=1$	$x=4$
$y=-2-(-2)^2$ $=-2-4$ $=-6$	$y=0-0^2$ $=0$	$y=1-1^2$ $=1-1$ $=0$	$y=4-4^2$ $=4-16$ $=-12$

7.8 **A.** The graph of $g(t)$ is symmetric about the y-axis, which means that $g(-t) = g(t)$ and the function is even. In case you are curious about the other answer choices, functions are odd if they are origin-symmetric, which $g(t)$ is not. Functions are one-to-one if they pass the horizontal line test, which means that a horizontal line intersects the graph exactly once. Horizontal lines between $t = -2$ and $t = 3$ intersect the graph at least twice, so the graph fails the horizontal line test.

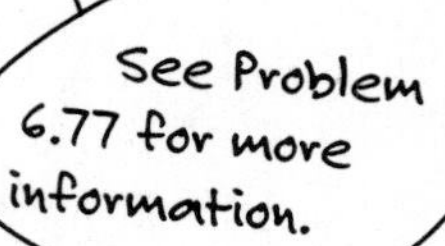

A function is monotonic if it only travels in one direction, either increasing or decreasing over its entire domain. This function increases, decreases, increases, and then finally decreases, so it is not monotonic. Finally, $g(t)$ does not have an inverse because functions with an inverse are both one-to-one and monotonic. As explained above, $g(t)$ is neither.

In fact, monotonic functions are one-to-one functions and vice versa. Choices C, D, and E are all related, and none of them describe g(t).

7.9 **D.** Remember, this problem refers to the graph of $g(t)$ above Problem 7.8. If $g(b) = 3$, then b is a location along the horizontal axis at which the function reaches a height of 3. There are two such locations along the horizontal axis: $b = -3$ and $b = 3$. If $g(0) = a$, then a is the height of the function at the position $t = 0$ along the horizontal axis. When $t = 0$, the function $g(t)$ has height $g(0) = -2$. Therefore, $a = -2$.

There are two possible values of $a + b$, one value when $b = -3$ and one value when $b = 3$.

$$a+b=-2+(-3) \qquad a+b=-2+(3)$$
$$=-5 \qquad\qquad\qquad =1$$

At $t = -5$, the height of the function is 0; therefore, $g(-5) = 0$. At $t = 1$, the function is below the x-axis, so $g(1)$ is negative, which means $g(1) < g(-5)$. The greatest value of $g(a + b)$ is $g(5) = 0$.

7.10 **C.** The line passing near (1, 17) and (7, 15), illustrated below, represents a fairly good line of best fit for the data.

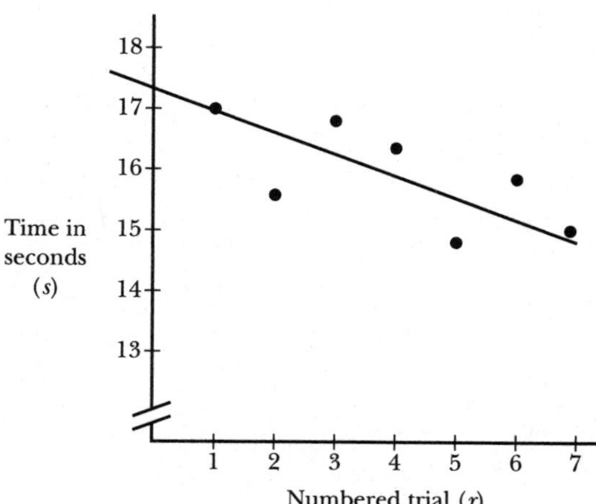

The equation representing the relationship between r and s should be similar to the equation of this line. Notice that this line slopes downward, so the slope of the linear equation must be negative. In other words, the coefficient of r must be negative. This is only true in two of the answer choices: B and C.

The y-intercept of the line is between 17 and 18. Notice that answer choice C has a y-intercept of 17.3, but choice B has no constant, which means its y-intercept is 0. Because $s(r) = -0.\bar{3}r + 17.3$ has a negative slope and a y-intercept between 17 and 18, it most closely represents the relationship between r and s.

7.11 **D.** Apply the DIY method to identify the correct answer. If $a = 1$, $b = 2$, and $c = 3$, only choice D is equal to b^2.

$$\begin{aligned} b^2 &= ac - a + c - 1 \\ 2^2 &= 1(3) - 1 + 3 - 1 \\ 4 &= 3 - 1 + 3 - 1 \\ 4 &= 4 \quad \textbf{\textit{True}} \end{aligned}$$

Why is this the correct answer? If a, b, and c are consecutive integers, then b is 1 more than a and 1 less than c. In other words, $b = a + 1$ and $b = c - 1$. Use these expressions to calculate b^2, as demonstrated below.

$$\begin{aligned} b^2 &= b \cdot b \\ &= (a+1)(c-1) \\ &= a(c) + a(-1) + 1(c) + 1(-1) \\ &= ac - a + c - 1 \end{aligned}$$

7.12 **D.** According to Problem 6.93, if p and q vary inversely, then the following equation is true, such that k is the constant of proportionality.

$$\begin{aligned} pq &= k \\ 4(6) &= k \\ 24 &= k \end{aligned}$$

Now calculate the value of p that corresponds to $q = 9$.

$$\begin{aligned} pq &= k \\ p(9) &= 24 \\ p &= \frac{24}{9} \\ p &= \frac{8}{3} \end{aligned}$$

7.13 **A.** This question defines the operation $x \nabla y = |x - y|$. Consider each of the statements independently. Statement I is true because $x - y$ and $y - x$ are opposites, so their values are the same if you take the absolute values: $|x - y| = |y - x|$. For example, if you apply the DIY technique, substituting $x = 2$ and $y = -3$ into the statement, the equation is true.

$$\begin{aligned} 2 \nabla -3 &= -3 \nabla 2 \\ |2 - (-3)| &= |-3 - 2| \\ |2 + 3| &= |-5| \\ 5 &= 5 \end{aligned}$$

Statement II is not true. In the following counterexample, the DIY technique is applied using $x = 1$, $y = 2$, and $z = -3$.

$$\begin{aligned} x\nabla(y\nabla z) &\neq (x\nabla y)\nabla z \\ 1\nabla(2\nabla 3) &\neq (1\nabla 2)\nabla 3 \\ 1\nabla|2-3| &\neq |1-2|\nabla 3 \\ 1\nabla|-1| &\neq |-1|\nabla 3 \\ 1\nabla 1 &\neq 1\nabla 3 \\ |1-1| &\neq |1-3| \\ 0 &\neq 2 \end{aligned}$$

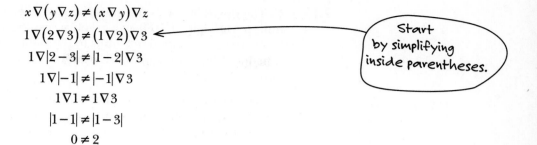

Start by simplifying inside parentheses.

Statement III is also false. In the counterexample below, $x = 1$, $y = 2$, and $a = -5$.

$$\begin{aligned} ax\nabla ay &\neq a\left|x\nabla y\right| \\ -5(1)\nabla -5(2) &\neq -5|1-2| \\ -5\nabla -10 &\neq -5|-1| \\ |-5-(-10)| &\neq -5(1) \\ |-5+10| &\neq -5 \\ |5| &\neq -5 \\ 5 &\neq -5 \end{aligned}$$

Therefore, only statement I is true.

7.14 **B.** Isolate the variable expressions containing j and k on the left side of the equation and move the constants to the right side of the equation.

$$\begin{aligned} j-4k+3 &= 6j+k+2 \\ (j-6j)+(-4k-k) &= 2-3 \\ -5j-5k &= -1 \end{aligned}$$

Factor –5 out of the expression left of the equal sign and solve for the quantity $j + k$.

$$\begin{aligned} -5(j+k) &= -1 \\ j+k &= \frac{-1}{-5} \\ j+k &= \frac{1}{5} \end{aligned}$$

7.15 **C.** The graph of $h(x)$ is the graph of $g(x)$ reflected across the x-axis and moved 2 units to the right. To reflect a graph across the x-axis, you multiply the function by -1. To move a function 2 units to the right, you subtract 2 from x within the function. When you apply both transformations, the result is $h(x) = -g(x-2)$, answer choice C.

To review graphical transformations during the SAT, you can use your calculator to graph simple transformations of basic graphs. For example, the graph of $y = x^2$ is similar to the graph of $g(x)$. The graph of $y = -(x-2)^2$ is similar to the graph of $h(x)$.

7.16 **.5 or 1/2.** There are four equal segments per whole unit, so each of the units represents a length of 0.25, or 1/4. The marked segment represents the value 2.25, or 9/4. According to the diagram, it is also equal to $(s+1)^2$. Set these values equal and solve for s.

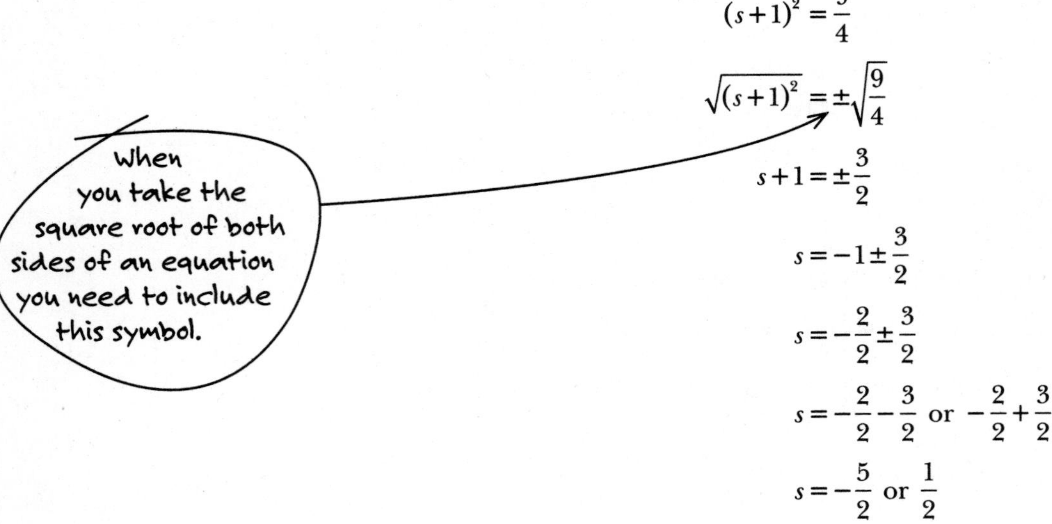

When you take the square root of both sides of an equation you need to include this symbol.

$$(s+1)^2 = \frac{9}{4}$$

$$\sqrt{(s+1)^2} = \pm\sqrt{\frac{9}{4}}$$

$$s+1 = \pm\frac{3}{2}$$

$$s = -1 \pm \frac{3}{2}$$

$$s = -\frac{2}{2} \pm \frac{3}{2}$$

$$s = -\frac{2}{2} - \frac{3}{2} \text{ or } -\frac{2}{2} + \frac{3}{2}$$

$$s = -\frac{5}{2} \text{ or } \frac{1}{2}$$

There are two possible values of s, but you cannot record a negative grid-in response. Therefore, the only appropriate responses are $s = 1/2$ or $s = 0.5$.

7.17 **26.** The graph of $f(x)$ is a transformation of the function $y = |x|$, if it is moved one unit to the left and three units up. Therefore, $f(x) = |x+1| + 3$, which means $a = 1$ and $b = 3$. Substitute these values into the expression you are asked to evaluate.

$$\begin{aligned} b^3 - a^2 &= 3^3 - 1^2 \\ &= 27 - 1 \\ &= 26 \end{aligned}$$

Hard problems
start here.

7.18 **2.** All three of the equations contain y. If you solve the first equation for x and the second equation for z, you can express all three variables in terms of y.

$$\begin{aligned} 6y+1&=x+3 \\ 6y+1-3&=x \\ 6y-2&=x \end{aligned} \qquad \begin{aligned} 4y&=\frac{1}{2}z \\ 2(4y)&=2\left(\frac{1}{2}z\right) \\ 8y&=z \end{aligned}$$

Substitute $x = 6y - 2$ and $z = 8y$ into the last equation of the system and solve for y.

$$\begin{aligned} yz&=2x \\ y(8y)&=2(6y-2) \\ 8y^2&=12y-4 \\ 8y^2-12y+4&=0 \end{aligned}$$

Divide the terms of the equation by 4 and factor the quadratic expression.

$$\begin{aligned} \frac{8y^2}{4}-\frac{12y}{4}+\frac{4}{4}&=\frac{0}{4} \\ 2y^2-3y+1&=0 \\ (2y-1)(y-1)&=0 \end{aligned}$$

Set each factor equal to 0 and solve.

$$\begin{aligned} 2y-1&=0 \\ 2y&=1 \\ y&=\frac{1}{2} \end{aligned} \qquad \begin{aligned} y-1&=0 \\ y&=1 \end{aligned}$$

There are two possible values for y, so there are two solutions to the system of equations. Calculate the corresponding values of x and z for each by substituting $y = 1/2$ and $y = 1$ into the equations you previously solved for x and z.

Solution containing $y=\frac{1}{2}$		Solution containing $y=1$	
$x=6y-2$	$z=8y$	$x=6y-2$	$z=8y$
$=6\left(\frac{1}{2}\right)-2$	$=8\left(\frac{1}{2}\right)$	$=6(1)-2$	$=8(1)$
$=3-2$	$=4$	$=6-2$	$=8$
$=1$		$=4$	

There are two solutions to the system of equations: $(x, y, z) = (1, 1/2, 4)$ and $(4, 1, 8)$. Hence, there are also two values of $x_1 \cdot y_1 \cdot z_1$: $1(1/2)(4) = 2$ and $4(1)(8) = 32$. The least of those two products is 2.

$g^{-1}(3)$ asks, "What value of x has a function value of –3?" On the other hand, g(1) asks, "What is the value of g(x) when x = 1?"

7.19 **5.** Use the table to evaluate $g^{-1}(3)$ and $g(1)$. Because $g(-1) = -3$, you know that $g^{-1}(-3) = -1$.

$$g^{-1}\left(g^{-1}(-3)\right)+g\left(g(1)\right)=g^{-1}(-1)+g(2)$$

Now evaluate $g^{-1}(-1)$ and $g(2)$. Note that $g(0) = -1$, so $g^{-1}(-1) = 0$.

$$g^{-1}(-1) + g(2) = 0 + 5$$

Therefore, $g^{-1}\left(g^{-1}(-3)\right)+g\left(g(1)\right)=5$.

7.20 **4.** Apply the technique demonstrated in Problems 6.51–6.52 to solve the quadratic inequality. Note that you can factor the quadratic: $(x+4)(x-3) > 0$. Therefore, the critical numbers are $x = -4$ and $x = 3$. These critical numbers divide the number line into three segments: $x < -4$, $-4 < x < 3$, and $x > 3$. Only the intervals $x < -4$ and $x > 3$ satisfy the inequality.

Now solve the absolute value inequality, applying the technique demonstrated in Problem 6.54.

$$\begin{aligned}|3x-4| &< 11\\ -11 < 3x-4 &< 11\\ -11+4 < 3x &< 11+4\\ -7 < 3x &< 15\\ -\frac{7}{3} < \frac{3x}{3} &< \frac{15}{3}\\ -\frac{7}{3} < x &< 5\end{aligned}$$

x = 4 is greater than 3, which satisfies the quadratic; x = 4 is also between –7/3 and 5.

To satisfy this compound inequality, x has to be greater than –7/3 and less than 5. To satisfy the previous inequality, x had to be less than –4 or greater than 3. The only integer that satisfies both is $x = 4$.

7.21 **4/3 or 1.33.** A perfect square is a product of one integer multiplied times itself. For example, a^2 is a perfect square because $a \cdot a = a^2$. Similarly, 9 is a perfect square because it is the product of 3 multiplied times itself: $3 \cdot 3 = 9$. Let x^2 represent one of the unknown perfect squares and let y^2 represent the other. Note that the square root of x^2 is x and the square root of y^2 is y.

Therefore you know that $x^2 - y^2 = 12$. A difference of perfect squares can be factored, as explained in Problem 6.18: $(x+y)(x-y) = 12$. You are told that the sum of the square roots is 9, so $x + y = 9$. Substitute this into the factored version of the difference of perfect squares, and solve for $x - y$, the difference of the square roots.

$$\begin{aligned}(x+y)(x-y)&=12\\ 9(x-y)&=12\\ x-y&=\frac{12}{9}\\ x-y&=\frac{4}{3}\end{aligned}$$

This solution assumes that x^2 is the larger of the two perfect squares. The problem asks for the absolute value of the difference of the square roots, because it does not indicate the order in which you should subtract. Note that $|x-y| = |y-x| = 4/3$.

7.22 **3/2 or 1.5.** Evaluate the function $p(x) = x^2 - 3x$ for $x = 1 + y$ and $x = y - 1$ and set the results equal.

$$\begin{aligned} p(1+y) &= p(y-1) \\ (1+y)^2 - 3(1+y) &= (y-1)^2 - 3(y-1) \\ 1+2y+y^2-3-3y &= y^2-2y+1-3y+3 \\ y^2+(2y-3y)+(1-3) &= y^2+(-2y-3y)+(1+3) \\ y^2-y-2 &= y^2-5y+4 \end{aligned}$$

Group like terms and solve for y.

$$\begin{aligned} (y^2-y^2)+(-y+5y) &= 4+2 \\ 4y &= 6 \\ y &= \frac{6}{4} \\ y &= \frac{3}{2} \end{aligned}$$

7.23 **.916 or .917.** Translate the given information into an equation, writing 80 percent as $0.80 = 80/100 = 4/5$.

$$\begin{aligned} 2x+3y &= \frac{4}{5}(8x) \\ 2x+3y &= \frac{32}{5}x \end{aligned}$$

This solution uses fractions, but you could use decimals if you prefer.

Solve the equation for one of the variables. For example, in the solution below, you express y in terms of x.

$$\begin{aligned} 3y &= \frac{32}{5}x - 2x \\ 3y &= \frac{32}{5}x - \frac{10}{5}x \\ 3y &= \frac{22}{5}x \\ y &= \left(\frac{1}{3}\right)\left(\frac{22}{5}x\right) \\ y &= \frac{22x}{15} \end{aligned}$$

The question asks you to evaluate $(5y)/(8x)$. Substitute $y = (22x)/15$ into that fraction.

$$\frac{5y}{8x}=\frac{5\left(\frac{22x}{15}\right)}{8x}$$
$$=\frac{\frac{22x}{3}}{8x}$$
$$=\frac{22x}{3}\div 8x$$
$$=\frac{22x}{3}\cdot\frac{1}{8x}$$
$$=\frac{22\cancel{x}}{24\cancel{x}}$$
$$=\frac{11}{12}$$
$$\approx 0.91\overline{6}$$

Note that you cannot enter the fraction 11/12 into the answer grid, because it would require 5 columns instead of 4. You must grid your answer as a rounded or truncated decimal.

Be careful with the units! The particle is 1 MILLIMETER left of center at $t = 1$ but is 3.5 CENTIMETERS right of center at $t = 10$, so $p(10) = 35$, not 3.5. Remember the function outputs distances in millimeters, and there are 10 millimeters in a centimeter.

7.24 **5.** In order to solve this problem, you have to determine the values of a and b in the position function $p(t)$. You know that $p(1) = -1$ and $p(10) = 35$. Substitute $t = 1$ into the function and set the expression equal to -1. Similarly, substitute $t = 10$ into the function and set the expression equal to 35. The result is two equations with two variables, a and b—a system of equations.

$$\begin{aligned} p(1)&=1^2-a(1)+b \\ -1&=1-a+b \\ -2&=-a+b \\ a-2&=b \end{aligned} \qquad \begin{aligned} p(10)&=10^2-a(10)+b \\ 35&=100-10a+b \\ 10a-b&=100-35 \\ 10a-b&=65 \end{aligned}$$

You can solve the system using the substitution method, substituting $b = a - 2$ into the equation $10a - b = 65$.

$$\begin{aligned} 10a-(a-2)&=65 \\ 10a-a+2&=65 \\ 9a&=65-2 \\ 9a&=63 \\ a&=\frac{63}{9} \\ a&=7 \end{aligned}$$

If $a = 7$, then $b = a - 2 = 5$. You now know the values of the constants in the position equation: $p(t) = t^2 - 7t + 5$. Evaluate $p(7)$ to identify the position of the particle, with respect to the center of the wire, at time $t = 7$ seconds.

$$\begin{aligned} p(7)&=7^2-7(7)+5 \\ &=49-49+5 \\ &=5 \end{aligned}$$

The particle is 5 millimeters right of the center when $t = 7$.

7.25 **13.** The function ♦a♦ is only defined for integer values of a that are greater than 4. To calculate ♦a♦, you divide a by 3 and by 4, creating a fraction whose numerator and denominator are the whole number remainders, respectively. For example, to calculate ♦5♦, you note that $5 \div 3 = 1$ remainder 2 and $5 \div 4 = 1$ remainder 1. Divide the first remainder by the second: ♦5♦ $= 2/1 = 2$.

The function value will equal 1 when dividing an integer by 3 and dividing it by 4 produce the same remainder. You can identify this integer using trial and error, calculating consecutive values of the function beginning with $a = 5$.

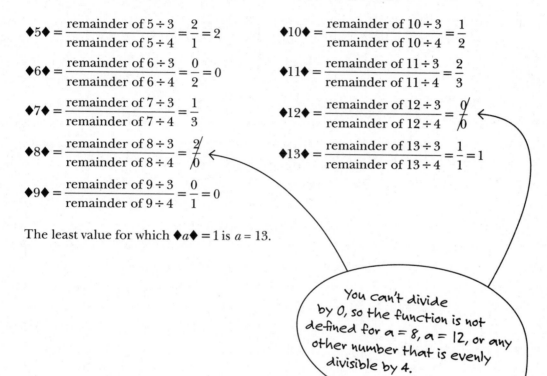

$$\blacklozenge 5 \blacklozenge = \frac{\text{remainder of } 5 \div 3}{\text{remainder of } 5 \div 4} = \frac{2}{1} = 2$$

$$\blacklozenge 6 \blacklozenge = \frac{\text{remainder of } 6 \div 3}{\text{remainder of } 6 \div 4} = \frac{0}{2} = 0$$

$$\blacklozenge 7 \blacklozenge = \frac{\text{remainder of } 7 \div 3}{\text{remainder of } 7 \div 4} = \frac{1}{3}$$

$$\blacklozenge 8 \blacklozenge = \frac{\text{remainder of } 8 \div 3}{\text{remainder of } 8 \div 4} = \frac{2}{0}$$

$$\blacklozenge 9 \blacklozenge = \frac{\text{remainder of } 9 \div 3}{\text{remainder of } 9 \div 4} = \frac{0}{1} = 0$$

$$\blacklozenge 10 \blacklozenge = \frac{\text{remainder of } 10 \div 3}{\text{remainder of } 10 \div 4} = \frac{1}{2}$$

$$\blacklozenge 11 \blacklozenge = \frac{\text{remainder of } 11 \div 3}{\text{remainder of } 11 \div 4} = \frac{2}{3}$$

$$\blacklozenge 12 \blacklozenge = \frac{\text{remainder of } 12 \div 3}{\text{remainder of } 12 \div 4} = \frac{0}{0}$$

$$\blacklozenge 13 \blacklozenge = \frac{\text{remainder of } 13 \div 3}{\text{remainder of } 13 \div 4} = \frac{1}{1} = 1$$

The least value for which ♦a♦ $= 1$ is $a = 13$.

You can't divide by 0, so the function is not defined for $a = 8$, $a = 12$, or any other number that is evenly divisible by 4.

Chapter 8

MATH REVIEW: GEOMETRY AND MEASUREMENT

28% of the SAT

Geometry questions are the easiest to identify on the SAT. Usually accompanied by a diagram that contains lines, segments, angles, polygons (like triangles and rectangles)—a diagram riddled with angle measurements, lengths, variables, and who knows what else—geometry questions can be very intimidating. There is good news: You do not need to memorize every geometric theorem in order to succeed. In fact, the most important formulas are listed for you at the beginning of each mathematics section on the test. However, you do need to be familiar with all sorts of geometric figures, and studying them is the only way to prepare. In this chapter, you will review all of the important geometric theorems, principles, formulas, and concepts you need to know for the SAT.

Now that you have completed the SAT algebra review by working through Chapters 6 and 7, it's time to turn your attention to the second largest set of questions on the SAT, geometry. If you skipped ahead to this chapter, beware! You will need to apply some of the algebraic techniques reviewed in the previous two chapters in order to solve problems in this chapter! For example, in order to calculate a perimeter, you might need to solve a linear equation.

Lines and Angles

Angle classifications and parallel lines

8.1 Given the diagram below in which $l \parallel m$, identify the pairs of angles that belong to each of the following classifications and describe the relationship between the pairs of angles within each classification.

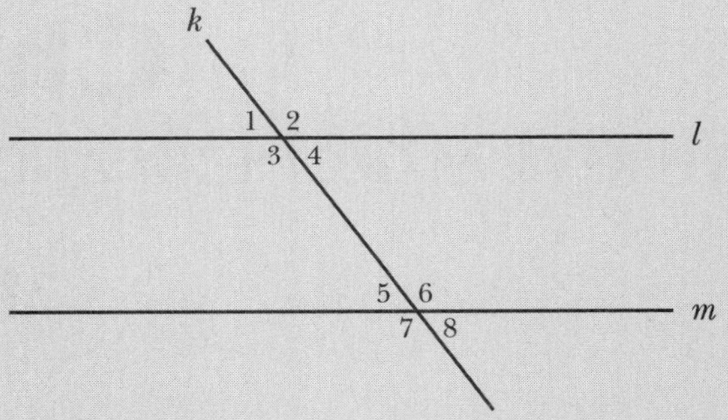

k
1 2
3 4
l
5 6
7 8
m

- Vertical angles
- Corresponding angles
- Same-side interior angles
- Alternate-interior angles
- Same-side exterior angles
- Alternate-exterior angles

For example, there are many pairs of vertical angles, so list them all. Then, complete this sentence: "Vertical angles are …." Complete these two tasks for each of the angle classifications.

A pair of parallel (non-intersecting) lines that is intersected by a third line called a transversal forms pairs of angles that have very specific names and properties.

- Vertical angles are formed by the intersection of two lines. In this diagram, angles 1 and 4 are vertical angles, as are angles 2 and 3. Similarly, angles 5 and 8 are vertical angles, as are angles 6 and 7. Vertical angles are congruent, which means they have the same measure.
- Corresponding angles appear in the same corresponding location where the parallel lines are intersected by the transversal, *k*. Angles 1 and 5 are corresponding angles, as are 2 and 6, 3 and 7, and 4 and 8. Corresponding angles are congruent.
- Same-side interior angles appear between the parallel lines on the same side of the transversal. In this diagram, angles 3 and 5 are same-side interior angles, as are angles 4 and 6. Same-side interior angles are supplementary, which means the sum of the two angle measures is 180°.
- Alternate-interior angles appear on opposite sides of the transversal and between the parallel lines. In this diagram, 3 and 6 are alternate-interior angles, as are 4 and 5. Alternate-interior angles are congruent.
- Same-side exterior angles appear on the same side of the transversal and are not between the parallel lines. In this diagram, 1 and 7 are same-side exterior angles, as are 2 and 8. Same-side exterior angles are supplementary.

Technically, the two lines have to be in the same plane. Unless you are specifically told otherwise, you can assume that each geometric drawing consists of all coplanar elements—all the lines, angles, polygons, circles, etc. lie in the same plane in each drawing.

- Alternate-exterior angles appear on opposite sides of the transversal and are not between the parallel lines. In this diagram, 1 and 8 are alternate-exterior angles, as are 2 and 7. Alternate-exterior angles are congruent.

Note: Problems 8.2–8.4 refer to the diagram below, in which $l \parallel m$ and $s \perp t$. Lines m, s, and t intersect at point Q.

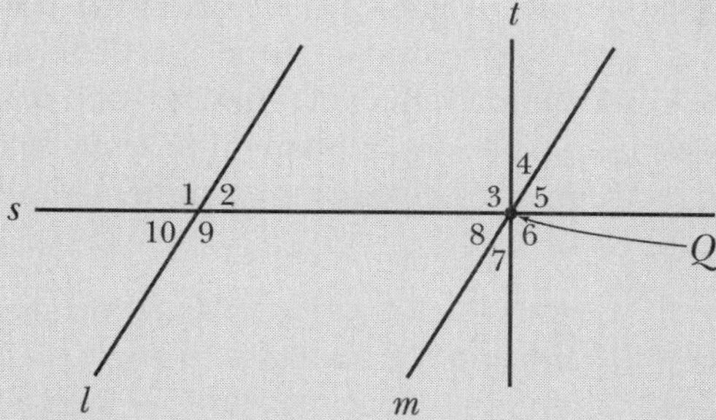

8.2 Classify each of the numbered angles as acute, obtuse, or right.

Perpendicular lines intersect at 90° angles, called right angles. Angles whose measures are less than 90° are called acute angles, and angles whose measures are between 90° and 180° are called obtuse angles. Incidentally, the angle measure of a straight line is 180°, so angles measuring 180° are called straight angles.

In this drawing, lines s and t are perpendicular, so they intersect at right angles. Therefore, angles 3 and 6 are right angles. Angles 4, 5, 7, 8, 2, and 10 are smaller than right angles, so they are acute. Angles 1 and 9 are larger than right angles, so they are obtuse.

Note: Problems 8.2–8.4 refer to the diagram in Problem 8.2, in which $l \parallel m$ and $s \perp t$. Lines m, s, and t intersect at point Q.

8.3 Identify all numbered angles that are complementary to angle 4.

Two angles are complementary if the sum of their measures is 90°. If those angles are adjacent, complementary angles will, when combined, form a right angle. In this diagram, angles 4 and 5 form a right angle when combined, so the sum of their angles is 90° and they are complementary.

Notice that angles 5 and 8 are vertical angles, so their angle measures are equal. Therefore, 5 is not the only angle that is a complement of 4. Angle 8, which is congruent to angle 5, is also a complement of angle 4.

All of the angles on the SAT are measured in degrees. You may have heard of radians, or see that "RAD" is an option on your calculator, but you won't find radians on the SAT.

Note: Problems 8.2–8.4 refer to the diagram in Problem 8.2, in which l || m and s ⊥ t. Lines m, s, and t intersect at point Q.

8.4 Identify all of the numbered angles that are supplementary to angle 9.

When combined, angles 9 and 10 form a straight line, a 180° angle, so angles 9 and 10 are supplementary. Angles 9 and 2 also form a straight line, so they are supplementary as well. Notice that angles 8 and 9 are same-side interior angles, which are also supplementary. Because angles 8 and 5 are vertical angles, they are congruent, so if angle 8 is a supplement to angle 9, then angle 5 is also. In conclusion, there are four angles that are supplementary to angle 9: angles 2, 5, 8, and 10.

8.5 In the figure below, $l \perp m$ and the measure of angle 3 is four times the measure of angle 2. Calculate the sum of the measures of angles 1 and 2.

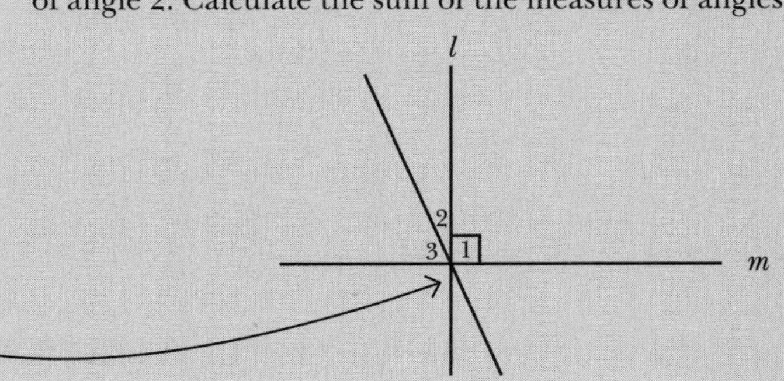

This symbol (which looks like a tiny square) indicates a right angle, which makes sense in this problem because lines l and m are perpendicular.

If l and m are perpendicular lines, then they intersect at 90° angles. Therefore, the measure of angle 1 is 90° and the sum of the measures of angles 2 and 3 is 90°: $m\angle 2 + m\angle 3 = 90°$. In other words, angles 2 and 3 are complementary.

The little "m" before the angle symbol stands for "the measure of." In other words, m∠2 is read "the measure of angle 2."

The problem states that angle 3 is four times as large as angle 2: $m\angle 3 = 4m\angle 2$. Use this information to rewrite the equation above, which states that angles 2 and 3 are complementary, and solve for the measure of angle 2.

$$\begin{aligned} m\angle 2 + m\angle 3 &= 90° \\ m\angle 2 + (4m\angle 2) &= 90° \\ 5m\angle 2 &= 90° \\ m\angle 2 &= \frac{90°}{5} \\ m\angle 2 &= 18° \end{aligned}$$

Angle 3 is four times as large as angle 2, so the measure of angle 3 is equal to 4(18°) = 72°. Note that 18° and 72° are the measures of complementary angles, because 18° + 72° = 90°. Finally, calculate the sum of the measures of angles 1 and 2, as directed by the problem.

$$\begin{aligned} m\angle 1 + m\angle 2 &= 90° + 18° \\ &= 108° \end{aligned}$$

8.6 Given the diagram below, in which r and s are parallel lines, express x in terms of y.

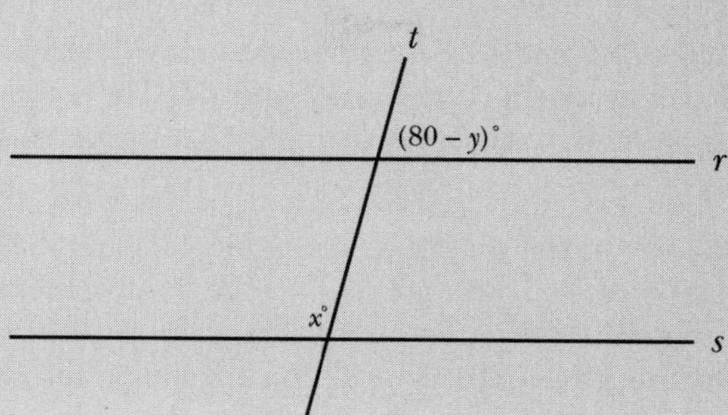

The problem asks you to express x in terms of y, so the final answer will begin with "$x =$" and the expression following the equal sign will contain the variable y. Remember that vertical angles are congruent, and based on that information alone, you know that at least one other angle in the diagram has a measure of $(80 - y)°$, as illustrated below.

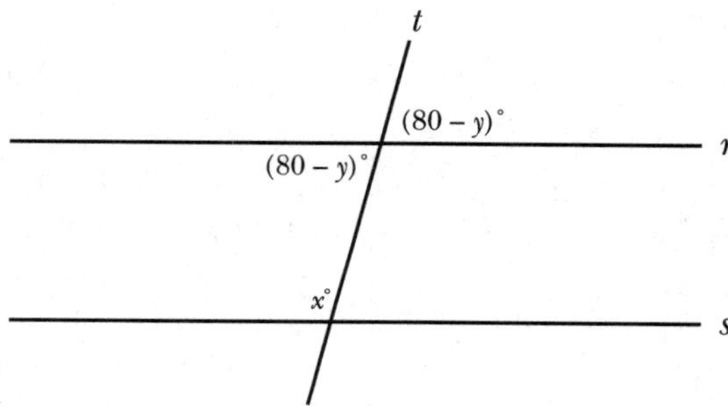

The newly labeled angle and the angle with measure $x°$ are same-side interior angles, so they are supplementary angles; their measures have a sum of 180°.

$$x + (80 - y) = 180$$

Solve the equation for x to express x in terms of y.

$$x - y = 180 - 80$$
$$x - y = 100$$
$$x = y + 100$$

8.7 In the diagram below, $\overrightarrow{ZF}$ bisects $\angle AZE$. What is the measure of $\angle DZE$?

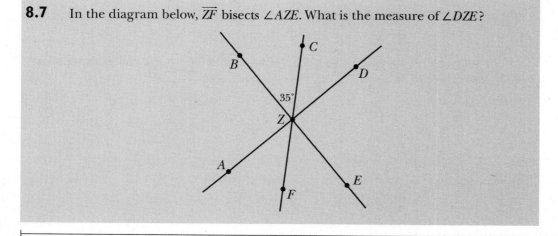

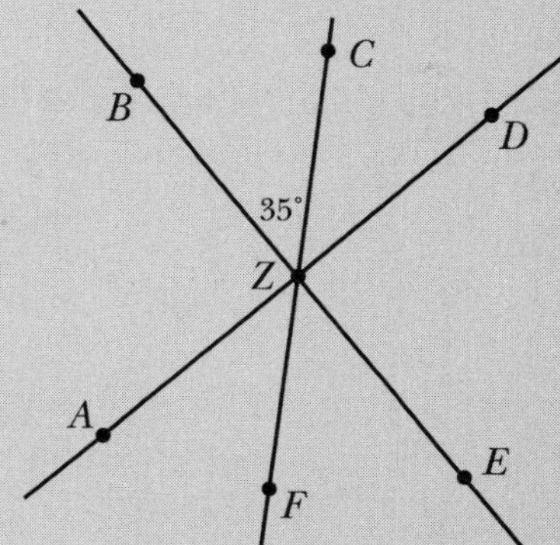

An angle bisector is a ray that divides one angle into two smaller, congruent angles. In other words, it divides an angle in half. If $\overrightarrow{ZF}$ bisects $\angle AZE$, then you can conclude that $m\angle AZF = m\angle EZF$.

Notice that angles *BZC* and *EZF* are vertical angles. Therefore, they both measure 35°. Remember that angles *AZF* and *EZF* are equal halves of angle *AZE*, so $m\angle AZF = 35°$ as well. Because angle *CZD* is a vertical angle of *AZF*, it measures 35°.

There are six angles in the diagram, and four of them measure 35°. The remaining two angles, *AZB* and *DZE*, are congruent vertical angles. Let *x* represent the angle measure they share. The sum of all six angle measures is 360°, because the angles circle around point *Z* completely. Set up an equation stating this fact and solve for *x*.

A note about notation: $\overline{RV}$ refers to the geometric segment, because it has a bar over the endpoints. When you leave off the bar, you are referring to length, so RV is the length of segment $\overline{RV}$.

$$\begin{aligned} m\angle AZB + m\angle BZC + m\angle CZD + m\angle DZE + m\angle EZF + m\angle FZA &= 360 \\ x + 35 + 35 + x + 35 + 35 &= 360 \\ 2x + 140 &= 360 \\ 2x &= 220 \\ x &= 110 \end{aligned}$$

The measure of angles *AZB* and *DZE* is 110°.

8.8 Points *R*, *S*, *T*, and *V* lie on a line in that order, such that *T* is the midpoint of $\overline{SV}$ and *S* is the midpoint of $\overline{RV}$. If $RT = 18$, what is the value of *TV*?

Construct a diagram that represents the information given. Begin by drawing a line and then label four points *R*, *S*, *T*, and *V* on the line. Because *T* is the midpoint of $\overline{SV}$, $ST = TV$. Similarly, because *S* is the midpoint of $\overline{RV}$, $RS = SV$.

You don't have to draw the diagram accurately, making sure all the segments that are supposed to be equal are actually equal. The diagram works as long as you label the information correctly, even if your actual drawing is not to scale.

2*x* *x* *x*
R *S* *T* *V*

Let *x* represent the equal lengths *ST* and *TV*. Because $RS = SV$, you can conclude that $RS = x + x = 2x$. According to the problem, $RT = 18$. Note that $RT = RS + ST$. Use the following equation to calculate *x*.

$$\begin{aligned} RS + ST &= 18 \\ 2x + x &= 18 \\ 3x &= 18 \\ x &= \frac{18}{3} \\ x &= 6 \end{aligned}$$

The problem asks you to calculate *TV*, which has length *x* in the diagram above. Therefore, $TV = x = 6$.

8.9 Line segment $\overline{XY}$ lies on a number line, such that the coordinates of its endpoints are $X = -6$ and $Y = 11$. If A and B are the coordinates of points that split the segment into three congruent, adjacent segments such that $A < B$, what is the value of $A \cdot B$?

The points with coordinates A and B trisect the line segment, creating three segments of equal length as illustrated below.

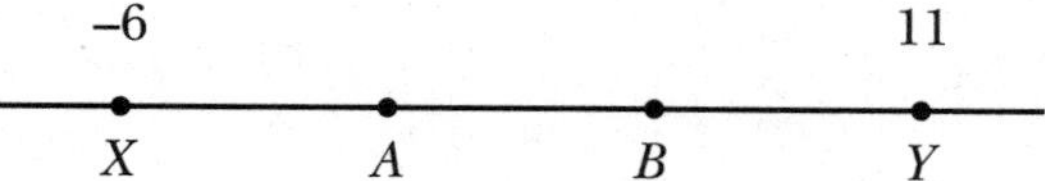

To calculate the lengths of the three equal segments, begin by calculating XY, the length of the original segment. Compute the absolute value of the difference of the coordinates of the endpoints.

$$\begin{aligned} XY &= |Y - X| \\ &= |11-(-6)| \\ &= |11+6| \\ &= 17 \end{aligned}$$

Divide that length by 3; each of the three congruent line segments has length 17/3. To calculate A, add 17/3 to X. Add 17/3 again to calculate B.

$$\begin{aligned} A &= X + \frac{17}{3} \\ &= -6 + \frac{17}{3} \\ &= -\frac{18}{3} + \frac{17}{3} \\ &= -\frac{1}{3} \end{aligned} \qquad \begin{aligned} B &= A + \frac{17}{3} \\ B &= -\frac{1}{3} + \frac{17}{3} \\ B &= \frac{16}{3} \end{aligned}$$

The problem asks you calculate $A \cdot B$.

$$\begin{aligned} A \cdot B &= \left(-\frac{1}{3}\right)\left(\frac{16}{3}\right) \\ &= -\frac{16}{9} \end{aligned}$$

Triangles

Including the Pythagorean theorem

8.10 Given the diagram below, calculate the value of x.

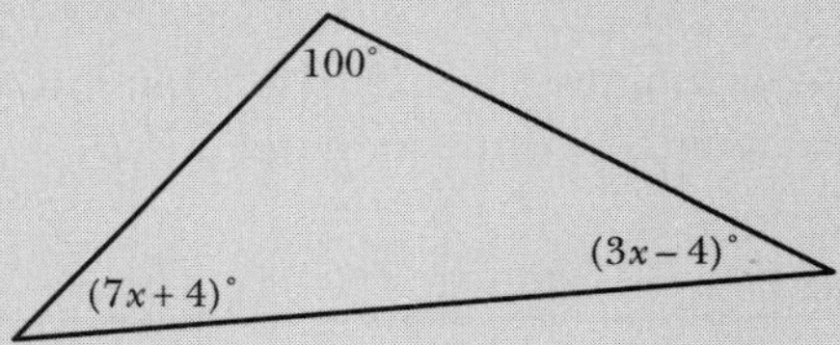

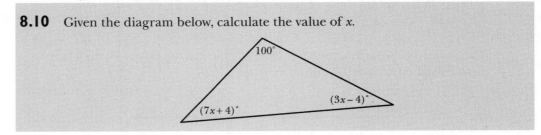

The sum of the interior angles of a triangle is 180°. Therefore, you can add the given angle measurements, set the sum equal to 180°, and solve for x.

$$\begin{aligned}100+(7x+4)+(3x-4)&=180\\(7x+3x)+(100+4-4)&=180\\10x+100&=180\\10x&=80\\x&=\frac{80}{10}\\x&=8\end{aligned}$$

8.11 Given the figure below, in which $\overleftrightarrow{AC} \parallel \overleftrightarrow{BD}$ and $\overrightarrow{BC}$ bisects $\angle ABD$, express the measure of angle BAC in terms of x.

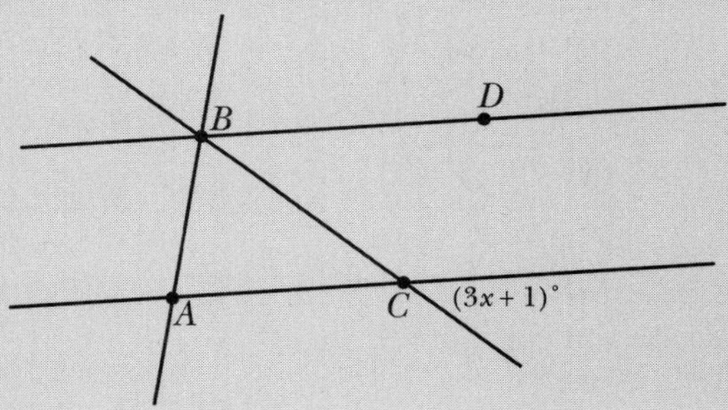

The angle measuring $(3x + 1)°$ and angle CBD are corresponding angles formed by parallel lines that are intersected by a transversal. Therefore, the angles have the same measure: $m\angle DBC = (3x+1)°$. The problem states that $\overrightarrow{BC}$ bisects angle ABD, so angles ABC and DBC have the same measure: $m\angle ABC = (3x+1)°$.

Furthermore, the angle in the original diagram that measures $(3x + 1)°$ and angle BCA are vertical angles, so they have the same measure as well: $m\angle BCA = (3x+1)°$. The sum of the angle measures of triangle ABC is equal to 180°.

$$m\angle BAC + m\angle ABC + m\angle BCA = 180°$$

You now know the measures of two of those angles. Substitute them into the equation and solve for $m\angle BAC$.

$$\begin{aligned}m\angle BAC+(3x+1)+(3x+1)&=180\\m\angle BAC+6x+2&=180\\m\angle BAC&=178-6x\end{aligned}$$

8.12 Given triangle XYZ below, in which $XY = YZ$, calculate all possible values of a. Note that the figure may not be drawn to scale.

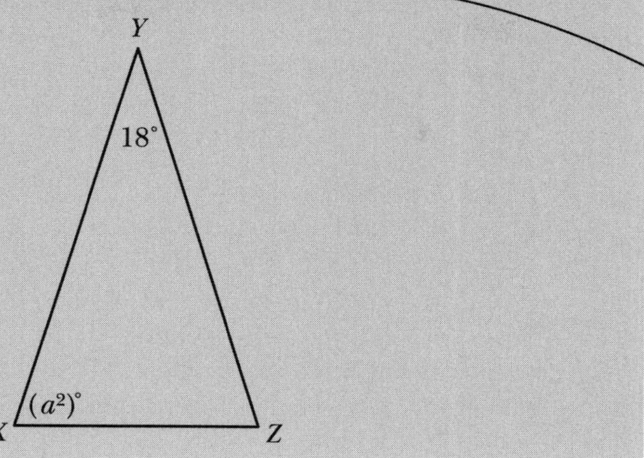

The fact that the diagram is not to scale should not affect how you approach the problem. It just means you shouldn't make any assumptions about how big or small the angles might be based on the diagram alone.

A triangle with at least two congruent sides is called an isosceles triangle, so triangle XYZ is isosceles. According to the isosceles triangle theorem, the angles of an isosceles triangle that are opposite the congruent sides have the same angle measure. In this diagram, angle Z is opposite side $\overline{XY}$ and angle X is opposite side $\overline{YZ}$. Therefore, angles X and Z have the same measure. Recall that the sum of the angles of a triangle is 180°.

$$m\angle X + m\angle Y + m\angle Z = 180°$$

The diagram states that angle Y has a measure of 18°, and because angles X and Z are congruent, they both have a measure of $(a^2)°$. Substitute the angle measurements into the equation and solve for a.

$$\begin{aligned}(a^2)+18+(a^2)&=180\\ 2a^2+18&=180\\ 2a^2&=162\\ a^2&=\frac{162}{2}\\ a^2&=81\end{aligned}$$

To solve for a, take the square root of both sides of the equation. Note that you need to include a "±" sign to the right of the equal sign whenever you apply an even root to both sides of an equation.

Like a square root

$$\begin{aligned}\sqrt{a^2}&=\pm\sqrt{81}\\ a&=\pm 9\end{aligned}$$

There are two possible values of a: –9 and 9. For either value, the measures of the congruent angles in the isosceles triangle are $(a^2)° = 81°$.

8.13 Given the diagram below, in which A and C are complementary angles, calculate AB.

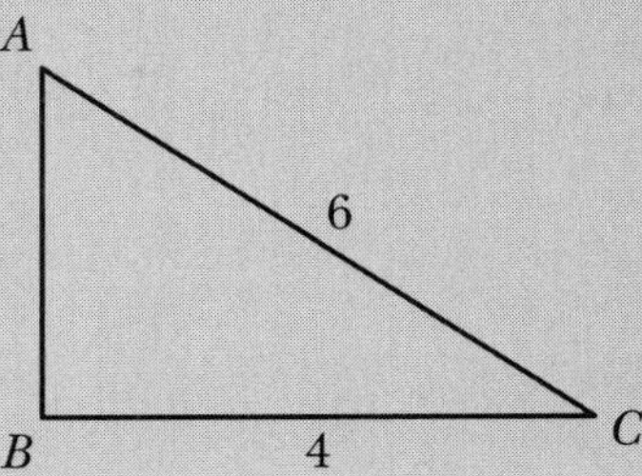

If A and B are complementary angles, then their measures have a sum of 90°. The sum of all three angle measures in the triangle is 180°. Therefore, the measure of angle B is a right angle, because it measures 90°, as demonstrated below.

$$\begin{aligned}(m\angle A + m\angle C) + m\angle B &= 180^\circ\\(90^\circ) + m\angle B &= 180^\circ\\m\angle B &= 180^\circ - 90^\circ\\m\angle B &= 90^\circ\end{aligned}$$

If two acute angles in a triangle are complementary, then the triangle is a right triangle, and vice versa.

Triangle ABC contains a right angle, so it is called a right triangle. This is important because a special relationship exists between the lengths of the sides of a right triangle. No, not *they-go-on-long-walks-together-and-share-meaningful-glances* special—just *describable-by-an-equation* special. Specifically, the lengths of the sides of a right triangle satisfy the Pythagorean theorem.

According to the Pythagorean theorem, if the lengths of the sides of a right triangle are x, y, and z (such that z is the longest side, called the hypotenuse), then $x^2 + y^2 = z^2$. The sum of the squares of the legs' lengths is equal to the square of the hypotenuse's length. Note that the hypotenuse is always the side opposite the right angle; the remaining sides of the right triangle are called the legs.

In triangle ABC, the hypotenuse is side $\overline{AC}$, because it is opposite angle B, the right angle. The legs are $\overline{AB}$ and $\overline{BC}$. Apply the Pythagorean theorem and solve for AB.

$$\begin{aligned}(AB)^2 + (BC)^2 &= (AC)^2\\(AB)^2 + 4^2 &= 6^2\\(AB)^2 + 16 &= 36\\(AB)^2 &= 36 - 16\\(AB)^2 &= 20\\\sqrt{(AB)^2} &= \sqrt{20}\\AB &= \sqrt{4\cdot 5}\\AB &= 2\sqrt{5}\end{aligned}$$

Geometric lengths are positive values, so you do not need to insert a "±" symbol into the equation when you take the square root of both sides; a length cannot be negative.

8.14 If the length of the hypotenuse in a certain right triangle is two and a half times the length of the short leg in the triangle, what is the length of the long leg of the right triangle in terms of the short leg?

Let s, l, and h represent the lengths of the short leg, long leg, and hypotenuse, respectively. The problem states that the hypotenuse is 2.5 times, or 5/2 times, the length of the short leg. Therefore, $h = (5/2)s$. Apply the Pythagorean theorem, replacing h with the equivalent value $(5/2)s$ and solve for l.

$$s^2 + l^2 = h^2$$

$$s^2 + l^2 = \left(\frac{5}{2}s\right)^2$$

$$s^2 + l^2 = \frac{25}{4}s^2$$

$$l^2 = \frac{25}{4}s^2 - s^2$$

$$l^2 = \frac{25}{4}s^2 - \frac{4}{4}s^2$$

$$l^2 = \frac{21}{4}s^2$$

$$\sqrt{l^2} = \sqrt{\frac{21}{4}s^2}$$

$$l = \frac{\sqrt{21}}{\sqrt{4}}\sqrt{s^2}$$

$$l = \frac{\sqrt{21}}{2}s$$

Like in Problem 8.13, you can leave off the "±" symbol when you take the square root of both sides in this step because geometric lengths are positive.

8.15 List the four unique Pythagorean triples that represent right triangles with the least perimeter. Omit triples that are multiples of other triples.

Pythagorean triples are sets of three numbers that satisfy the Pythagorean theorem. In other words, they are lengths of sides of right triangles. It is useful to memorize the most common Pythagorean triples, because recognizing them on a test means you can calculate side lengths automatically, without having to apply the Pythagorean theorem. The following four triples contain the smallest numbers (and therefore represent right triangles with the least area).

3, 4, 5   5, 12, 13   8, 15, 17   7, 24, 25

The first two numbers in each triple represent leg lengths, and the final number in each triple represents the length of the hypotenuse. For example, if you encounter a right triangle with leg lengths 3 and 4, based on the first of the Pythagorean triples listed above, you could automatically conclude that the length of the hypotenuse is 5.

Note that multiplying any triple by a real number produces another Pythagorean triple. For example, if you multiply each number in the triple 3, 4, 5 by 2, the result is Pythagorean triple 6, 8, 10.

8.16 In the diagram below, a wire stretches from the ground to the top of a tower. If the height of the tower is 180 feet and the slope of the wire is $\frac{5}{4}$, what is the value of d, the distance between the base of the tower and the point at which the wire is anchored to the ground, in feet?

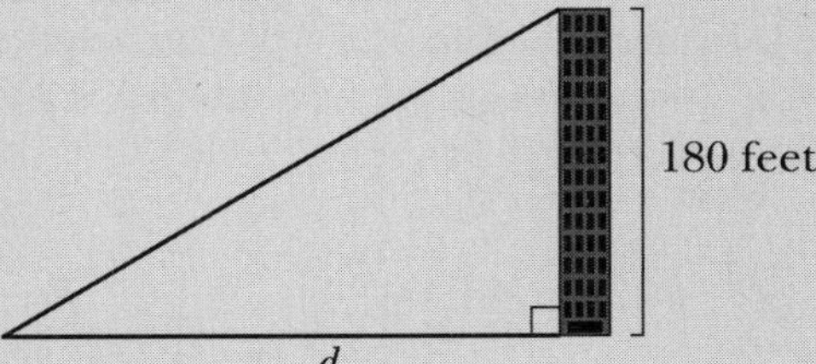

Although this looks like it might be a Pythagorean theorem problem (because it contains a right triangle and side lengths d and 180), it is not. Instead, it is a simple slope problem. In Chapter 6, you review the concept of slope, the value that describes the steepness of a line as a fraction, the change in vertical distance (y) divided by the change in horizontal distance, x.

The slope of the wire in the diagram is 5/4, which means that for every 5 feet the wire travels vertically, it travels 4 feet horizontally. Divide the total vertical change (180 feet from the ground to the roof) by 5, the vertical component of the slope: $180 \div 5 = 36$.

Imagine the entire journey along the wire, from the ground to the roof, as a series of 36 consecutive steps on a staircase. In each step you are traveling 5 feet higher than the step before. According to the slope, you are also traveling a horizontal change of 4 feet to the right in each step, for a total horizontal change of $36 \cdot 4 = 144$ feet.

8.17 Is it possible to form a triangle with sides of length 5, 8, and 12 inches? Why or why not?

According to the triangle inequality theorem, the sum of the lengths of any two sides of a triangle must always be greater than the length of the remaining side. For example, if a triangle has side lengths a, b, and c, then the following must be true: $a + b > c$, $a + c > b$, and $b + c > a$. You can form a triangle with sides of length 5, 8, and 12, because the sum of any two side lengths is greater than the remaining side length, as demonstrated below.

$$\begin{array}{ccc} 5+8>12 & 5+12>8 & 8+12>5 \\ 13>12 & 17>8 & 20>5 \end{array}$$

8.18 A certain triangle has lengths a, b, and c, such that $c > b > a$. Given $a = 4$, $c = 10$, and b is an integer, list all possible values of b.

The length of side b is an integer greater than 4 but less than 10. In other words, b could be equal to 5, 6, 7, 8, or 9. However, the side lengths must abide by the rules of the triangle inequality theorem, as described in Problem 8.17. If $b = 5$, then $a + b = 4 + 5$ is not greater than the remaining side length, $c = 10$. Similarly, if $b = 6$, then $a + b = 10$, which is not greater than 10 either. Therefore, there are only three possible values of b: 7, 8, and 9.

8.19 Given the diagram below, calculate *BE*.

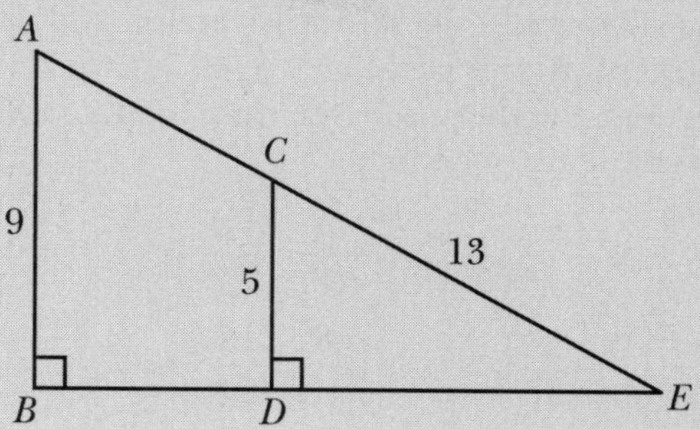

This diagram contains two overlapping triangles, *CED* and *AEB*, that share angle *E*. Note that both triangles also contain right angles: triangle *CED* contains right angle *CDE* and triangle *AEB* contains right angle *B*. When two triangles have two congruent angles, then the triangles are similar.

If two triangles have two angles in common, then the two triangles actually have all three angles in common. The third angle is equal to 180 minus the sum of the two angles they share.

Similar triangles have the same shape but different sizes—one is a larger version of the other. Corresponding sides of similar triangles are in the same proportion. In this problem, triangle *CED* is a smaller version of triangle *AEB*, and you are given corresponding leg lengths $AB = 9$ and $CD = 5$. Therefore, the sides of triangle *AEB* are $AB/CD = 9/5$ times as long as the corresponding sides in triangle *CED*. This value is called the scale factor.

Technically, congruent triangles are also similar, but you should focus your attention on similar triangles that are different sizes and how to deal with them.

To calculate any length of the larger triangle, multiply the length of the corresponding side in the smaller triangle by 9/5. Alternately, you could calculate lengths in the smaller triangle by multiplying corresponding lengths in the larger triangle by 5/9. You are asked to calculate *BE*, which is 9/5 as large as the corresponding side length *DE* in the smaller right triangle.

The length *DE* is not given in the diagram, but it is easy to calculate. *CED* is a right triangle, and two of its side lengths are values in the Pythagorean triple 5, 12, 13. Therefore, $DE = 12$. (If you did not recognize the Pythagorean triple, you can still apply the Pythagorean theorem to calculate *DE*.) If $DE = 12$, then the corresponding side *BE* is 9/5 as long.

$$\begin{aligned} BE &= \frac{9}{5}(DE) \\ &= \frac{9}{5}(12) \\ &= \frac{108}{5} \end{aligned}$$

8.20 Given the diagram below, calculate *VW*.

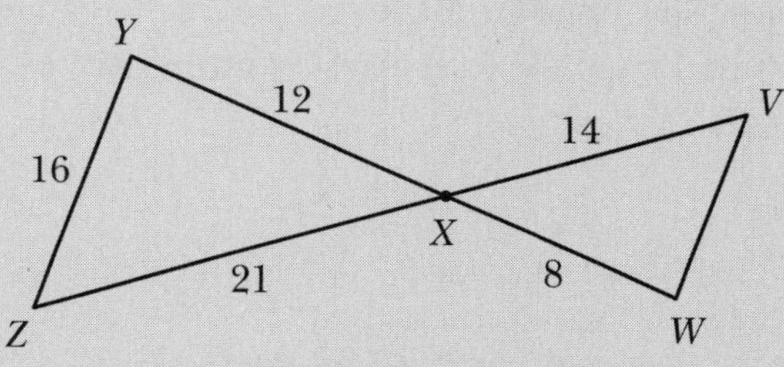

Note: Figure not drawn to scale.

You can prove triangles are similar using:
(1) Two pairs of corresponding congruent angles (as described in Problem 8.19). Also called angle-angle similarity theorem.
(2) Side-angle-side similarity theorem (described here).
(3) Side-side-side similarity theorem (when all corresponding sides are in the same proportion).

The two triangles in the diagram are similar according to the side-angle-side similarity theorem, which states that two triangles are similar if two corresponding sides are in the same proportion and the angles formed by those sides are congruent. In this problem, angles *YXZ* and *WXV* are vertical angles, so they are congruent. Furthermore, the ratio of *WX* to *XY* is the same as the ratio of *VX* to *XZ*, as demonstrated below.

$$\frac{WX}{XY} = \frac{VX}{XZ}$$
$$\frac{8}{12} = \frac{14}{21}$$
$$\frac{8 \div 4}{12 \div 4} = \frac{14 \div 7}{21 \div 7}$$
$$\frac{2}{3} = \frac{2}{3}$$

Therefore, every side length in triangle *VXW* is 2/3 the corresponding side length in triangle *ZXY*, and *VW* is 2/3 as large as *YZ*.

$$\begin{aligned} VW &= \frac{2}{3}(YZ) \\ &= \frac{2}{3}(16) \\ &= \frac{32}{3} \end{aligned}$$

8.21 Given the diagram below, list the angles in order, from the angle with the least measure to the angle with the greatest measure.

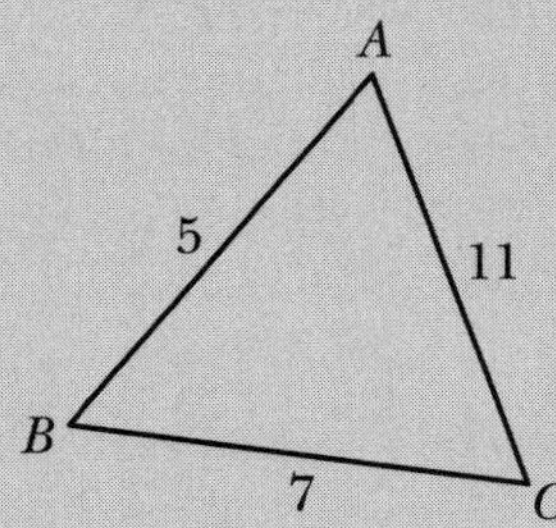

Note: Figure not drawn to scale.

The largest angle in a triangle is opposite the longest side, and vice versa. The longest side in this triangle is $\overline{AC}$, which has length 11. Because angle *B* is opposite side $\overline{AC}$, angle *B* is the largest angle in the triangle. Similarly, because $\overline{AB}$ is the smallest side, opposite angle *C* is the smallest angle in the triangle. The angles, in order from the least measure to the greatest measure, are $\angle C$, $\angle A$, and $\angle B$.

Note: Problems 8.22–8.23 refer to the diagram below, isosceles right triangle LMN.

8.22 Apply the Pythagorean theorem to calculate LM if $MN = 8$.

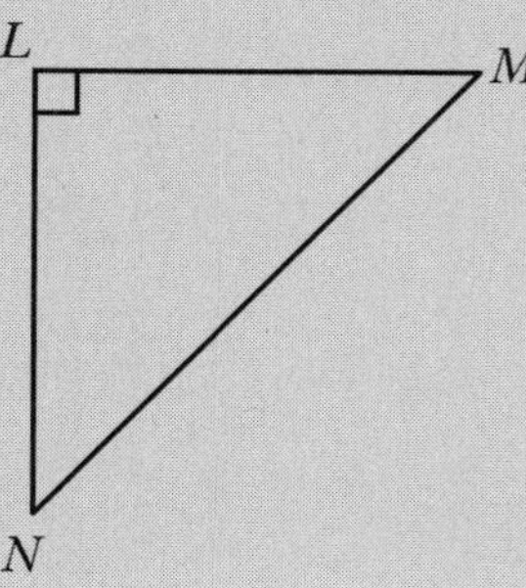

The problem states that LMN is an isosceles right triangle, so $LM = LN$. Note that the legs of a right triangle are equal when the right triangle is isosceles—the length of the hypotenuse is always greater than the lengths of the legs, so MN cannot be equal to either of the leg lengths. Let l represent the length of the legs in the isosceles right triangle: $l = LM = LN$. Apply the Pythagorean theorem to calculate l.

$$\begin{aligned}(LM)^2 + (LN)^2 &= (MN)^2\\ l^2 + l^2 &= 8^2\\ 2l^2 &= 64\\ l^2 &= \frac{64}{2}\\ l^2 &= 32\\ \sqrt{l^2} &= \sqrt{32}\\ l &= \sqrt{16 \cdot 2}\\ l &= 4\sqrt{2}\end{aligned}$$

You conclude that $LM = LN = 4\sqrt{2}$.

Note: Problems 8.22–8.23 refer to the diagram in Problem 8.22, isosceles right triangle LMN.

8.23 Apply the properties of 45°–45°–90° triangles to calculate MN if $LN = 9\sqrt{2}$.

If a triangle is isosceles, then the angles opposite the congruent sides are also congruent. In right triangle LMN, sides $\overline{LM}$ and $\overline{LN}$ are congruent, so the angles opposite those sides are also congruent: $\angle M \cong \angle N$. As Problem 8.13 explains, the acute angles of a right triangle are complementary, and if those complementary angles are equal, they both must measure 45°. (The only two angles of equal measure that add to 90° are 45° angles.) Because of the angles this triangle contains, it is called a 45°–45°–90° triangle.

This symbol means "congruent." If two geometric things are congruent, they are basically the same. In other words, segments are congruent if they are the same length, angles are congruent if they have the same measure, triangles are congruent if they are copies of each other, etc.

And the length of a side is equal to the length of the hypotenuse DIVIDED by $\sqrt{2}$.

The side lengths of 45°–45°–90° triangles adhere to this rule: The length of the hypotenuse of the triangle is $\sqrt{2}$ times the length of a leg. In this problem, you are given the length of a leg, so multiply that length by $\sqrt{2}$ to calculate the *MN*, the length of the hypotenuse.

$$\begin{aligned} MN &= \sqrt{2}(LN) \\ &= \sqrt{2}\left(9\sqrt{2}\right) \\ &= 9\sqrt{4} \\ &= 9\cdot 2 \\ &= 18 \end{aligned}$$

8.24 Given the diagram below, calculate *XY* and *YZ*.

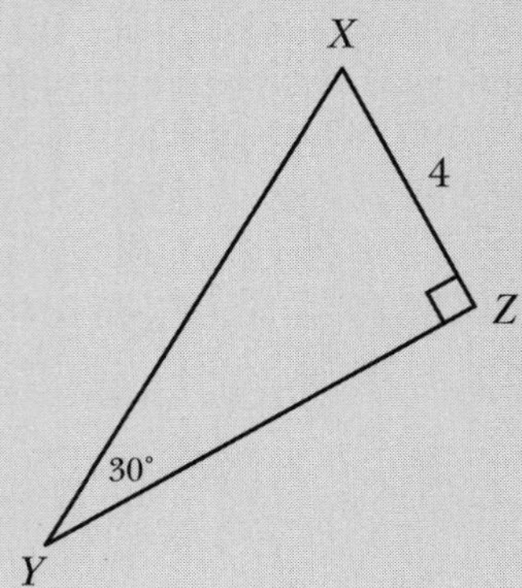

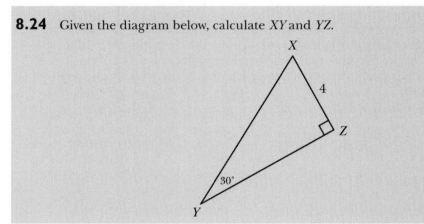

In Problem 8.23, you investigate 45°–45°–90° right triangles, whose side lengths follow specific rules. Similarly, 30°–60°–90° right triangles (also named for the angle measurements within the triangle) have side lengths that follow specific rules. Let *s* represent the length of the shortest side of the triangle, the short leg opposite the smallest angle (which measures 30°). If *h* is the length of the hypotenuse, then $h = 2s$. In other words, the hypotenuse is twice as long as the short leg. The long leg of the triangle is $\sqrt{3}$ times as long as the short leg, so if l represents the length of the long leg, then $l = \sqrt{3}s$.

In this triangle, the shortest side has length $s = XZ = 4$. The hypotenuse must be twice as long.

$$\begin{aligned} h &= 2s \\ XY &= 2(4) \\ XY &= 8 \end{aligned}$$

The long leg of the triangle is $\sqrt{3}$ times as long as the short leg.

$$\begin{aligned} l &= \sqrt{3}s \\ YZ &= \sqrt{3}(4) \\ YZ &= 4\sqrt{3} \end{aligned}$$

8.25 Given the diagram below, in which angles *CBD* and *CDB* are congruent and $m\angle CBA = 105°$, express *AD* in terms of *x*.

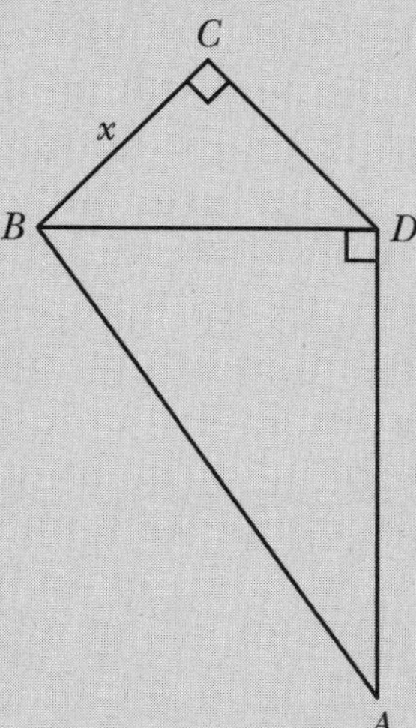

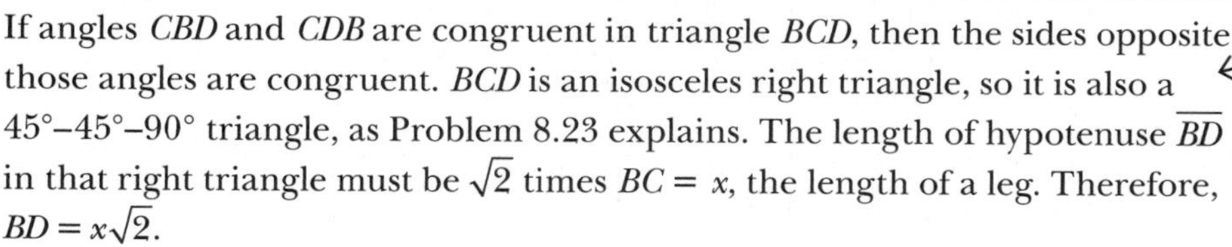

If angles *CBD* and *CDB* are congruent in triangle *BCD*, then the sides opposite those angles are congruent. *BCD* is an isosceles right triangle, so it is also a 45°–45°–90° triangle, as Problem 8.23 explains. The length of hypotenuse $\overline{BD}$ in that right triangle must be $\sqrt{2}$ times $BC = x$, the length of a leg. Therefore, $BD = x\sqrt{2}$.

The isosceles triangle theorem states that when two sides of a triangle are congruent, the angles opposite those sides are congruent. In this problem, you are applying the converse, which is also true: If two angles in a triangle are congruent, the opposite sides are congruent.

Not only is $\overline{BD}$ the hypotenuse of right triangle *BCD*, it is also a leg of right triangle *ABD*. You are given the measure of angle *CBA*, and you have determined that the measure of angle *CBD* is 45°. Use this information to calculate the measure of angle *ABD*.

$$\begin{aligned} m\angle CBA &= m\angle CBD + m\angle ABD \\ 105° &= 45° + m\angle ABD \\ 105° - 45° &= m\angle ABD \\ 60° &= m\angle ABD \end{aligned}$$

Recall that the three angle measures within a triangle must have a sum of 180°, so the measure of angle *A* must be 30°, as calculated below.

$$\begin{aligned} m\angle DBA + m\angle A + m\angle ADB &= 180° \\ 60° + m\angle A + 90° &= 180° \\ 150° + m\angle A &= 180° \\ m\angle A &= 180° - 150° \\ m\angle A &= 30° \end{aligned}$$

Therefore, right triangle *ABD* is a 30°–60°–90° triangle, and the length of its long leg $l = AD$ must be $\sqrt{3}$ times the length of short leg $s = BD$.

$$\begin{aligned} l &= \sqrt{3}s \\ AD &= \sqrt{3}(BD) \\ AD &= \sqrt{3}\left(x\sqrt{2}\right) \\ AD &= x\sqrt{6} \end{aligned}$$

8.26 List the five postulates or theorems used to prove triangles are congruent.

There are five techniques you can apply to prove that a pair of triangles is congruent:

- Angle-side-angle: If the triangles contain two congruent, corresponding angles and the included sides (the sides between those angles) are congruent, then the triangles are congruent.
- Side-angle-side: If the triangles contain two congruent, corresponding sides and the angles formed by those sides are congruent, then the triangles are congruent.
- Side-side-side: If the triangles have the same corresponding side lengths, the triangles are congruent.
- Angle-angle-side: If the triangles have two congruent, corresponding angles and the corresponding lengths of two non-included sides are congruent, then the triangles are congruent.
- Hypotenuse-leg: If two right triangles have congruent hypotenuses and one pair of corresponding legs is congruent, then the right triangles are congruent.

Non-included means "not between." The congruent sides can't be sandwiched between the congruent angles in this theorem, unlike the angle-side-angle method.

Once you prove two triangles are congruent, you know that all of the corresponding angles and sides must be congruent as well.

8.27 If the opposite sides of the quadrilateral below are parallel, verify that $\angle BAC \cong \angle CDB$ by demonstrating that they are corresponding parts of congruent triangles.

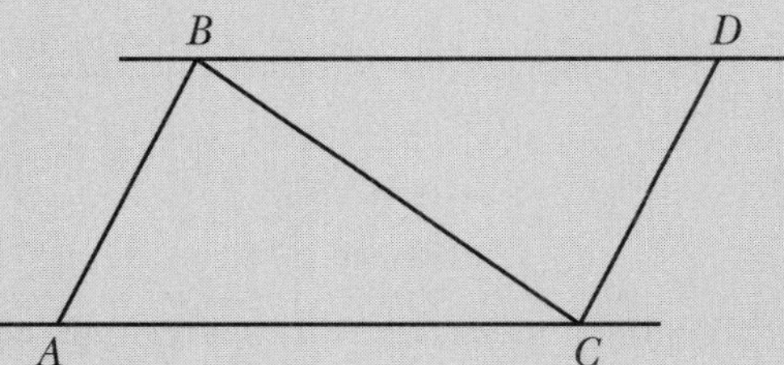

The parallel lines are cut by a transversal, so the alternate-interior angles formed are congruent (as explained in Problem 8.1). Consider the diagram below, which illustrates the congruent alternate interior angles using arc notation: $\angle ABC \cong \angle DCB$ and $\angle BCA \cong \angle CBD$.

B D A C

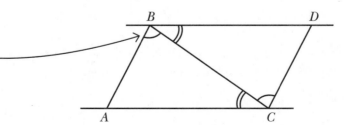

The single-arc angles are congruent to each other, and the double-arc angles are congruent to each other. The angles with the same arc markings are in the same corresponding position in each triangle.

Note that triangles *ABC* and *DCB*, also formed by the parallel lines and the transversal, have two congruent, corresponding angles. Furthermore, the triangles share overlapping side $\overline{BC}$, which is located between the congruent pairs of angles in each triangle. Therefore, the triangles are congruent according to the angle-side-angle technique described in Problem 8.26.

Because triangles *ABC* and *DCB* are congruent, all of their corresponding parts are congruent. For example, the unmarked angles in the triangles (angles *BAC* and *CDB*) are corresponding angles, so they are congruent.

Quadrilaterals and other Polygons

Figures with more than three sides

Note: Problems 8.28–8.29 refer to the diagram below, in which ABCD and EFCG are rectangles.

8.28 Calculate *AC*.

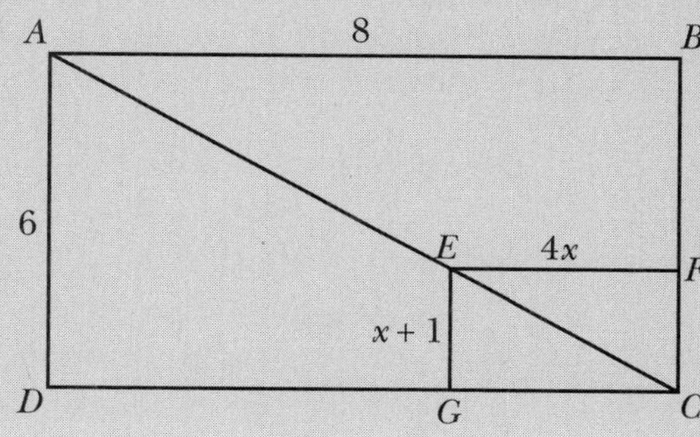

Rectangles are quadrilaterals (four-sided polygons) whose adjacent sides meet at 90° angles. Their opposite sides are congruent, as are their diagonals. Therefore, $AD = BC = 6$, and triangle *ABC* is a right triangle because $m\angle B = 90°$. Apply the Pythagorean theorem to calculate *AC*.

$$\begin{aligned}(AB)^2 + (BC)^2 &= (AC)^2 \\ 8^2 + 6^2 &= (AC)^2 \\ 64 + 36 &= (AC)^2 \\ 100 &= (AC)^2 \\ \sqrt{100} &= \sqrt{(AC)^2} \\ 10 &= AC\end{aligned}$$

You could have avoided using the Pythagorean theorem altogether if you had noticed that 6, 8, 10 is a Pythagorean triple, a multiple of the 3, 4, 5 triple created by doubling each of the values.

Note: Problems 8.28–8.29 refer to the diagram in Problem 8.28, in which ABCD and EFCG are rectangles.

8.29 Identify the value of x.

Rectangles *ABCD* and *EFCG* are similar, so all corresponding sides are in the same proportion. Note that all rectangles are not automatically similar (unlike squares and circles, which are). How do you know that these particular rectangles are similar? Triangles *ABC* and *CDA* are congruent right triangles according to the hypotenuse-leg technique described in Problem 8.26. They share the same hypotenuse, and not only one but both pairs of corresponding legs are congruent. For the same reasons, triangles *EFC* and *CGE* are congruent.

Therefore, they are also congruent according to the side-side-side technique. If you need more practice with congruent triangles or any geometric concept discussed in this book, check out The Humongous Book of Geometry Problems.

Furthermore, triangles *ECG* and *ACD* are similar triangles (because two of their corresponding angles are congruent), as are triangles *ABC* and *EFC*. Because the larger rectangle is comprised of two congruent right triangles that are similar to the two congruent right triangles that form the smaller rectangle, the rectangles are similar.

The rectangles are similar because they are made up of triangles that are similar.

Because the rectangles are similar, their corresponding sides are in the same proportion. In other words, the ratio of $x + 1$ to 6 is the same as the ratio of $4x$ to 8. Create this proportion and solve it for x.

$$\frac{x+1}{6}=\frac{4x}{8}$$
$$8(x+1)=6(4x)$$
$$8x+8=24x$$
$$8=24x-8x$$
$$8=16x$$
$$\frac{8}{16}=x$$
$$\frac{1}{2}=x$$

Note: Problems 8.30–8.31 refer to the diagram below, in which the sides of square QRST have midpoints W, X, Y, and Z.

8.30 Determine whether quadrilateral *WXYZ* is a square and explain your answer.

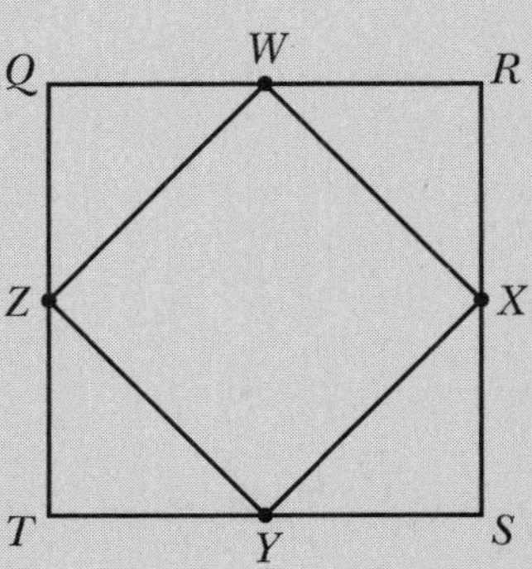

The sides of a square are congruent, so $QR = RS = ST = QT$. Points W, X, Y, and Z divide those congruent sides into halves, which are also congruent.

$$QW = RW = RX = SX = SY = TY = TZ = QZ$$

Consider the small right triangles QWZ, RWX, SYX, and TYZ. Each of those triangles is isosceles—its legs are each half the length of the side of the square. Furthermore, the angles formed by the congruent legs in each triangle are also congruent; they each measure 90°. Therefore, all four of the triangles are congruent according to the side-angle-side technique. Because corresponding parts of congruent triangles are congruent, the hypotenuses of the right triangles are congruent: $WX = XY = YZ = WZ$.

As Problem 8.23 explains, isosceles right triangles are also 45°–45°–90° triangles, so the acute angles in each of the right triangles measure 45°. Notice that the straight angle QWR is composed of three angles, two of which measure 45°. Use this information to calculate the measure of angle XWZ.

$$\begin{aligned} m\angle QWR &= m\angle QWZ + m\angle XWZ + m\angle RWX \\ 180° &= 45° + m\angle XWZ + 45° \\ 180° &= 90° + m\angle XWZ \\ 180° - 90° &= m\angle XWZ \\ 90° &= m\angle XWZ \end{aligned}$$

By the same logic, angles WXY, XYZ, and WZY also measure 90°. Therefore, $WXYZ$ is, indeed, a square, because each of its sides is the same length and each of its angles measure 90°.

Note: Problems 8.30–8.31 refer to the diagram in Problem 8.30, in which the sides of square QRST have midpoints W, X, Y, and Z.

8.31 Let A and B be the midpoints of $\overline{WZ}$ and $\overline{WX}$, respectively, and let s represent the side length of square $QRST$. Express AB in terms of s.

Begin by constructing a diagram that contains points A and B.

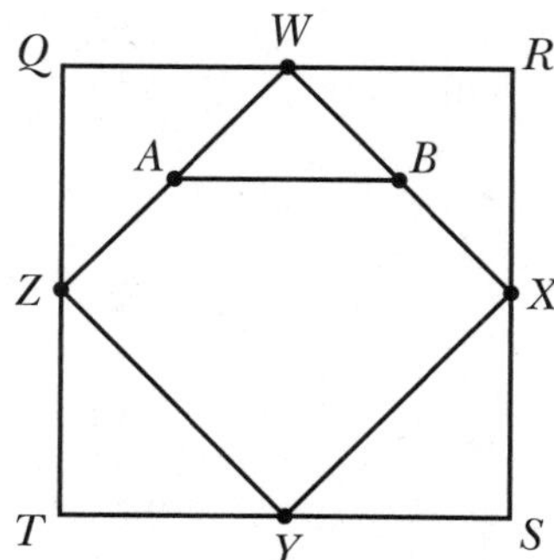

If the length of a side of $QRST$ is s, then $QZ = QW = s/2$, because Z and W are the midpoints of sides of the square and divide those sides in half. As Problem 8.30 explains, triangle QWZ is a 45°–45°–90° triangle, so the length of its hypotenuse is $\sqrt{2}$ times the length of a leg.

$$\begin{aligned} WZ &= \sqrt{2}(QZ) \\ &= \sqrt{2}\left(\frac{s}{2}\right) \\ &= \frac{\sqrt{2}}{2}s \end{aligned}$$

Point A divides $\overline{WZ}$ in half, so AW is half of WZ.

$$\begin{aligned} AW &= \frac{1}{2}(WZ) \\ &= \frac{1}{2}\left(\frac{\sqrt{2}}{2}s\right) \\ &= \frac{\sqrt{2}}{4}s \end{aligned}$$

By all of the same logic, side $\overline{BW}$ has the same length, so triangle AWB is isosceles. According to Problem 8.30, angle AWB is a right angle, so triangle AWB is an isosceles right triangle, another 45°–45°–90° triangle, whose leg length you have already calculated. Thus, hypotenuse $\overline{AB}$ of right triangle AWB is $\sqrt{2}$ times as long as a leg.

$$\begin{aligned} AB &= \sqrt{2}(AW) \\ &= \sqrt{2}\left(\frac{\sqrt{2}}{4}s\right) \\ &= \frac{\sqrt{4}}{4}s \\ &= \frac{2}{4}s \\ &= \frac{1}{2}s \end{aligned}$$

8.32 Given the diagram below, in which parallelogram $ABCD$ has diagonal length $BD = 12$, calculate the value of z.

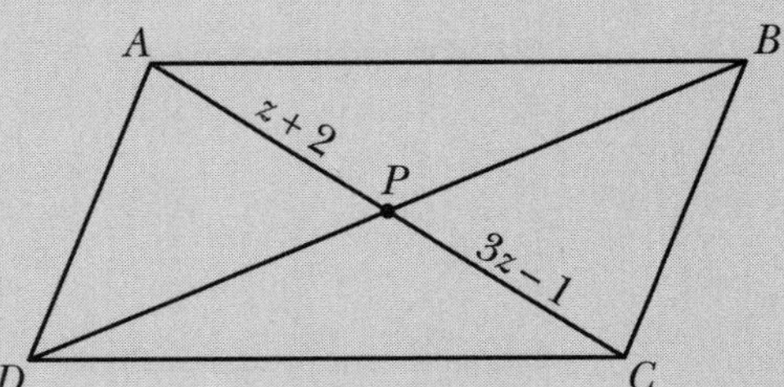

Note: Figure not drawn to scale.

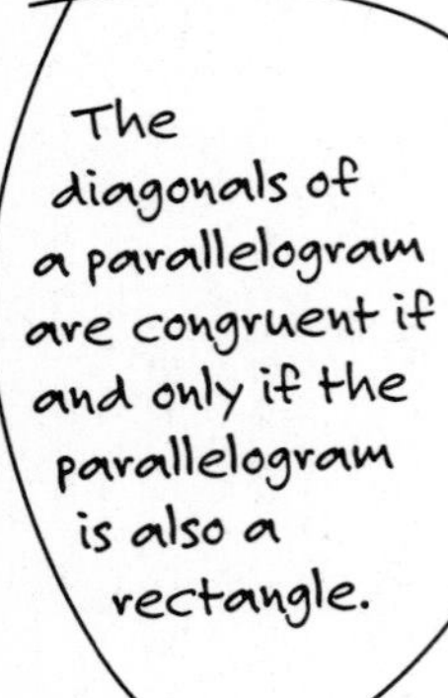

The diagonals of a parallelogram are congruent if and only if the parallelogram is also a rectangle.

The diagonals of a parallelogram bisect each other. In other words, they divide each other into two segments of equal length. However, the diagonals of a parallelogram are *not necessarily* equal to each other. You are told that $BD = 12$, but that does not mean $AC = 12$. In fact, that information is irrelevant and will not affect how you solve the problem.

Because the diagonals of a parallelogram bisect each other, you know that $AP = CP$. Set the lengths of the segments equal to each other and solve for z.

$$\begin{aligned} z+2 &= 3z-1 \\ 2+1 &= 3z-z \\ 3 &= 2z \\ \frac{3}{2} &= z \end{aligned}$$

8.33 Given the parallelogram in the diagram below, calculate the values of a and b.

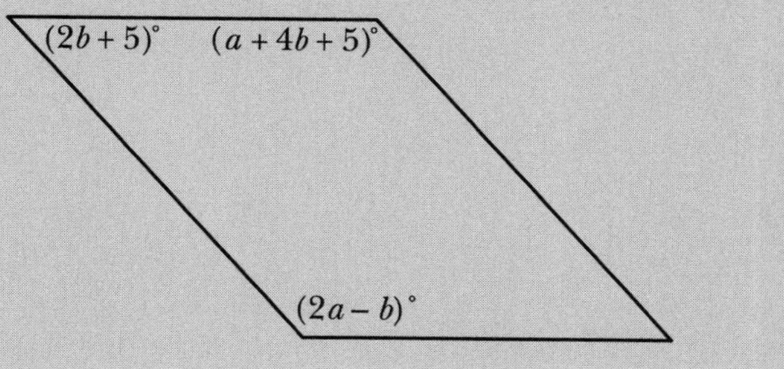

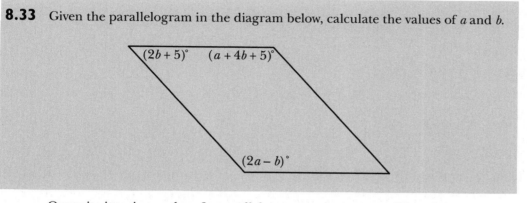

Opposite interior angles of a parallelogram are congruent. Therefore, $2a - b = a + 4b + 5$. Combine like terms to simplify the equation.

$$\begin{aligned} 2a-a-b-4b &= 5 \\ a-5b &= 5 \end{aligned}$$

Adjacent interior angles of a parallelogram are supplementary, so $(2a - b) + (2b + 5) = 180$. Once again, combine like terms to simplify the equation.

Or $(2b + 5) + (a + 4b + 5) = 180$

$$\begin{aligned} 2a+2b-b+5 &= 180 \\ 2a+b &= 180-5 \\ 2a+b &= 175 \end{aligned}$$

You now have a system of two equations containing two variables.

$$\begin{cases} a-5b=5 \\ 2a+b=175 \end{cases}$$

Solving the first equation for a allows you to apply the substitution method to solve the system. Substitute $a = 5b + 5$ into the second equation, $2a + b = 175$.

$$\begin{aligned} 2(5b+5)+b &= 175 \\ 10b+10+b &= 175 \\ 11b &= 175-10 \\ b &= \frac{165}{11} \\ b &= 15 \end{aligned}$$

Substitute $b = 15$ into the equation you solved for a to calculate the corresponding value of a.

$$\begin{aligned} a &= 5b+5 \\ &= 5(15)+5 \\ &= 75+5 \\ &= 80 \end{aligned}$$

You conclude that $a = 80$ and $b = 15$.

8.34 Given the diagram below, in which $\overline{AD} \parallel \overline{XY} \parallel \overline{BC}$ and X and Y are the midpoints of the sides on which they lie, express XY in terms of s.

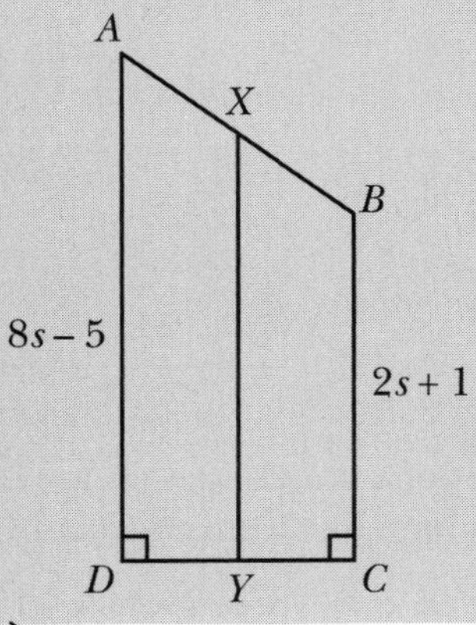

Quadrilateral $ABCD$ is a trapezoid, because exactly one pair of opposite sides are parallel. Those parallel sides are called the bases; the non-parallel sides are called the legs. A segment, like $\overline{XY}$, that connects the midpoints of the legs of a trapezoid is called the median. It is not only parallel to the bases of that trapezoid, its length is the average of the lengths of the bases. In other words, to calculate XY, the length of the median, add the lengths of the bases ($2s + 1$ and $8s - 5$) and divide by 2.

Parallelograms have two sets of opposite parallel sides. Trapezoids have only one set of opposite parallel sides.

$$\begin{aligned} XY &= \frac{(2s+1)+(8s-5)}{2} \\ &= \frac{10s-4}{2} \\ &= \frac{10s}{2} - \frac{4}{2} \\ &= 5s-2 \end{aligned}$$

8.35 What is the sum of the interior angles of a hexagon?

The sum of the interior angles of any polygon is equal to $180(n - 2)$, where n is the number of sides. A hexagon has $n = 6$ sides, so substitute this value into the formula.

$$\begin{aligned} \text{sum of interior angles} &= 180(6-2) \\ &= 180(4) \\ &= 720 \end{aligned}$$

The sum of the interior angles of a hexagon is 720°.

8.36 Given the diagram below, in which side $\overline{CD}$ of regular pentagon *ABCDE* lies on line *l*, calculate the value of *x*.

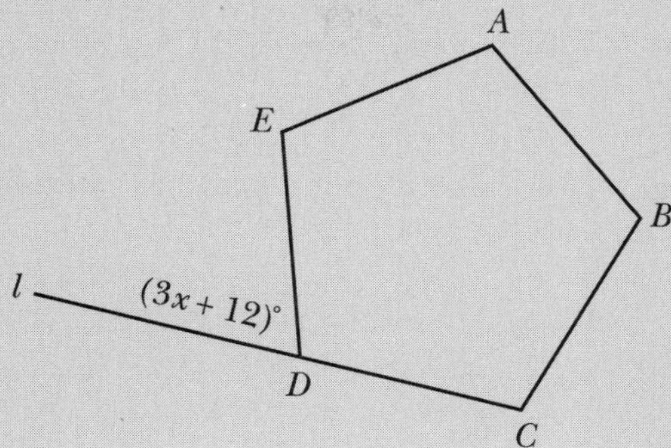

A regular polygon has congruent sides and congruent interior angles. A square, for example, is the only regular, four-sided polygon. You are given the measure of an exterior angle of a regular pentagon: $(3x + 12)°$. The exterior angle is supplementary to the interior angle of the polygon; together they form a straight line.

According to Problem 8.35, the sum of the interior angles of an n-sided polygon is $180(n - 2)$. Calculate the sum of the interior angles of a pentagon, which has $n = 5$ sides.

$$\begin{aligned}180(n-2) &= 180(5-2)\\ &= 180(3)\\ &= 540\end{aligned}$$

Because all five of the angles are congruent in a regular pentagon, you can divide the sum by 5 to calculate the measure of each interior angle: $540° \div 5 = 108°$. If each interior angle of the pentagon measures 108°, then each exterior angle measures 72°, the supplement of 108°. Set $3x + 12$ equal to 72 and solve for x.

$$\begin{aligned}3x+12 &= 72\\ 3x &= 72-12\\ 3x &= 60\\ x &= \frac{60}{3}\\ x &= 20\end{aligned}$$

8.37 A regular, n-sided polygon has exterior angles that each measure 30°. What is the value of n?

As Problem 8.36 explains, the measure of one interior angle of a regular polygon is equal to $180(n - 2) \div n$ degrees. The measure of each exterior angle of any n-sided regular polygon is equal to $360 \div n$ degrees. In this case, you are given the measure of an exterior angle and are asked to calculate n. Apply the exterior angle formula and cross-multiply to solve the ensuing proportion.

The sum of the interior angles $180(n - 2)$ divided by the number of angles, n.

You could have solved Problem 8.36 by calculating the measure of an exterior angle of a regular pentagon: $360° \div 5 = 72°$. Then you could set 72° equal to the angle measure you're given, $3x + 12$.

$$\frac{360}{n} = 30$$
$$30n = 360$$
$$n = \frac{360}{30}$$
$$n = 12$$

A 12-sided polygon (also called a dodecagon) has exterior angles that measure 30°.

Circles

Figures with no sides!

8.38 If the radius of a circle C is $x - 2$ and the diameter of the circle is $x + 1$, what is the value of x?

The radius of a circle is the distance between the center and any point on the circle. The diameter of a circle is the length of a segment that passes through the center whose endpoints lie on the circle. The diameter of a circle is always twice the radius, because a diameter is composed of two radii.

$$\text{diameter} = 2(\text{radius})$$

Substitute the given values for the radius and diameter into this equation.

$$x + 1 = 2(x - 2)$$
$$x + 1 = 2x - 4$$
$$1 + 4 = 2x - x$$
$$5 = x$$

8.39 Let x represent the radius of circle O in the diagram below. If the radius of circle A is $\frac{4}{3}$ the radius of circle C, what is the diameter of circle B in terms of x?

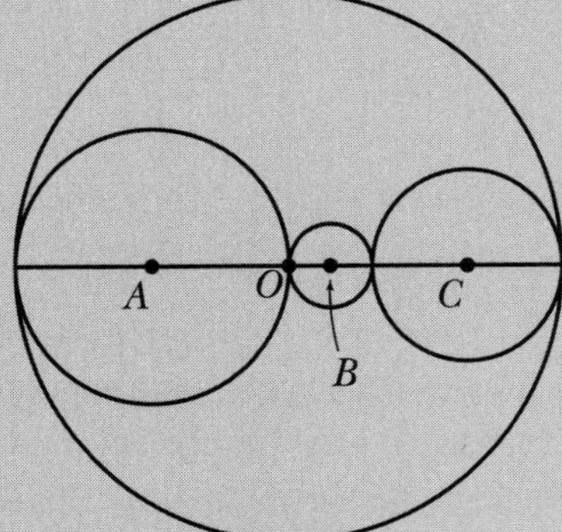

The large circle in the diagram has center O and radius x. Note that the radius of circle O is equal to the diameter of circle A. Therefore, the radius of circle A is $x/2$. According to the problem, the radius of circle A is $4/3$ times the radius of circle C. Calculate the radius of circle C in terms of x.

$$\text{radius of circle } A = \frac{4}{3}(\text{radius of circle } C)$$
$$\frac{x}{2} = \frac{4}{3}(\text{radius of circle } C)$$
$$\left(\frac{3}{4}\right)\left(\frac{x}{2}\right) = \text{radius of circle } C$$
$$\frac{3x}{8} = \text{radius of circle } C$$

Your goal is to calculate the diameter of circle *B*. Note that its length is equal to the diameter of circle *O* minus the diameters of circles *A* and *C*.

$$\begin{aligned}\text{diameter of } B &= \text{diameter of } O - \text{diameter of } A - \text{diameter of } C\\ &= 2(x) - 2\left(\frac{x}{2}\right) - 2\left(\frac{3x}{8}\right)\\ &= 2x - x - \frac{6x}{8}\\ &= x - \frac{3x}{4}\\ &= \frac{4x}{4} - \frac{3x}{4}\\ &= \frac{x}{4}\end{aligned}$$

The diameter of each circle is equal to the radius of each circle multiplied by 2.

8.40 If the measure of angle *AXB* is less than 180° in the diagram below, what can you conclude about $\overline{AX}$, $\overline{BX}$, the numbered angles, and the measure of $\overset{\frown}{AB}$?

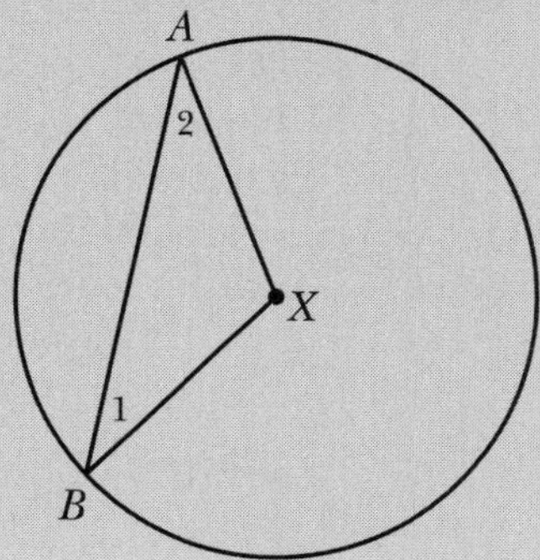

$\overline{AX}$ and $\overline{BX}$ are radii of the circle, so they have the same length. That makes triangle *AXB* isosceles, so the base angles—numbered angles 1 and 2 opposite the congruent sides of the triangle—are congruent.

On the SAT, arcs are measured in degrees, based on their central angles. The central angle of $\overset{\frown}{AB}$ is angle *AXB*, because its vertex *X* lies on the center of the circle and the sides of the angles pass through the endpoints of the arc, *A* and *B*. Mathematically speaking, central angle *AXB* "subtends" the arc. The measure of $\overset{\frown}{AB}$ is equal to the measure of its central angle *AXB*.

When an angle subtends an arc, its sides intersect the circle at the endpoints of the arc. In this case, the sides of angle AXB intersect circle X at points A and B, the endpoints of $\overset{\frown}{AB}$.

Note: Problems 8.41–8.42 refer to the diagram below, in which $\overline{WY}$ and $\overline{XZ}$ are diameters of circle O and m$\overset{\frown}{YZ}$ = 100°.

8.41 Determine whether the following statement is true and explain your answer: $\overleftrightarrow{WZ} \parallel \overleftrightarrow{XY}$.

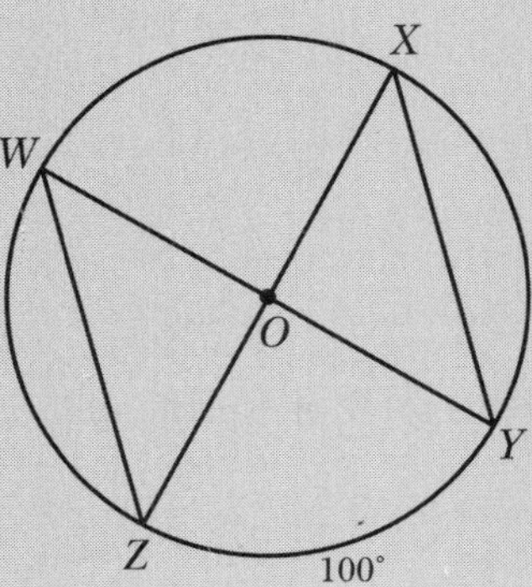

Because $\overline{OW}$, $\overline{OX}$, $\overline{OY}$, and $\overline{OZ}$ are radii of the circle, the segments are all the same length. Note that angles *WOZ* and *XOY* are vertical angles, so they have the same measures. Therefore, triangles *WOZ* and *XOY* are congruent according to the side-angle-side technique. Furthermore, triangles *WOZ* and *XOY* are isosceles triangles, so all four base angles are congruent: angles *W*, *X*, *Y*, and *Z*.

Notice that angles *W* and *Y* are alternate-interior angles formed when lines $\overleftrightarrow{WZ}$ and $\overleftrightarrow{XY}$ are intersected by transversal $\overleftrightarrow{WY}$. Because those alternate-interior angles are congruent, you can conclude that $\overleftrightarrow{WZ} \parallel \overleftrightarrow{XY}$.

See Problem 8.1.

Note: Problems 8.41–8.42 refer to the diagram in Problem 8.41, in which $\overline{WY}$ and $\overline{XZ}$ are diameters of circle O and m$\overset{\frown}{YZ}$ = 100°.

8.42 Calculate the measure of angle *WZX*.

The central angle of $\overset{\frown}{YZ}$ is *YOZ*. Therefore, the measures are equal. Note that angles *YOZ* and *WOX* are vertical angles, so they are congruent. Because angle *WOX* is the central angle of $\overset{\frown}{WX}$, their measures are also equal. Thus, the measures of these angles and the arcs they subtend are all 100°, as illustrated below.

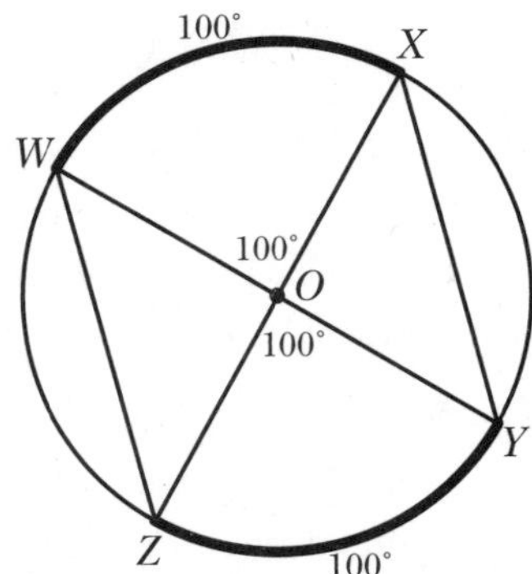

Angle *WOX* is not the only angle that subtends $\overset{\frown}{WX}$. Notice that angle *WZX* also subtends the arc, because its sides intersect the circle at *W* and *X*, the endpoints of the arc. Unlike a central angle, which has a vertex located at the center of a circle, angle *WZX* has a vertex that lies on the circle itself, so it is called an inscribed angle.

The measures of inscribed angles are half the measures of the arcs they subtend. In this problem, inscribed angle *WZX* subtends an arc with measure 100°, so the measure of angle *WZX* is 50°.

8.43 Let *ABCDEFGH* be a regular octagon inscribed in a circle *O*. Calculate the measure of obtuse angle *COH*.

Construct a diagram of an octagon—an eight-sided regular polygon—inscribed in a circle. Each vertex of the octagon lies on the circle. Imagine that the circle is divided into eight congruent arcs, segmented by the vertices of the octagon. A circle contains a total of 360°, so each of those arcs (and their corresponding central angles) has a measure of $360° \div 8 = 45°$.

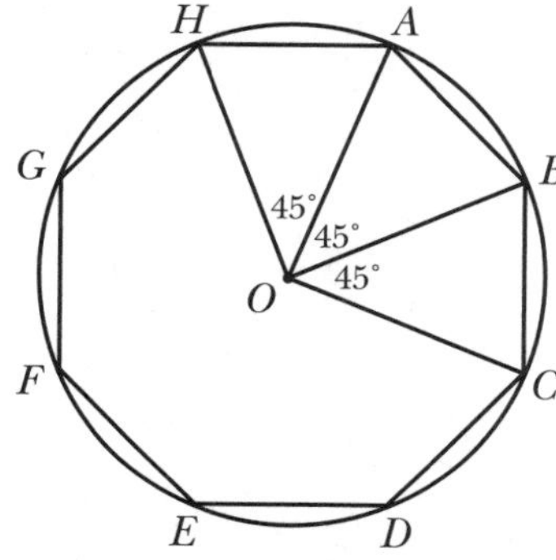

Angle *COH* is composed of three angles (*AOH*, *AOB*, and *BOC*) that each measure 45°, so the measure of angle *COH* is $3(45°) = 135°$.

8.44 Given the diagram below, in which circle *C* has radius *r*, the measure of angle *MLO* is 60°, and $LM = LO$, express the length *LO* in terms of *r*.

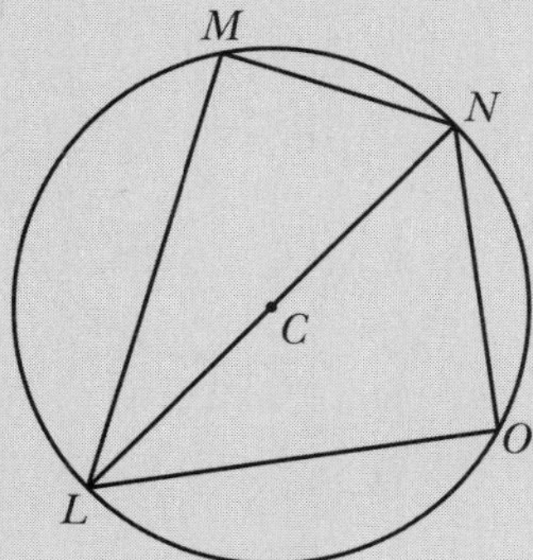

Note: Figure not drawn to scale.

Notice that angles *LMN* and *LON* subtend semicircles $\widehat{LON}$ and $\widehat{LMN}$, respectively, because they intersect the circle at the endpoints of a diameter. Semicircles represent one-half of a circle, so they measure 180°. The vertices of angles *LMN* and *LON* lie on circle *C*, so they are inscribed angles. As Problem 8.43 explains, the measures of angles *LMN* and *LON* are half the measures of the arcs they subtend. Thus, the angles both measure $180° \div 2 = 90°$.

When an arc measures 180° or more, you use three letters to represent it. The first and last letters are endpoints and the middle letter is a point that lies on the arc somewhere between those endpoints. For example, in this diagram, major arc $\widehat{MLN}$ (called a major arc because it measures more than 180°) is the entire circle C except for the minor arc $\widehat{MN}$.

You now know that triangles *LMN* and *LON* are right triangles, and the problem tells you that the triangles have a pair of corresponding, congruent legs: $LM = LO$. The triangles share the same hypotenuse, $\overline{LN}$. Thus, the right triangles are congruent according to the hypotenuse-leg technique, and all corresponding parts of both triangles are congruent. This includes corresponding angles *MLN* and *OLN*. Together, the angles measure 60°, so that means they must each measure 30°.

If $m\angle OLN = 30°$ and $m\angle LON = 90°$, then $m\angle LNO = 60°$, because the sum of the interior angles of any triangle is 180°. Thus, triangle *LON* is a 30°–60°–90° triangle. As Problem 8.24 explains, the length of short leg $\overline{NO}$ is half the length of the hypotenuse $\overline{LN}$, which is also a diameter of the circle. Because the radius of circle *C* is *r*, the diameter of the circle is $LN = 2r$.

$$\begin{aligned} NO &= \frac{1}{2}(LN) \\ &= \frac{1}{2}(2r) \\ &= r \end{aligned}$$

The length of the long leg $\overline{LO}$ is $\sqrt{3}$ times the length of the short leg.

$$\begin{aligned} LO &= \sqrt{3}(NO) \\ &= \sqrt{3}(r) \\ &= r\sqrt{3} \end{aligned}$$

8.45 In the diagram below, $\overleftrightarrow{AC}$ is tangent to circle *O* at point *A* and $\overleftrightarrow{BC}$ is tangent to circle *O* at point *B*. Can you conclude that *AOBC* is a square? Why or why not?

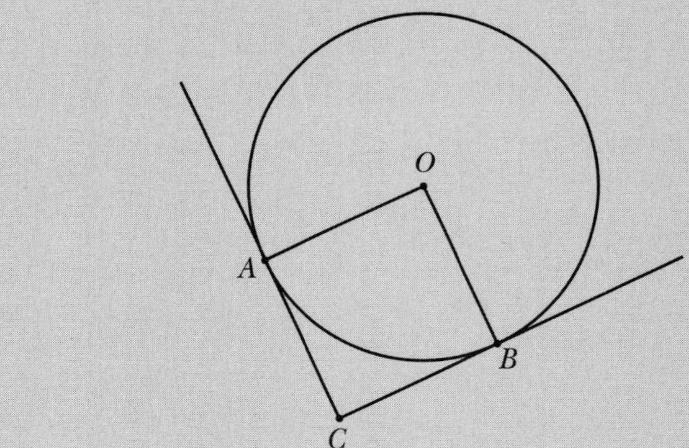

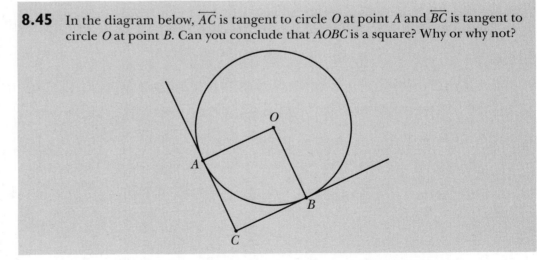

The radius connecting the center of a circle to a point of tangency is perpendicular to that tangent line. In this diagram, $\overline{OA} \perp \overleftrightarrow{AC}$ and $\overline{OB} \perp \overleftrightarrow{BC}$. Therefore, angles *A* and *B* of quadrilateral *AOBC* are right angles. You also know that two sides of the quadrilateral are congruent: $\overline{AO} \cong \overline{BO}$ because they are two radii of the same circle. Unfortunately, that is not enough information to ensure that *AOBC* is a square.

Consider the following diagram, in which the radii are still perpendicular to the lines of tangency but points A and B are closer together on the circle. Angles OAC and OBC are still right angles, and the two sides that are radii of the circle are still congruent. Furthermore, all of the restrictions described in the original problem are also true here. Clearly, $AOBC$ is not a square.

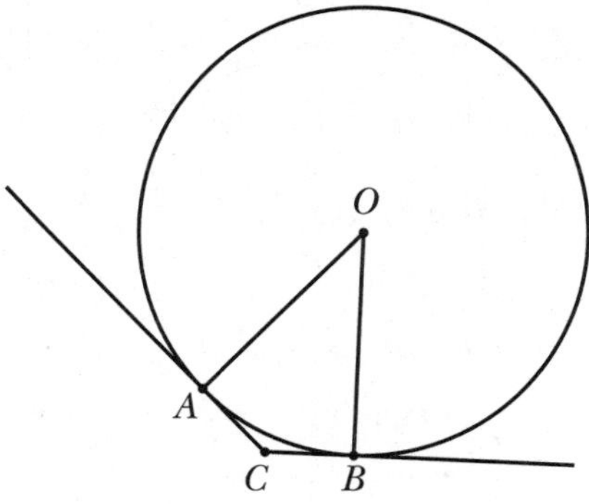

Just because $AOBC$ looks like a square in the original diagram, you cannot assume that it is. You cannot prove that sides $\overline{AC}$ and $\overline{BC}$ are congruent to $\overline{AO}$ and $\overline{BO}$.

The lesson here is not to make too many assumptions based on a diagram alone.

Area and Perimeter

(For two-dimensional figures)

Note: Problems 8.46–8.47 refer to a rectangle ABCD in which AB = 13 and AD = 4.

8.46 Calculate the area of $ABCD$.

The area of a rectangle is the product of its length and width ($a = l \cdot w$). Typically, the length of a rectangle is defined as the longer of its two side lengths (unless the side lengths of the rectangle are all equal, in which case the rectangle is also a square). Therefore, the length of rectangle $ABCD$ is $AB = 13$ and the width is $AD = 4$, as illustrated below.

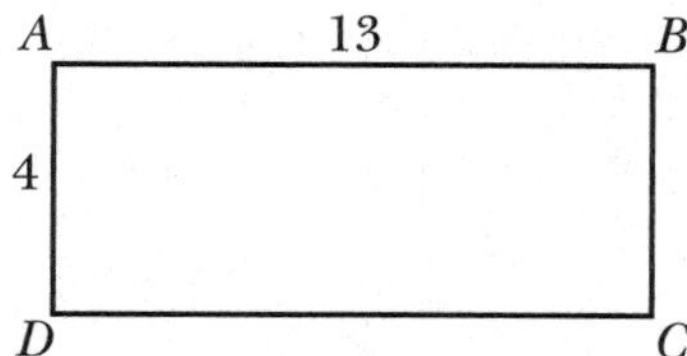

Multiply the length and width to calculate the area of $ABCD$.

$$\begin{aligned}\text{area}\left(\text{rectangle}\right) &= l \cdot w \\ \text{area}\left(\text{rectangle } ABCD\right) &= (13)(4) \\ \text{area}\left(\text{rectangle } ABCD\right) &= 52\end{aligned}$$

Note: Problems 8.46–8.47 refer to a rectangle ABCD in which AB = 13 and AD = 4.

8.47 Calculate the perimeter of *ABCD*.

The perimeter of a figure is the sum of the lengths of its sides. Note that rectangle *ABCD*, like all parallelograms, has congruent opposite sides. Therefore, $AB = CD = 13$ and $AD = BC = 4$. As Problem 8.46 explains, the length of the rectangle is $l = 13$ and the width is $w = 4$. Most textbooks use the following formula for the perimeter of a rectangle, the sum of twice the length and twice the width of the rectangle.

Because there are four sides of a rectangle, two that are L long and two that are W long.

$$\text{perimeter}(\text{rectangle}) = 2l + 2w$$
$$\text{perimeter}(ABCD) = 2(13) + 2(4)$$
$$\text{perimeter}(ABCD) = 26 + 8$$
$$\text{perimeter}(ABCD) = 34$$

Note: Problems 8.48–8.49 refer to a rhombus with side length 6 and altitude 5.

8.48 Calculate the perimeter of the rhombus.

A rhombus is a parallelogram with four congruent sides. All four sides of this rhombus have length 6, so multiply that length by 4 to calculate the perimeter of the figure.

$$\begin{aligned}\text{perimeter}(\text{rhombus}) &= 4(\text{side length})\\ &= 4(6)\\ &= 24\end{aligned}$$

You do not need to know the altitude of a rhombus to calculate its perimeter.

Note: Problems 8.48–8.49 refer to a rhombus with side length 6 and altitude 5.

8.49 Calculate the area of the rhombus.

The altitude of a rhombus is a line segment connecting two opposite sides that is perpendicular to both of the sides. As its name suggests, an altitude describes the "height" of a rhombus. Construct a diagram of a rhombus that has side length 6 and an altitude of 5.

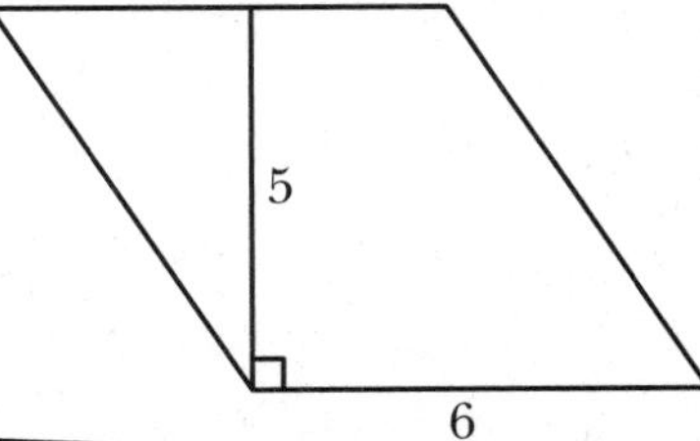

A parallelogram is a quadrilateral whose opposite sides are parallel. Rectangles, squares, and rhombi are all parallelograms but trapezoids are not.

The area of a rhombus, and indeed the area of any parallelogram, is equal to the base b (the length of one side of the parallelogram) multiplied by the height

h (the length of the altitude perpendicular to that side). In this problem, an altitude of length 5 is perpendicular to a side of length 6.

$$\begin{aligned}\text{area}(\text{parallelogram}) &= b \cdot h \\ &= (6)(5) \\ &= 30\end{aligned}$$

8.50 Given a square *WXYZ* with side length 6, calculate *WY*.

This problem asks you to calculate the length of the diagonal of a square, a segment connecting non-adjacent vertices of the figure, illustrated below. Remember that all four interior angles of a square measure 90°.

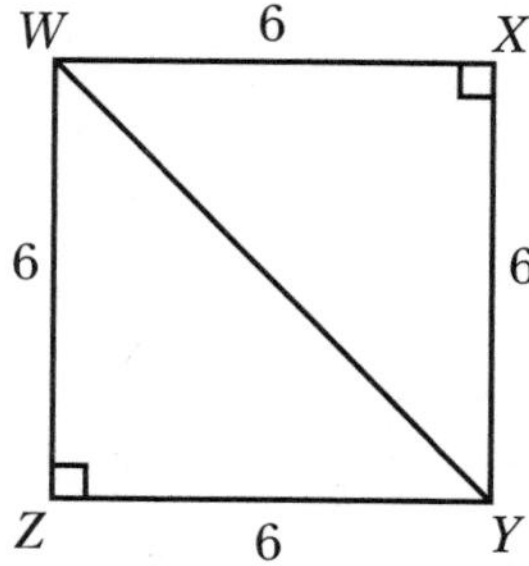

The diagonal divides the square into two congruent isosceles right triangles, *WXY* and *WZY*. Therefore, *WXY* and *WZY* are 45°–45°–90° right triangles, and the length of their shared hypotenuse $\overline{WY}$ is $\sqrt{2}$ times the length of a leg—a side of the square. Therefore, $WY = 6\sqrt{2}$.

Note: Problems 8.51–8.52 refer to a right triangle with hypotenuse length 13 and a leg with length 5.

8.51 Calculate the perimeter of the triangle.

You are given the lengths of two sides of the right triangle, illustrated below.

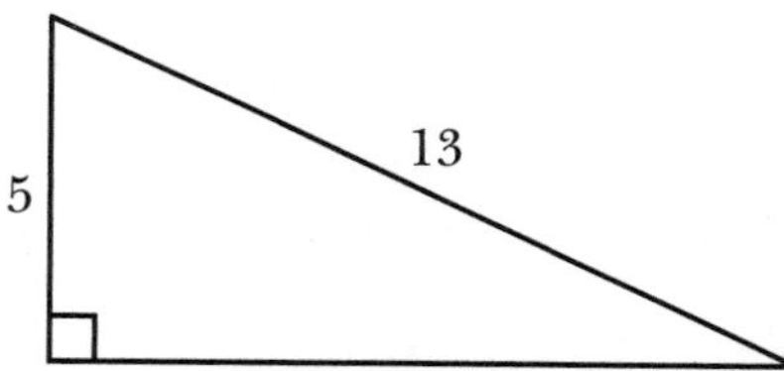

Notice that the lengths of the sides are part of the Pythagorean triple 5, 12, 13. Therefore, the length of the remaining side is 12. (If you did not recognize the Pythagorean triple, you can apply the Pythagorean theorem to calculate the length instead.) The perimeter of the triangle is the sum of the lengths of its three sides.

$$\text{perimeter} = 5 + 12 + 13 = 30$$

Note: Problems 8.51–8.52 refer to a right triangle with hypotenuse length 13 and a leg with length 5.

8.52 Calculate the area of the triangle.

To calculate the area of a triangle, you apply a formula very similar to the parallelogram area formula presented in Problem 8.49. In both area formulas, you multiply the length of one side b (the base) by the length of an altitude h (the height). Any side of the figure can serve as the base, but the altitude must be perpendicular to that base. One notable difference: In the triangle area formula, you then multiply that product by 1/2.

area of a parallelogram: $a = bh$ area of a triangle: $a = \frac{1}{2}bh$

Consider the diagram below, the right triangle described in this problem including the length of the leg calculated in Problem 8.51.

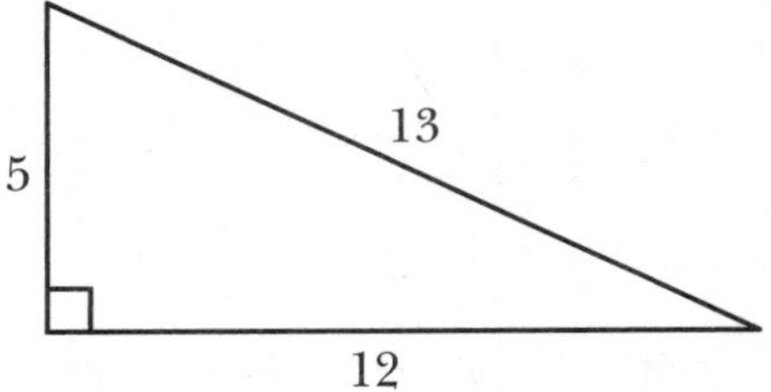

The legs of right triangles are perpendicular to each other. Therefore, one leg can serve as the base and the other can serve as the altitude. Thus, $b = 12$ and $h = 5$ (or vice versa).

Altitudes are usually not sides of a triangle. This is only true for right triangles.

$$\begin{aligned} \text{area} &= \frac{1}{2}bh \\ &= \frac{1}{2}(12)(5) \\ &= \frac{12}{2}(5) \\ &= 6(5) \\ &= 30 \end{aligned}$$

Note: Problems 8.53–8.54 refer to the diagram below.

8.53 Calculate the perimeter of triangle BDC.

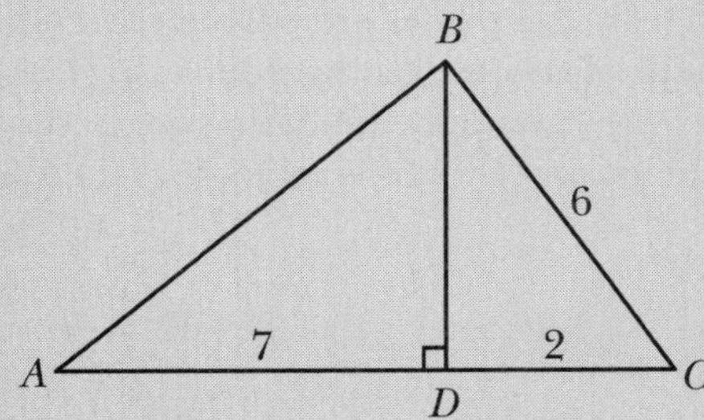

Apply the Pythagorean theorem to calculate length BD in right triangle BDC.

$$
\begin{aligned}
(BD)^2 + (CD)^2 &= (BC)^2 \\
(BD)^2 + 2^2 &= 6^2 \\
(BD)^2 + 4 &= 36 \\
(BD)^2 &= 32 \\
\sqrt{(BD)^2} &= \sqrt{32} \\
BD &= \sqrt{16 \cdot 2} \\
BD &= 4\sqrt{2}
\end{aligned}
$$

Now add the lengths of the sides of triangle BDC to calculate its perimeter.

$$
\begin{aligned}
\text{perimeter}(\text{triangle } BDC) &= BD + CD + BC \\
&= 4\sqrt{2} + 2 + 6 \\
&= 8 + 4\sqrt{2}
\end{aligned}
$$

Note: Problems 8.53–8.54 refer to the diagram in Problem 8.53.

8.54 Calculate the area of triangle ABC.

Let $\overline{AC}$ represent the base of triangle ABC and $\overline{BD}$ represent an altitude perpendicular to that base. The length of the base AC is $AD + DC = 9$. According to Problem 8.53, the height of triangle ABC is $h = BD = 4\sqrt{2}$. Substitute h and b into the triangle area formula.

$$
\begin{aligned}
\text{area}(\text{triangle } ABC) &= \frac{1}{2}bh \\
&= \frac{1}{2}(9)\left(4\sqrt{2}\right) \\
&= \frac{36\sqrt{2}}{2} \\
&= 18\sqrt{2}
\end{aligned}
$$

Note: Problems 8.55–8.56 refer to the region illustrated below, in which all of the interior angles are right angles.

8.55 Calculate the perimeter of the region.

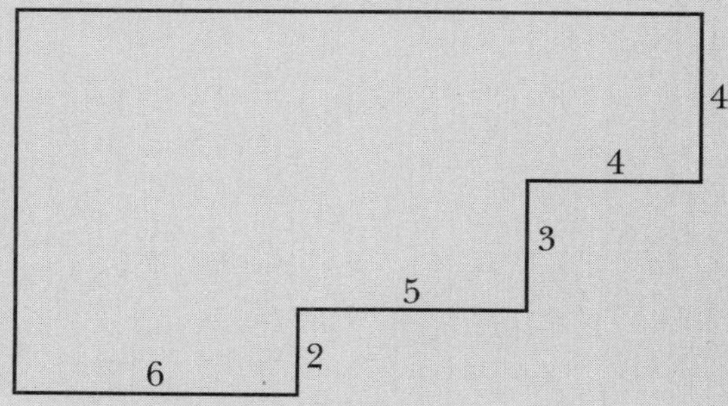

Divide the region into rectangles in order to calculate the lengths of the top and left sides of the figure, as illustrated below.

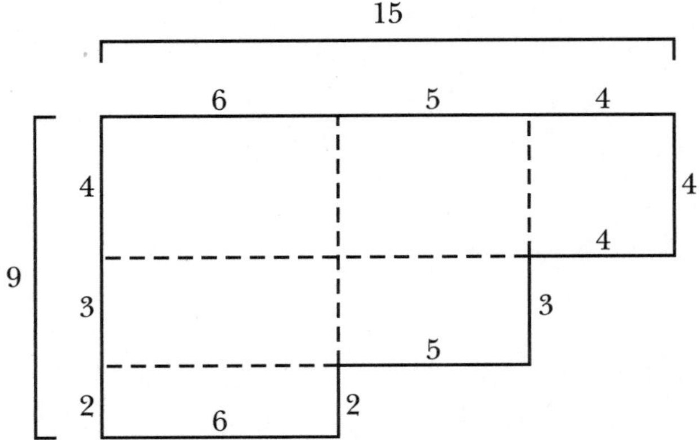

Add the lengths of the sides of the figure—the solid segments in the diagram, not the dotted internal segments. The sum below begins with the length of the left side of the figure and proceeds counterclockwise.

$$\begin{aligned} \text{perimeter} &= 9+6+2+5+3+4+4+15 \\ &= 48 \end{aligned}$$

Note: Problems 8.55–8.56 refer to the region illustrated in Problem 8.55, in which all of the interior angles are right angles.

8.56 Calculate the area of the region.

Divide the region into rectangles and then add the areas of those rectangles. In the diagram below, the region is divided into three rectangles: *A*, *B*, and *C*.

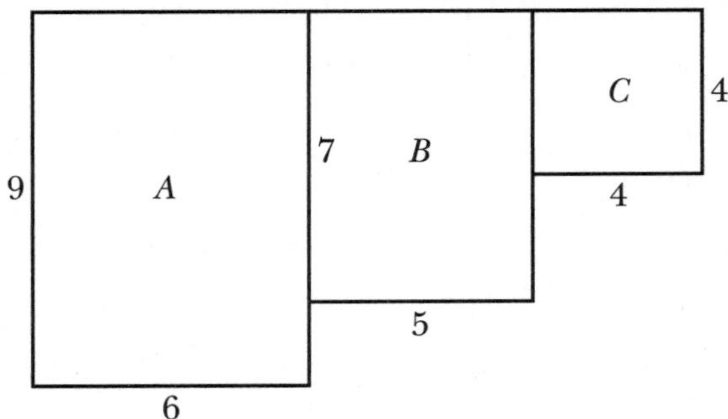

$$\begin{aligned} \text{area} &= \text{area}(\text{rectangle } A) + \text{area}(\text{rectangle } B) + \text{area}(\text{rectangle } C) \\ &= (9 \cdot 6) + (7 \cdot 5) + (4 \cdot 4) \\ &= 54 + 35 + 16 \\ &= 105 \end{aligned}$$

8.57 Calculate the area and circumference of a circle with radius 4.

The circumference of a circle is the length of the entire arc of the circle; it is, essentially, the perimeter of the circle, but the word perimeter is usually applied to polygons. Both the area and circumference formulas, presented below, require only the radius r of the circle. In this problem, $r = 4$.

$$\begin{aligned} \text{area} &= \pi r^2 \\ &= \pi(4)^2 \\ &= 16\pi \end{aligned} \qquad \begin{aligned} \text{circumference} &= 2\pi r \\ &= 2\pi(4) \\ &= 8\pi \end{aligned}$$

8.58 In the diagram below, square *KLMN* has side length s and is inscribed in circle *O*. Express the area of the shaded region in terms of s.

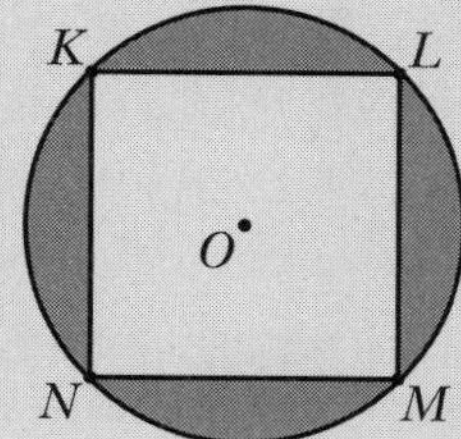

The area of the shaded region is the area of circle *O* minus the area of square *KLMN*. According to the problem, square *KLMN* has side length s, and the area of a square is equal to the square of its side length, s^2.

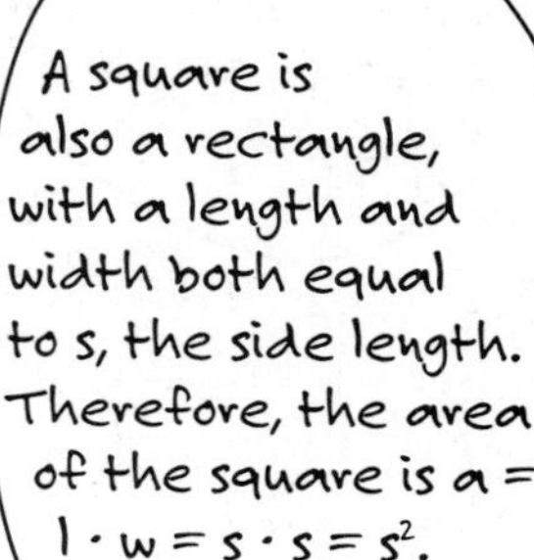

A square is also a rectangle, with a length and width both equal to s, the side length. Therefore, the area of the square is a = l · w = s · s = s².

To calculate the area of the circle, you need to identify its radius. Consider the diagram below, which uses one diagonal of the square to form triangle *LMN*. Because angle *LMN* is an inscribed angle that subtends a semicircle, it is also a right angle. Therefore, triangle *LMN* is an isosceles right triangle, a 45°–45°–90° triangle, whose hypotenuse is $\sqrt{2}$ times the length of a leg: $LN = s\sqrt{2}$.

See Problem 8.44 for more information.

$$\begin{aligned} LN &= \sqrt{2}(LM) \\ &= \sqrt{2}(s) \\ &= s\sqrt{2} \end{aligned}$$

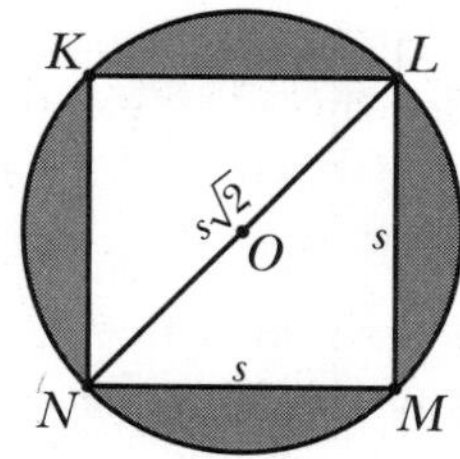

Note that $\overline{LN}$ is both the hypotenuse of triangle *LMN* and the diameter of circle *O*. Therefore, the radius r of circle *O* is half of *LN*.

$$\begin{aligned} r &= \frac{1}{2}(LN) \\ r &= \frac{1}{2}\left(s\sqrt{2}\right) \\ &= \frac{s\sqrt{2}}{2} \end{aligned}$$

Apply the area formula for a circle.

$$\begin{aligned}\text{area}(\text{circle } O) &= \pi r^2 \\ &= \pi\left(\frac{s\sqrt{2}}{2}\right)^2 \\ &= \pi\left(\frac{s^2 \cdot 2}{4}\right) \\ &= \pi\left(\frac{s^2}{2}\right) \\ &= \frac{\pi}{2}s^2\end{aligned}$$

Calculate the area of the shaded region.

$$\begin{aligned}\text{area}(\text{shaded region}) &= \text{area}(\text{circle } O) - \text{area}(\text{square } KLMN) \\ &= \frac{\pi}{2}s^2 - s^2 \\ &= s^2\left(\frac{\pi}{2} - 1\right)\end{aligned}$$

Both of these are acceptable answers. You don't have to factor s^2 out of both terms at the end.

8.59 Given the diagram below, a circle with radius 2 whose center is marked, calculate the total area of the shaded regions.

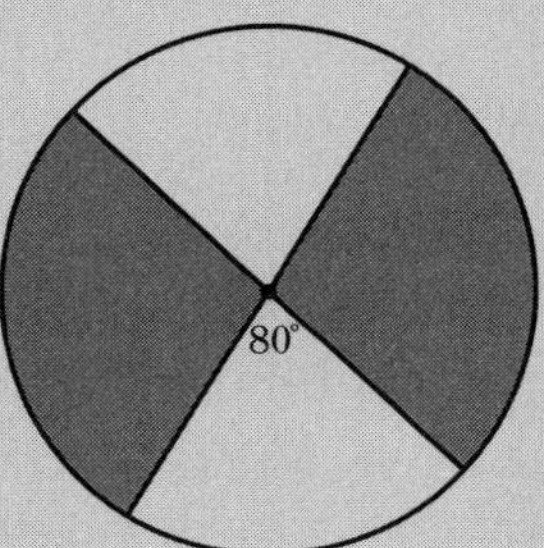

The shaded regions are two sectors of the circle. Sectors are portions of the circle bounded by radii; they resemble pieces of pie. To calculate the area of a sector, multiply the area of the entire circle by $a/360$, where a is the measure of the sector's central angle.

Note that the two unshaded sectors of the circle have congruent, vertical angles for central angles that measure 80°. Together, the two unshaded sectors account for 2(80°) = 160° of the circle. Therefore, the shaded sectors account for 360° – 160° = 200° of the circle. Multiply the area of the entire circle by the fraction 200/360 (or 5/9) to determine how much of the circle's area corresponds to the shaded region.

$$\begin{aligned}\text{shaded area} &= \frac{5}{9}(\pi r^2) \\ &= \frac{5}{9}(\pi \cdot 2^2) \\ &= \frac{5}{9}(4\pi) \\ &= \frac{20\pi}{9}\end{aligned}$$

In other words, 5/9 of this circle is shaded, so the area of the shaded region is 5/9 of the circle's total area.

Note: Problems 8.60–8.61 refer to the diagram below, in which the diameter of the circle is 16.

8.60 Calculate the area of the shaded region.

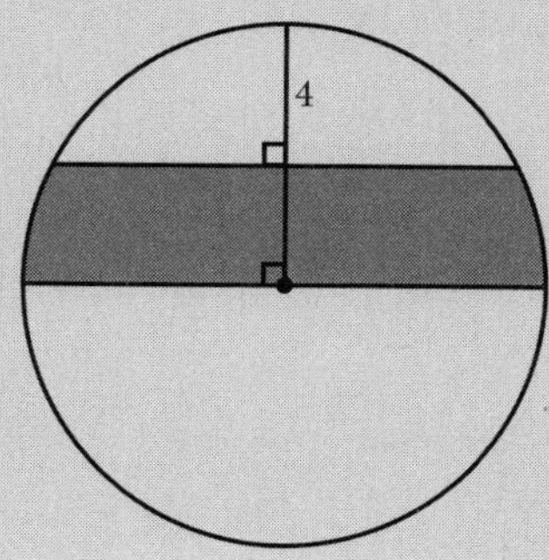

The diameter of the circle is 16, so its radius is 8. One radius is illustrated in the diagram. The top portion of that radius has length 4, so in order for the entire radius to have length 8, the bottom portion of the radius (lying in the shaded region) must also have length 4. The shaded region is illustrated in more detail below.

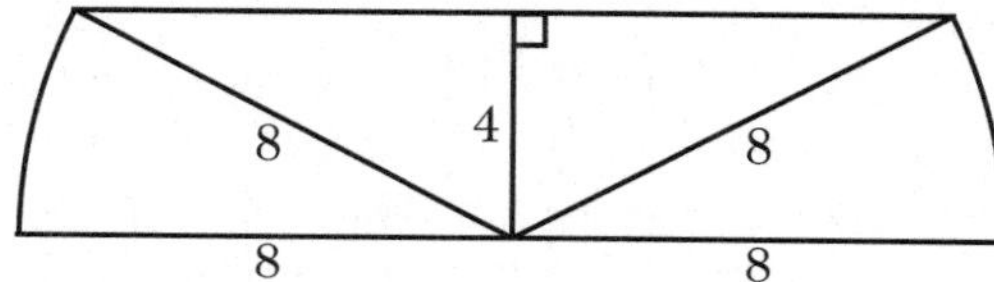

Notice that the shaded region has been divided into four smaller regions by radii drawn from the center of the circle to the upper corners of the shaded region. This produces two right triangles and two sectors. All four of the radii have length 8.

The right triangles in the figure each have one leg whose length is half the length of their hypotenuses. Therefore, the right triangles are 30°–60°–90° triangles, as explained in Problem 8.24, and the lengths of the long legs of those triangles are $\sqrt{3}$ times the length of the short legs: $4\sqrt{3}$. In the following diagram, the angle measures of the right triangles are indicated, as is the length of the missing side. Note that the angles adjacent to the 60° angles must measure 30°, because the vertical segment is perpendicular to the diameter of the circle.

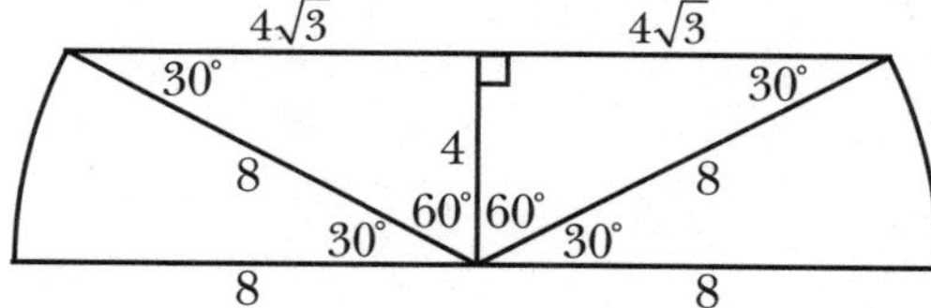

The shaded region is composed of two congruent right triangles and two congruent sectors. Calculate the area of each.

$$\text{area of one right triangle} = \frac{1}{2}bh = \frac{1}{2}(4\sqrt{3})(4) = (2\sqrt{3})(4) = 8\sqrt{3}$$

$$\text{area of one sector} = \frac{30}{360}(\pi r^2) = \frac{1}{12}(\pi \cdot 8^2) = \frac{1}{12}(64\pi) = \frac{16}{3}\pi$$

The area of the shaded region is equal to the sum of the areas of two right triangles and two sectors.

$$\begin{aligned}\text{area of shaded region} &= 2(\text{area of one right triangle}) + 2(\text{area of one sector})\\ &= 2(8\sqrt{3}) + 2\left(\frac{16}{3}\pi\right)\\ &= 16\sqrt{3} + \frac{32}{3}\pi\end{aligned}$$

Note: Problems 8.60–8.61 refer to the diagram in Problem 8.60, in which the diameter of the circle is 16.

8.61 Calculate the perimeter of the shaded region.

The perimeter is composed of two arcs and two parallel lines. One of the parallel lines is a diameter with length 16 and the other is formed by two legs of two right triangles, as explained in Problem 8.60. The length of that segment is $4\sqrt{3} + 4\sqrt{3} = 8\sqrt{3}$.

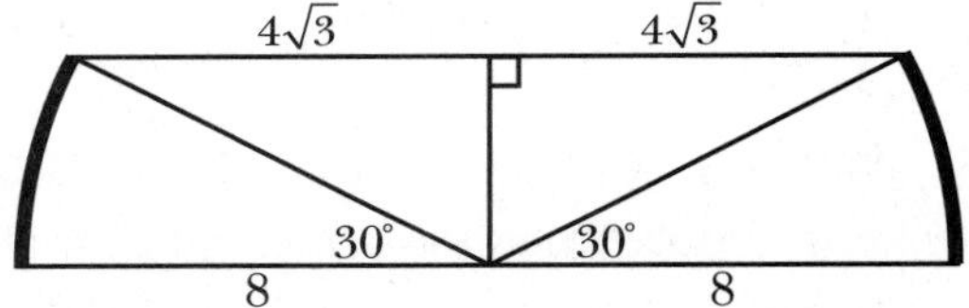

To calculate the lengths of the arcs, apply a method similar to the technique described in Problem 8.59. Rather than multiply the area of the circle by $a/360$, where a is the central angle of a sector, multiply the *circumference* of the circle by the same fraction. In other words, the two sectors in this circle, together, account for $30° + 30° = 60°$ of the $360°$ in the circle, or $60/360 = 1/6$. Therefore, the arcs of those sectors account for $1/6$ of the total circumference of the circle. Recall that the radius of the circle is $r = 8$, as the diameter is 16.

$$\begin{aligned}\text{arc length} &= \frac{1}{6}(2\pi r)\\ &= \frac{1}{6}(2\pi \cdot 8)\\ &= \frac{1}{6}(16\pi)\\ &= \frac{8}{3}\pi\end{aligned}$$

The total perimeter is $\frac{8}{3}\pi + 16 + 8\sqrt{3}$.

Solid Geometry

Volume and surface area

Note: Problems 8.62–8.63 refer to the right rectangular prism illustrated below.

8.62 Calculate the volume of the solid.

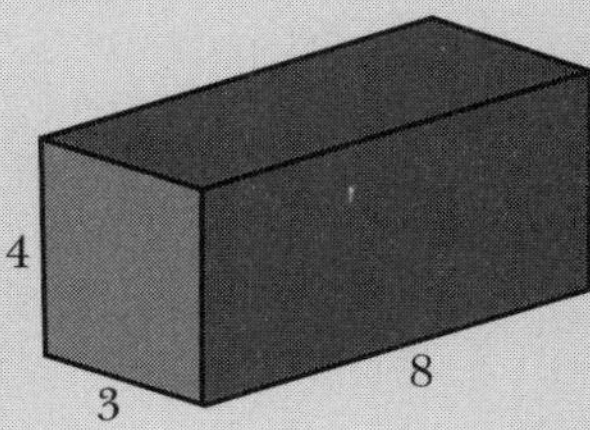

The volume of a box, which in mathematical language is called a right rectangular prism, is the product of its length, width, and height.

$$\begin{aligned} \text{volume} &= 4 \cdot 3 \cdot 8 \\ &= 96 \end{aligned}$$

Note: Problems 8.62–8.63 refer to the right rectangular prism illustrated in Problem 8.62.

8.63 Calculate the surface area of the solid.

The three-dimensional solid has six rectangular faces, the sides that form the "shell" of the solid. Two of those faces are 3×4 rectangles (each with area $3 \cdot 4 = 12$), and four of the faces are 4×8 rectangles (each with area $4 \cdot 8 = 32$), as illustrated below.

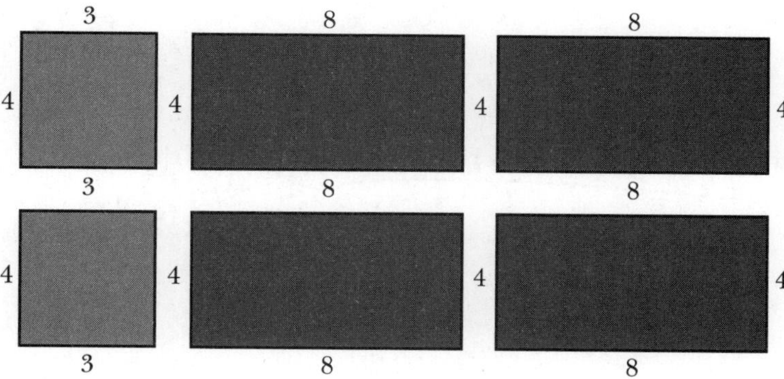

Add the areas of all six rectangles.

$$\begin{aligned} \text{total surface area} &= 2(\text{area of } 3 \times 4 \text{ rectangle}) + 4(\text{area of } 4 \times 8 \text{ rectangle}) \\ &= 2(12) + 4(32) \\ &= 152 \end{aligned}$$

Note: Problems 8.64–8.66 refer to the diagram below, a right prism with right triangle bases.

8.64 Calculate the area of a base of the solid.

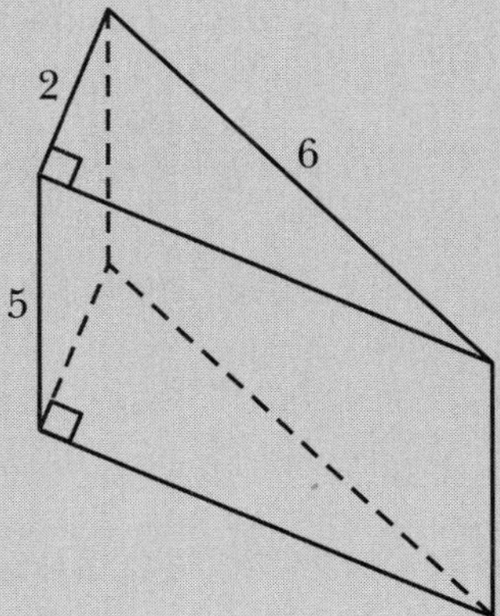

A right prism has two sides, or faces, called bases that are parallel to each other. In this diagram, the bases are congruent right triangles. The other sides of the prism, called the lateral faces, are rectangles that join corresponding sides of the congruent bases.

The base of this prism has a leg with length 2 and a hypotenuse of length 6. Apply the Pythagorean theorem to calculate the length x of the other leg in the right triangle.

$$\begin{aligned} 2^2 + x^2 &= 6^2 \\ 4 + x^2 &= 36 \\ x^2 &= 32 \\ x &= \sqrt{16 \cdot 2} \\ x &= 4\sqrt{2} \end{aligned}$$

Calculate the area A of the right triangular base of the prism.

This base b is part of the triangle area formula, so in this context "base" means "base of the triangle," not "base of the prism."

$$\begin{aligned} A &= \frac{1}{2}bh \\ &= \frac{1}{2}\left(4\sqrt{2}\right)(2) \\ &= \frac{2}{2}\left(4\sqrt{2}\right) \\ &= 4\sqrt{2} \end{aligned}$$

Note: Problems 8.64–8.66 refer to the diagram in Problem 8.64, a right prism with right triangle bases.

8.65 Calculate the volume of the solid.

The volume of a prism is equal to the area B of the base multiplied by the height h of the prism. The height of a right prism is the length of a lateral edge, one of the segments that connects, and is perpendicular to, the bases. The height of this right prism is 5.

According to Problem 8.64, the area of the base is $4\sqrt{2}$. Calculate the volume V of the solid.

$$\begin{aligned} V &= B \cdot h \\ &= 4\sqrt{2}(5) \\ &= 20\sqrt{2} \end{aligned}$$

Note: Problems 8.64–8.66 refer to the diagram in Problem 8.64, a right prism with right triangle bases.

8.66 Calculate the surface area of the solid.

This solid has five faces, two bases of area $4\sqrt{2}$ each and three lateral rectangular faces, as illustrated in the following diagram.

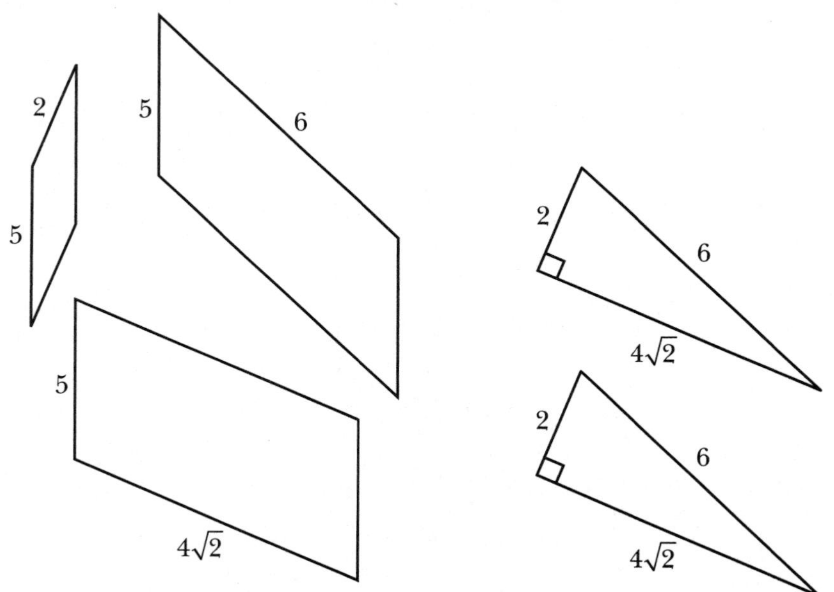

The areas of the lateral faces are $5 \cdot 2 = 10$, $5 \cdot 6 = 30$, and $5 \cdot 4\sqrt{2} = 20\sqrt{2}$. Add these to the areas of the bases to calculate the total surface area s of the prism.

$$\begin{aligned} s &= 10 + 30 + 20\sqrt{2} + 4\sqrt{2} + 4\sqrt{2} \\ &= 40 + 28\sqrt{2} \end{aligned}$$

8.67 A sheet of rectangular paper with length 10 inches and width 6 inches is rolled into a right circular cylinder by joining opposite sides of the rectangle as illustrated below. What is the volume of the cylinder?

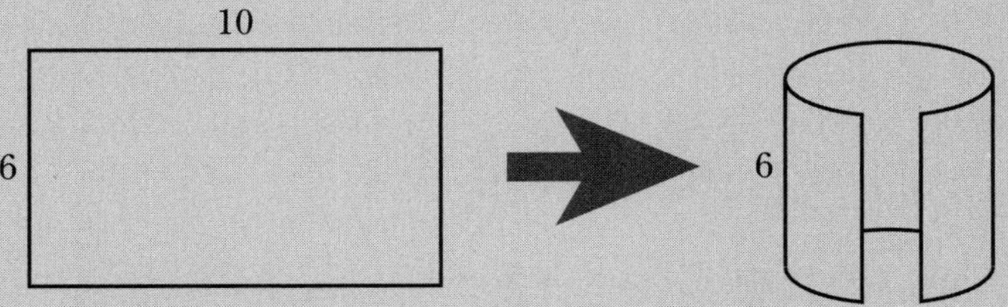

The length 10 of the rectangle becomes the circumference of the circular base of the cylinder. The height of the cylinder is $h = 6$. As Problem 8.65 explains, the area of any right prism (including a right rectangular cylinder, which is a right prism with circular bases) is equal to the area of the base B times the height h: $V = B \cdot h$.

To calculate the area of the circular base, you must first calculate its radius, based on the known value of the circumference, $C = 10$.

$$\begin{aligned} C &= 2\pi r \\ 10 &= 2\pi r \\ \frac{10}{2\pi} &= r \\ \frac{5}{\pi} &= r \end{aligned}$$

Now that you know the radius of the circular base, calculate the area B of the base.

$$\begin{aligned} B &= \pi r^2 \\ &= \pi\left(\frac{5}{\pi}\right)^2 \\ &= \pi\left(\frac{25}{\pi^2}\right) \\ &= \frac{25}{\pi} \end{aligned}$$

Calculate the volume of the cylinder.

$$\begin{aligned} V &= B \cdot h \\ &= \frac{25}{\pi}(6) \\ &= \frac{150}{\pi} \end{aligned}$$

Note: Problems 8.68–8.69 refer to the diagram below, a right circular cylindrical solid with a base of radius r = $\sqrt{6}$ and height 12 after a square hole with side length 2 is drilled through the length of the solid.

8.68 Calculate the volume of the solid.

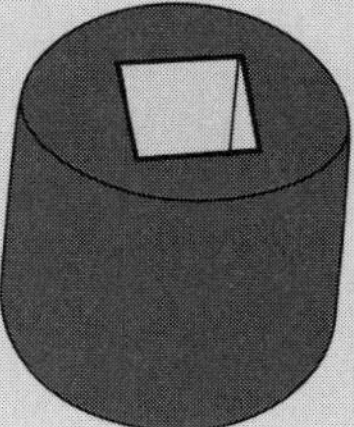

You need to calculate the volume of the right cylinder and subtract the volume of the hole, which is a right rectangular prism. The base of the cylinder is a circle, and you are given its radius *r*. Use this information to calculate the area *B* of the circular base.

$$\begin{aligned} B &= \pi r^2 \\ &= \pi\left(\sqrt{6}\right)^2 \\ &= 6\pi \end{aligned}$$

As Problem 8.67 explains, the volume V_1 of the cylinder is the product of the height $h = 12$ and the area of the base $B = 6\pi$.

$$\begin{aligned} V_1 &= B \cdot h \\ &= 6\pi(12) \\ &= 72\pi \end{aligned}$$

V_1 represents the volume of the entire cylinder (if there were no hole in it), and V_2 will represent the volume of the hole. Therefore, the volume of the solid is $V_1 - V_2$.

The area of the right rectangular prism hole is the product of its length, width, and height. Like any right prism, you can also calculate the area by multiplying the area of its base (in this case a square with side $s = 2$) by the height $h = 12$. Using either technique, the volume V_2 of the hole in the solid is $V_2 = 2 \cdot 2 \cdot 12 = 48$.

Calculate the volume V of the solid in this problem by subtracting the volume V_2 of the hole from the volume V_1 of the cylinder.

$$V = 72\pi - 48$$

Note: Problems 8.68–8.69 refer to the diagram in Problem 8.68, a right circular cylindrical solid with a base of radius r = $\sqrt{6}$ and height 12 after a square hole with side length 2 is drilled through the length of the solid.

8.69 Calculate the surface area of the solid, including the bases.

The surface area of the solid consists of two circles (from which a square region with side length 2 is excluded) and a rectangle with height 12 and length equal to the circumference *C* of the circle. Note that this is similar to the cylinder you explored in Problem 8.67. The components of the surface area are illustrated in the following diagram. You calculate the area of each individually and then add them together to compute the total surface area.

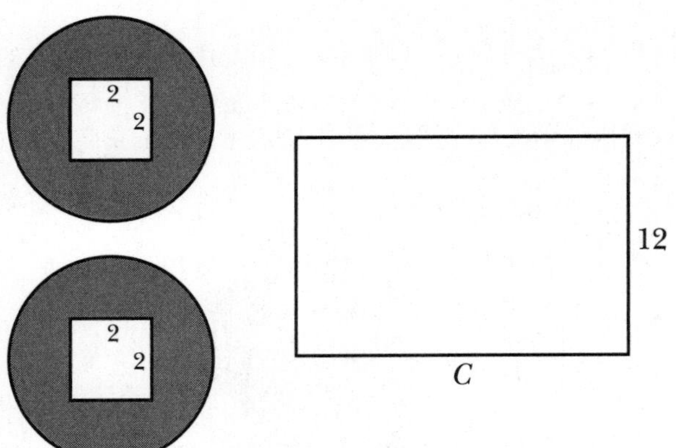

According to Problem 8.68, each circular base would have an area of 6π if it were not for the square hole. The side length of the square hole is 2, so the area of the hole is $2 \cdot 2 = 4$. Thus, the area of each base is $6\pi - 4$.

To calculate the area of the rectangle, you must first calculate the circumference C of the circular base. You are given the radius of the circle.

$$\begin{aligned} C &= 2\pi r \\ &= 2\pi\left(\sqrt{6}\right) \\ &= 2\pi\sqrt{6} \end{aligned}$$

The area of the rectangle is the product of its length and width: $12 \cdot 2\pi\sqrt{6} = 24\pi\sqrt{6}$. Calculate the total surface area by adding the areas of the two bases and the area of the rectangle.

$$\begin{aligned} \text{surface area} &= \text{area}(\text{base 1}) + \text{area}(\text{base 2}) + \text{area}(\text{rectangle}) \\ &= (6\pi - 4) + (6\pi - 4) + \left(24\pi\sqrt{6}\right) \\ &= 12\pi - 8 + 24\pi\sqrt{6} \\ &= 4\left(3\pi - 2 + 6\pi\sqrt{6}\right) \end{aligned}$$

8.70 If a cube with side length 3 is inscribed in a sphere, what is the diameter of the sphere?

If a cube is inscribed in a sphere, each of the cube's vertices intersects the surface of the circle, as illustrated in the following diagram.

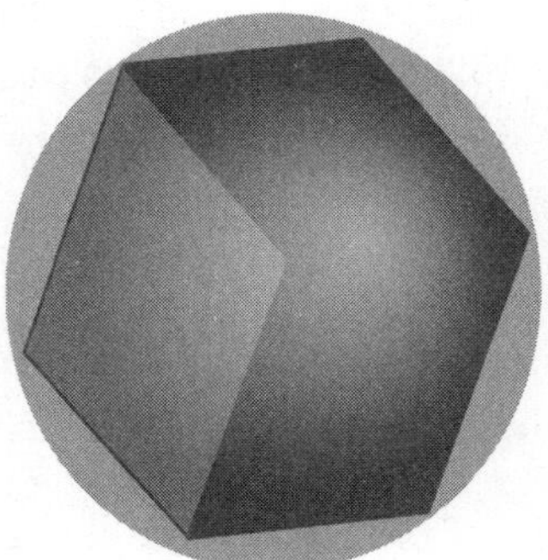

The trickiest part of this problem is visualizing what a diameter of the sphere is in relation to the inscribed cube. A line segment that connects a vertex of one side of the cube that extends to the opposite vertex on the opposite side of the cube is also a diameter of the circle, the segment with length d in the following diagram. As the diagram illustrates, you can construct a right triangle whose hypotenuse is the diameter and has legs of length y (one side of the cube, so $y = 3$) and length x, which is a diagonal of a square face.

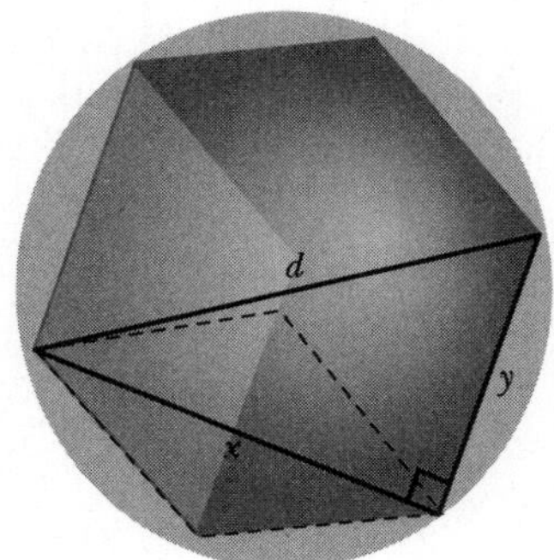

The segment with length x is a hypotenuse of a 45°–45°–90° triangle whose legs are the sides of the square base. Thus, x is $\sqrt{2}$ times as long as the side length: $x = 3\sqrt{2}$.

Apply the Pythagorean theorem to calculate the length d of the diameter.

$$\begin{aligned}
x^2 + y^2 &= d^2 \\
\left(3\sqrt{2}\right)^2 + 3^2 &= d^2 \\
9(2) + 9 &= d^2 \\
18 + 9 &= d^2 \\
27 &= d^2 \\
\sqrt{27} &= d \\
\sqrt{9 \cdot 3} &= d \\
3\sqrt{3} &= d
\end{aligned}$$

Coordinate Geometry

Geometry with (x,y) coordinates

Problems 8.71–8.73 refer to points A = (–2,3) and B = (v,w).

8.71 If point B lies on the same horizontal line as point A, what can you conclude about the values of v and w?

Point A lies on vertical line $x = -2$ and horizontal line $y = 3$. If point B lies on the same horizontal line, then you can conclude it has the same y-coordinate as point A. You conclude that $w = 3$. You cannot make any assumptions about v, the x-coordinate of point B.

Problems 8.71–8.73 refer to points A = (–2,3) and B = (v,w).

8.72 If $w = -7$ and the slope of $\overrightarrow{AB}$ is $\frac{4}{3}$, what is the value of v?

The slope of the line passing through two points is the difference of their y-coordinates divided by the difference of their x-coordinates. In other words, the slope m of the line passing through points (x_1, y_1) and (x_2, y_2) is calculated according to the following formula.

This is also discussed in Problem 6.40, because the algebra section of the SAT deals with equations of lines, which also require you to understand slope.

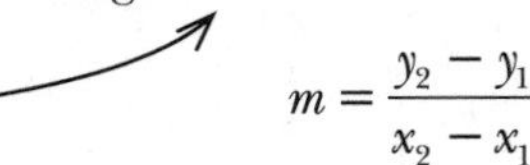

$$m = \frac{y_2 - y_1}{x_2 - x_1}$$

In this problem, $m = 4/3$, $x_1 = -2$, $y_1 = 3$, $x_2 = v$, and $y_2 = w = -7$.

$$\frac{4}{3} = \frac{-7-3}{v-(-2)}$$
$$\frac{4}{3} = \frac{-10}{v+2}$$

Cross-multiply and solve for v.

$$\begin{aligned} 4(v+2) &= 3(-10) \\ 4v+8 &= -30 \\ 4v &= -30-8 \\ 4v &= -38 \\ v &= -\frac{38}{4} \\ v &= -\frac{19}{2} \end{aligned}$$

Problems 8.71–8.73 refer to points A = (–2,3) and B = (v,w).

8.73 If $v = 1$ and $\overrightarrow{AB}$ is perpendicular to the line with equation $4x - 6y = 11$, what is the value of w?

Perpendicular lines have slopes that are opposite reciprocals, so the slope of $\overrightarrow{AB}$ is the opposite reciprocal of the slope of the line with equation $4x - 6y = 11$. One way to calculate the slope of $4x - 6y = 11$ is to apply the technique demonstrated in Problem 6.37: Solve for y, and the slope is the coefficient of x. A quicker way to calculate the slope is to apply the shortcut presented in Problem 6.65: The slope of linear equation $Ax + By = C$ is $-A/B$. Therefore, the slope of $4x - 6y = 11$ is $-(4/-6) = +2/3$.

The slope of the line passing through points A and B is $-3/2$, the opposite reciprocal of $+2/3$. Substitute $m = -3/2$, $x_1 = -2$, $y_1 = 3$, $x_2 = v = 1$, and $y_2 = w$ into the slope formula.

$$m = \frac{y_2 - y_1}{x_2 - x_1}$$
$$-\frac{3}{2} = \frac{w-3}{1-(-2)}$$
$$-\frac{3}{2} = \frac{w-3}{1+2}$$
$$-\frac{3}{2} = \frac{w-3}{3}$$

Cross-multiply and solve for w.

$$-3(3) = 2(w-3)$$
$$-9 = 2w - 6$$
$$-3 = 2w$$
$$\frac{-3}{2} = w$$

Note: Problems 8.74–8.76 refer to the diagram below.

8.74 Identify the coordinates of point C.

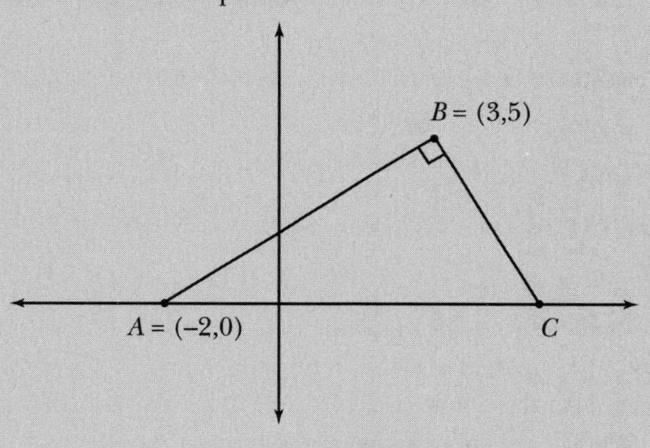

Note: Figure not drawn to scale.

Triangle ABC in the diagram is a right triangle because its legs $\overline{AB}$ and $\overline{BC}$ form a right angle. Therefore, the legs are perpendicular. Recall that perpendicular lines have opposite reciprocal slopes. Calculate the slope m_1 of $\overline{AB}$.

$$m_1 = \frac{y_2 - y_1}{x_2 - x_1}$$
$$= \frac{5-0}{3-(-2)}$$
$$= \frac{5}{3+2}$$
$$= \frac{5}{5}$$
$$= 1$$

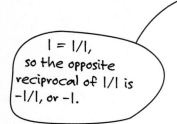

1 = 1/1, so the opposite reciprocal of 1/1 is −1/1, or −1.

The opposite reciprocal of 1 is –1. Therefore, the slope m_2 of $\overline{BC}$ is –1. Because point C lies on the x-axis, like point A, its y-coordinate is 0. Let x represent its x-coordinate. Therefore, $C = (x,0)$. Apply the slope formula for $\overline{BC}$, setting $m = -1$.

$$m = \frac{y_2 - y_1}{x_2 - x_1}$$
$$-1 = \frac{0-5}{x-3}$$
$$-1 = \frac{-5}{x-3}$$

Cross-multiply and solve for x.

$$-1(x-3) = -5$$
$$-x+3 = -5$$
$$-x = -8$$
$$x = 8$$

The coordinates of point C are $(x,0) = (8,0)$.

Note: Problems 8.74–8.76 refer to the diagram in Problem 8.74.

8.75 Calculate BC.

To calculate the distance d between two points (x_1, y_1) and (x_2, y_2) on the coordinate plane, apply the distance formula, $d = \sqrt{(x_2 - x_1)^2 + (y_2 - y_1)^2}$.

You identify these coordinates in Problem 8.74.

In this problem, you are asked to calculate BC, which is both the length of segment $\overline{BC}$ and the distance between points B and C. Therefore, you can apply the distance formula such that $(x_1, y_1) = (3,5)$ are the coordinates of B and $(x_2, y_2) = (8,0)$ are the coordinates of C.

$$d = \sqrt{(x_2 - x_1)^2 + (y_2 - y_1)^2}$$
$$BC = \sqrt{(8-3)^2 + (0-5)^2}$$
$$BC = \sqrt{5^2 + (-5)^2}$$
$$BC = \sqrt{25+25}$$
$$BC = \sqrt{50}$$
$$BC = \sqrt{25 \cdot 2}$$
$$BC = 5\sqrt{2}$$

Note: Problems 8.74–8.76 refer to the diagram in Problem 8.74, right triangle ABC.

8.76 Calculate the area of triangle *ABC*.

Consider the diagram below, in which an altitude $\overline{BD}$ is drawn, connecting vertex B to base $\overline{AC}$ of the triangle. Note that B and D both lie on the same vertical line, $x = 3$, and the same horizontal line as points A and C, $y = 0$. Therefore, the coordinates of D are $(x,y) = (3,0)$.

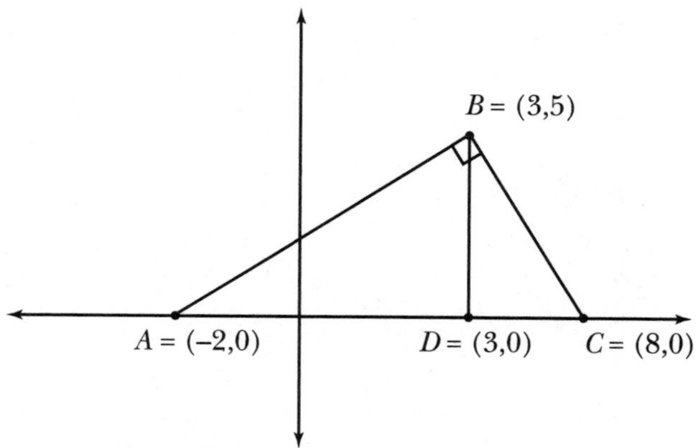

Note: Figure not drawn to scale.

Because B and D lie on the same vertical line, the length BD is equal to the absolute value of the difference of the y-coordinates: $BD = |5 - 0| = 5$. Similarly, A and C lie on the same horizontal line, so $AC = |8 - (-2)| = 10$. The area of triangle ABC is one-half the product of base AC and height BD.

$$\begin{aligned} A &= \frac{1}{2}bh \\ &= \frac{1}{2}(10)(5) \\ &= 25 \end{aligned}$$

Note: In Problems 8.77–8.78, points R = (–3,1), S = (8,1), and T = (8,–4) are vertices of rectangle RSTU.

8.77 Calculate the area and perimeter of *RSTU*.

Plot the three known vertices on the coordinate plane. The coordinates of the fourth vertex must be $U = (-3,-4)$.

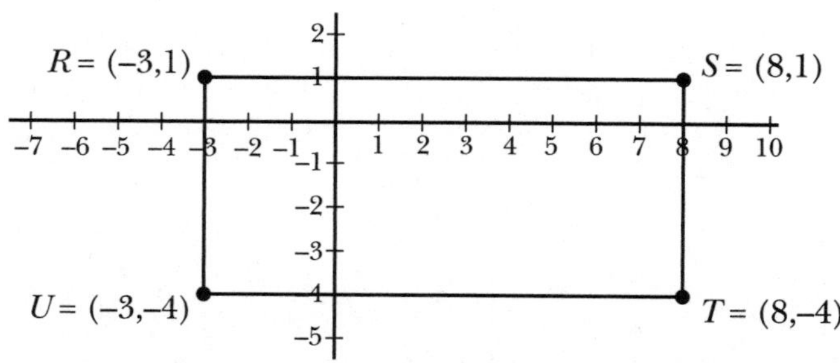

As Problem 8.76 explains, you do not need to apply the distance formula to calculate lengths of horizontal and vertical segments. The length RS is equal to the absolute value of the difference of the x-coordinates of R and S: $RS = |8-(-3)| = 11$. Similarly, ST is equal to the absolute value of the difference of the y-coordinates of S and T: $ST = |-4-1| = 5$. The length and width of $RSTU$ are, respectively, $l = 11$ and $w = 5$. Substitute these values into the area and perimeter formulas for a rectangle, presented in Problems 8.46–8.47.

$$\begin{aligned} \text{area} &= l \cdot w \\ &= 11(5) \\ &= 55 \end{aligned} \qquad \begin{aligned} \text{perimeter} &= 2l + 2w \\ &= 2(11) + 2(5) \\ &= 22 + 10 \\ &= 32 \end{aligned}$$

Note: In Problems 8.77–8.78, points R = (–3,1), S = (8,1), and T = (8,–4) are vertices of rectangle RSTU.

8.78 Calculate the coordinates of M, the point at which $\overline{RT}$ and $\overline{SU}$ intersect.

Segments $\overline{RT}$ and $\overline{SU}$ are the diagonals of the rectangle. Recall that the diagonals of all parallelograms bisect each other, so the diagonals of the rectangle intersect, and overlap, at their shared midpoint. To calculate the midpoint M of a segment with endpoints (x_1, y_1) and (x_2, y_2), apply the midpoint formula below.

Whereas the distance formula gives you a numeric answer (a distance), the midpoint formula gives you a coordinate answer (a point). The x-coordinate of the midpoint is the average of the x-coordinates of the endpoints, and the y-coordinate of the midpoint is the average of the y-coordinates of the endpoints.

$$M = \left(\frac{x_1 + x_2}{2}, \frac{y_1 + y_2}{2} \right)$$

In this problem, the coordinates of the endpoints of $\overline{RT}$ are $(x_1, y_1) = (-3,1)$ and $(x_2, y_2) = (8,-4)$. Apply the midpoint formula.

$$\begin{aligned} M &= \left(\frac{-3+8}{2}, \frac{1+(-4)}{2} \right) \\ &= \left(\frac{5}{2}, -\frac{3}{2} \right) \end{aligned}$$

The diagonal $\overline{RT}$ has midpoint $M = (5/2, -3/2)$. Note that applying the midpoint formula to the other diagonal, $\overline{SU}$, produces the same midpoint M.

Note: Problems 8.79–8.80 refer to the diagram below, in which $\overline{XY}$ is a diameter of the circle and $\overleftrightarrow{WY}$ is tangent to the circle at point Y.

8.79 Calculate the area of the circle.

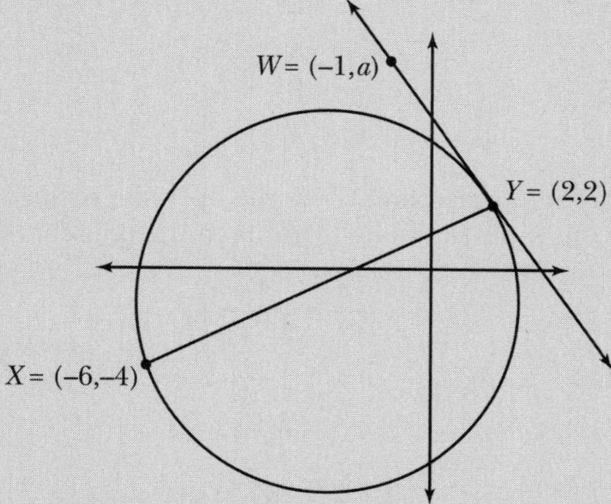

Note: Figure not drawn to scale.

The diameter has endpoints $X = (-6,-4)$ and $Y = (2,2)$. Apply the distance formula to calculate the length d of the diameter.

$$\begin{aligned} d &= \sqrt{(x_2 - x_1)^2 + (y_2 - y_1)^2} \\ &= \sqrt{(2-(-6))^2 + (2-(-4))^2} \\ &= \sqrt{(2+6)^2 + (2+4)^2} \\ &= \sqrt{8^2 + 6^2} \\ &= \sqrt{64+36} \\ &= \sqrt{100} \\ &= 10 \end{aligned}$$

The diameter of the circle is 10, so the radius r is half as long: $r = 5$. Apply the area formula for a circle.

$$\begin{aligned} A &= \pi r^2 \\ &= \pi (5)^2 \\ &= 25\pi \end{aligned}$$

The area of the circle is 25π.

Note: Problems 8.79–8.80 refer to the diagram in Problem 8.79, in which $\overline{XY}$ is a diameter of the circle and $\overleftrightarrow{WY}$ is tangent to the circle at point Y.

8.80 Calculate the value of a, the y-coordinate of point W.

According to Problem 8.45, a radius (and, by extension, a diameter) is perpendicular to the tangent line of a circle at the point of tangency. Because $\overleftrightarrow{WY}$ is tangent to the circle at point Y, it is perpendicular to $\overline{XY}$. Calculate the slope m_1 of the diameter $\overline{XY}$.

$$\begin{aligned} m_1 &= \frac{2-(-4)}{2-(-6)} \\ &= \frac{2+4}{2+6} \\ &= \frac{6}{8} \\ &= \frac{3}{4} \end{aligned}$$

The slope m_2 of the tangent line $\overleftrightarrow{WY}$ is the opposite reciprocal of the diameter's slope: $m_2 = -4/3$. Apply the slope formula to the tangent line.

$$\begin{aligned} -\frac{4}{3} &= \frac{2-a}{2-(-1)} \\ -\frac{4}{3} &= \frac{2-a}{2+1} \\ -\frac{4}{3} &= \frac{2-a}{3} \end{aligned}$$

Cross-multiply and solve for a.

$$\begin{aligned} -4(3) &= 3(2-a) \\ -12 &= 6-3a \\ -18 &= -3a \\ \frac{-18}{-3} &= a \\ 6 &= a \end{aligned}$$

Transformations and Geometric Perception

Visualizing and manipulating objects

Note: Problems 8.81–8.82 refer to the figure below, a pentagon ABCDE that is transformed to create a congruent pentagon A'B'C'D'E', such that A corresponds to A', B corresponds to B', etc.

8.81 If $ABCDE$ is reflected across the x-axis, what are the coordinates of D'?

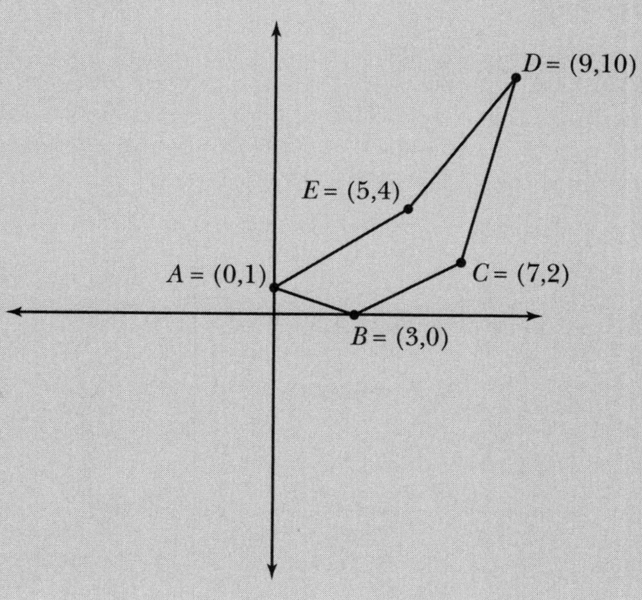

Reflecting $ABCDE$ across the x-axis does not affect the x-coordinate of each point. However, the y-coordinate of each vertex in the original pentagon is the opposite of each corresponding vertex in the reflected polygon. For example, when vertex $A = (0,1)$ is reflected across the x-axis, the result is the corresponding vertex $A' = (0,-1)$. In the figure below, the entire pentagon is reflected across the x-axis and each of the new vertices is labeled.

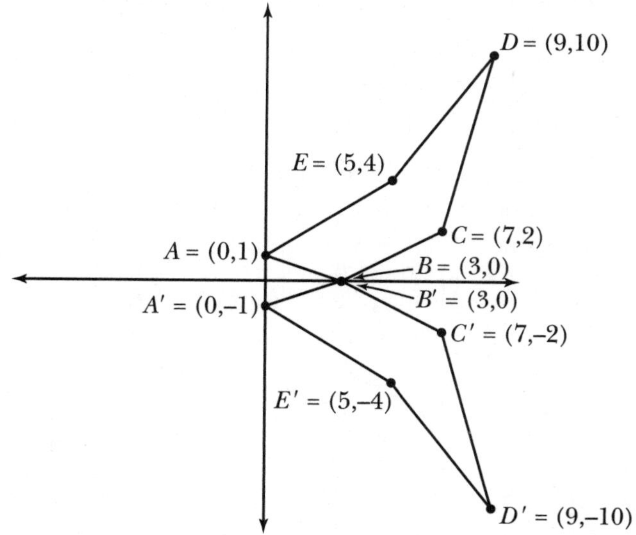

The coordinates D' are $(x,y) = (9,-10)$.

Note: Problems 8.81–8.82 refer to the figure in Problem 8.81, a pentagon ABCDE that is transformed to create a congruent pentagon A'B'C'D'E', such that A corresponds to A', B corresponds to B', etc.

8.82 If *ABCDE* is reflected across the line $x = -2$, what are the coordinates of C'?

Reflecting the pentagon across the vertical line $x = -2$ does not affect the y-coordinate of each vertex. To determine how it affects each x-coordinate, you have to calculate the horizontal distance between $x = -2$ and the x-coordinate of each point.

To calculate the distance, subtract the x-coordinate 7 from -2 (or vice versa) and take the absolute value. The distance is 9.

For example, the coordinates of vertex C are $(x,y) = (7,2)$. That point is 9 units right of the vertical line $x = -2$. Thus, the corresponding vertex C' must have an x-coordinate that is 9 units left of (and therefore less than) $x = -2$. In other words, its x-coordinate is $-2 - 9 = -11$. The y-coordinates of points C and C' are both 2, as illustrated below.

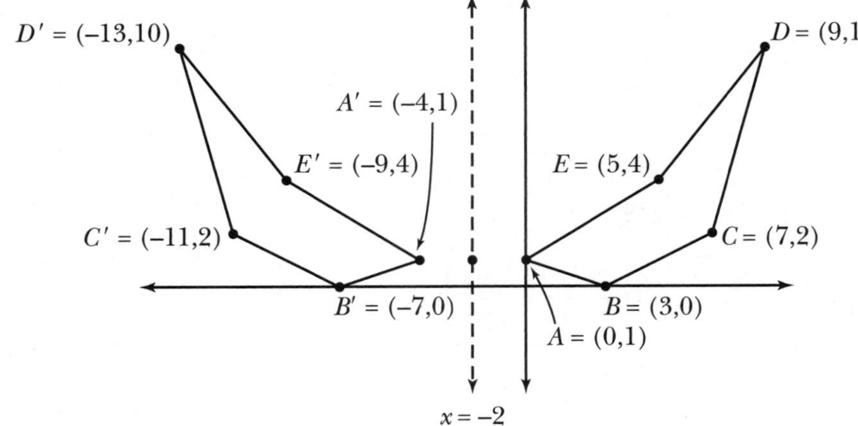

You conclude that the coordinates of C' are $(x,y) = (-11,2)$.

8.83 Identify all lines of symmetry for the figure below.

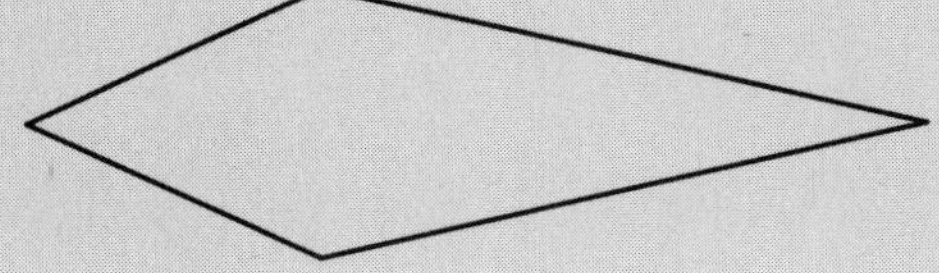

A line of symmetry divides a figure into two congruent halves, such that reflecting one-half of the figure across the line of symmetry produces the other half of the figure. In other words, a line of symmetry acts like a mirror, where the portions of the figure on either side of the line are reflections of each other. This figure has only one line of symmetry, the dotted horizontal line illustrated below.

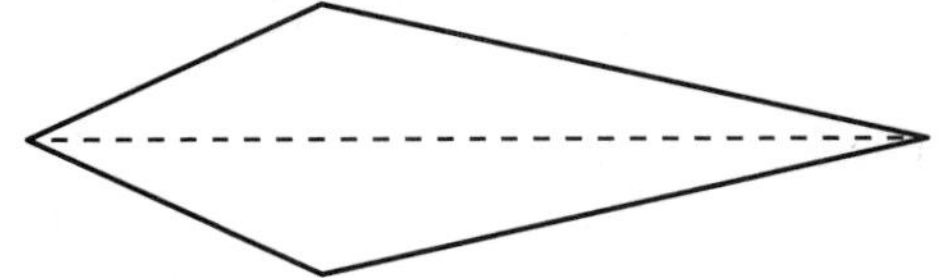

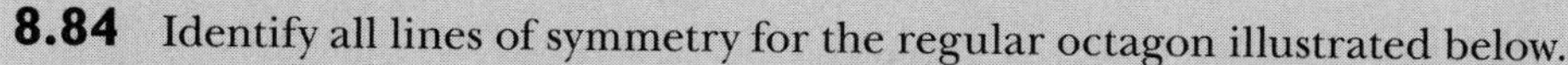

8.84 Identify all lines of symmetry for the regular octagon illustrated below.

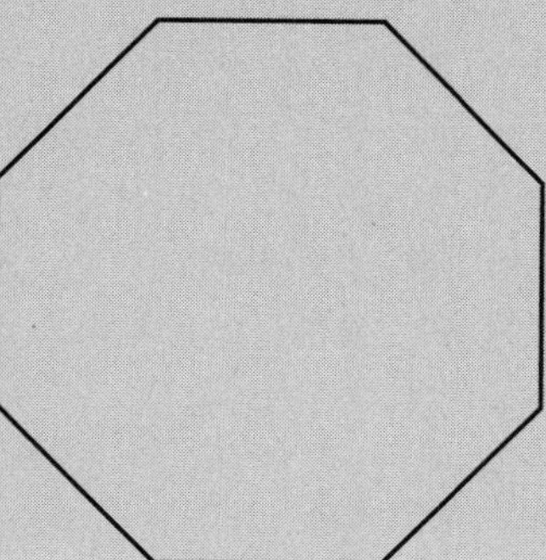

Initially, you may be tempted to conclude that there are four lines of symmetry, the lines that pass through opposite vertices of the octagon. However, the lines passing through the center of the octagon and bisecting each of the sides are also lines of symmetry. For example, the vertical line dividing the octagon in half from top to bottom is a line of symmetry, as is the horizontal line that divides the octagon in half from left to right. You can draw eight different lines of symmetry for this figure, as illustrated below.

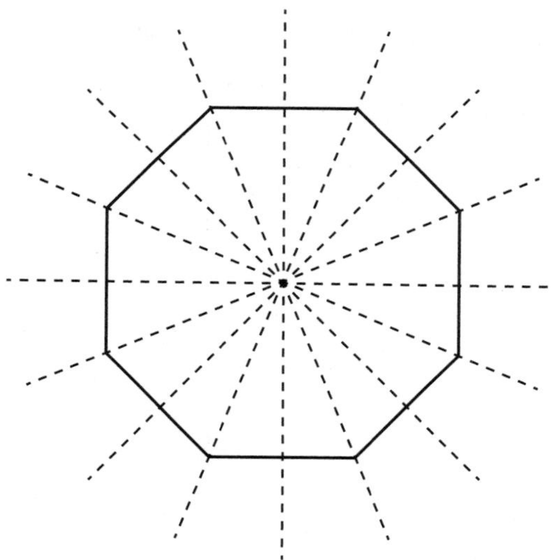

8.85 Identify a rotation that describes the transformation below.

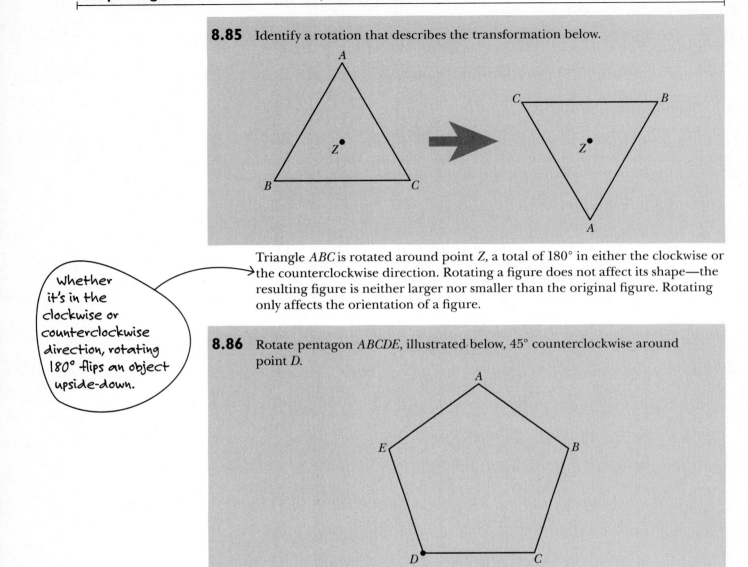

Triangle *ABC* is rotated around point *Z*, a total of 180° in either the clockwise or the counterclockwise direction. Rotating a figure does not affect its shape—the resulting figure is neither larger nor smaller than the original figure. Rotating only affects the orientation of a figure.

Whether it's in the clockwise or counterclockwise direction, rotating 180° flips an object upside-down.

8.86 Rotate pentagon *ABCDE*, illustrated below, 45° counterclockwise around point *D*.

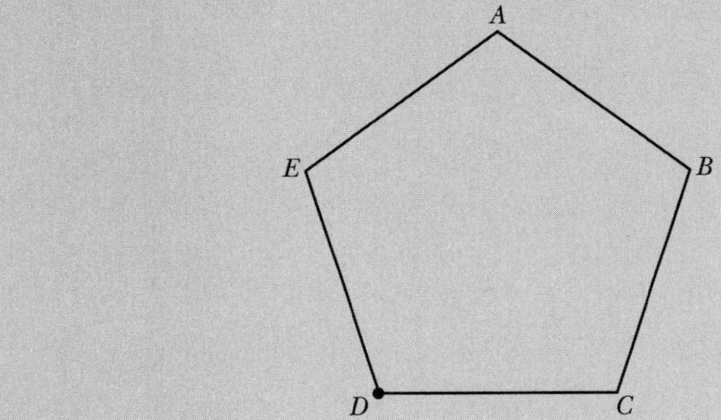

Think of point *D* as an anchor point for the pentagon. It is the only point whose location will not change as a result of the rotation. To rotate the pentagon 45° around point *D*, draw two new sides for the pentagon, each one forming a 45° angle with one of the sides of angle *D* and each the same length as the sides of the regular pentagon, as illustrated below. Make sure the new sides are placed counterclockwise with respect to the original sides.

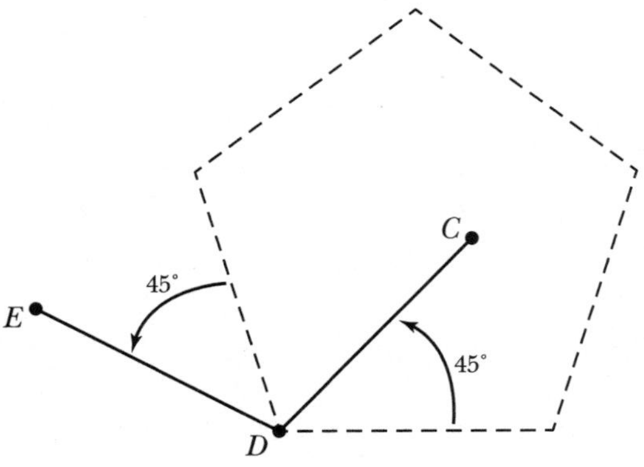

Once you have constructed the new angle D, copy the remainder of the pentagon, maintaining the angle and side measurements.

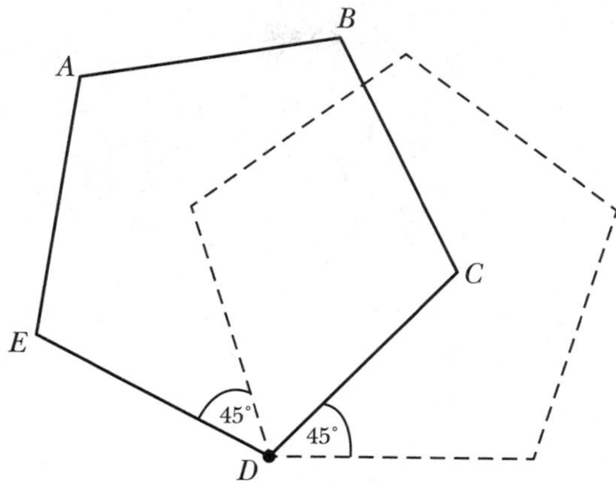

Note: In Problems 8.87–8.88, you determine whether it is possible to create a geometric shape using all three of the pieces below. You are permitted to rotate and flip the pieces, but the pieces may not overlap.

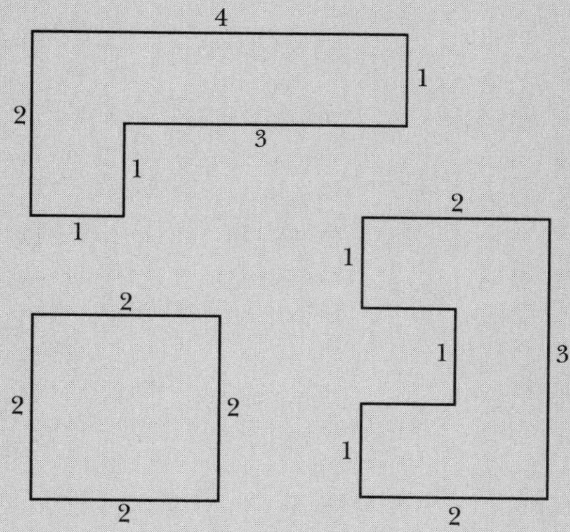

8.87 Can you create the figure below using the pieces? Explain your answer.

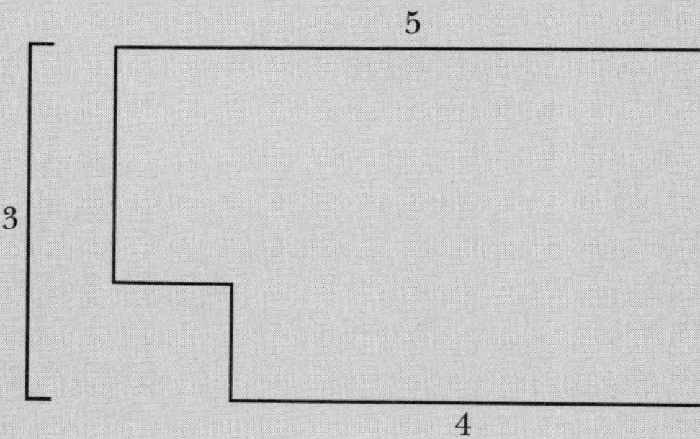

You can create the figure using the three pieces, as illustrated below.

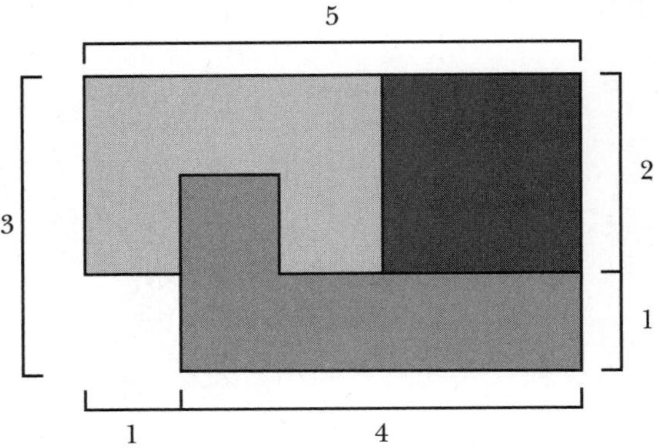

Note: In Problems 8.87–8.88, you determine whether it is possible to create a geometric shape using all three of the pieces below. You are permitted to rotate and flip the pieces, but the pieces may not overlap.

8.88 Can you create the figure below using the pieces? Explain your answer.

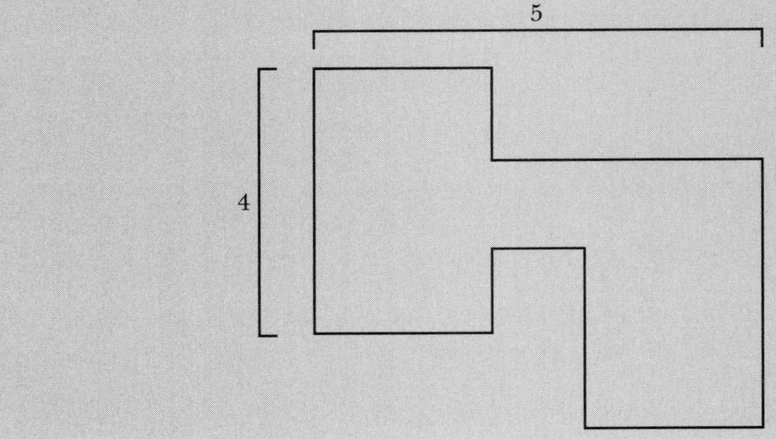

You cannot create the figure using the three pieces. The illustration below is as close as you can get, but the long L-shaped piece would need to overlap one of the other pieces.

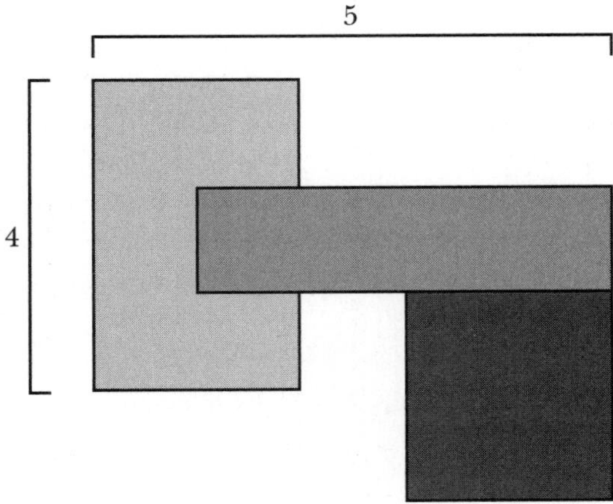

8.89 The figure below is an overhead view of the track for an amusement park ride, including a dark rectangle representing the point at which riders enter and exit.

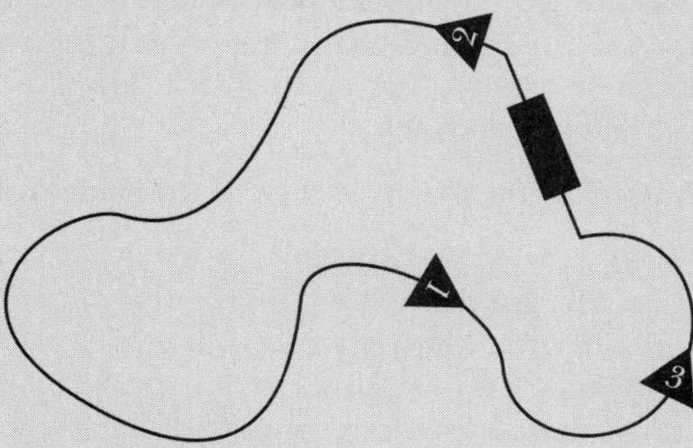

Riders drive cars along the track, but the cars (numbered 1, 2, and 3 in the diagram with an arrow indicating the direction of each car) are on rails so they cannot pass each other. Which of the following track diagrams could represent an overhead view of the same track at some moment during the same day?

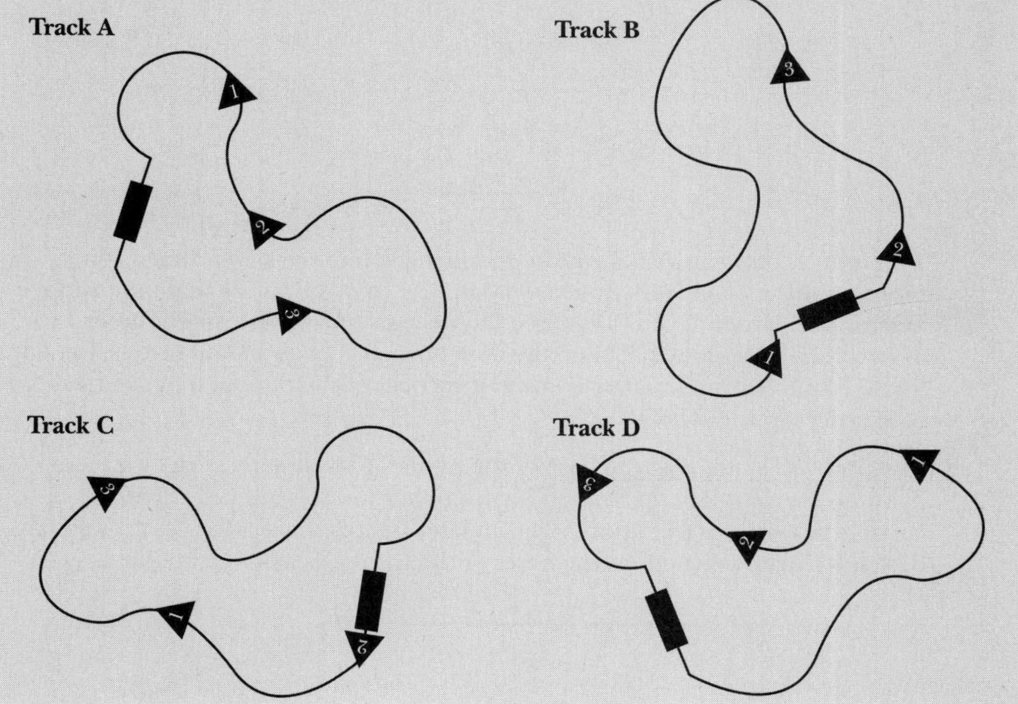

Note the order of the cars in the original overhead view of the track: Car 2 follows Car 1, which follows Car 3. According to the problem, the cars cannot pass each other, so that order must remain constant throughout the day. Therefore, Tracks B and D are not possible overhead views of the track. In both of those choices, Car 1 follows Car 2. Thus, the correct answer is either Track A or Track C, which preserve the order of the cars established in the original diagram.

Imagine that you are looking down on the tracks from above, perhaps from a roller coaster or an observation area in the park. Depending upon where you are in the park, you may see the track from a different perspective. The track is the same shape, but from your point of view it may be rotated from its original position.

Track A is a rotation of the original track, but Track C is a reflection of the original track, fundamentally changing it. In fact, Tracks C and D are essentially vertical reflections of each other. Examine the dark rectangle in the original track. Car 2 has just left the enter/exit rectangle and makes an immediate left to stay on the track. In Track C, Car 2 has just left the enter/exit rectangle and makes an immediate right to stay on the track. Thus, Track C is fundamentally different than the original track.

Tracks A, B, and D are all rotations of the original track. Only C is a reflection.

Only Track A is a possible overhead view of the same track.

8.90 The solid in the diagram below is a right regular pyramid with square base *ABCD* and vertex *E*. Draw a two-dimensional representation of the three-dimensional pyramid that you could cut out of the page and fold to form the solid. The representation should be one solid shape so no taping or gluing is required. Use dotted lines to indicate folds and label the vertices.

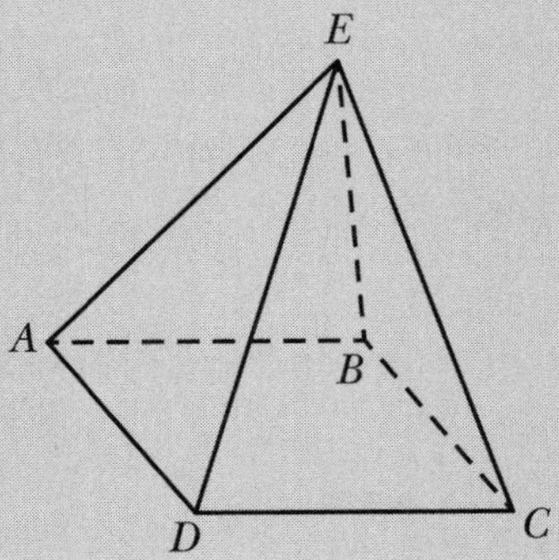

This is an exercise in visualization, requiring you to see a two-dimensional representation of a three-dimensional figure. In order to calculate the surface area of this pyramid, you would need to deconstruct the pyramid into its two-dimensional sides and calculate the area of each. This is a similar exercise, but it also requires you to visualize how the vertices of the flattened figure relate to the vertices in the solid.

The problem states that the base of the pyramid is a square. A right regular pyramid rises straight up from its base and has four lateral faces that are congruent isosceles triangles. Adjacent lateral faces share a leg, and the square base has sides congruent to the bases of the triangles, as illustrated below.

In other words, the vertex E is directly above the middle of the base.

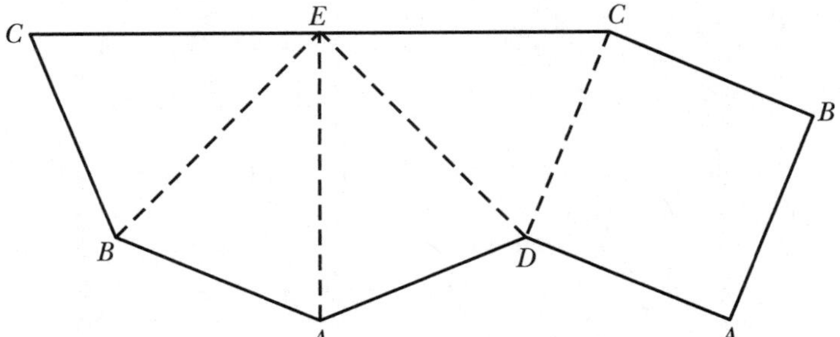

There is no single correct answer to this problem, but you can check your answer by cutting it out of a piece of paper and folding it to verify the three-dimensional pyramid shape. Note that points *A*, *B*, and *C* appear more than once in the two-dimensional drawing. Points with the same label should be matched together as you construct the solid.

Chapter 9

SAT PRACTICE: GEOMETRY AND MEASUREMENT

Test the skills you practiced in Chapter 8

This chapter contains 25 SAT-style questions to help you practice the skills and concepts you explored in Chapter 8. The questions are clustered by difficulty—early questions are easier than later questions. A portion of a student test booklet is reproduced on the following page; use it to record your answers. Full answer explanations are located at the end of the chapter.

Try not to peek at the answers until you have worked through all 25 questions. If you want to practice SAT pacing, try to finish all of the questions in no more than 30-35 minutes.

Practice Questions: Geometry and Measurement

9.1 ⒶⒷⒸⒹⒺ
9.2 ⒶⒷⒸⒹⒺ
9.3 ⒶⒷⒸⒹⒺ
9.4 ⒶⒷⒸⒹⒺ
9.5 ⒶⒷⒸⒹⒺ

9.6 ⒶⒷⒸⒹⒺ
9.7 ⒶⒷⒸⒹⒺ
9.8 ⒶⒷⒸⒹⒺ
9.9 ⒶⒷⒸⒹⒺ
9.10 ⒶⒷⒸⒹⒺ

9.11 ⒶⒷⒸⒹⒺ
9.12 ⒶⒷⒸⒹⒺ
9.13 ⒶⒷⒸⒹⒺ
9.14 ⒶⒷⒸⒹⒺ
9.15 ⒶⒷⒸⒹⒺ

9.16

9.17

9.18

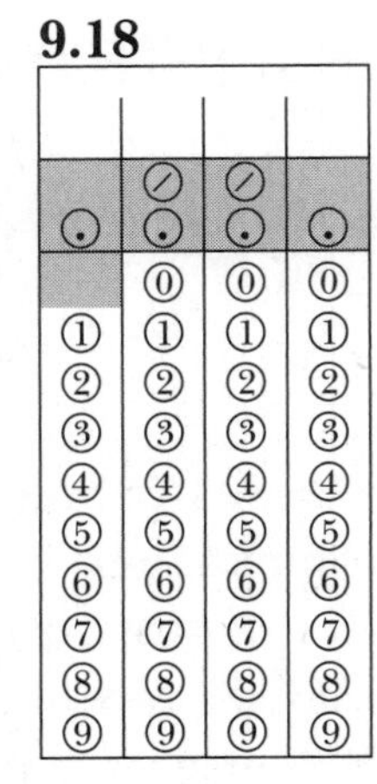

9.19

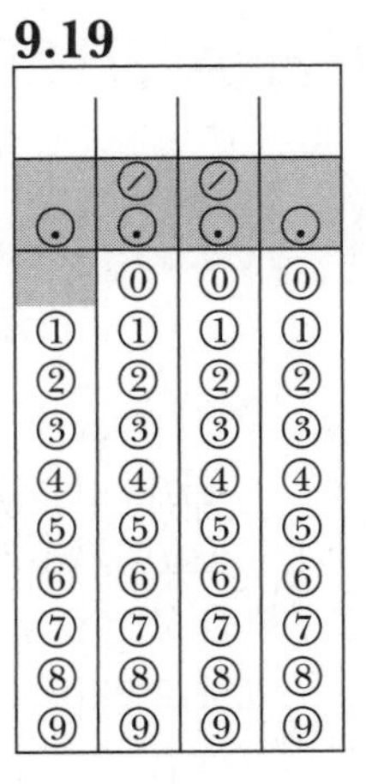

9.20

9.21

9.22

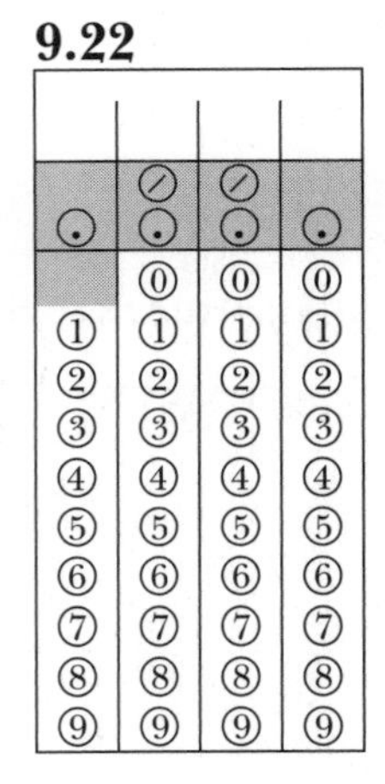

9.23

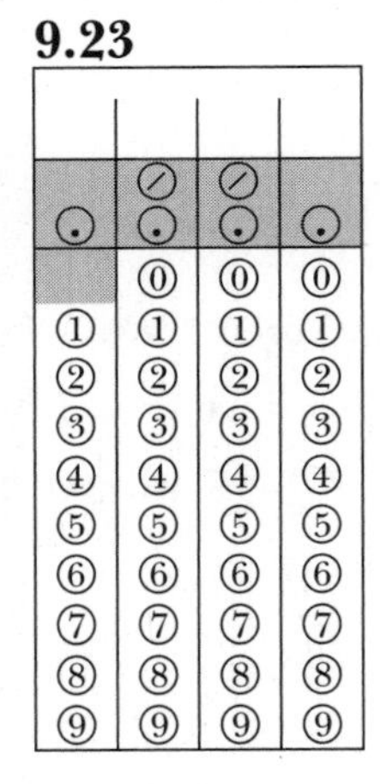

9.24

9.25

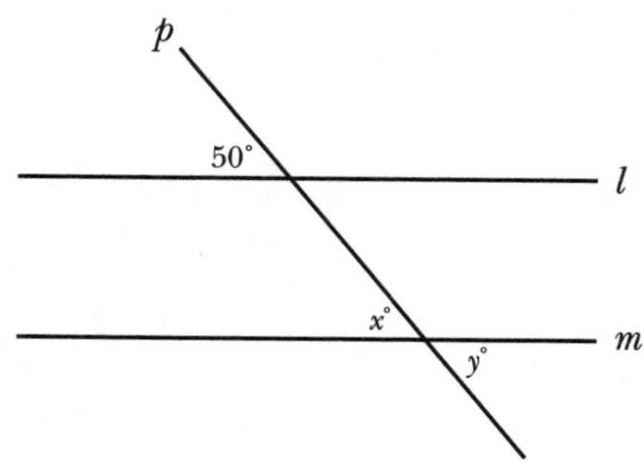

9.1 In the figure above, if lines l and m are parallel, what is the value of $x + y$?

(A) 80
(B) 100
(C) 130
(D) 180
(E) 260

9.2 If the base angles of an isosceles triangle XYZ measure 45°, which of the following statements <u>must</u> be true?

I. $XZ = \sqrt{2}(XY)$
II. $\overline{XY} \perp \overline{YZ}$
III. Triangle XYZ is a right triangle

(A) II only
(B) III only
(C) I and II only
(D) II and III only
(E) I, II, and III

<u>Questions 9.3 and 9.4 refer to the region below, which consists of a square with side length 8 and a semicircle.</u>

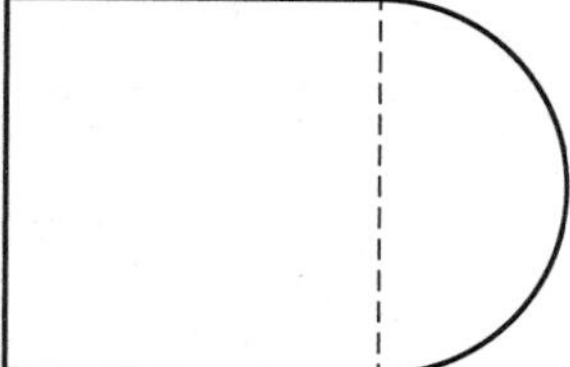

9.3 What is the area of the region?

(A) $4(\pi + 8)$
(B) $8(\pi + 8)$
(C) $16(\pi + 1)$
(D) $16(\pi + 4)$
(E) $16(\pi + 16)$

9.4 What is the perimeter of the region?

(A) $4(\pi + 6)$
(B) $4(\pi + 8)$
(C) $8(\pi + 2)$
(D) $8(\pi + 4)$
(E) $8(\pi + 8)$

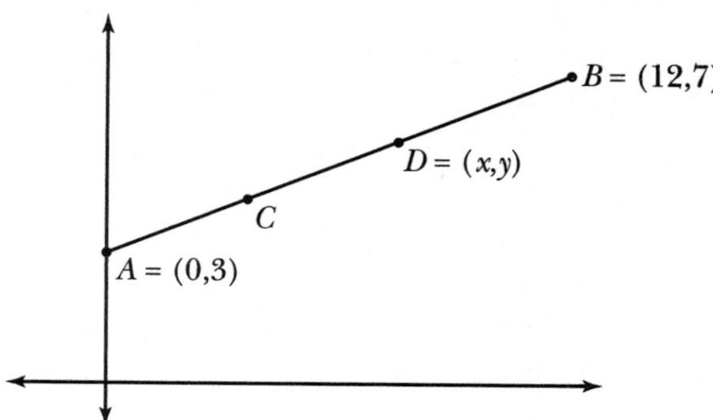

<u>Note:</u> Figure not drawn to scale.

9.5 In the figure above, C is the midpoint of $\overline{AB}$ and D is the midpoint of $\overline{BC}$. What is the value of $x + y$?

(A) 11
(B) 12.5
(C) 13.5
(D) 14
(E) 15

9.6 How much larger is one internal angle of a regular nine-sided polygon than one interior angle of a regular hexagon?

(A) 20°
(B) 25°
(C) 30°
(D) 35°
(E) 40°

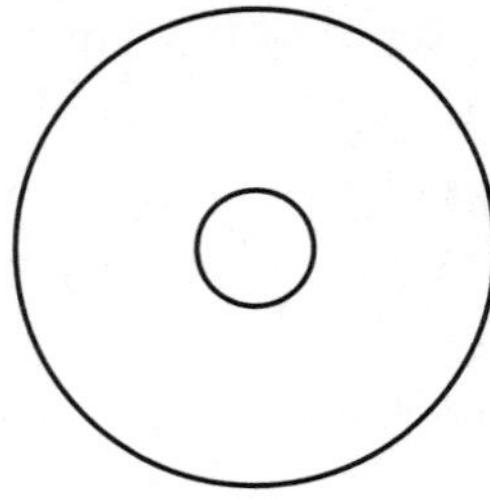

9.7 If the ratio of the areas of the concentric circles in the figure above is 1:16, and the radius of the small circle is 1.5, what is the circumference of the large circle?

(A) 6π
(B) 12π
(C) 16π
(D) 18π
(E) 36π

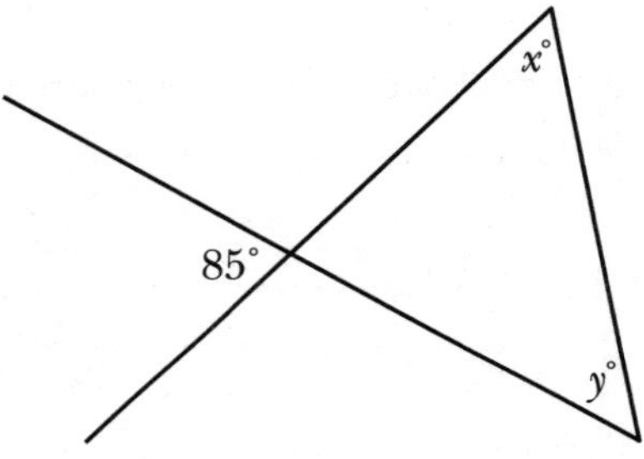

9.8 Given the figure above, what is the value of y in terms of x?

(A) $y = 95 - x$
(B) $y = 85 - x$
(C) $y = 5x + 85$
(D) $y = 95 - 5x$
(E) $y = 180 - (x + 95)$

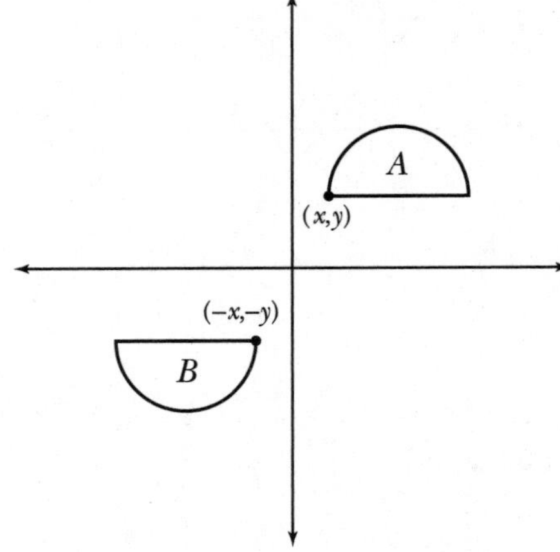

9.9 Given the diagram above, which of the following transforms semicircular region A into region B?

I. Reflecting A across both the x- and y-axes
II. Rotating A 180° around the origin
III. Reflecting A across the line $y = -x$

(A) I only
(B) I and II only
(C) I and III only
(D) II and III only
(E) I, II, and III

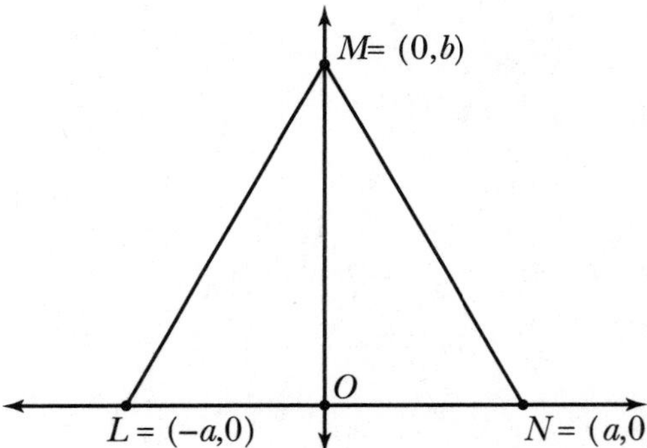

9.10 Given equilateral triangle LMN with side length 10 in the figure above, what is the value of b?

(A) 5
(B) $5\sqrt{2}$
(C) $5\sqrt{3}$
(D) 10
(E) $10\sqrt{2}$

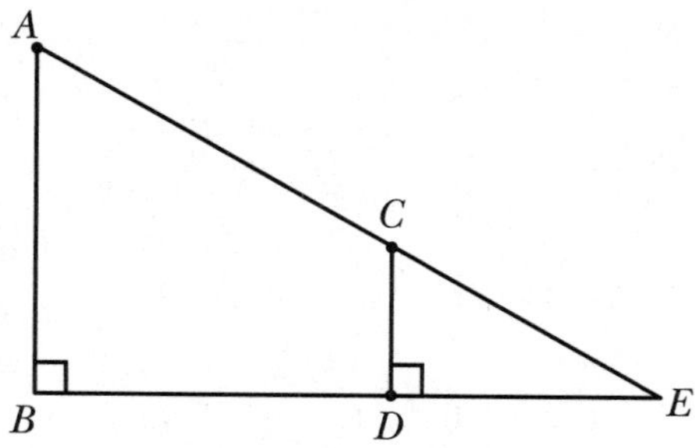

9.11 If $DE = 5$, $CE = 6$ and $BE = 18$ in the figure above what is the value of AB?

(A) $\frac{36}{5}$

(B) $3\sqrt{11}$

(C) $\sqrt{11}+13$

(D) $\frac{18\sqrt{11}}{5}$

(E) $\frac{36\sqrt{11}}{5}$

9.12 If a circle with radius 6 is circumscribed around a square, what is the perimeter of the square?

(A) $6\sqrt{2}$

(B) $12\sqrt{2}$

(C) 18

(D) $18\sqrt{2}$

(E) $24\sqrt{2}$

9.13 In a certain triangle, the length of the shortest side is one-half the length of one side and one-third the length of the other side. Which of the following represents a possible value for the perimeter of the triangle, if the lengths of all three sides are integers?

(A) 552

(B) 553

(C) 554

(D) 555

(E) 556

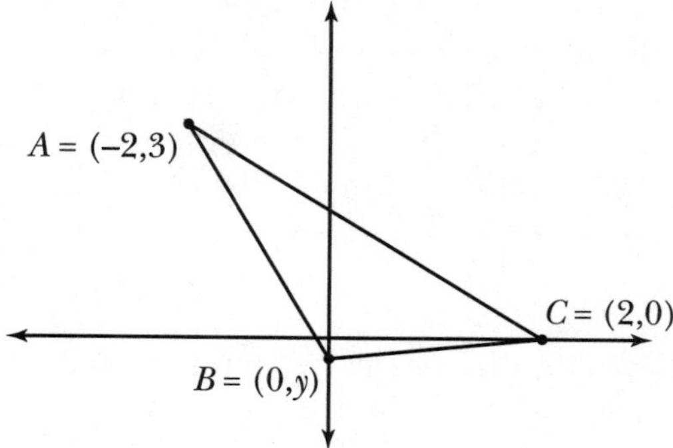

Note: Figure not drawn to scale.

9.14 In the figure above, if $\overline{AB} \perp \overline{BC}$ and $y < 0$, what is the value of y?

(A) -4

(B) -2

(C) -1

(D) $-\frac{1}{2}$

(E) $-\frac{1}{4}$

9.15. If x and y are consecutive integers such that $4 < x < y$, which of the following could represent the lengths of the sides of a triangle?

I. $x, y, 2y - x$

II. $x, y - 1, y - x$

III. $2x, 2y - 5, 3y$

(A) II only

(B) III only

(C) I and II only

(D) I and III only

(E) Not enough information is provided

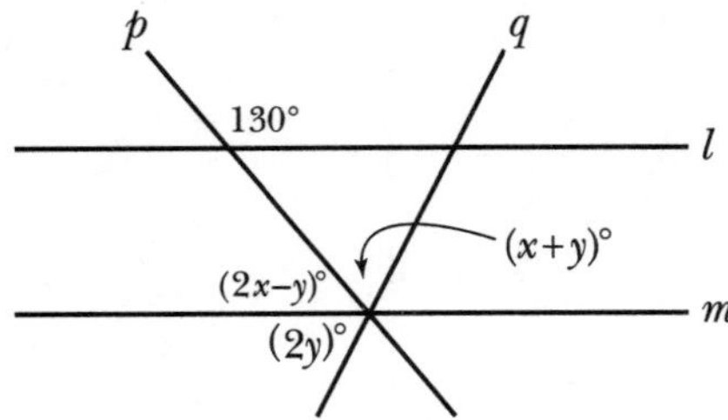

9.16 Given the diagram above, in which l and m are parallel lines, what is the value of y?

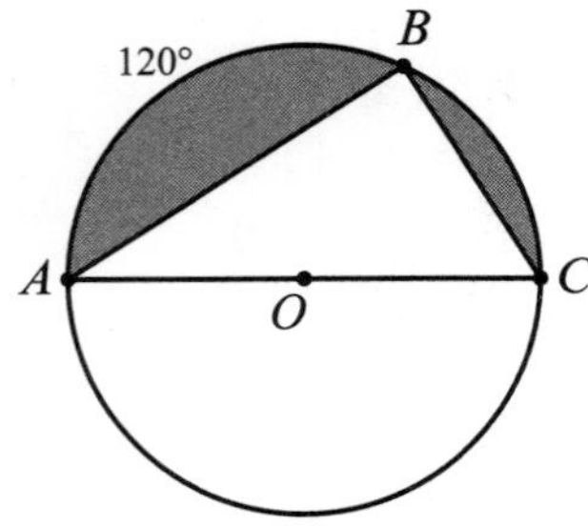

9.17 In the diagram above, triangle ABC is inscribed in circle O, which has radius 2. If the measure of $\overparen{AB}$ is 120°, what is the area of the shaded region?

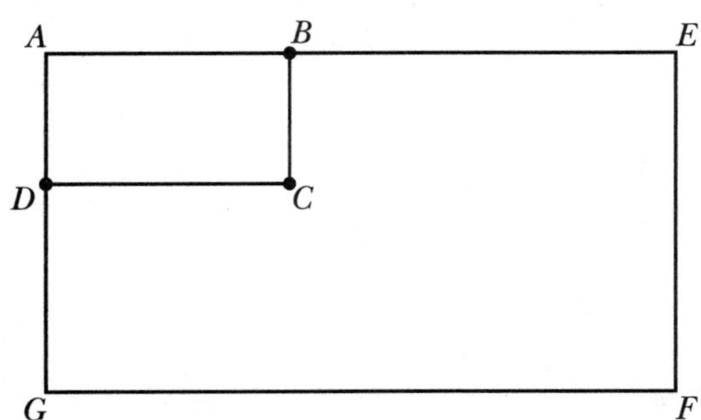

Note: Figure not drawn to scale.

9.18 In the diagram above, $ABCD$ and $AEFG$ are similar rectangles. The ratio of AB to BE is 1:4. What is the ratio of the area of $ABCD$ to the area of $AEFG$?

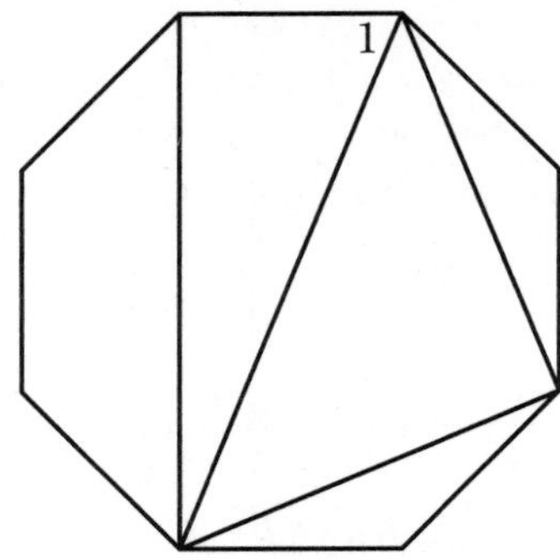

9.19 Given the regular octagon above, what is the measure of angle 1?

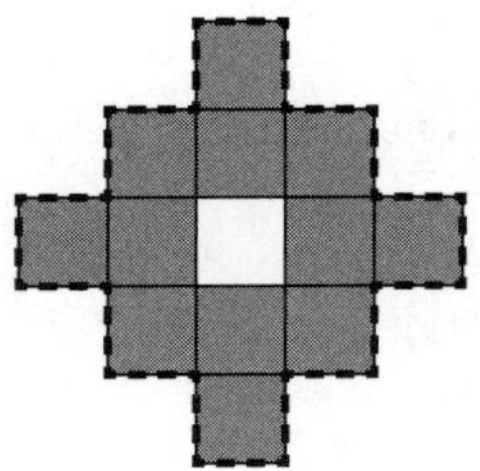

9.20 The shaded region above is composed of squares. If the area of the region is 36, what is the total length of the dotted perimeter of the region?

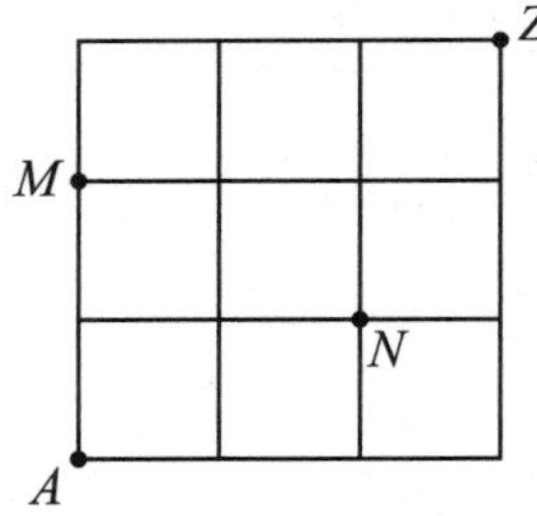

9.21 In the figure above, you may travel either up or right along the linear paths. How many paths exist from point A to point Z that do not pass through either of the points M and N?

9.22 Mike and Carla work in the same building. At 4:00 p.m., Carla leaves work and heads due north to walk home, at a constant rate of 2 meters per second. Ten minutes later, Mike leaves and walks due east at a constant rate. At precisely 4:25 p.m., they are both still walking in the same directions at the same rates, and they are 5 kilometers apart. What is Mike's constant rate of speed, in meters per second?

9.23 A wooden cube with side length 6 is painted on all sides. Once the paint dries, the cube is cut into smaller cubes of side length 2 such that no wood is lost. What is the ratio of painted to unpainted sides for the entire set of smaller cubes?

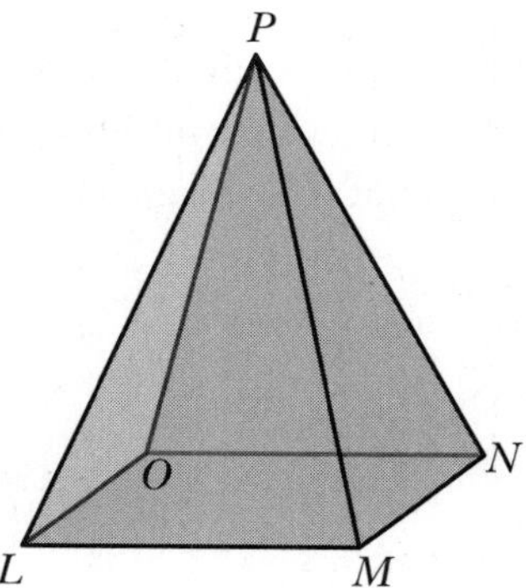

9.24 In the diagram above, the right regular pyramid has an altitude of 4 and a square base with area 49. What is the value of MP, the slant height of the solid?

9.25 A 48-foot-long piece of lumber in the shape of a right circular cylinder has volume 75π cubic feet. If the lumber is cut into 40 discs of equal width such that no lumber is lost, what is the total surface area of each disc, rounded to the nearest square inch?

Solutions

9.1 **B**. Parallel lines *l* and *m* are cut by a transversal. According to Problem 8.1, they form congruent corresponding angles, so $x° = 50°$. Note that the angles with measures $x°$ and $y°$ are vertical angles, so they are also congruent: $y° = 50°$. You are asked to calculate $x + y$.

$$x + y = 50 + 50 = 100$$

9.2 **B**. Draw a sketch of the information given. The base angles of an isosceles triangle are the angles opposite the legs (congruent sides). In the diagram below, the legs of the triangle are $\overline{XZ}$ and $\overline{XY}$, but your diagram may differ.

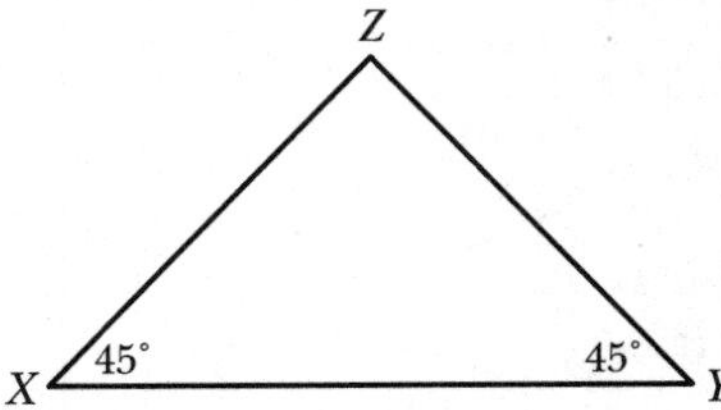

The interior angles of a triangle must total 180°. Use this information to calculate the measure of angle *Z*.

$$\begin{aligned} m\angle X + m\angle Y + m\angle Z &= 180° \\ 45° + 45° + m\angle Z &= 180° \\ 90° + m\angle Z &= 180° \\ m\angle Z &= 90° \end{aligned}$$

You conclude that angle *Z* is a right angle, so *XYZ* is a right triangle and statement III is true. Because *XYZ* is a 45°–45°–90° triangle, its hypotenuse must be $\sqrt{2}$ times the length a leg. Statement I is only true if *XY* is the length of a leg and *XZ* is the length of the hypotenuse. This is not true in the diagram above. Similarly, statement II is only true if $\overline{XY}$ and $\overline{YZ}$ are legs of the triangle. Once again, the diagram demonstrates that this is not necessarily true.

Statements I and II would both be true if you could GUARANTEE that $\overline{XZ}$ is the hypotenuse of the right triangle, but it doesn't have to be, as the diagram demonstrates.

9.3 **B**. The area of the region is equal to the area of the square plus the area of the semicircle. The square has side length 8, so its area is $l \cdot w = 8 \cdot 8 = 64$. Notice that the diameter of the semicircle is also a side of the square. Therefore, the semicircle has diameter 8 and radius $8 \div 2 = 4$.

Calculate the area *A* of a circle with radius 4.

$$A = \pi r^2 = \pi(4)^2 = 16\pi$$

Remember that the figure contains a semicircle, not a circle. The area of the semicircle is half the area of the circle: $16\pi \div 2 = 8\pi$. You conclude that the area of the region is $8\pi + 64$. Factor 8 out of this expression to get $8(\pi + 8)$.

9.4 **A**. According to Problem 9.3, the radius of the semicircle is $r = 4$. The perimeter of this region is the perimeter of the semicircle plus three sides of the square with side length 8. The three sides of the square have length $3(8) = 24$. Note that the circumference *C* of a semicircle is equal to half the circumference of the circle with the same radius.

$$\begin{aligned} C &= \frac{1}{2}(2\pi r) \\ &= \pi r \\ &= \pi(4) \\ &= 4\pi \end{aligned}$$

The perimeter of the region is $24 + 4\pi$. Factor 4 out of the terms to get $4(\pi + 6)$.

9.5 **E**. Apply the midpoint formula to calculate the x- and y-coordinates of C.

$$\begin{aligned} C &= \left(\frac{x_1 + x_2}{2}, \frac{y_1 + y_2}{2}\right) \\ &= \left(\frac{0+12}{2}, \frac{3+7}{2}\right) \\ &= \left(\frac{12}{2}, \frac{10}{2}\right) \\ &= (6,5) \end{aligned}$$

Now that you know the coordinates of C, you can apply the midpoint formula again to calculate the midpoint of $\overline{BC}$.

$$\begin{aligned} D &= \left(\frac{6+12}{2}, \frac{5+7}{2}\right) \\ (x,y) &= \left(\frac{18}{2}, \frac{12}{2}\right) \\ (x,y) &= (9,6) \end{aligned}$$

You conclude that $x + y = 9 + 6 = 15$.

9.6 **A**. The sum of the interior angles of an n-sided polygon is $180°(n - 2)$. You divide that value by n to calculate one interior angle measure of a regular n-sided polygon. Calculate the measures of one interior angle of a regular nine-sided polygon and a regular six-sided polygon (a hexagon).

Regular nine-sided polygon

$$\begin{aligned} \frac{180°(n-2)}{n} &= \frac{180°(9-2)}{9} \\ &= \frac{180°(7)}{9} \\ &= \frac{1{,}260°}{9} \\ &= 140° \end{aligned}$$

Regular hexagon

$$\begin{aligned} \frac{180°(n-2)}{n} &= \frac{180°(6-2)}{6} \\ &= \frac{180°(4)}{6} \\ &= \frac{720°}{6} \\ &= 120° \end{aligned}$$

One internal angle of a nine-sided regular polygon is $140° - 120° = 20°$ larger than one internal angle of a regular hexagon.

9.7 **B**. Let r_1 represent the radius of the small circle and A_1 represent the area of the small circle. Similarly, r_2 and A_2 represent the radius and area of the large circle. According to the problem, $A_1{:}A_2 = 1{:}16$. Express this as a proportion.

$$\frac{A_1}{A_2} = \frac{1}{16}$$

$\pi \div \pi = 1$, so you can reduce the fraction by eliminating these factors.

According to the area formula for a circle, $A_1 = \pi(r_1)^2$ and $A_2 = \pi(r_2)^2$. The question states that $r_1 = 1.5$.

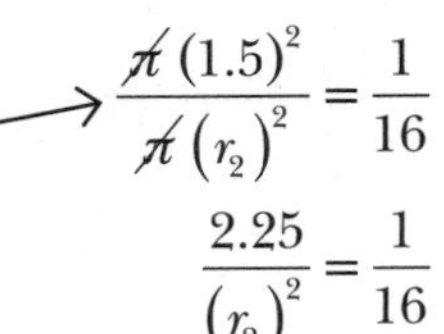

$$\frac{\pi(1.5)^2}{\pi(r_2)^2} = \frac{1}{16}$$
$$\frac{2.25}{(r_2)^2} = \frac{1}{16}$$

Cross-multiply and solve for r_2.

$$\begin{aligned} 2.25(16) &= (r_2)^2 \\ 36 &= (r_2)^2 \\ \sqrt{36} &= r_2 \\ 6 &= r_2 \end{aligned}$$

The radius of the large circle is $r_2 = 6$, so its circumference is $2\pi r_2 = 2\pi(6) = 12\pi$.

9.8 **A**. The angles within the triangle must have measures that sum to 180°. Two of the triangles measures are given: $x°$ and $y°$. The third angle of the triangle is a vertical angle of an angle measuring 85°, so it has a measure of 85° as well.

$$x + y + 85 = 180$$

Solve the equation for y to express y in terms of x.

$$\begin{aligned} y &= 180 - 85 - x \\ &= 95 - x \end{aligned}$$

9.9 **B**. Statement I is true. To understand why, consider the point (x,y) identified in region A. Reflecting A across the x-axis transforms this point into $(x,-y)$. Reflecting that region across the y-axis transforms the point into $(-x,-y)$, which is the point in region B that corresponds to the original point (x,y) in region A.

The slope of the line from the origin to point (x,y) is y/x. The slope of the line from the origin to point (y,–x) is –x/y. Those slopes are negative reciprocals, so the lines are perpendicular, and you have rotated the region 90°.

Statement II is true. Consider the diagram below, in which region A is rotated clockwise around the origin, 90° at a time. After one rotation, region A now appears in the fourth quadrant and point (x,y) is transformed into point $(y,-x)$. Rotating the region again produces region B in the third quadrant.

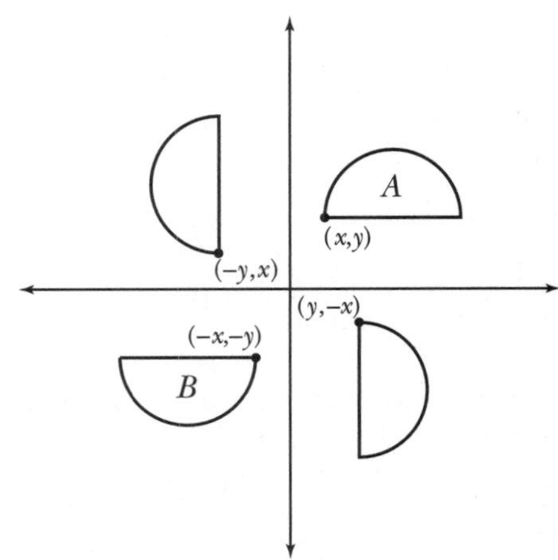

Statement III looks plausible, but it is false. When you reflect a point across a line, the perpendicular distance between the original point and the line must be equal to the perpendicular distance between the reflected point and the line. In the following diagram, region B is illustrated with a dotted boundary. The actual region that results when A is reflected across the line $y = -x$ is depicted with a solid boundary. The segments extending from line $y = -x$ to the points (x,y) and $(-y,-x)$ are congruent and perpendicular to $y = -x$.

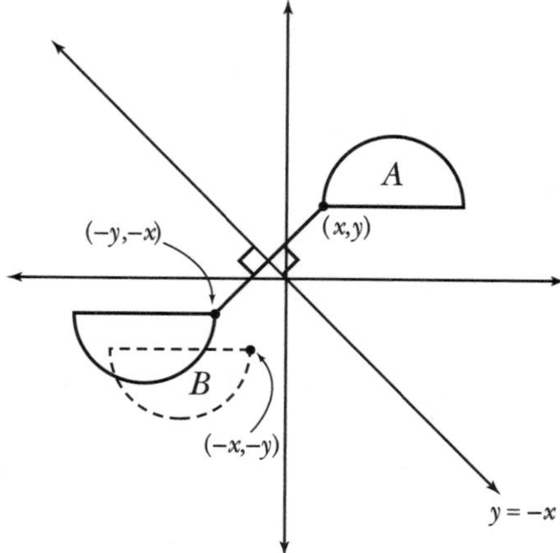

Reflecting points across the line $y = -x$ reverses the x- and y-coordinates and changes their signs. Therefore, reflecting (x,y) across $y = -x$ produces the point $(-y,-x)$, not $(-x,-y)$.

9.10 **C**. The altitude of the triangle (a segment of the y-axis) forms a 90° angle with the x-axis. Let point O represent the origin. The altitude splits the equilateral triangle into two congruent right triangles, LOM and NOM. An equilateral triangle has congruent sides and three congruent interior angles. Therefore, angles L, M, and N all measure 60°. Angles LMO and NMO each measure half of 60°, or 30°. That makes LOM and NOM 30°–60°–90° triangles with hypotenuse length $LM = MN = 10$.

Consider triangle NMO. Angle NMO measures 30°, so the opposite side has a length one-half the length of the hypotenuse: $NO = 5$. Angle MNO measures 60°, so the opposite side has length $MO = \sqrt{3}(NO) = \sqrt{3}(5)$.

9.11 **D**. Right triangles ABE and CDE both contain a right angle and share angle E. Because they contain two congruent corresponding angles, they are similar (as Problem 8.19 explains). Thus, corresponding sides of the triangles are in the same proportion. You are given the lengths of two corresponding sides: $DE = 5$ and $BE = 18$, so each side in triangle ABE is 18/5 the length of the corresponding side of triangle CDE.

You are asked to calculate AB, which is 18/5 times CD, a length not yet known. Apply the Pythagorean theorem to calculate CD.

$$(CD)^2 + (DE)^2 = (CE)^2$$
$$(CD)^2 + 5^2 = 6^2$$
$$(CD)^2 + 25 = 36$$
$$(CD)^2 = 36 - 25$$
$$(CD)^2 = 11$$
$$CD = \sqrt{11}$$

Recall that AB is 18/5 times that length.

$$\begin{aligned} AB &= \frac{18}{5}(CD) \\ &= \frac{18}{5}\sqrt{11} \\ &= \frac{18\sqrt{11}}{5} \end{aligned}$$

9.12 **E**. Draw a diagram to illustrate the information given. If a circle is circumscribed around a square, then the square is inscribed in the circle; the four vertices of the square lie on the circle. Notice that the diameter of the circle (with length 6 + 6 = 12) is also a diagonal of the square, which has side length s.

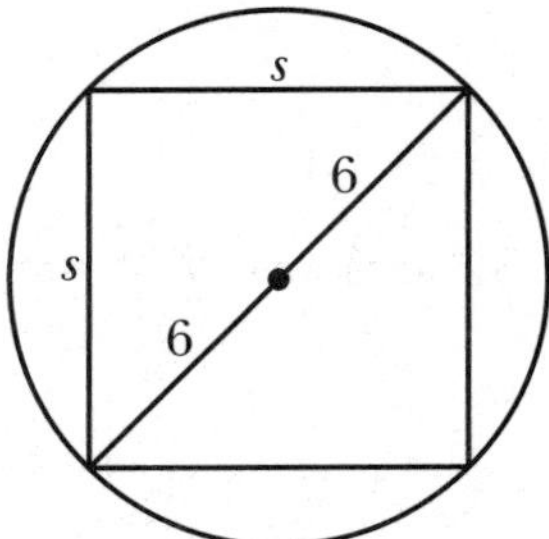

Or properties of 45°-45°-90° triangles

The diameter/diagonal divides the square into two isosceles right triangles, each with legs of length s and hypotenuse length 12. Apply the Pythagorean theorem to calculate s.

$$\begin{aligned} s^2 + s^2 &= 12^2 \\ 2s^2 &= 144 \\ s^2 &= \frac{144}{2} \\ s^2 &= 72 \\ s &= \sqrt{72} \\ s &= \sqrt{36 \cdot 2} \\ s &= 6\sqrt{2} \end{aligned}$$

The perimeter of a square is four times the length of a side: $4(6\sqrt{2}) = 24\sqrt{2}$.

9.13 **A**. Let x represent the length of the shortest side. It is one-half as long as a side with length $2x$, and it is one-third as long as a side with length $3x$. The perimeter of the triangle is $x + 2x + 3x = 6x$. If the lengths are all integers, the perimeter must be a multiple of 6. Of the choices given, only 552 is evenly divisible by 6.

9.14 **C**. If two lines (or line segments) are perpendicular, then their slopes are opposite reciprocals. Apply the slope formula to express the slopes of $\overline{AB}$ and $\overline{BC}$ in terms of y.

$$\text{slope of } \overline{AB} = \frac{y-3}{0-(-2)} = \frac{y-3}{2} \qquad \text{slope of } \overline{BC} = \frac{0-y}{2-0} = -\frac{y}{2}$$

The slopes must be opposite reciprocals, so their product is –1.

$$\left(\frac{y-3}{2}\right)\left(-\frac{y}{2}\right)=-1$$
$$\frac{-y(y-3)}{4}=-1$$
$$\frac{-y^2+3y}{4}=-1$$

For example, 5/7 and -7/5 are opposite reciprocals, and their product is -1:

$$\frac{5}{7}\left(-\frac{7}{5}\right)=-\frac{35}{35}=-\frac{1}{1}$$

Cross-multiply and solve for y.

$$-y^2+3y=-4$$
$$0=y^2-3y-4$$

Factor to solve the quadratic equation.

$$(y-4)(y+1)=0$$
$$y=4, y=-1$$

The question states that $y < 0$, so the correct answer is $y = -1$.

9.15 C. If x and y are consecutive integers and $x < y$, then $y = x + 1$. This question requires you to apply the triangle inequality theorem, which states that the sum of any two side lengths in a triangle must be greater than the length of the remaining side. Consider the first statement, substituting $y = x + 1$ into the list.

$$x,\ y,\ 2y-x=x,\ x+1,\ 2(x+1)-x$$
$$=x,\ x+1,\ 2x+2-x$$
$$=x,\ x+1,\ x+2$$

Note that x must be greater than 4. If you add any two sides, such as x and $(x + 1)$, the result is greater than the third side.

$$x+(x+1)>x+2$$
$$2x+1>x+2$$

How can you be certain? Substitute an x-value greater than 4 into the expression.

$$2(5)+1>5+2$$
$$11>7$$

Of course, you also need to verify that $x + (x + 2) > x + 1$ and $(x + 1) + (x + 2) > x$, but those inequalities are true as well. Statement I satisfies the triangle inequality theorem. To check statements II and III, apply the same procedure, substituting $y = x + 1$ to express them in terms of x.

Statement II

$$x,\ y-1,\ y-x=x,\ (x+1)-1,\ (x+1)-x$$
$$=x,\ x,\ 1$$

Statement III

$$2x,\ 2y-5,\ 3y=2x,\ 2(x+1)-5,\ 3(x+1)$$
$$=2x,\ 2x+2-5,\ 3x+3$$
$$=2x,\ 2x-3,\ 3x+3$$

Substitute an x-value greater than 4, such as $x = 5$, into the expressions and verify that they satisfy the triangle inequality theorem. Statement II is true, but statement III is not. Note that $2x + (2x - 3)$ is not greater than $3x + 3$ when $x = 5$.

$$\begin{aligned} 2(5)+(2\cdot 5-3) &> 3\cdot 5+3 \\ 10+7 &> 15+3 \\ 17 &> 18 \quad \textbf{False} \end{aligned}$$

Only statements I and II are true.

9.16 **30**. Combining the three angles measuring $(2y)°$, $(2x - y)°$, and $(x + y)°$ produces a straight line measuring 180°.

$$\begin{aligned} 2y+(2x-y)+(x+y) &= 180 \\ (2x+x)+(2y-y+y) &= 180 \\ 3x+2y &= 180 \end{aligned}$$

This is one of two equations you will need to solve this problem. Now consider the diagram below, which states that 50° and 130° are supplementary angles. Furthermore, the angles measuring 50° and $(2x - y)°$ are alternate-interior angles formed by parallel lines cut by a transversal.

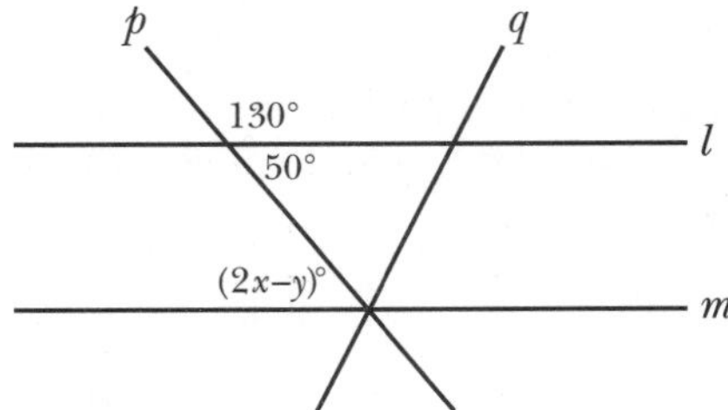

The alternate-interior angles are congruent, so $2x - y = 50$. This is the second equation you need to solve the problem. You now have a system of two linear equations in two variables.

$$\begin{cases} 3x+2y=180 \\ 2x-y=50 \end{cases}$$

Apply the elimination technique to solve the system; multiply the top equation by 2 and the bottom equation by –3 to eliminate the x-terms.

$$\begin{cases} 2(3x+2y=180) \\ -3(2x-y=50) \end{cases} \rightarrow \begin{cases} 6x+4y=360 \\ -6x+3y=-150 \end{cases} \rightarrow 7y=210$$

Solve the equation for y.

$$\begin{aligned} y &= \frac{210}{7} \\ y &= 30 \end{aligned}$$

9.17 **2.81** or **2.82**. The area of the shaded region is equal to the area of the semicircle defined by $\overset{\frown}{ABC}$ minus the area of triangle ABC. The radius of circle O is $r = 2$, so the area of the semicircle is half of $A = \pi r^2$.

$$\begin{aligned}\text{area of semicircle} &= \frac{1}{2}\pi(2)^2 \\ &= \frac{1}{2}(4\pi) \\ &= 2\pi\end{aligned}$$

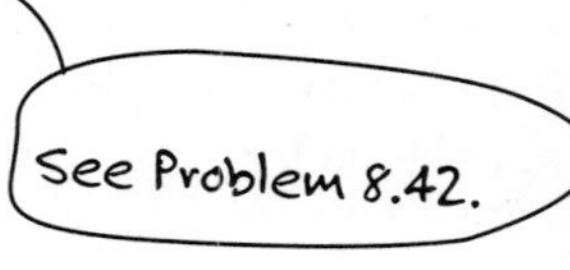

To calculate the area of the triangle, draw the altitude $\overline{CX}$, as illustrated below, and note that the measure of angle ABC is 60°, because it is an inscribed angle that subtends an arc measuring 120°. Similarly, the measure of angle ACB is 90°, so the altitude splits the large 30°–60°–90° triangle into two, smaller 30°–60°–90° triangles.

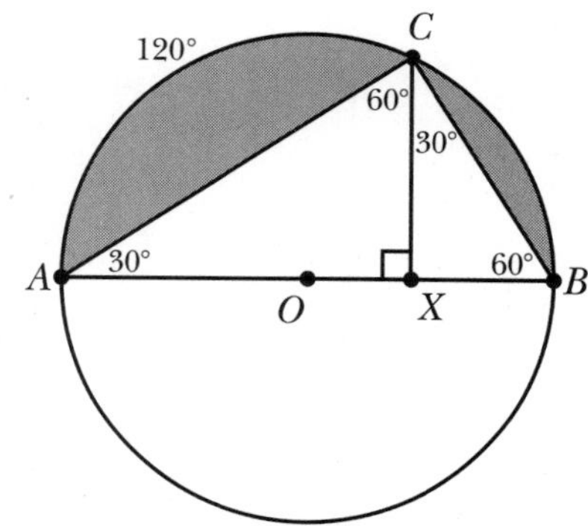

Because $\overline{AB}$ is a diameter, it has length 4. The short leg of triangle ABC is half as long: $BC = 2$. That leg is also the hypotenuse of right triangle BCX, so the short leg of that triangle has length $BX = 1$ and the long leg has length $CX = \sqrt{3}(1) = \sqrt{3}$. Thus, the base of triangle ABC is 4 and its height is $CX = \sqrt{3}$. Apply the area formula for a triangle.

$$\begin{aligned}\text{Area of triangle } ABC &= \frac{1}{2}bh \\ &= \frac{1}{2}(4)\left(\sqrt{3}\right) \\ &= \frac{4}{2}\sqrt{3} \\ &= 2\sqrt{3}\end{aligned}$$

Calculate the area of the shaded region.

$$\begin{aligned}\text{area of shaded region} &= \text{area of semicircle} - \text{area of triangle} \\ &= 2\pi - 2\sqrt{3} \\ &\approx 2.81908\end{aligned}$$

The question does NOT say that AB:AE = 1:4. You need to calculate the scale factor between the rectangles.

9.18 **1/25** or **0.04**. The question states that the ratio of *AB* to *BE* is 1:4. In other words, *BE* is four times the length of *AB*, so *BE* = 4(*AB*). This allows you to express *AE*, the length of rectangle *AEFG* in terms of *AB*.

$$\begin{aligned} AB + BE &= AE \\ AB + 4(AB) &= AE \\ 5(AB) &= AE \end{aligned}$$

Each side of the large rectangle is five times as long as the corresponding side of the smaller rectangle: *AE* = 5(*AB*) and *AG* = 5(*AD*). Calculate the ratio of the areas of the rectangles, noting that *AB* and *AD* are, respectively, the length and width of rectangle *ABCD*; *AE* and *AG* are, respectively, the length and width of rectangle *AEFG*.

$$\begin{aligned} \frac{\text{area of } ABCD}{\text{area of } AEFG} &= \frac{AB \cdot AD}{AE \cdot AG} \\ &= \frac{\cancel{AB} \cdot \cancel{AD}}{5(\cancel{AB}) \cdot 5(\cancel{AD})} \\ &= \frac{1}{5 \cdot 5} \\ &= \frac{1}{25} \end{aligned}$$

9.19 **67.5**. Calculate the measure of one interior angle of the regular octagon.

$$\begin{aligned} \frac{180(n-2)}{n} &= \frac{180(8-2)}{8} \\ &= \frac{180(6)}{8} \\ &= \frac{1{,}080}{8} \\ &= 135 \end{aligned}$$

Consider the following figure, which includes two triangles with a 135° angle. Those triangles are isosceles because their legs are sides of the regular octagon. The measures of the angles of a triangle have a sum of 180°, so the congruent base angles of the triangles each measure 180° – 135° = 45° ÷ 2 = 22.5°.

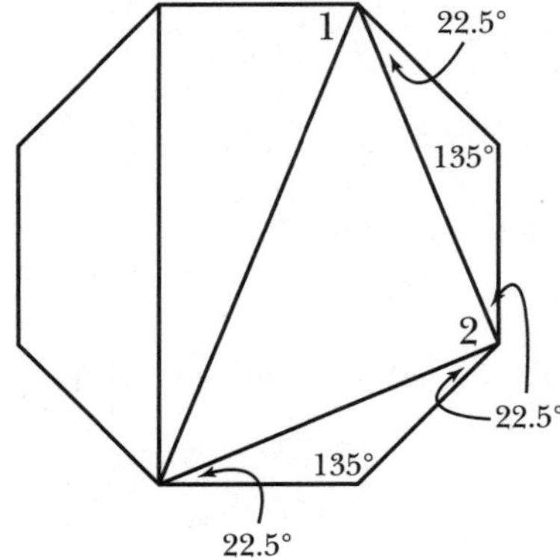

These two triangles are congruent, according to the side-angle-side technique, so their bases, the diagonals of the octagon that form numbered angle 2 in the diagram, are congruent. That makes numbered angle 2 the vertex angle of an isosceles triangle. To calculate the measure of angle 2, notice that its measure plus the measures of the small 22.5° angles together form an interior angle of the octagon, measuring 135°.

$$m\angle 2 + 22.5^\circ + 22.5^\circ = 135^\circ$$
$$m\angle 2 + 45^\circ = 135^\circ$$
$$m\angle 2 = 135^\circ - 45^\circ$$
$$m\angle 2 = 90^\circ$$

If angle 2 accounts for 90° of the 180° total degrees in the shaded isosceles triangle below, then the congruent base angles must each measure 45°.

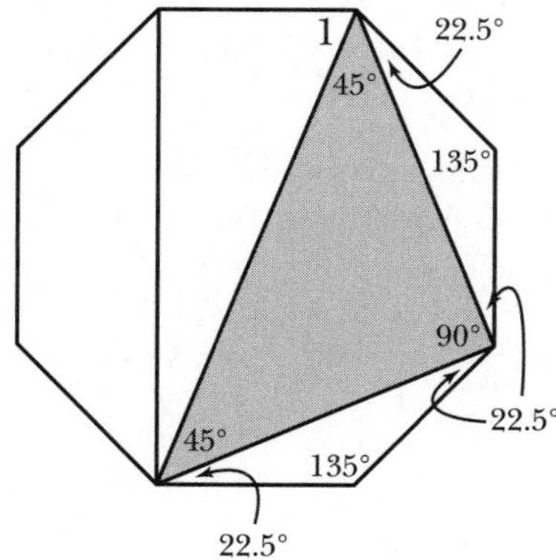

Finally, note that angle 1, a 45° angle, and a 22.5° angle form an interior angle of the octagon, totaling 135°. Use this information to calculate the measure of angle 1.

$$m\angle 1 + 45^\circ + 22.5^\circ = 135^\circ$$
$$m\angle 1 + 67.5^\circ = 135^\circ$$
$$m\angle 1 = 135^\circ - 67.5^\circ$$
$$m\angle 1 = 67.5^\circ$$

9.20 **34.6**. The region consists of 12 squares with an unknown side length, x. The area of each square is $A = x^2$, so the total area of the region is $12 \cdot x^2 = 36$. Solve the equation for x.

$$12x^2 = 36$$
$$x^2 = \frac{36}{12}$$
$$x^2 = 3$$
$$x = \sqrt{3}$$

Now state the perimeter in terms of x. As the following diagram illustrates, the dotted boundary is composed of 20 segments of length x.

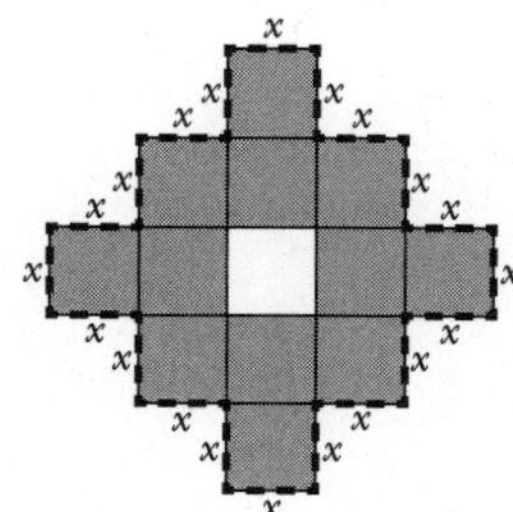

Thus, the perimeter is $20 \cdot x = 20\sqrt{3} \approx 34.641$.

9.21 **7**. You can only travel along the gridlines, and you can only travel up or right as you pass from A to Z. If you cannot pass through points M or N, then you cannot travel along the gridlines above, below, left, or right of M and N. In the diagram below, the impassable gridlines are dotted.

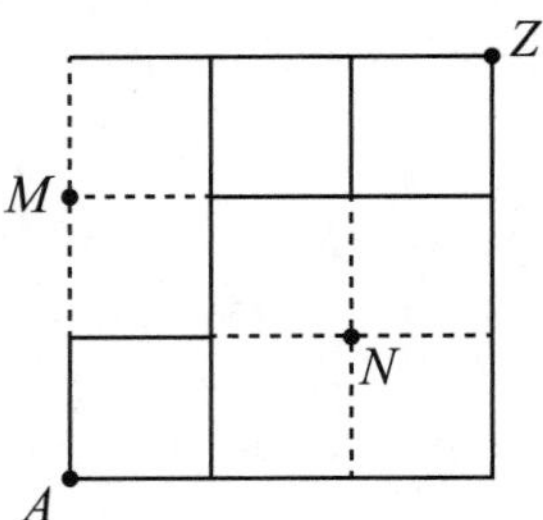

There are seven ways to pass from A to Z without passing through points M or N, as illustrated in the following diagram.

Think in an orderly fashion. In these solutions, you travel up as far as you can before ever traveling right. In the first row of solution paths, the first move you make is up. In the second row, the first move you make is right, once again traveling as far up as you can before turning right.

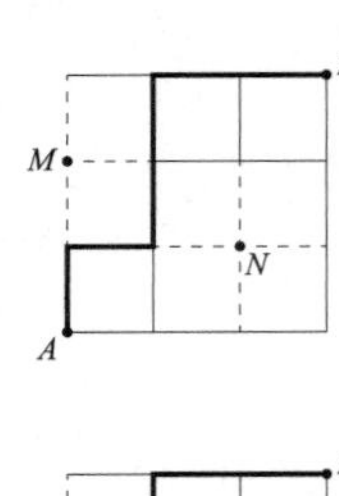

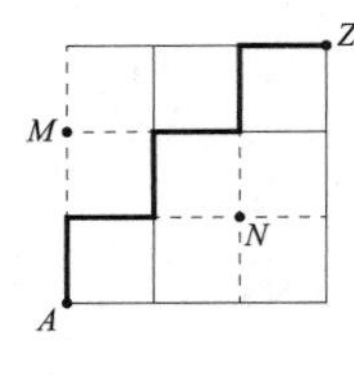

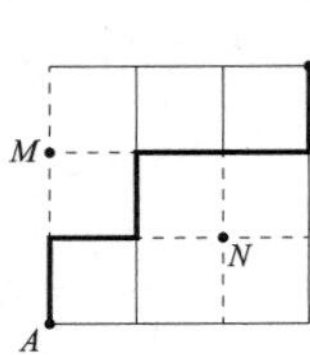

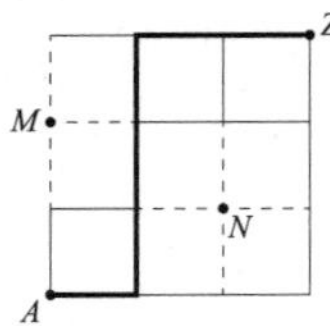

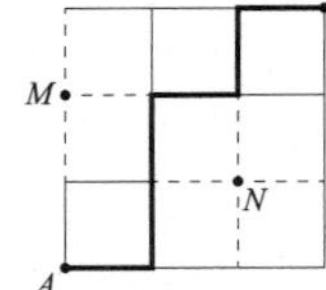

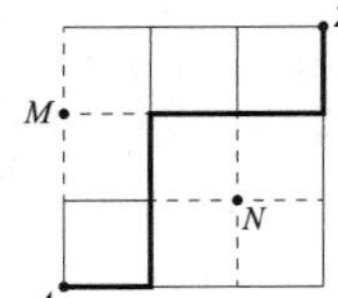

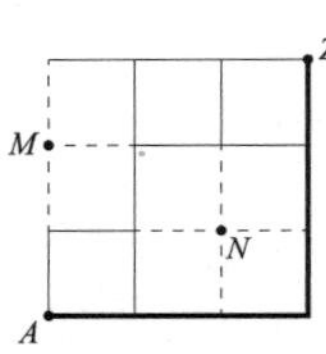

9.22 **40/9** or **4.44**. Carla walks for 25 minutes at a rate of 2 meters per second. That means she walks $2(60) = 120$ meters per minute over a 25 minute period, or $120(25) = 3{,}000$ meters. Recall that there are 1,000 meters in a kilometer, so Carla walks 3 kilometers. Mike walks for 15 minutes at an unknown (but constant) rate of speed. The figure below shows their relative positions at precisely 4:25 p.m. In the diagram, M represents Mike's position, C represents Carla's position, and W is the building at which they work.

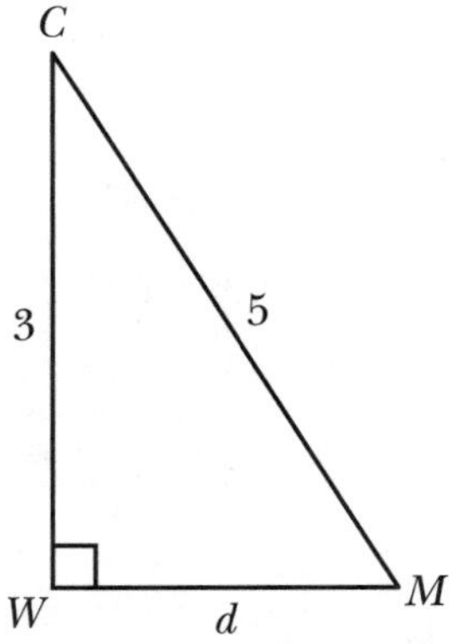

Notice that *CWM* is a right triangle. Apply the Pythagorean theorem to calculate *d*, or notice that 3 and 5 are part of the Pythagorean triple 3, 4, 5. Therefore, Mike has traveled 4 kilometers, or 4,000 meters, in 15 minutes. There are 15(60) = 900 seconds in 15 minutes, so Mike has traveled 4,000 meters in 900 seconds, which is equal to $4{,}000 \div 900 = 4.\overline{4}$ meters per second.

9.23 **.333** or **1/3**. A cube with side length 6 can be cut into 27 cubes of side length 2, as illustrated below.

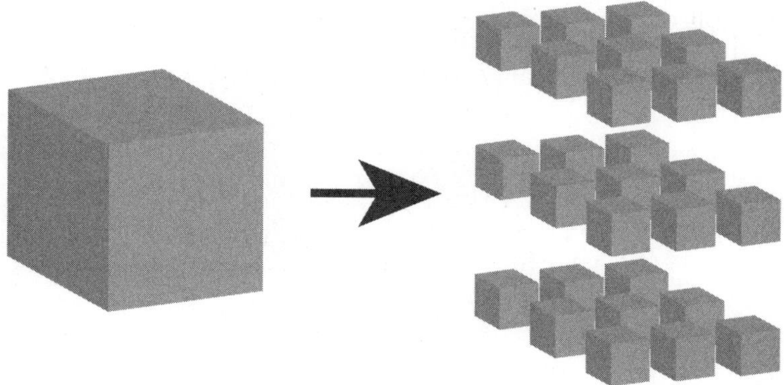

Each cube has 6 sides, so all 27 cubes have a total of 6(27) = 162 sides. Each side of the large cube is painted, and each of the 6 large painted sides represents 9 sides of the smaller cubes. Therefore, 6(9) = 54 of the smaller sides are painted. The ratio of painted to unpainted sides is 54/162 = 1/3.

9.24 **6.36**. Consider right triangle *PQM* in the diagram below, such that *Q* is the center of the square (the point at which the square's diagonals intersect), the altitude is $\overline{PQ}$, and $\overline{MQ}$ connects points *Q* and *M*. This right triangle contains $\overline{MP}$, the segment whose length you are asked to calculate.

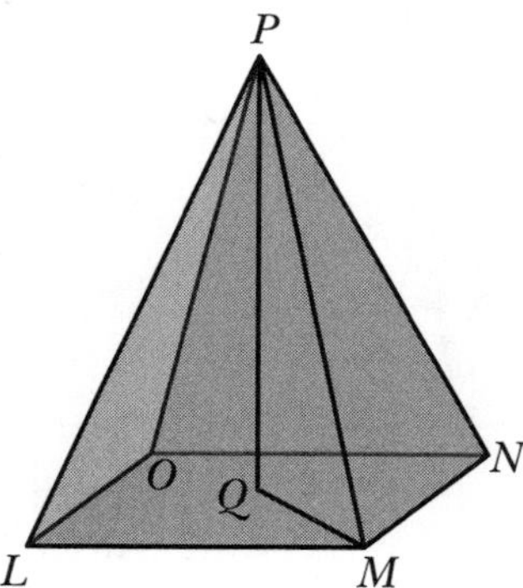

You are given *PQ* = 4, but you need to calculate *MQ* in order to apply the Pythagorean theorem to right triangle *PQM*. The diagram below gives an overhead view of the base of the pyramid, square *LMNO*. Notice that *LMO* is an isosceles right triangle, or a 45°–45°–90° triangle. Therefore, $MO = 7\sqrt{2}$ and the two halves of that hypotenuse are each half as long: $MQ = OQ = 7\sqrt{2} / 2$.

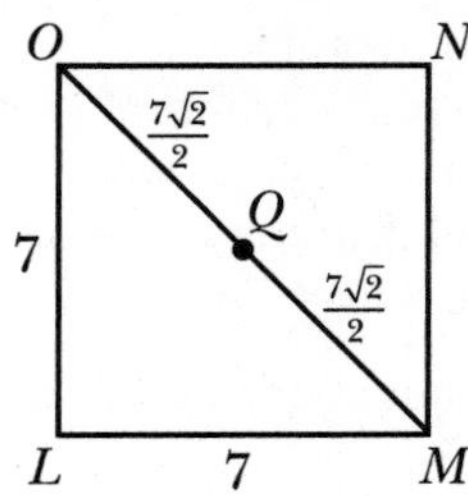

Now apply the Pythagorean theorem to right triangle *PQM*.

$$(PQ)^2 + (MQ)^2 = (MP)^2$$

$$4^2 + \left(\frac{7\sqrt{2}}{2}\right)^2 = (MP)^2$$

$$16 + \frac{49(2)}{4} = (MP)^2$$

$$16 + \frac{49}{2} = (MP)^2$$

$$\frac{32}{2} + \frac{49}{2} = (MP)^2$$

$$\frac{81}{2} = (MP)^2$$

$$\sqrt{\frac{81}{2}} = \sqrt{(MP)^2}$$

$$\frac{9}{\sqrt{2}} = MP$$

$$6.363961 \approx MP$$

9.25 **2771**. The diagram below illustrates the information given.

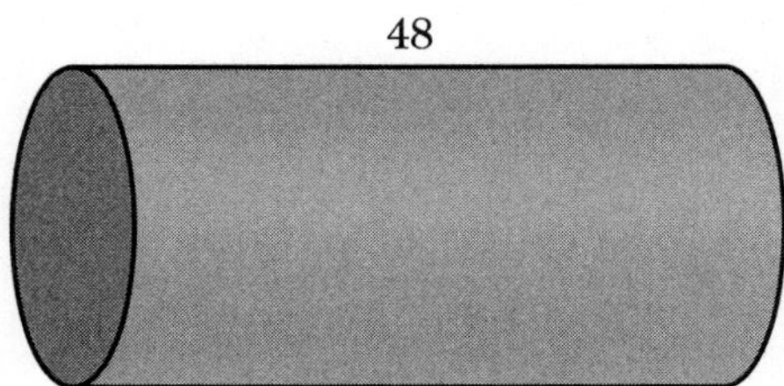

You are given $h = 48$ feet and $V = 75\pi$ cubic feet. You need to calculate the radius of the cylinder. Note that the measurements are reported in feet. Apply the formula for the volume of a cylinder.

But the answer needs to be in square INCHES. After you calculate the radius, convert it to inches.

$$V = \pi r^2 h$$
$$75\pi = \pi r^2 (48)$$
$$\frac{75\cancel{\pi}}{48\cancel{\pi}} = r^2$$
$$\frac{25}{16} = r^2$$
$$\sqrt{\frac{25}{16}} = \sqrt{r^2}$$
$$\frac{5}{4} = r$$

The radius is 5/4, or 1.25, feet. This is the equivalent of 12(1.25) = 15 inches. The question states that the cylinder is cut into 40 equal pieces, so the height (or length) of each disc is 48/40 =1.2 feet. This is the equivalent of 12(1.2) = 14.4 inches. The surface area of each disc includes the areas of the bases, which are circles with a 15-inch radius.

$$\begin{aligned}\text{area of one circular base} &= \pi r^2 \\ &= \pi(15)^2 \\ &= 225\pi \text{ square inches}\end{aligned}$$

The area of both circular bases is $2(225\pi) = 450\pi$ square inches. The surface area also includes the area of the side of the cylinder. As Problem 8.67 explains, the side of the cylinder is a rectangle whose length is the circumference of the base: $C = 2\pi r = 2\pi(15) = 30\pi$ inches. The width of the rectangle is $w = 14.4$ inches.

$$\begin{aligned}\text{rectangular area of side of cylinder} &= l \cdot w \\ &= 30\pi \cdot 14.4 \\ &= 432\pi \text{ square inches}\end{aligned}$$

The total surface area is $450\pi + 432\pi = 882\pi = 2{,}770.88472$ square inches. Round the answer to the nearest square inch: 2,771.

Chapter 10

MATH REVIEW: DATA ANALYSIS, STATISTICS, AND PROBABILITY

13% of the SAT

The final section of mathematical content tested on the SAT constitutes the smallest percentage of the test, by far. Some of the concepts explored in this chapter are fairly straightforward, including line graphs, pie charts, and arithmetic means (or averages). Be careful—just because a question looks simple does not mean it actually is. Besides, SAT writers always find a way to make things tricky.

Make sure to spend time reviewing the principles of probability at the end of this chapter, because they are just different enough from the other questions on the SAT to stagger you mentally and ruin your momentum on test day if you are not adequately prepared for them.

This is it—your math review is almost done. This chapter is shorter than the others, but don't speed through it. Take your time. Once you're finished practicing these concepts in Chapter 11, it's time to pull the training wheels off of this math bicycle and move on to the final part of the book: problem sets that look and feel like actual SAT tests.

Graphs and Data Interpretation

Pie charts, line graphs, bar graphs, pictographs

Note: Problems 10.1–10.3 refer to the graph below, which records the temperature (in degrees Fahrenheit) of an office, measured in two-hour intervals over a 24-hour period.

10.1 At what time is the lowest temperature recorded?

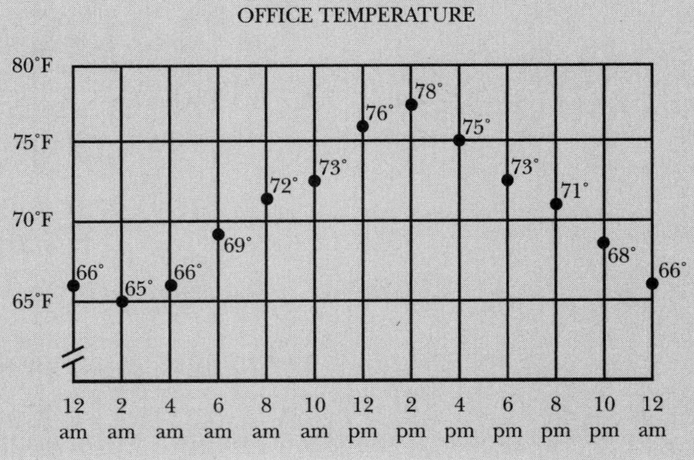

The lowest temperature in the office is 65°, recorded at 2 a.m.

Note: Problems 10.1–10.3 refer to the graph in Problem 10.1, which records the temperature (in degrees Fahrenheit) of an office, measured in two-hour intervals over a 24-hour period.

10.2 If the temperature changed at a constant rate between 4:00 p.m. and 8:00 p.m., what is the temperature at 4:45 p.m.?

If the temperature changes at a constant rate during the four hours specified by the problem, then the graph connecting 75° at 4 p.m. and 71° at 8 p.m. is a straight line. Over the four-hour period, the temperature decreases four degrees. In other words, it decreases one degree per hour, or 0.25 degrees every 15 minutes. Therefore, from 4 p.m. to 4:45 p.m. it decreases $3(0.25) = 0.75$ degrees. The temperature at 4:45 p.m. is $75° - 0.75° = 74.25°$.

Note: Problems 10.1–10.3 refer to the graph in Problem 10.1, which records the temperature (in degrees Fahrenheit) of an office, measured in two-hour intervals over a 24-hour period.

10.3 What is the average rate at which the temperature changes between 4:00 a.m. and 2:00 p.m.?

The temperature rises from 66° at 4 a.m. to 78° at 2 p.m., a total increase of 12° in a 10-hour period. The average rate of increase is the temperature change divided by the time change.

In other words, the slope of the line passing through the points.

$$\text{average rate of change} = \frac{12°}{10 \text{ hours}}$$
$$= \frac{12}{10} \text{ degrees per hour}$$
$$= \frac{6}{5} \text{ or } 1.2 \text{ degrees per hour}$$

Note: Problems 10.4–10.6 refer to the graph below, which records the average precipitation (in inches) for three states during the four seasons of the year.

10.4 What is Florida's annual average precipitation?

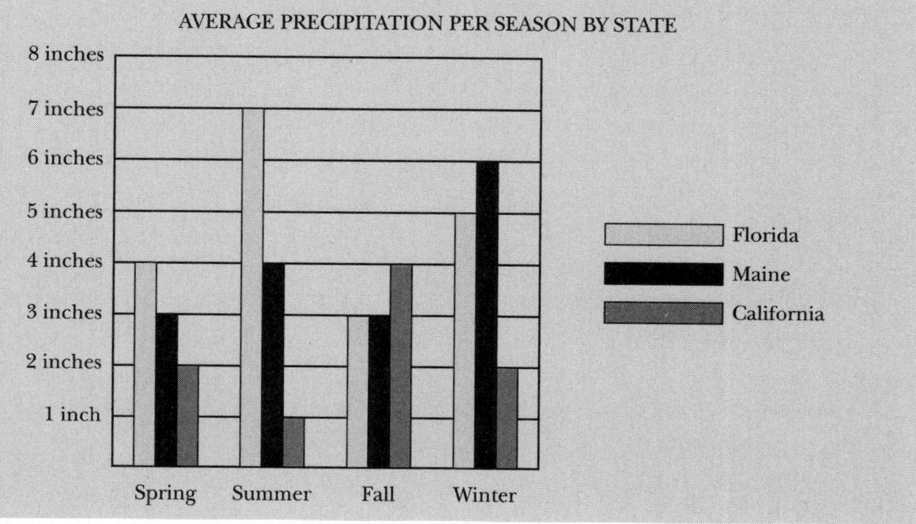

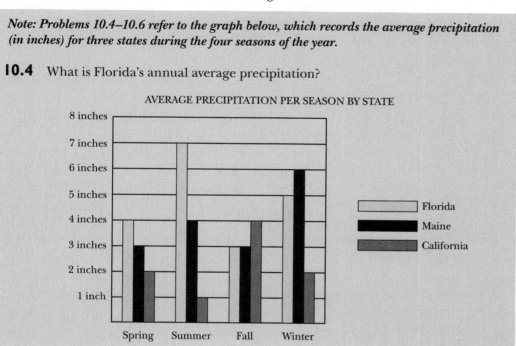

Florida's annual average precipitation is the total of its precipitation averages for all four seasons.

$$\text{annual average} = \text{spring average} + \text{summer average} + \text{fall average} + \text{winter average}$$
$$= 4 + 7 + 3 + 5$$
$$= 19 \text{ inches}$$

Note: Problems 10.4–10.6 refer to the graph in Problem 10.4, which records the average precipitation (in inches) for three states during the four seasons of the year.

10.5 How much more precipitation would California need per month during the winter and spring in order to match Maine's precipitation during the same seasons?

Maine receives 6 + 3 = 9 inches of precipitation during the six months of winter and spring. California receives 2 + 2 = 4 inches. Thus, California has 9 – 4 = 5 fewer inches of precipitation. In order to match Maine's total, California would need to accumulate those 5 additional inches of precipitation over the 6-month period, an extra $5 / 6 = 0.8\overline{3}$ inches of rain per month.

The 12-month year is divided into four seasons of equal length, so each season lasts 12 ÷ 4 = 3 months.

Note: Problems 10.4–10.6 refer to the graph in Problem 10.4, which records the average precipitation (in inches) for three states during the four seasons of the year.

10.6 In which seasons does the precipitation in California represent less than 20% of the total precipitation in all three states during those seasons?

Calculate the total precipitation for all three states each season.

$$\begin{aligned}&\text{spring:} && 4+3+2 &&= 9 \text{ inches}\\ &\text{summer:} && 7+4+1 &&= 12 \text{ inches}\\ &\text{fall:} && 3+3+4 &&= 10 \text{ inches}\\ &\text{winter:} && 5+6+2 &&= 13 \text{ inches}\end{aligned}$$

Now calculate the percentage of each total attributed to California.

$$\begin{aligned}&\text{spring:} && 2/9 &&= 22.\overline{2}\%\\ &\text{summer:} && 1/12 &&= 8.\overline{3}\%\\ &\text{fall:} && 4/10 &&= 40\%\\ &\text{winter:} && 2/13 &&\approx 15.4\%\end{aligned}$$

California accounts for less than 20% of the total precipitation in all three states during the summer and winter seasons.

Note: Problems 10.7–10.8 refer to the pictograph below, which lists the average salaries for various occupations in the U.S.

10.7 Which occupation's average salary is twice the average salary of an elementary school teacher?

Occupation	Average Salary
civil engineer	$ $ $ $ $ $
elementary school teacher	$ $ $ ½
flight attendant	$ $ $ $
licensed practical nurse	$ $ $
pharmacist	$ $ $ $ $ $ $
real estate attorney	$ $ $ $ $ $ $ $ $ $ ½
sales representative	$ $ $ $ $

$ = $10,000

The pictograph lists 3.5 money bags next to elementary school teacher. According to the diagram, each bag represents $10,000. Thus, the average salary is 3.5($10,000) = $35,000. Twice this salary is 2($35,000) = $70,000. Note that the pharmacist lists 7 money bags as the average salary, which is equal to 7($10,000) = $70,000. You conclude that a pharmacist's average salary is twice that of an elementary school teacher.

Note: Problems 10.7–10.8 refer to the pictograph in Problem 10.7, which lists the average salaries for various occupations in the U.S.

10.8 What two occupations' average salaries have the same sum as the average salaries of a civil engineer and a sales representative?

The average salary of a civil engineer is \$60,000, and the average salary of a sales representative is \$50,000. The sum of those salaries is \$110,000. The only two other average salaries that have the same sum are the pharmacist (\$70,000) and the flight attendant (\$40,000).

Note: Problems 10.9–10.10 refer to the graph below, which illustrates the total number of wins per season for two professional basketball teams over an 11-year period.

10.9 During which season was the difference of the two teams' total wins the least? In which season was the difference the greatest?

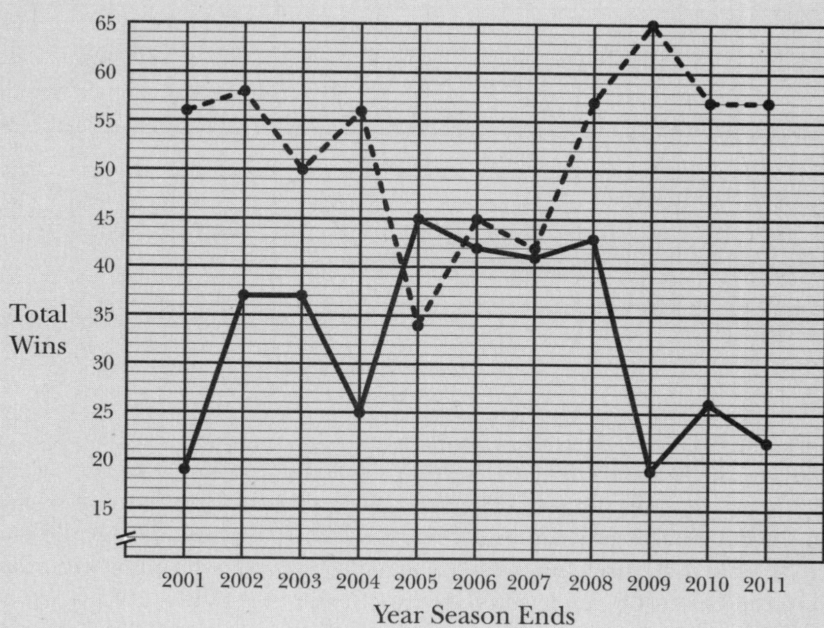

The two line graphs are closest in the 2007 season, so that year represents the smallest difference in total wins. In that year, the Lakers won one more game than the Wizards, 42 wins versus 41 wins.

In other words, the 2006–2007 NBA season.

The graphs are farthest apart in 2009, when the Lakers record their highest number of wins and the Wizards tie their worst performance: 65 wins versus 19 wins.

Note: Problems 10.9–10.10 refer to the graph in Problem 10.9, which illustrates the total number of wins per season for two professional basketball teams over an 11-year period.

10.10 If both teams continue to win or lose at the same average rate defined by the last three seasons in the graph, how many years will it take the Wizards to win more games than the Lakers in a single season?

The average win or loss rate is the slope of the line connecting the points at the start and end of the period. The problem states that the period spans the final three seasons, so on both line graphs, connect the points representing the 2009 and 2011 seasons, and extend them beyond the edge of the graph, as illustrated below.

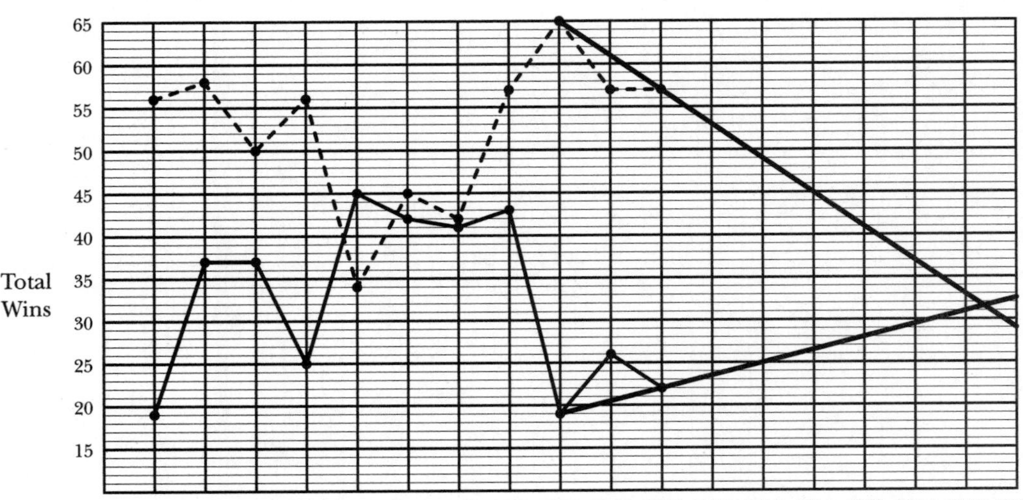

Between 2009 and 2011, the Lakers dropped from 65 wins to 57 wins, a decrease of 8 wins in two years. Therefore, they would continue to lose an average of $8 \div 2 = 4$ games per year. The Wizards improved from 19 to 22 wins, an improvement of 3 games over a 2-year period. Therefore, they would continue to win $3 \div 2 = 1.5$ more games each year. The Wizards will finally win more games than the Lakers during the 2018 season. It will take $2018 - 2011 = 7$ years.

In those 7 years, the Lakers will go from 57 to $57 - 7(4) = 29$ wins. The Wizards will go from 22 to approximately $22 + 7(1.5) = 32.5$ wins.

Note: Problems 10.11–10.12 refer to the table below, which lists the number of different types of pizzas sold at a certain restaurant during a one-week period.

10.11 Construct a pie chart (circle graph) to illustrate the data.

A pizza pie chart!

pizza variety	total sold
cheese	80
pepperoni	50
sausage	40
bacon	25
other	5

First calculate the total number of pizzas sold: 80 + 50 + 40 + 25 + 5 = 200. Now calculate the percentage of the 200 pizzas that each variety represents.

cheese: $\frac{80}{200} = 0.4 = 40\%$

pepperoni: $\frac{50}{200} = 0.25 = 25\%$

sausage: $\frac{40}{200} = 0.2 = 20\%$

bacon: $\frac{25}{200} = 0.125 = 12.5\%$

other: $\frac{5}{200} = 0.025 = 2.5\%$

To construct a circle graph, you need to calculate the central angle for the sector representing each variety of pizza. Recall that a circle contains 360°, so the central angle representing the cheese sector is 40 percent of the circle: $360°(0.4) = 144°$. Repeat this process for the other pizza varieties.

pepperoni: $360°(0.25) = 90°$

sausage: $360°(0.2) = 72°$

bacon: $360°(0.125) = 45°$

other: $360°(0.025) = 9°$

Finally, construct a circle with the given central angle measurements. It does not have to be precise, so do not bother using a protractor unless you wish it to be extremely accurate. Make sure to label each of the sectors.

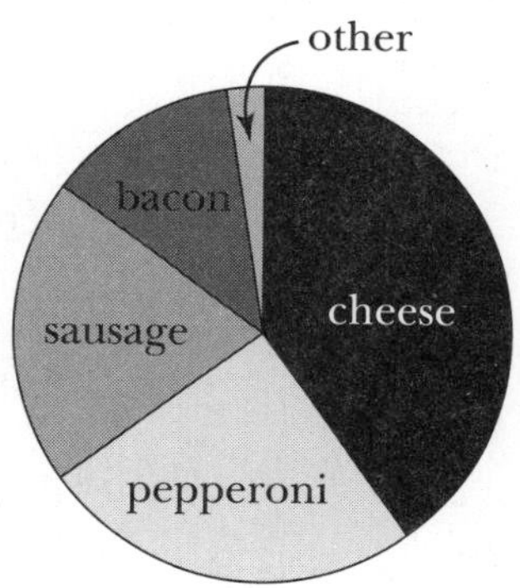

Note: Problems 10.11–10.12 refer to the table in Problem 10.11, which lists the number of different types of pizzas sold at a certain restaurant during a one-week period.

10.12 If the restaurant sells 312 pizzas the following week, in the same ratio as the week described by the chart, approximately how many bacon pizzas were sold over the two-week period?

According to Problem 10.11, 12.5% of the pizzas sold were bacon. Therefore, 12.5% of the 312 pizzas sold the following week were bacon: 312(0.125) = 39. A total of 25 + 39 = 64 bacon pizzas were sold over the two-week period.

This is the number of bacon pizzas sold in the first of the two weeks, the number from the bacon row of the chart in Problem 10.11.

10.13 The diagram below reports the colors of the cars sold at a particular dealership during a one-month period in which the sales of blue and black cars were equal. If 61 of the 404 total cars sold that month were black, how many brown cars were sold? Note that the graph includes two diameters of the circle.

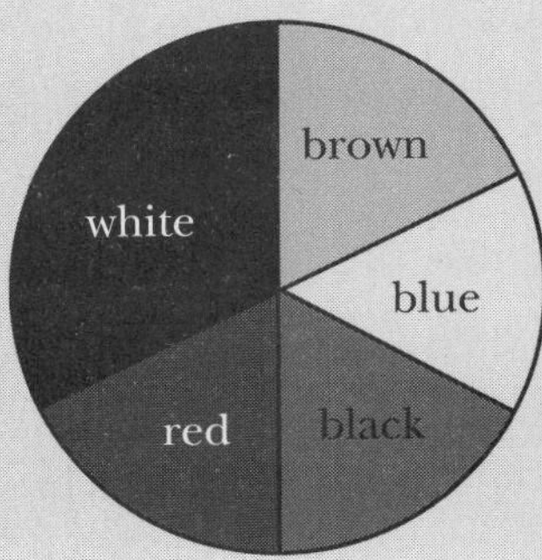

Consider the diagram below, which replaces the shading and labels of the diagram with numbered angles.

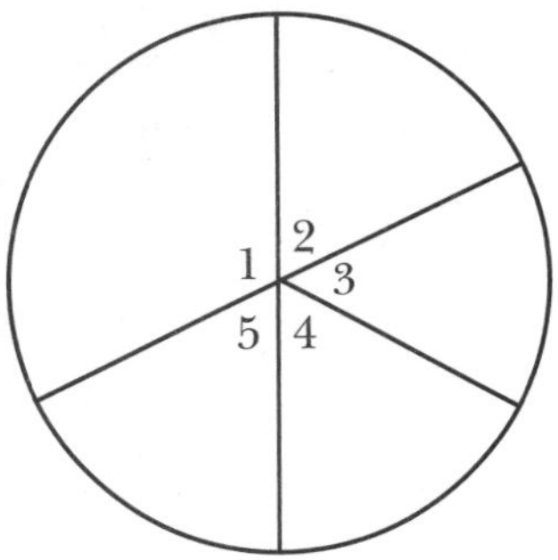

The problem states that the sales of blue and black cars were equal, so angles 3 and 4 are congruent. 61 black cars and 61 blue cars were sold, for a total of 122 cars. The problem states that the two segments that look like diameters actually are diameters. Therefore, angle 1 and the angle formed by combining 3 and 4 are vertical angles. They have the same central angle, so they represent the same number of cars sold. In other words, 122 white cars were sold.

This leaves 404 – 122 – 122 = 160 cars that are either brown or red, angles 2 and 5 in the diagram. The central angles of those regions are also vertical angles, so they are congruent—the same number of red and brown cars were sold. Thus, 160 ÷ 2 = 80 red cars and (the solution to the problem) 80 brown cars were sold.

Statistical Measures of Central Tendency

Mean, median, mode, weighted averages

Problems 10.14–10.17 refer to the following data values: 3, –3, –6, 12, –5, 4, –1, 0, 4.

10.14 What is the average (arithmetic mean) of the data?

An average is also called a mean, but because there are different types of means (such as a geometric mean) the specific term "arithmetic mean" is used to describe an average. To compute the average, add the values and divide by the number of values you added. In this data set, there are 9 values. Calculate their sum.

Zero is one of the data values. Even though it doesn't affect the sum, you do count it as a data value, so divide by 9, NOT 8.

$$3 + (-3) + (-6) + 12 + (-5) + 4 + (-1) + 0 + 4 = 8$$

The average of the data is the sum, 8, divided by the number of data values: $8 \div 9 = 0.\overline{8}$.

Problems 10.14–10.17 refer to the following data values: 3, –3, –6, 12, –5, 4, –1, 0, 4.

10.15 What is the mode of the data?

The mode of a data set is the value that occurs most often. A data set may contain more than one mode. In this data set, 4 is the only value that occurs more than once. Thus, the mode is 4. Note that 3 is not the mode; 3 and –3 are not the same values—they are opposites—so each of those values only occurs once.

Problems 10.14–10.17 refer to the following data values: 3, –3, –6, 12, –5, 4, –1, 0, 4.

10.16 What is the median of the data?

The median of the data is the "middle" number once all of the data values have been listed in order, from least to greatest. Begin by arranging the terms in increasing order.

$$-6, -5, -3, -1, 0, 3, 4, 4, 12$$

The middle number, and therefore the median, of this data set is 0.

Zero is the middle number because there are four numbers that are less than 0 and four numbers that are greater than 0.

Problems 10.14–10.17 refer to the following data values: 3, –3, –6, 12, –5, 4, –1, 0, 4.

10.17 If you add the data value 14 to the set, what are the new average (arithmetic mean) and median values?

According to Problem 10.14, the nine existing data values have a sum of 8, so adding the data value 14 brings the total sum to 8 + 14 = 22. Divide by 10, the number of elements in the new set, to calculate the new average: 22 ÷ 10 = 2.2.

To calculate the median, list the data values from least to greatest, as you did in Problem 10.16, but include the new data value 14.

$$-6, -5, -3, -1, 0, 3, 4, 4, 12, 14$$

There is no "middle value" of this data set, because there is an even number of data values. When this occurs, you should take the two middle numbers, in this case 0 and 3, and calculate their average.

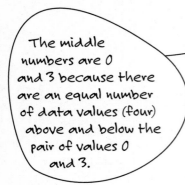

The middle numbers are 0 and 3 because there are an equal number of data values (four) above and below the pair of values 0 and 3.

$$\begin{aligned}\text{median} &= \frac{0+3}{2}\\ &= \frac{3}{2}\\ &= 1.5\end{aligned}$$

The median of the new data set is 1.5. Note that the median of an odd number of data values, like the median you calculated in Problem 10.16, is always a data value of the set. The median of an even number of data values may not be, as this problem demonstrates.

Problems 10.18–10.20 refer to the data values below, in which c is a positive integer.

$$c-2,\ c,\ c+3,\ c-4,\ c+2,\ c-1,\ c-5,\ -2(1-c)-c$$

10.18 What is the mode of the data?

Begin by simplifying the final data value.

$$\begin{aligned}-2(1-c)-c &= -2(1)-2(-c)-c\\ &= -2+2c-c\\ &= c-2\end{aligned}$$

Each of the terms occurs once, except $c-2$. Therefore, the mode is $c-2$.

Problems 10.18–10.20 refer to the data values below, in which c is a positive integer.

$$c-2,\ c,\ c+3,\ c-4,\ c+2,\ c-1,\ c-5,\ -2(1-c)-c$$

10.19 What is the median of the data?

List the data values in order, from least to greatest. Remember that $-2(1-c)-c = c-2$, according to Problem 10.18.

$$c-5,\ c-4,\ c-2,\ c-2,\ c-1,\ c,\ c+2,\ c+3$$

There is an even number of data values, so the median is the average of the two middle values: $c-2$ and $c-1$.

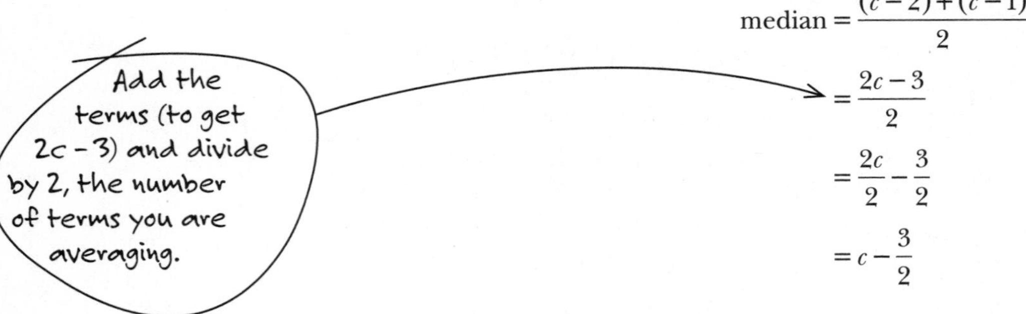

Add the terms (to get 2c - 3) and divide by 2, the number of terms you are averaging.

$$\begin{aligned}\text{median} &= \frac{(c-2)+(c-1)}{2}\\ &= \frac{2c-3}{2}\\ &= \frac{2c}{2}-\frac{3}{2}\\ &= c-\frac{3}{2}\end{aligned}$$

Problems 10.18–10.20 refer to the data values below, in which c is a positive integer.

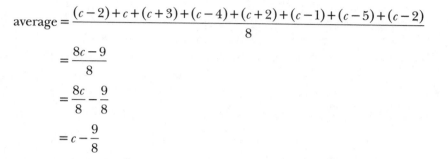

$$c-2,\ c,\ c+3,\ c-4,\ c+2,\ c-1,\ c-5,\ -2(1-c)-c$$

10.20 What is the average (arithmetic mean) of the data?

According to Problem 10.18, the final data value is $-2(1-c)-c=c-2$. Add the data values and divide by 8, the number of data values.

$$\begin{aligned}\text{average} &= \frac{(c-2)+c+(c+3)+(c-4)+(c+2)+(c-1)+(c-5)+(c-2)}{8}\\ &= \frac{8c-9}{8}\\ &= \frac{8c}{8}-\frac{9}{8}\\ &= c-\frac{9}{8}\end{aligned}$$

Problems 10.21–10.23 refer to the scatterplot below, which reports the number of hours a group of students at different grade levels spends doing homework in one week.

10.21 What is the average (arithmetic mean) of the time spent per week on homework for the entire group of students?

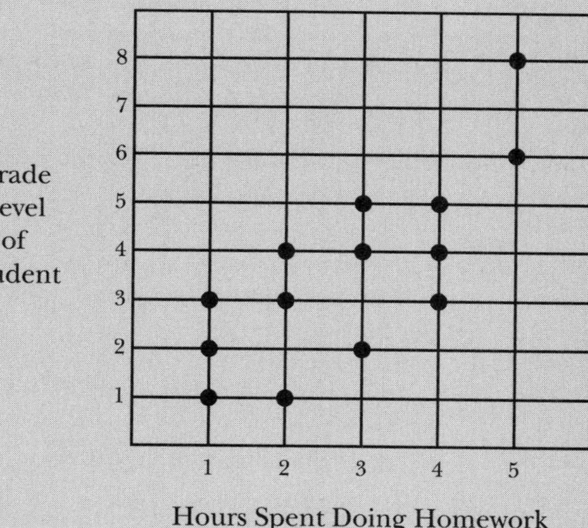

Each point on the scatterplot represents two pieces of data. For example, the point at (5,8) indicates that an 8th grade student spent 5 hours doing homework in one week. This problem asks you to focus on the time spent per week, the first number in the coordinate pair of each point.

Three students spent 1 hour, because there are three dots on the vertical line representing 1 hour. (Those students were in Grades 1, 2, and 3.) Similarly, there are three students who spent 2 hours, three students who spent 3 hours, three students who spent 4 hours, and two students who spent 5 hours. List the values of the data set, such that each value represents a period of hours spent on homework by one of the students.

$$1,\ 1,\ 1,\ 2,\ 2,\ 2,\ 3,\ 3,\ 3,\ 4,\ 4,\ 4,\ 5,\ 5$$

Calculate the average.

$$\begin{aligned}\text{average} &= \frac{1+1+1+2+2+2+3+3+3+4+4+4+5+5}{14}\\ &= \frac{40}{14}\\ &= \frac{20}{7}\\ &\approx 2.857\end{aligned}$$

Problems 10.21–10.23 refer to the scatterplot in Problem 10.21, which reports the number of hours a group of students at different grade levels spends doing homework in one week.

10.22 What is the median of the average time spent doing homework for the entire group of students?

According to Problem 10.21, you can list the times spent by students as the following set of data values.

1, 1, 1, 2, 2, 2, 3, 3, 3, 4, 4, 4, 5, 5

There is an even number of data values, so the median is the average of the two middle numbers, the seventh and eighth data values, both of which are 3. Therefore, the median of the data set is 3.

Problems 10.21–10.23 refer to the scatterplot in Problem 10.21, which reports the number of hours a group of students at different grade levels spends doing homework in one week.

10.23 Which is greater, the average (arithmetic mean) or the median grade levels of the students?

List the grade levels of the students in the same way you listed the hours they spent doing homework in Problem 10.21. The list should include, for example, two students in Grade 1 because there are two dots on the horizontal line for Grade 1. Similarly, there are three Grade 4 students and one Grade 6 student, to name a few.

1, 1, 2, 2, 3, 3, 3, 4, 4, 4, 5, 5, 6, 8

If there are N data values and N is even, then the median of the N data values is the average of the N/2 term and the (N/2) + 1 term. In this case, there are 14 values, so the median is the average of the 14/2 = 7th term and the 7 + 1 = 8th term.

The two middle terms of the data are, once again, the seventh and eighth terms in the list: 3 and 4. The median is the average of the two values: 3.5.

The average of the grade values is the sum of the grades divided by 14, the total number of data values.

$$\begin{aligned}\text{average} &= \frac{1+1+2+2+3+3+3+4+4+4+5+5+6+8}{14}\\ &= \frac{51}{14}\\ &\approx 3.643\end{aligned}$$

The average is greater than the median: 3.643 > 3.5.

10.24 If four consecutive odd integers have an arithmetic mean of 6, what is the sum of the integers?

If you're not sure why, flip back to Problems 4.2 and 4.3.

Let x, $x + 2$, $x + 4$, and $x + 6$ represent the four consecutive odd integers. If the average (arithmetic mean) of the numbers is 6, then the sum of the numbers divided by 4 is equal to 6.

$$\frac{x+(x+2)+(x+4)+(x+6)}{4}=6$$

$$\frac{4x+12}{4}=6$$

$$\frac{4x}{4}+\frac{12}{4}=6$$

$$x+3=6$$

$$x=6-3$$

$$x=3$$

The four odd integers are $x = 3$, $x = 5$, $x = 7$, and $x = 9$. To solve this problem without performing any arithmetic, you could note that 3 and 5 are the odd integers closest to *and less than* 6, whereas 7 and 9 are the odd integers closest to *and greater than* 6. Consider the diagram below, in which the four integers and their average are plotted.

Imagine that the four integers represent four equal weights in the diagram. The average, 6, is the balance point for those values, like the fulcrum of a seesaw balancing four children, two on each side.

Note: Problems 10.25–10.27 refer to the chart below, the shoe sizes of a group of children at an elementary school.

10.25 What is the median shoe size for the group?

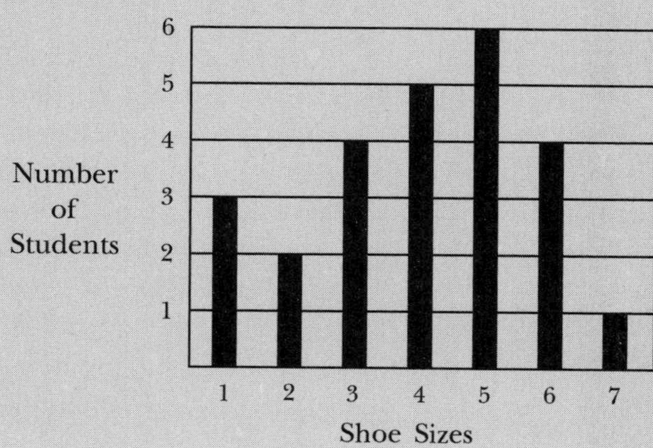

The bar chart is a frequency distribution; it expresses how many of each shoe size are data values in the set. In other words, three students have shoe size 1, two students have shoe size 2, four students have shoe size 3, and so on. Note that there are 3 + 2 + 4 + 5 + 6 + 4 + 1 = 25 students in the class. List all of the sizes in order from least to greatest, in an orderly fashion. The median is the middle number, the boxed data value below.

$$\begin{array}{ccccc} 1 & 1 & 1 & 2 & 2 \\ 3 & 3 & 3 & 3 & 4 \\ 4 & 4 & \boxed{4} & 4 & 5 \\ 5 & 5 & 5 & 5 & 5 \\ 6 & 6 & 6 & 6 & 7 \end{array}$$

The median of the data is 4.

Note: Problems 10.25–10.27 refer to the chart in Problem 10.25, the shoe sizes of a group of children at an elementary school.

10.26 What is the average shoe size?

While you could add all 25 of the shoe sizes listed in Problem 10.25, it is more efficient to apply a weighted average. Instead of writing three shoe sizes of 1, you would write 3(1), indicating that the value 1 is repeated three times. Similarly, you would write 6(5), indicating that six students have a shoe size of 5.

$$\begin{aligned} \text{average} &= \frac{3(1)+2(2)+4(3)+5(4)+6(5)+4(6)+1(7)}{\text{total number of students}} \\ &= \frac{3+4+12+20+30+24+7}{25} \\ &= \frac{100}{25} \\ &= 4 \end{aligned}$$

The average shoe size is 4.

Note: Problems 10.25–10.27 refer to the chart in Problem 10.25, the shoe sizes of a group of children at an elementary school.

10.27 One student was absent on the day the data was collected. When her shoe size is included, it lowers the overall average by approximately 0.115. What is her shoe size?

The original 25 students had an average shoe size of 4. Notice that the sum of the original shoe sizes is 100, a number that you can easily reproduce by multiplying the number of original students by their average shoe size: 25(4) = 100. To calculate the new average, add all 26 shoe sizes (using this shortcut) and divide by 26, the number of students. Let x represent the unknown shoe size.

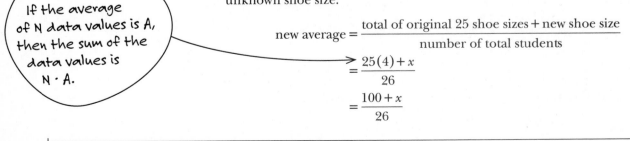

If the average of N data values is A, then the sum of the data values is N · A.

$$\begin{aligned} \text{new average} &= \frac{\text{total of original 25 shoe sizes} + \text{new shoe size}}{\text{number of total students}} \\ &= \frac{25(4)+x}{26} \\ &= \frac{100+x}{26} \end{aligned}$$

The problem states that adding the student decreases the original average of 4 by approximately 0.115. The new average is approximately 4 – 0.115 = 3.885.

$$3.885 = \frac{100+x}{26}$$

Cross-multiply to solve the proportion.

$$\begin{aligned} 3.885(26) &= 100+x \\ 101.01 &= 100+x \\ 101.01-100 &= x \\ 1.01 &= x \end{aligned}$$

Round the answer to the nearest shoe size—remember that the change in averages was stated as *approximate*. The 26th student's shoe size was 1.

Note: In Problems 10.28–10.29, Sam intends to ship 9 books in the same box. All of the books are the same weight except for one, which is 5 times as heavy as one of the lighter books.

10.28 If x represents the weight shared by 8 of the books, what is the average weight of all the books, in terms of x?

The eight books of equal weight each have weight x. The heavy book has weight $5x$. Calculate the average weight of all nine books using a weighted average.

$$\begin{aligned} \text{average weight} &= \frac{\text{weight of light books} + \text{weight of heavy book}}{\text{total number of books}} \\ &= \frac{8(x)+5x}{9} \\ &= \frac{13}{9}x \end{aligned}$$

The average weight of all 9 books is $(13/9)x$.

Note: In Problems 10.28–10.29, Sam intends to ship 9 books in the same box. All of the books are the same weight except for one, which is 5 times as heavy as one of the lighter books.

10.29 If the heavy book weighs 7 pounds and Sam adds one additional book weighing 2 pounds to the shipment, what is the average weight of all the books?

If the heavy book weighs 7 pounds, then each of the lighter books weighs one-fifth as much: $(1/5)(7) = 1.4$ pounds.

The tenth book in the shipment weighs 2 pounds. Calculate the average weight of all 10 books.

$$\begin{aligned} \text{average weight} &= \frac{\text{weight of 8 books} + \text{7-pound book} + \text{2-pound book}}{10} \\ &= \frac{8(1.4)+7+2}{10} \\ &= \frac{20.2}{10} \\ &= 2.02 \text{ pounds} \end{aligned}$$

Note: Problems 10.30–10.31 refer to different groups of randomly selected people. Each group contains exactly 7 people, each person in the group is at least 10 years old, and each age is an integer.

10.30 If the average age of the group members is 14, what is the highest possible age of the oldest group member?

If the average age of the 7 group members is 14, then you know that the sum of their ages is 7(14) = 98. In order to generate the largest possible age for one group member, you need to subtract from that total the six smallest possible ages. The problem states that the group members are no younger than 10.

If six of the group members were 10 years old, that represents 6(10) = 60 of the sum calculated above, 98. Therefore, the remaining person in the group would be 98 – 60 = 38 years old.

If you were to pick any age other than 10 for the six younger members of the group, it would leave less of the 98 total years for the oldest person. Therefore, the highest possible age of the oldest group member is 38.

Note: Problems 10.30–10.31 refer to different groups of randomly selected people. Each group contains exactly 7 people, each person in the group is at least 10 years old, and each age is an integer.

10.31 If the group contains a 17-year-old member and a 33-year-old member, and the average age of the group is 20, what is the average of the unknown ages?

Remember that the average age is equal to the sum of the ages divided by the total number of ages. You are given two of the ages: 17 and 33. Let x represent the average of the 5 unknown ages.

$$\text{average age} = \frac{17+33+5(x)}{7}$$

The problem states that the average age is 20. Substitute this into the equation and cross-multiply to solve for x.

$$\begin{aligned}
20 &= \frac{50+5x}{7} \\
20(7) &= 50+5x \\
140 &= 50+5x \\
140-50 &= 5x \\
90 &= 5x \\
18 &= x
\end{aligned}$$

The average of the 5 unknown ages is 18. To check your work, note that the average of 17, 33, 18, 18, 18, 18, and 18 is 20.

Note: In Problems 10.32–10.33, a group of 50 hotel customers was asked to rate their most recent stay on a scale from 1 star to 5 stars, such that 5 stars is the best possible rating. The average rating was 4.2 stars.

10.32 How many consecutive 3-star ratings would it take to lower the average to 3.5 stars?

Let x represent the number of 3-star ratings it would take to lower the total average from 4.2 stars to 3.5 stars. Calculate a weighted average.

$$\text{average rating} = \frac{\text{sum of total ratings}}{\text{total number of customers surveyed}}$$

$$3.5 = \frac{\text{sum of original ratings} + \text{sum of consecutive 3-star ratings}}{\text{number of current customers} + \text{number of new customers}}$$

$$3.5 = \frac{50(4.2) + x(3)}{50 + x}$$

$$3.5 = \frac{210 + 3x}{50 + x}$$

Cross-multiply and solve the proportion for x.

$$\begin{aligned} 3.5(50 + x) &= 210 + 3x \\ 175 + 3.5x &= 210 + 3x \\ 3.5x - 3x &= 210 - 175 \\ 0.5x &= 35 \\ x &= \frac{35}{0.5} \\ x &= 70 \end{aligned}$$

It would take 70 consecutive 3-star ratings to lower the average rating from 4.2 to 3.5 stars.

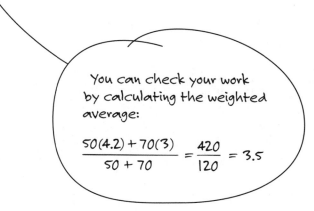

You can check your work by calculating the weighted average:

$$\frac{50(4.2) + 70(3)}{50 + 70} = \frac{420}{120} = 3.5$$

Note: In Problems 10.32–10.33, a group of 50 hotel customers was asked to rate their most recent stay on a scale from 1 star to 5 stars, such that 5 stars is the best possible rating. The average rating was 4.2 stars.

10.33 How many consecutive 5-star ratings would it take to raise the 4.2-star average to 4.5 stars?

Let x equal the number of consecutive 5-star ratings it would take to raise the average to 4.5 stars. Apply the technique demonstrated in Problem 10.32.

$$\begin{aligned}
\text{average rating} &= \frac{\text{sum of total ratings}}{\text{total number of customers surveyed}} \\
4.5 &= \frac{\text{sum of original ratings} + \text{sum of consecutive 5-star ratings}}{\text{number of current customers} + \text{number of new customers}} \\
4.5 &= \frac{50(4.2) + x(5)}{50 + x} \\
4.5 &= \frac{210 + 5x}{50 + x} \\
4.5(50 + x) &= 210 + 5x \\
225 + 4.5x &= 210 + 5x \\
225 - 210 &= 5x - 4.5x \\
15 &= 0.5x \\
\frac{15}{0.5} &= x \\
30 &= x
\end{aligned}$$

If the hotel receives 30 consecutive 5-star ratings, the average rating will increase from 4.2 stars to 4.5 stars.

Basic Probability

What are the odds?

Note: In Problems 10.34–10.37, marbles are drawn randomly from a bag that contains 8 white, 4 green, and 6 blue marbles.

10.34 What is the probability of selecting a blue marble from the bag?

When you draw a marble from the bag, there are 8 + 4 + 6 = 18 possible outcomes, because there are 18 marbles in the bag. Only 6 of those marbles are blue, so in 6 of the 18 outcomes, you draw a blue marble. The probability is 6/18 or 1/3.

You can report probability as a fraction (1/3), a decimal (1/3 = 0.333333...), or a percent (33.3333333...%).

Note: In Problems 10.34–10.37, marbles are drawn randomly from a bag that contains 8 white, 4 green, and 6 blue marbles.

10.35 What is the probability of *NOT* selecting a green marble from the bag?

Of the 18 marbles in the bag, 8 white + 6 blue = 14 of them are not green. Therefore, the probability of *not* drawing a green marble is 14/18 = 7/9.

Note: In Problems 10.34–10.37, marbles are drawn randomly from a bag that contains 8 white, 4 green, and 6 blue marbles.

10.36 Assume that each time Mabel draws a marble from the bag, she returns it to the bag before drawing again. What is the probability that the first two marbles she draws are green?

"Mabel draws a marble." Try to say that 10 times fast.

There are two events occurring here, and they are independent events. In other words, because she is drawing the marbles randomly, the marble color she chooses first does not affect the marble color she chooses second; each draw is completely random. According to the multiplication law of probability, if there is an x probability of one event occurring and a y probability of an independent event occurring, then the probability of them both occurring is $x \cdot y$.

There are 4 green marbles in the bag, so there is a 4/18 = 2/9 probability that the first marble she draws is green. Mabel then drops the marble she chose back into the bag, so the total number of marbles within is 18 once again. Just like the first time, there is a 2/9 probability of drawing a green marble. To determine the probability of both events occurring, multiply the probabilities.

$$\frac{2}{9} \cdot \frac{2}{9} = \frac{4}{81}$$

There is a 4/81 ≈ 4.94% probability that the first two marbles Mable draws will be green.

Note: In Problems 10.34–10.37, marbles are drawn randomly from a bag that contains 8 white, 4 green, and 6 blue marbles.

10.37 Each time Lisa draws a marble, she removes it from the bag. If Lisa has already drawn 2 green, 4 blue, and 1 white marble, what is the probability that the next two marbles she draws are white?

The first order of business is figuring out what is left in the bag. There are 4 – 2 = 2 green marbles, 6 – 4 = 2 blue marbles, and 8 – 1 = 7 white marbles left, a total of 2 + 2 + 7 = 11 marbles. There is a 7/11 chance that the first marble she draws will be white. However, once drawn, it is removed from the bag. This leaves 6 white marbles out of 10 total marbles in the bag. Therefore, there is a 6/10 = 3/5 probability that the second marble will be white. Apply the multiplication law of probability.

$$\frac{7}{11} \cdot \frac{3}{5} = \frac{21}{55}$$

There is a 21/55 ≈ 38.18% probability that the next two marbles she draws are white.

Note: In Problems 10.38–10.39, a player rolls two fair dice.

10.38 What is the probability that the dice will show the same value, such as ⚀⚀ or ⚅⚅?

You can apply the universal counting principle described in Problem 4.60 to calculate the number of possible rolls. The first die has one of six possible values, as does the second. Therefore, there are $6 \cdot 6 = 36$ possible ways to roll the dice, each of which is illustrated below.

You may be thinking, "Isn't a 1,4 roll the same thing as a 4,1 roll?" Not technically. They may have the same sum, but those are different rolls. If you counted only unique rolls of the dice, where 4,1 and 1,4 are the same, there are 21 possible rolls. However, those 21 rolls are not equally likely. For example, there are two ways to roll 1,4 (as explained already) but only one way to roll 6,6. Check out the big diagram o' dice to see why.

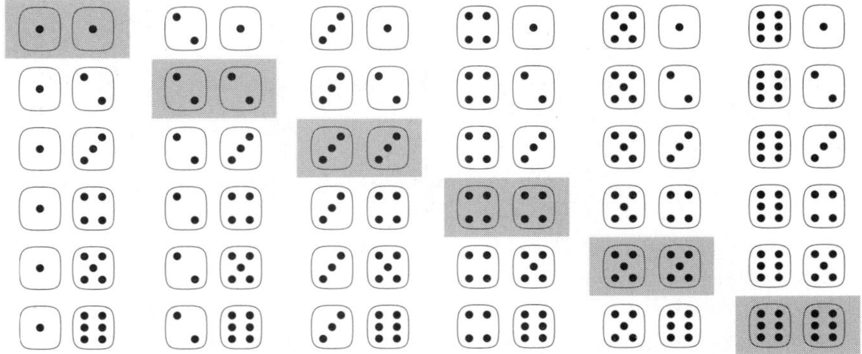

Of the 36 possible outcomes, exactly six rolls represent dice of the same value. Therefore, the probability of rolling the same number on the pair of dice is $6/36 = 1/6$.

Note: In Problems 10.38–10.39, a player rolls two fair dice.

10.39 What is the probability that the sum of the values rolled on the dice equals 8?

Refer to the diagram below, which illustrates all 36 possible rolls and highlights the rolls that have a sum of 8.

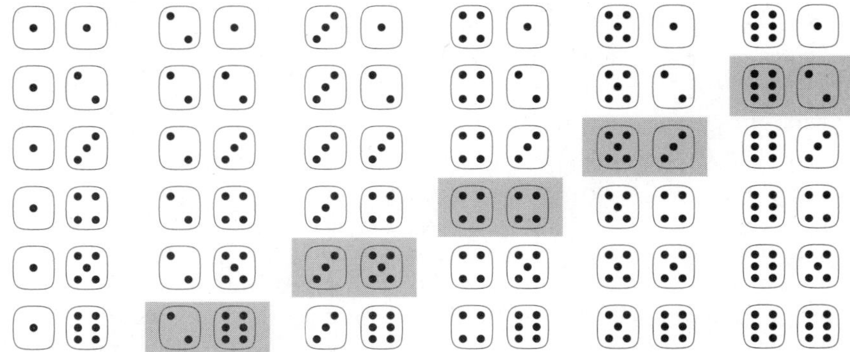

There are 5 ways to roll dice with a sum of 8, so the probability is $5/36$.

Note: Problems 10.40–10.41 describe a game in which players roll a normal, six-sided die on the square game board illustrated below, which contains a white square region, a black square region, and two shaded rectangular regions. Assume that the die is equally likely to fall anywhere on the game board.

10.40 What is the probability that a roll will land in a rectangular shaded region?

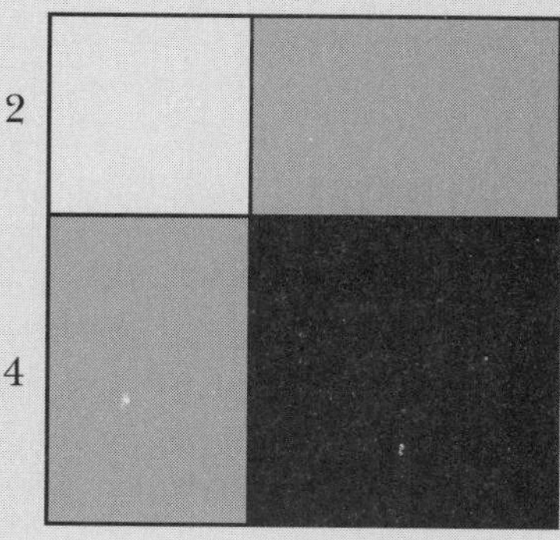

The square game board has sides of length $2 + 4 = 6$, so the area of the game board is $l \cdot w = 6 \cdot 6 = 36$. There are two shaded rectangular regions, each with area $l \cdot w = 2 \cdot 4 = 8$. Thus, the rectangular shaded area within the game board is $2(8) = 16$. The rectangular shaded regions account for $16/36 = 4/9$ of the game board. The probability that a roll will land in one of the shaded rectangles is $4/9$.

Note: Problems 10.40–10.41 describe a game in which players roll a normal, six-sided die on the square game board illustrated in Problem 10.40, which contains a white square region, a black square region, and two shaded rectangular regions. Assume that the die is equally likely to fall anywhere on the game board.

10.41 What is the probability that a player will roll a 4 that lands in the white square region?

There are two separate and independent events in this problem, the numerical outcome of the roll and the position of the roll. The probability of rolling a 4 with a fair die is $1/6$; rolling a 4 is 1 of 6 possible outcomes.

The white square region has an area of $l \cdot w = 2 \cdot 2 = 4$. According to Problem 10.40, the area of the game board is 36. Therefore, the white square region accounts for $4/36 = 1/9$ of the game board. The probability of a die landing in the white square region is $1/9$.

Multiply the two probabilities together to calculate the probability that a die will show 4 and land in the white square region.

$$\frac{1}{6} \cdot \frac{1}{9} = \frac{1}{54}$$

Note: Problems 10.42–10.46 refer to a standard dartboard, illustrated below. Its diameter is 17.75 inches, and it is divided into 20 congruent radial regions with varying point values, as indicated. The bull's-eye at the center of the dartboard has a 0.5-inch diameter; it is worth 50 points. The surrounding ring, called the bull, has a diameter of 1.25 inches and is worth 25 points.

There are also two concentric bands 0.35 inches wide. The inner band, the light shaded region of the illustration, grants triple the points in the region and has an outer edge that is 4.2 inches from the center. The outer band, the darker shaded region in the diagram, grants double the points of the region, and its outer ring is 6.7 inches from the center. Any darts falling outside the double ring do not score any points.

In these problems, assume that a thrown dart will strike the surface of the board, but it may not strike within the scoring area. In other words, it may land outside the double band. Also assume that a throw is equally likely to strike any point on the surface of the board. Though the regions of a dartboard are often bounded by wires, for the purposes of this problem, ignore the presence of the wires on the dartboard; assume the wires occupy no area.

10.42 What is the probability of hitting the bull's-eye on a single throw? Express your answer as a percentage rounded to the thousandths place.

The diameter of the dartboard is 17.75 inches, so its radius is 17.75 ÷ 2 = 8.875 inches. Calculate the area of the dartboard's surface.

$$\begin{aligned} A_{\text{dartboard}} &= \pi r^2 \\ &= \pi(8.875)^2 \\ &= 78.765625\pi \text{ square inches} \end{aligned}$$

The bull's-eye has a 0.5-inch diameter, so its radius is 0.5 ÷ 2 = 0.25 inches. Calculate its area.

$$\begin{aligned} A_{\text{bull's-eye}} &= \pi r^2 \\ &= \pi(0.25)^2 \\ &= 0.0625\pi \text{ square inches} \end{aligned}$$

The geometric probability of hitting the bull's-eye with a single throw is equal to the quotient of the bull's-eye area and the dartboard area.

$$\begin{aligned}\text{probability of hitting the bull's-eye} &= \frac{A_{\text{bull's-eye}}}{A_{\text{dartboard}}}\\ &= \frac{0.0625\not{\pi}}{78.765625\not{\pi}}\\ &\approx 0.0007935\end{aligned}$$

There is a 0.079% probability of hitting the bull's-eye in a single throw.

Note: Problems 10.42–10.46 refer to a standard dart board, as illustrated and described in Problem 10.42.

10.43 What is the probability of hitting one specific radial region of the dartboard—one of the 20 numbered regions—including the double and triple bands within that region? Express your answer as a percentage rounded to the thousandths place.

Any of the 20 radial regions extending from the center of the dartboard has an inner boundary defined by the outside of the bull and an outer boundary defined by the outside of the double band, as illustrated below. The bull has a diameter of 1.25 inches, so its radius is 1.25 ÷ 2 = 0.625 inches. Problem 10.42 states that the outer boundary of the double ring is 6.7 inches from the center.

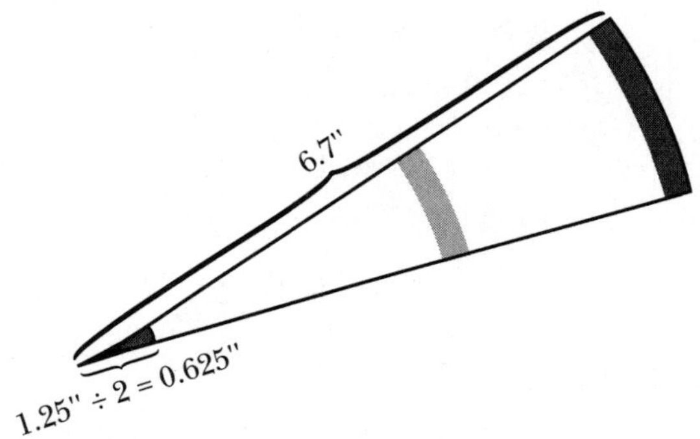

To calculate the area A of the radial region, compute the area A_1 of a circle with radius 6.7 inches and subtract the area A_2 of a circle with radius 0.625 inches. This represents the scoring area of the dartboard outside the bull. Then divide that area by 20, because all 20 radial regions are congruent.

$$\begin{aligned}A &= \frac{A_1 - A_2}{20}\\ &= \frac{\pi(6.7)^2 - \pi(0.625)^2}{20}\\ &= \frac{44.89\pi - 0.390625\pi}{20}\\ &= \frac{44.499375}{20}\pi\\ &\approx 2.22496875\pi \text{ square inches}\end{aligned}$$

The probability of throwing a dart into one specific region is equal to the area A divided by the area of the dartboard, 78.765625π according to Problem 10.42.

$$\begin{aligned}\text{probability of hitting a specific radial region} &= \frac{A}{A_{\text{dartboard}}}\\ &= \frac{2.22496875\not{\pi}}{78.765625\not{\pi}}\\ &\approx 0.028248\end{aligned}$$

The probability of throwing a dart in one specific radial region is 2.825%.

Note: Problems 10.42–10.46 refer to a standard dart board, as illustrated and described in Problem 10.42.

10.44 What is the probability of hitting the triple-20 band, the highest scoring region of the board? Express your answer as a percentage rounded to the thousandths place.

To calculate the area of the triple band within a single radial region, you apply a technique similar to the technique demonstrated in Problem 10.43. That narrow band has an outer bound 4.2 inches from the center of the dartboard. The band is 0.35 inches wide, so its inner bound is 4.2 – 0.35 = 3.85 inches from the center of the dartboard, as illustrated below.

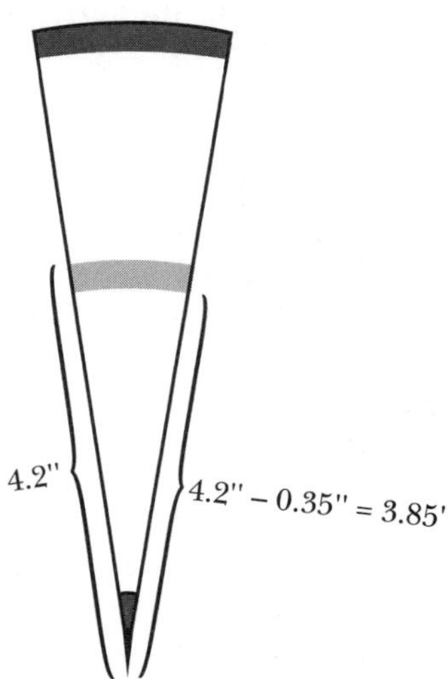

The area of the entire triple band for the whole dartboard is equal to the area A_1 of a circle with radius 4.2 inches minus the area A_2 of a circle with radius 3.85 inches. Exactly 1/20 of that area lies within a single radial scoring region of the dartboard, no matter which region that is, so divide $A_1 - A_2$ by 20.

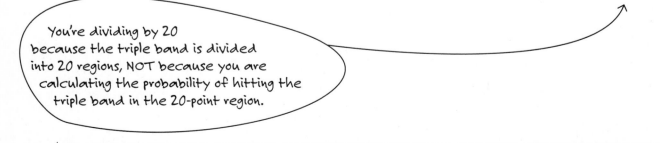

You're dividing by 20 because the triple band is divided into 20 regions, NOT because you are calculating the probability of hitting the triple band in the 20-point region.

$$\begin{aligned}\text{area of triple-20 region} &= \frac{A_1 - A_2}{20}\\ &= \frac{\pi(4.2)^2 - \pi(3.85)^2}{20}\\ &= \frac{17.64\pi - 14.8225\pi}{20}\\ &= \frac{2.8175}{20}\pi\\ &= 0.140875\pi \text{ square inches}\end{aligned}$$

The probability of hitting the triple-20 region is the area of that region divided by the area of the dartboard, 78.765625π.

$$\begin{aligned}\text{probability of hitting triple-20 region} &= \frac{A_{\text{triple-20 region}}}{A_{\text{dartboard}}}\\ &= \frac{0.140875\cancel{\pi}}{78.765625\cancel{\pi}}\\ &\approx 0.0017885\end{aligned}$$

The probability of hitting the highest-scoring region of the dartboard is approximately 0.179%.

Note: Problems 10.42–10.46 refer to a standard dart board, as illustrated and described in Problem 10.42.

10.45 What is the probability of scoring 38 points in a single throw? Express your answer as a percentage rounded to the thousandths place.

The only way to score 38 points in a single throw is to hit the double band of the 19-point radial region. All 20 radial regions have a congruent double band area. Apply a technique similar to the technique applied in Problem 10.44. The outer boundary of the double band is 6.7 inches from the center of the dartboard, and the band is 0.35 inches wide. Therefore, the inner boundary of the double band is 6.7 – 0.35 = 6.35 inches from the center of the dartboard.

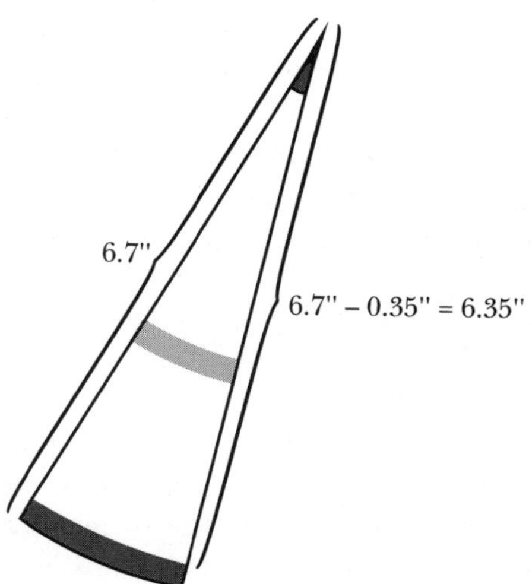

The area of the entire double band is equal to the area A_1 of a circle with radius 6.7 inches minus the area A_2 of a circle with radius 6.35 inches. The portion of the band lying within the 19-point radial region is 1/20 of that area.

$$\begin{aligned}\text{area of double-19 region} &= \frac{A_1 - A_2}{20}\\ &= \frac{\pi(6.7)^2 - \pi(6.35)^2}{20}\\ &= \frac{44.89\pi - 40.3225\pi}{20}\\ &= \frac{4.5675}{20}\pi\\ &= 0.228375\pi \text{ square inches}\end{aligned}$$

The probability of hitting the double-19 region is the area of that region divided by the area of the dartboard, 78.765625π.

$$\begin{aligned}\text{probability of hitting double-19 region} &= \frac{A_{\text{double-19 region}}}{A_{\text{dartboard}}}\\ &= \frac{0.228375\not{\pi}}{78.765625\not{\pi}}\\ &\approx 0.0028994\end{aligned}$$

The probability of hitting the double-19 region of the dartboard is approximately 0.290%.

Note: Problems 10.42–10.46 refer to a standard dart board, as illustrated and described in Problem 10.42.

10.46 What is the probability that you will score 14 points in a single throw? Express your answer as a percentage rounded to the hundredths place.

There are two ways to score 14 points in a single throw. First, you could toss a dart into the 14-point radial region, but not into the double and triple bands of that region. Second, you could toss a dart into the double band of the 7-point region. Notice that the double-7 region you are including is the same size as the double-14 region you are excluding. In other words, the probability of scoring 14 points is equal to the probability of hitting a radial region outside of the bull minus the probability of hitting one region of the triple band.

According to Problem 10.43, there is a 2.825% probability that you will hit a specific radial region (like 14). According to Problem 10.44, there is a 0.179% probability that you will hit any specific triple band within a region. Therefore, there is a 2.825% – 0.179% ≈ 2.65% probability that you will score 14 points in one throw.

You could omit the double band of the 14-point region, but you still have to include the double-band of the 7-point region. All of the small double band regions are the same size. It doesn't matter that the double-7 is located in a different place than the double-14 on the board—the area is the same, and that's all that matters in terms of probability.

Chapter 11

SAT PRACTICE: DATA ANALYSIS, STATISTICS, AND PROBABILITY

Test the skills you practiced in Chapter 10

This chapter contains 25 SAT-style questions to help you practice the skills and concepts you explored in Chapter 10. The questions are clustered by difficulty—early questions are easier than later questions. A portion of a student test booklet is reproduced on the following page; use it to record your answers. Full answer explanations are located at the end of the chapter.

Try not to peek at the answers until you have worked through all 25 questions. If you want to practice SAT pacing, try to finish all of the questions in no more than 30-35 minutes.

Practice Questions:
Data Analysis, Statistics, and Probability

11.1 Ⓐ Ⓑ Ⓒ Ⓓ Ⓔ	**11.6** Ⓐ Ⓑ Ⓒ Ⓓ Ⓔ	**11.11** Ⓐ Ⓑ Ⓒ Ⓓ Ⓔ
11.2 Ⓐ Ⓑ Ⓒ Ⓓ Ⓔ	**11.7** Ⓐ Ⓑ Ⓒ Ⓓ Ⓔ	**11.12** Ⓐ Ⓑ Ⓒ Ⓓ Ⓔ
11.3 Ⓐ Ⓑ Ⓒ Ⓓ Ⓔ	**11.8** Ⓐ Ⓑ Ⓒ Ⓓ Ⓔ	**11.13** Ⓐ Ⓑ Ⓒ Ⓓ Ⓔ
11.4 Ⓐ Ⓑ Ⓒ Ⓓ Ⓔ	**11.9** Ⓐ Ⓑ Ⓒ Ⓓ Ⓔ	**11.14** Ⓐ Ⓑ Ⓒ Ⓓ Ⓔ
11.5 Ⓐ Ⓑ Ⓒ Ⓓ Ⓔ	**11.10** Ⓐ Ⓑ Ⓒ Ⓓ Ⓔ	**11.15** Ⓐ Ⓑ Ⓒ Ⓓ Ⓔ

11.16

11.17

11.18

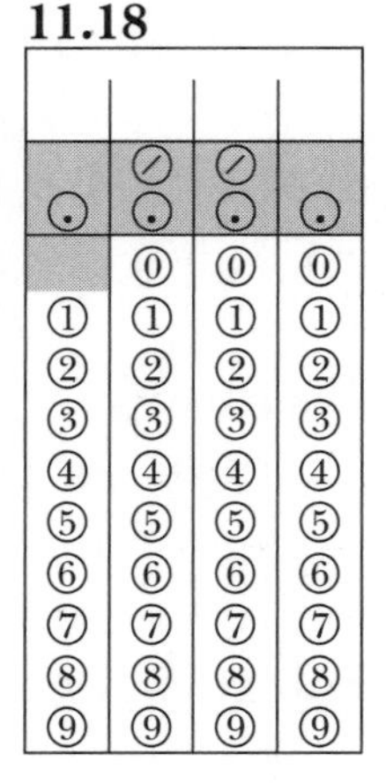

11.19

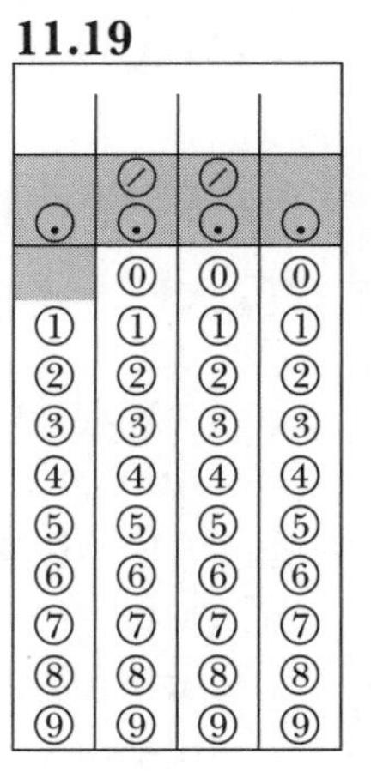

11.20

11.21

11.22

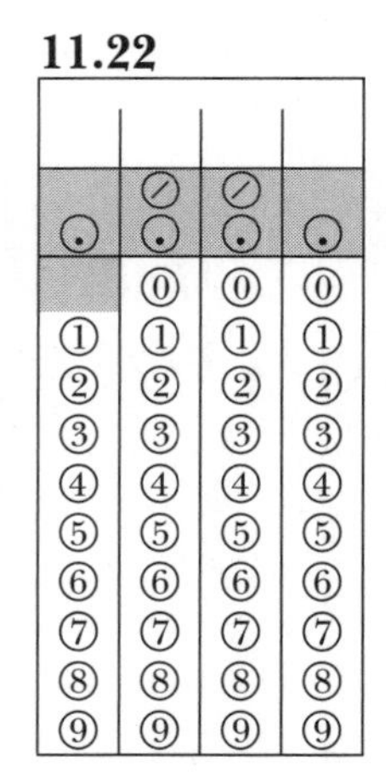

11.23

11.24

11.25

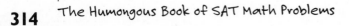

$$1, 7, 13, x - 1$$

11.1 If the average (arithmetic mean) of the four integers listed above is 9, what is the value of x?

(A) 15
(B) 16
(C) 17
(D) 18
(E) 20

Questions 11.2 and 11.3 refer to the graph below, which records the number of diaper changes required for two newborn twins during a seven-day period.

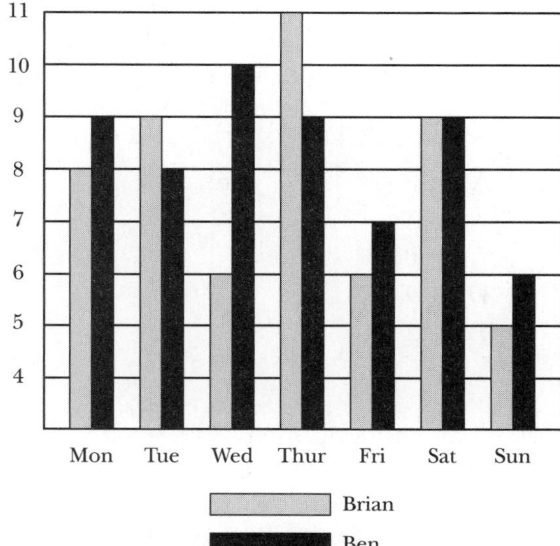

11.2 What is the median number of daily diaper changes for Brian?

(A) 6
(B) 7
(C) 7.5
(D) 8
(E) 9

11.3 How many more diaper changes did Ben require, on average, each day?

(A) $\frac{3}{7}$
(B) $\frac{4}{7}$
(C) 1
(D) 3
(E) 4

11.4 Jen keeps all of her shirts in the top drawer of her dresser. She owns 1 green shirt, 6 white shirts, and 5 yellow shirts. If she randomly selects one shirt from the drawer, what is the probability that she will select a white shirt?

(A) $\frac{1}{30}$
(B) $\frac{1}{12}$
(C) $\frac{1}{6}$
(D) $\frac{1}{5}$
(E) $\frac{1}{2}$

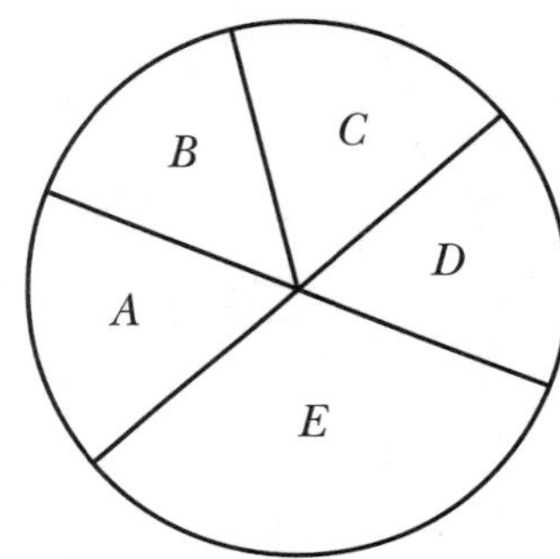

11.5 The circle graph above describes the multiple-choice answers selected by a class of students on a test question. According to the graph, which two choices were *most likely* chosen by the same number of students?

(A) A and B
(B) B and C
(C) C and D
(D) A and C
(E) A and D

11.6 A certain data set consists of integers from 1 to 10, such that each integer appears as many times as its value. For example, the set contains three 3's and eight 8's. What is the median of the set?

(A) 4
(B) 5
(C) 6
(D) 7
(E) 8

4, 2, 4, 3, 3, 1, 4, 2, 5, 3, 1

11.7 A recent survey asked a group of 12 students how many television shows they watched per day. Eleven of the responses are listed above. Unfortunately, one of the data values was lost. If the median (including the missing survey response) was 3 television shows, which of the following could be the missing response?

I. 2
II. 3
III. 4

(A) I only
(B) II only
(C) III only
(D) I and II only
(E) I, II, and III

11.8 At a certain middle school, 80 students are 11 years old, 95 students are 12 years old, 130 students are 13 years old, and 100 students are 14 years old. What is the average (arithmetic mean) age of the students?

(A) 12 years, 6 months
(B) 12 years, 7 months
(C) 12 years, 10 months
(D) 13 years, 1 month
(E) 13 years, 3 months

NUMBER OF DAYS PER MONTH
THE LUNCH SPECIAL IS FISH

September 🐟 🐟
October 🐟 🐟 🐟
November 🐟 🐟
December 🐟 🐟 🐟
January 🐟 🐟 🐟
February 🐟

11.9 A certain restaurant features a daily lunch special at a discounted price. The pictograph above records the number of days each month that the lunch special was some type of fish. If the restaurant's lunch special was fish a total of 24 times during the months of November, December, and January, how many days of October had lunch specials that were fish?

(A) 3
(B) 4
(C) 6
(D) 9
(E) 12

11.10 Three data values have an average (arithmetic mean) of 10. When a fourth data value, d, is included in the set, the average of the data decreases to 8. What is the value of d?

(A) 1.5
(B) 2
(C) 2.5
(D) 3
(E) 3.5

11.11 A certain data set has the same median and average (arithmetic mean), a. If one additional data value x is added to the set, such that $x > a$, which of the following statements must be true?

I. The average (arithmetic mean) increases
II. The median increases
III. The mode changes

(A) I only
(B) II only
(C) I and II only
(D) I and III only
(E) I, II, and III

11.12 What is the probability of rolling three consecutive, increasing numbers in a row using a single fair die?

(A) $\frac{1}{216}$

(B) $\frac{1}{54}$

(C) $\frac{1}{18}$

(D) $\frac{1}{9}$

(E) $\frac{2}{9}$

Questions 11.13 and 11.14 refer to the information below.

A jar contains 8 red and 4 green gumballs. The gumballs are selected randomly from the jar, and once selected, they are not returned to the jar.

11.13 What is the probability that, of the 12 gumballs in the jar, the first three gumballs Chris selects will be green?

(A) $\frac{3}{440}$

(B) $\frac{1}{55}$

(C) $\frac{8}{165}$

(D) $\frac{133}{165}$

(E) $\frac{3}{2}$

11.14 What is the probability that, of the 12 gumballs in the jar, the first three gumballs Donna selects will alternate in color?

(A) $\frac{28}{165}$

(B) $\frac{5}{27}$

(C) $\frac{8}{33}$

(D) $\frac{32}{55}$

(E) $\frac{77}{120}$

11.15 A set M contains 5 data values with a sum of k and an average of x. Set N contains 15 data values with a sum of $-k$ and an average of y. Assume that k is a positive real number. Which of the following statements must be true?

I. M contains more positive data values than N

II. The average of all 20 data values is 0

III. $x > y$

(A) II only
(B) III only
(C) I and II only
(D) II and III only
(E) I, II, and III

11.16 A pile containing only four cards of the same suit (the jack, queen, king, and ace of hearts) is shuffled. The cards are then placed face down on a table and revealed one at a time. What is the probability that the ace of hearts will be revealed after the king of hearts?

2004, 2010, 2013, 1998, 2009, 2007

11.17 Seven friends compare the model years of their cars and discover that all of the years are different. Six of the model years are listed above. When the seventh model year is included, the median year is 2009. What is the earliest possible model year of the seventh car?

2, 4, 6, –1, 9, 18, 3, 0, 1, 5

11.18 The numbers listed above are each written on separate slips of paper and placed into a box. One number is selected at random from the box. What is the probability that the number drawn is less than the average (arithmetic mean) of all the numbers?

11.19 The average age of Alex, Brenna, and Catie is 18. The average age of Alex, Brenna, Catie, Dawn, Ellie, and Fiona is 22. If Dawn, Ellie, and Fiona are at least 19 years old, what is Fiona's maximum age?

Questions 11.20 and 11.21 refer to the information below.

In a certain game, one turn consists of two parts: (1) drawing a ping pong ball at random from a bag and (2) dropping that ball into a tall rectangular box with a 2 foot × 8 foot base that is divided into four rectangular scoring regions, as illustrated below. Assume that each throw lands somewhere in the scoring box, but because the ball bounces, its final location will be random. That final location determines the points scored on each turn. The bag contains 20 ping pong balls, 15 that are white and 5 that are red. If a player draws a red ping pong ball on his or her turn, the points scored by that ball are doubled.

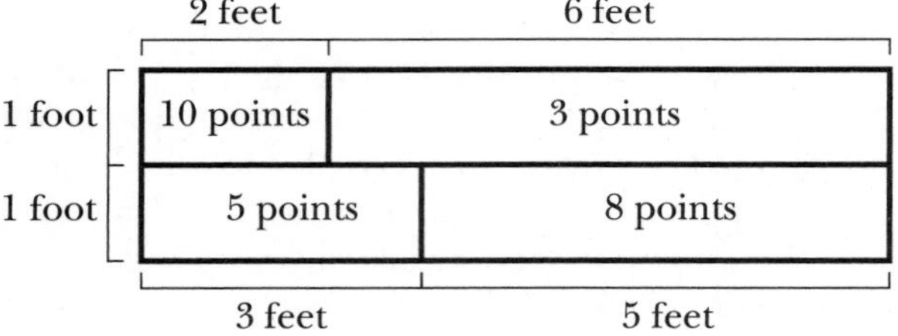

Sample turn: A player draws a red ball and drops it into the scoring box. The ball comes to rest in the lower-right, 8-point region of the box, so the player scores 16 points on that turn. Balls are not returned to the bag at the end of each turn, so the next player has one fewer ping pong ball from which to choose.

11.20 What is the probability that a player will score 3 points on the first turn of the game?

11.21 What is the probability of scoring 10 points on the 11th turn if 7 white ping pong balls have been drawn in the first 10 turns?

11.22 If the average (arithmetic mean) of real numbers x and y is 8, what is the average of x, y, and $\frac{1}{4}(x+y)$?

11.23 A certain high school surveys its graduating seniors about the types of courses they took between kindergarten and 12th grade. A students took an average of 10 mathematics courses, and B students took an average of 14 mathematics courses. Combined, the two groups took an average of 13 mathematics courses. You can conclude that B is k times as large as A for what value of k?

$$2x+2, \quad x^2+5, \quad x+3, \quad \frac{1}{2}(x+10),$$
$$3x+2, \quad x-1, \quad 18-x^2, \quad 2x$$

11.24 The data set above contains eight expressions with an average (arithmetic mean) of 9. What is the difference of the median and the mode of the data values?

11.25 Books A, B, C, D, and E are selected randomly and placed, one by one, on a bookshelf from left to right. What is the probability that Book A will be placed at the far left and Book E will be placed at the far right?

Solutions

11.1 **B**. The average of four numbers is the sum of the numbers divided by 4.

$$\text{average} = \frac{1+7+13+(x-1)}{4}$$

According to the question, the average is 9.

$$9 = \frac{21+x-1}{4}$$
$$9 = \frac{20+x}{4}$$

Cross-multiply and solve the proportion.

$$36 = 20 + x$$
$$36 - 20 = x$$
$$x = 16$$

11.2 **D**. During the seven-day time period, Brian needed 8, 9, 6, 11, 6, 9, and 5 daily diaper changes. Arrange these data values in order, from least to greatest: 5, 6, 6, 8, 9, 9, 11. The median is the number in the middle: 8. There are three data values less than, and three data values greater than, 8.

If you think this problem is weird and contrived, I still have graphs like this from shortly after my twins' birth. We also had one for feeding and another one for how much we didn't sleep.

11.3 **B**. Calculate the total diaper changes required for each of the twins during the seven-day period.

$$\text{Brian: } 8+9+6+11+6+9+5 = 54$$
$$\text{Ben: } 9+8+10+9+7+9+6 = 58$$

Ben needed 4 more diaper changes than Brian during the week. Over the course of 7 days, this was an average of 4/7 more diapers per day.

If you multiply 4/7 by 7 days, you get 4, which is the number of additional diapers Ben soiled.

11.4 **E**. Jen's dresser drawer contains 1 + 6 + 5 = 12 shirts. Of those 12 shirts, 6 are white. Therefore, there is a 6/12 = 1/2 probability that the shirt Jen draws will be white.

11.5 **E**. This is a very meta question—a graph about multiple choice answers assessed on a multiple choice question. The question asks you to judge, from the circle graph, which two sectors are most likely to be the same size. At first glance, sectors *A*, *B*, *C*, and *D* all look very similar, so you might be wondering how any one answer choice is more valid than the others.

The trick to this question is recognizing the vertical angles in the diagram. In the following illustration, two segments are darkened for emphasis. They, like all region boundaries of a circle graph, definitely pass through the center of the circle. They appear to be diameters of the circle. They form vertical angles, which are congruent, so regions *A* and *D* are most likely the same size.

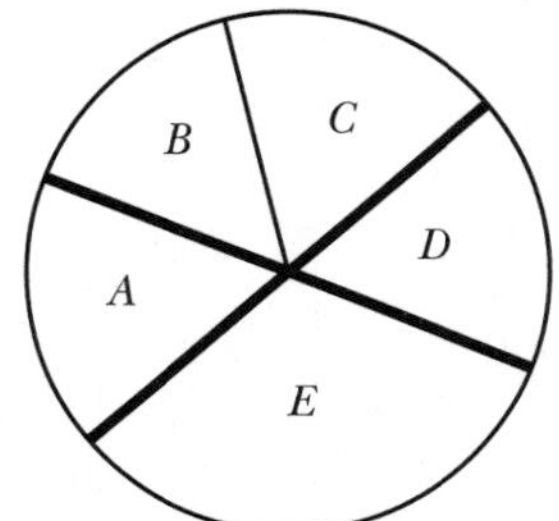

Even if the darkened segments in the diagram are *not* diameters—there's no statement guaranteeing that they are—those segments provide enough evidence for you to conclude that A and D are *most likely* the same size.

11.6 **D**. The data set contains one 1, two 2s, three 3s, four 4s, and so on. Rather than try to generate some sort of trick or clever technique to answer this question, it only takes a moment to list all 55 data values in the set. Even better, as you construct the data set, the data values are automatically arranged in order, from least to greatest.

1	2	2	3	3	3	4	4	4	4
5	5	5	5	5	6	6	6	6	6
6	7	7	7	7	7	7	7	8	8
8	8	8	8	8	8	9	9	9	9
9	9	9	9	9	10	10	10	10	10
10	10	10	10	10					

There are 55 total values. When there is an odd number n of data values, the median is the $(n + 1)/2$ term in the list. In other words, add 1 to n and divide the number by 2. In this problem, $n + 1 = 55 + 1 = 56$, and $56 \div 2 = 28$. Therefore, the median is the 28th number in the list, highlighted in the diagram below. The median is 7—the final 7 in the data set; exactly 27 data values appear before it and 27 data values appear after it.

1	2	2	3	3	3	4	4	4	4
5	5	5	5	5	6	6	6	6	6
6	7	7	7	7	7	7	(7)	8	8
8	8	8	8	8	8	9	9	9	9
9	9	9	9	9	10	10	10	10	10
10	10	10	10	10					

If there is an even number N of data values, then the median is the average (arithmetic mean) of the N/2 term and the next term, (N/2) + 1. (If the terms are written in order.) In this problem, there are N = 12 terms, so the median is the average of the 12/2 = 6th term and the 7th term.

11.7 **E**. Begin by listing the 11 known data values in order, from least to greatest.

1 1 2 2 3 3 3 4 4 4 5

Now test statement I by inserting the data value 2. The median of the 12 data values is the average of the two middle numbers—the average of data values 6 and 7—underlined below.

1 1 2 2 2 <u>3 3</u> 3 4 4 4 5

Each time you insert a new possible value for the missing survey response, start over with the 11 known responses. You should have 12 total data values.

The median of the data is 3, so the missing survey response could be 2. Now test statement II by adding data value 3 to the 11 known data values.

1 1 2 2 3 <u>3 3</u> 3 4 4 4 5

Once again, the median is 3, so 3 also could be the missing survey response. Finally, add 4 to the 11 known data values and identify the median.

1 1 2 2 3 <u>3 3</u> 4 4 4 4 5

The median is still 3, so all three statements (I, II, and III) represent possible missing survey responses. Whether that missing response is 2, 3, or 4 television shows, in all cases, the median of the responses is 3.

11.8 **B**. The average age of the students is equal to the total sum of the student ages divided by the number of students. Use a weighted average to quickly compute the sum of the ages; multiply each age by the number of students of that age.

$$\begin{aligned}\text{sum of ages} &= 80(11)+95(12)+130(13)+100(14)\\ &= 880+1{,}140+1{,}690+1{,}400\\ &= 5{,}110\end{aligned}$$

There are 80 + 95 + 130 + 100 = 405 total students. Compute the average age of the students.

$$\begin{aligned}\text{average age} &= \frac{\text{sum of ages}}{\text{number of students}}\\ &= \frac{5{,}110}{405}\\ &\approx 12.617\end{aligned}$$

The average student is 12.617 years old, which appears to be closest to answer choice A. You may be wondering, "How many months is 0.617 years?" There are 12 months in a year, so 12(0.617) ≈ 7.4 months. The average student age is approximately 12 years and 7 months.

11.9 **D**. Usually, pictographs contain some sort of legend or key, telling you how much each individual picture is worth. This pictograph does not. However, the question states that 24 of the daily specials were fish during the months of November, December, and January. Notice that those three months are labeled with a total of 2 + 3 + 3 = 8 fish pictures. Therefore, each of the pictures must represent 24 ÷ 8 = 3 lunch specials.

The month of October in the chart is labeled with three fish pictures. Each of those represents 3 lunch specials, so in October, 3(3) = 9 of the lunch specials were fish.

11.10 **B**. Three data values have an average of 10. Therefore, the sum of the values is 3(10) = 30. Add the fourth value, d, and divide by 4, the number of values in the data set. This represents the average of the four data values, and the question states that the average is equal to 8.

$$8=\frac{30+d}{4}$$

Cross-multiply and solve for d.

$$\begin{aligned}8(4) &= 30+d\\ 32 &= 30+d\\ 32-30 &= d\\ 2 &= d\end{aligned}$$

The fourth data value is $d = 2$.

11.11 **A**. Apply the DIY method to see how adding a data value to a data set affects the mean, median, and mode of the set. You need to construct a data set whose average (mean) is equal to its median. The easiest such set is a list of consecutive integers.

1 2 3 4 5 6 7

The average of the seven data values above is the sum of the data (28) divided by 7, the number of data values: $28 \div 7 = 4$. The median is the middle number of the data, 4. Because this set has the same median and average, you can use it to test the conditions described in this problem. You need to add one data value that is greater than 4. Below, the data value 8 is added to the set.

1 2 3 4 5 6 7 8

The average of this data set is $36 \div 8 = 4.5$, and the median is 4.5. Both the average and the median increased from 4 to 4.5. The mode was unaffected—no data value is repeated in either list—so statement III is incorrect. You may be tempted to conclude that statements I and II are true, but do not be too hasty. *Must* they both be true? No.

Consider the new data set below.

1 2 3 4 4 4 5 6 7

The average of the data is $36 \div 9 = 4$ and the median also equals 4. When you add the data value 8 to the data set, the new average is $44 \div 10 = 4.4$ but the median remains 4. Therefore, statement II is not always true.

Statement I is the only statement that *must* be true. Adding any data value that is greater than the average will increase the average of the data. Remember, the average is the balancing point of the data set, as explained in Problem 10.24. If you add a value greater than the average, it upsets the balance unless you move the fulcrum right, increasing the average.

11.12 **B**. Rolling a fair die produces six different outcomes, each equally likely to occur. The outcomes are the integers 1, 2, 3, 4, 5, and 6. According to Problem 10.38, there are $6 \cdot 6 = 36$ different outcomes when you roll two dice. Therefore, there are $6 \cdot 6 \cdot 6 = 216$ different outcomes when you roll a single die three times.

Or roll the same die two times. Same odds.

Of those 216 possible outcomes, there are only four ways to roll consecutive, increasing numbers, listed below.

1, 2, 3 2, 3, 4 3, 4, 5 4, 5, 6

There is a $4/216 = 1/54$ probability of rolling the die as described.

11.13 **B**. There are 4 green gumballs of the 12 total gumballs in the jar. Therefore, there is a $4/12 = 1/3$ probability of selecting a green gumball first. That leaves 3 green out of 11 gumballs in the jar, so there is a $3/11$ probability of drawing a green gumball second. Similarly, there is a $2/10 = 1/5$ probability of drawing a green gumball third. Multiply the three probabilities.

$$\left(\frac{1}{\not{3}}\right)\left(\frac{\not{3}}{11}\right)\left(\frac{1}{5}\right)=\frac{1}{11\cdot 5}$$
$$=\frac{1}{55}$$

The probability of selecting three green gumballs is 1/55.

11.14 **C**. There are 12 ways to select the first gumball, 11 ways to select the second, and 10 ways to select the third, so there are $12 \cdot 11 \cdot 10 = 1{,}320$ ways to select the first three.

If Donna's chosen gumballs alternate in color, then she chose either (A) red-green-red or (B) green-red-green. Take those possibilities one at a time. If she chose red-green-red, then there were 8 ways to choose the first red gumball, 4 ways to choose the green gumball, and then 7 ways to choose the second red gumball, a total of $8 \cdot 4 \cdot 7 = 224$ ways to accomplish the task.

There are fewer ways for Donna to choose gumballs in the green-red-green pattern. Specifically, there are 4 ways to select the first green gumball, 8 ways to select the red gumball, and then 3 ways to select the second green gumball, a total of $4 \cdot 8 \cdot 3 = 96$ different ways.

There are $224 + 96 = 320$ different ways for Donna to select gumballs of alternating color, so the probability is $320/1{,}320 = 8/33$.

There is no information about how sweaty or edible these overly handled gumballs may be, nor is there any information about why Chris and Donna are doing this in the first place. It seems entirely unsanitary.

11.15 **D**. Investigate each of the statements individually. Statement I is false. Just because a set has a positive sum—the question states that k is a positive real number—the majority of the numbers in the set need not be positive. For example, consider the set M of five data values below, which has a sum of $k = 1$.

$$M = \{-1, -2, -3, -4, 11\}$$

Now consider set N, which has 15 data values and a sum of $-k = -1$.

$$N = \{1, 1, 1, 1, 1, 1, 1, 1, 1, 1, 1, 1, 1, 1, -15\}$$

Although the sum of the data values in N is negative, N contains more positive data values than M.

Statement II is true. You may not know what the 20 data values are, but you know that the sum of the data values is $k + (-k) = 0$. Thus, the average of the 20 values is $0/20 = 0$.

Statement III is true. The sum of the data values in M is positive, and the average is $x = k/5$. The sum of the data values in N is negative, and the average is $y = -k/15$. A positive number x is always greater than a negative number y.

11.16 **.5 or 1/2**. Begin by calculating the number of ways you can reveal the cards. The first card could be any of the 4 cards, so there are 4 ways to select the first card. That leaves 3 possibilities for the second card, 2 possibilities for the third card, and only one possibility for the final card. Therefore, there are $4 \cdot 3 \cdot 2 \cdot 1 = 24$ ways to reveal the cards.

The last card is whatever's left over. There's no choice involved at all.

The cases in which the ace is revealed after the king are listed below. The cards are revealed in order, from left to right, so the ace is revealed after the king when the ace (A) appears right of the king (K).

J Q K A	Q J K A	K J Q A
J K Q A	Q K J A	K J A Q
J K A Q	Q K A J	K Q J A
		K Q A J
		K A J Q
		K A Q J

There are 12 ways to reveal the ace after the king, so the probability is 12/24 = 1/2, or 0.5.

11.17 **2011**. List the known model years from least to greatest.

1998, 2004, 2007, 2009, 2010, 2013

When the seventh model year is included, there will be an odd number of data values, so the median will be the middle number, the fourth number in the ordered list. The question states that the median is 2009.

1998, 2004, 2007, <u>2009</u>, 2010, 2013

At the moment, 2009 is not the middle value, so the missing model year must be greater than 2009; it needs to appear right of 2009 in the ordered list if 2009 is the middle value. You are asked to identify the *earliest* possible model year, but keep in mind that all of the model years are different. If the unknown model year were 2009 or 2010, the median would still be 2009, but those numbers are already present in the data.

The earliest possible model year is the first unique year greater than 2009. The correct answer is 2011.

In other words, a year not already in the list

11.18 **.6, 6/10, or 3/5**. This problem is difficult not because of the concepts involved—averages and simple probability—but because you need to combine those skills. Begin by calculating the average of the 10 numbers listed.

$$\begin{aligned} \text{average} &= \frac{2+4+6-1+9+18+3+0+1+5}{10} \\ &= \frac{47}{10} \\ &= 4.7 \end{aligned}$$

Six of the 10 numbers are less than the average: 2, 4, –1, 3, 0, and 1. Therefore the probability that a number less than the average will be drawn is 6/10 = 3/5 = 0.6.

11.19 **40**. The names of the girls in this problem begin with consecutive letters, so let their first initials represent their current ages. Therefore, the average of a, b, c, d, e, and f is 22.

$$22 = \frac{a+b+c+d+e+f}{6}$$

You know that the average of a, b, and c is 18, so the sum of their ages is $3(18) = 54$. Hence, $a + b + c = 54$. Replace $a + b + c$ with 54 in the equation above.

$$22 = \frac{54+d+e+f}{6}$$

Cross-multiply and isolate the remaining variables on the right side of the equation.

$$\begin{aligned} 22(6) &= 54+d+e+f \\ 132 &= 54+d+e+f \\ 132-54 &= d+e+f \\ 78 &= d+e+f \end{aligned}$$

Once again, you do not know the girls' individual ages, but you know that the sum of Dawn's, Ellie's, and Fiona's ages is 78. To calculate the maximum possible age for Fiona, Dawn and Ellie would have to have the minimum possible age. The problem states that all three of them are at least 19 years old. Substitute 19 for d and e to calculate the maximum possible value of f.

$$\begin{aligned} 78 &= 19+19+f \\ 78 &= 38+f \\ 78-38 &= f \\ 40 &= f \end{aligned}$$

Fiona's maximum age is 40.

11.20 **9/32 or .281**. Each turn is composed of two parts: drawing the ping pong ball and dropping it into the scoring box. In order to score 3 points on the first turn, the player would need to draw a white ping pong ball and drop it into the 1 foot × 6 foot, 3-point scoring region. Calculate both of the probabilities separately.

Because 15 of the 20 ping pong balls are white, there is a 15/20 = 3/4 probability that the player will select a white ball. How likely is it that the white ball will fall in the correct region? You need to calculate the geometric probability by dividing the area of the 3-point scoring region by the area of the entire rectangular scoring area.

$$\begin{aligned} \text{probability of landing in 3-point region} &= \frac{\text{area of 3-point region}}{\text{area of scoring box base}} \\ &= \frac{1 \text{ foot} \cdot 6 \text{ feet}}{2 \text{ feet} \cdot 8 \text{ feet}} \\ &= \frac{6}{16} \\ &= \frac{3}{8} \end{aligned}$$

The 3-point scoring region accounts for 3/8 of the entire scoring area, so the probability that the ping pong ball will come to rest in that region is 3/8. To complete the problem, multiply the two probabilities.

$$\frac{3}{4}\cdot\frac{3}{8}=\frac{9}{32}$$

11.21 **.137 or .138**. Remember that ping pong balls are not returned to the bag at the end of each turn. If 7 white ping pong balls were chosen in the first 10 turns, then 10 – 7 = 3 red ping pong balls were also chosen. That leaves 15 – 7 = 8 white and 5 – 3 = 2 red ping pong balls in the bag before the 11th turn.

Remember, the base of the scoring area is 1 + 1 = 2 feet by 2 + 6 = 8 feet. Its area is (2 feet)(8 feet) = 16 square feet.

There are two ways to score 10 points. The player could drop a white ball in the 10-point region or drop a red ball in the 5-point region. Calculate these two probabilities separately and add them together.

Case 1: White ball in 10-point region. The player has a 8/10 = 4/5 probability of drawing a white ball. The area of the 10-point region is 2(1) = 2, which is 2/16 = 1/8 of the scoring area. Multiply the two probabilities.

$$\begin{aligned}\text{Case 1 probability} &= (\text{white ball probability})(\text{10-point region probability})\\ &= \frac{4}{5}\cdot\frac{1}{8}\\ &= \frac{4}{40}\\ &= \frac{1}{10}\end{aligned}$$

Case 2: Red ball in a 5-point region. The player has a 2/10 = 1/5 chance of drawing a red ball. The area of the 5-point region is 1(3) = 3, so the probability that the ball lands in that region is 3/16. Multiply these probabilities.

$$\begin{aligned}\text{Case 2 probability} &= (\text{red ball probability})(\text{5-point region probability})\\ &= \frac{1}{5}\cdot\frac{3}{16}\\ &= \frac{3}{80}\end{aligned}$$

Add the probabilities calculated for both cases, because both result in a score of 10.

$$\begin{aligned}\text{Total probability} &= \text{Case 1 probability} + \text{Case 2 probability}\\ &= \frac{1}{10}+\frac{3}{80}\\ &= \frac{8}{80}+\frac{3}{80}\\ &= \frac{11}{80}\end{aligned}$$

You must grid the solution as a decimal, because 11/80 does not fit into the grid: 11/80 = 0.1375.

11.22 **20/3, 6.66, or 6.67**. You do not know the values of x and y, but you know their average. Set up an equation as though you were calculating the average, which you know is equal to 8.

$$\text{average of } x \text{ and } y = \frac{x+y}{2}$$
$$8 = \frac{x+y}{2}$$

Cross-multiply the proportion.

$$8 \cdot 2 = x + y$$
$$16 = x + y$$

The sum of x and y is 16. Use this information to calculate $(1/4)(x+y)$.

$$\frac{1}{4}(x+y) = \frac{1}{4}(16)$$
$$= \frac{16}{4}$$
$$= 4$$

Now calculate the average of x, y, and $(1/4)(x+y)$.

$$\text{average} = \frac{x+y+(1/4)(x+y)}{3}$$

Substitute $x + y = 16$ and $(1/4)(x+y) = 4$ into the equation.

$$\text{average} = \frac{16+4}{3}$$
$$= \frac{20}{3}$$

11.23 **3**. If A students took 10 math courses and B students took 14 math courses, then the total number of math courses taken by those two groups is $10A + 14B$. The total number of students in those two groups is $A + B$. According to the question, if you combine the groups, the average number of math courses per student is 13.

$$\text{combined average} = \frac{\text{total number of courses}}{\text{total number of students}}$$
$$13 = \frac{10A+14B}{A+B}$$

Cross-multiply and combine like terms.

$$13(A+B) = 10A + 14B$$
$$13A + 13B = 10A + 14B$$
$$13A - 10A = 14B - 13B$$
$$3A = B$$

B is 3 times as large as A.

11.24 **.5 or 1/2.** The average of the expressions is equal to the sum of the expressions divided by 8. Begin by calculating the sum of the expressions.

$$[2x+2]+[x^2+5]+[x+3]+\left[\frac{1}{2}(x+10)\right]+[3x+2]+[x-1]+[18-x^2]+[2x]$$
$$=2x+2+x^2+5+x+3+\frac{1}{2}x+5+3x+2+x-1+18-x^2+2x$$
$$=(x^2-x^2)+\left(2x+x+\frac{1}{2}x+3x+x+2x\right)+(2+5+3+5+2-1+18)$$
$$=\left(9x+\frac{1}{2}x\right)+(34)$$
$$=\frac{18}{2}x+\frac{1}{2}x+34$$
$$=\frac{19}{2}x+34$$

Divide the sum by 8 to get the average of 9.

$$9=\frac{(19/2)x+34}{8}$$

Cross-multiply and solve for x.

$$9(8)=\frac{19}{2}x+34$$
$$72=\frac{19}{2}x+34$$
$$72-34=\frac{19}{2}x$$
$$38=\frac{19}{2}x$$
$$\left(\frac{2}{19}\right)\left(\frac{38}{1}\right)=\left(\frac{2}{19}\right)\left(\frac{19}{2}x\right)$$
$$\frac{76}{19}=x$$
$$4=x$$

Now that you know the value of x, you can evaluate all of the expressions for $x=4$ to calculate the actual data values.

$2x+2 = 8+2 = 10$

$x^2+5 = 16+5 = 21$

$x+3 = 4+3 = 7$

$\frac{1}{2}(x+10) = \frac{1}{2}(14) = 7$

$3x+2 = 12+2 = 14$

$x-1 = 4-1 = 3$

$18-x^2 = 18-16 = 2$

$2x = 2(4) = 8$

Only one data value appears twice; the mode is 7. To calculate the median, list the data values in order from least to greatest.

2, 3, 7, 7, 8, 10, 14, 21

The median is the average of the middle values: $(7 + 8) \div 2 = 7.5$. Therefore, the difference of the median and mode is $7.5 - 7 = 0.5$.

11.25 **.05 or 1/20**. This is a probability problem combined with the fundamental counting principle. First, calculate the number of ways you can arrange a set of 5 books. The illustration below uses a square to represent the position of each book on the shelf. If you are truly selecting books at random, then there are 5 ways to choose the first (leftmost) book, 4 ways to choose the second book, 3 ways to choose the third, and so on.

[5] [4] [3] [2] [1]

There are $5 \cdot 4 \cdot 3 \cdot 2 \cdot 1 = 120$ possible ways to arrange the books. Now you need to determine how many of those arrangements satisfy the conditions described in the question. If Book *A* must appear at the far left, then there is only one way to choose the first book. Similarly, there is only one way to choose Book *E*.

[1] [ ] [ ] [ ] [1]

There are 3 choices for the second position on the bookshelf—only Books *B*, *C*, and *D* could occupy that spot. This leaves 2 choices for the third position and only 1 choice for the fourth position.

[1] [3] [2] [1] [1]

There are $1 \cdot 3 \cdot 2 \cdot 1 \cdot 1 = 6$ ways to arrange the books such that Book *A* is drawn first and Book *E* is drawn last. Therefore, the probability of this occurring is $6/120 = 1/20$.

Part Three

PRACTICE SAT TESTS

Save These for Last

This section of the book contains three full SAT math practice tests. Recall that each SAT contains three graded math sections that are defined in a specific way:

- 20 multiple-choice questions for which you have 25 minutes
- 8 multiple-choice and 10 grid-in questions, again with a 25-minute time limit
- 16 multiple-choice questions, this time with a 20-minute time limit

The practice tests in this book present the sections in the order listed above, which is typically how they appear on the SAT. Keep in mind that the sections may (and occasionally do) appear in a different order, but that should not affect the way you pace yourself or approach the problems.

Important Note: Some SAT preparatory books include charts that "calculate" an SAT math score based on their practice tests. Unless that book is written by the College Board and contains actual, retired SAT tests, those score approximations are not only misleading, they are invalid. Actual SAT scores are based on your performance, the performance of others taking the same test as you, the comparative benchmarked historical performance of specific questions, and complex statistical manipulations. None of that is possible on a practice test, let alone a practice test written by people other than actual SAT test writers. This test is intended to help you prepare for the SAT, not to predict your future SAT score.

Make sure to replicate the SAT testing circumstances as much as possible. When you take a practice test, complete the entire thing—all three sections—at once. Stick to the stated time limits, and apply the strategies you learned in Chapters 1–3.

Some practice books give you a false sense of security with questions that are far too easy and not representative of the SAT at all. Other books have ridiculously hard questions you'd never see on the SAT. Almost all SAT prep books give you a scaled SAT score for the practice tests they include. At best, those scaled scores are wild guesses. At worst, they are outright fabrications. Worry about your score on SAT day. For now, worry about your content knowledge.

Chapter 12

SAT PRACTICE TEST 1

Good luck!

This chapter is designed to simulate the mathematics portion of the SAT. Record your answers on the student test booklet reproduced on the following page. The solutions are located at the end of the chapter. The following formulas are provided at the beginning of each SAT mathematics section, and you can refer to them during the practice test.

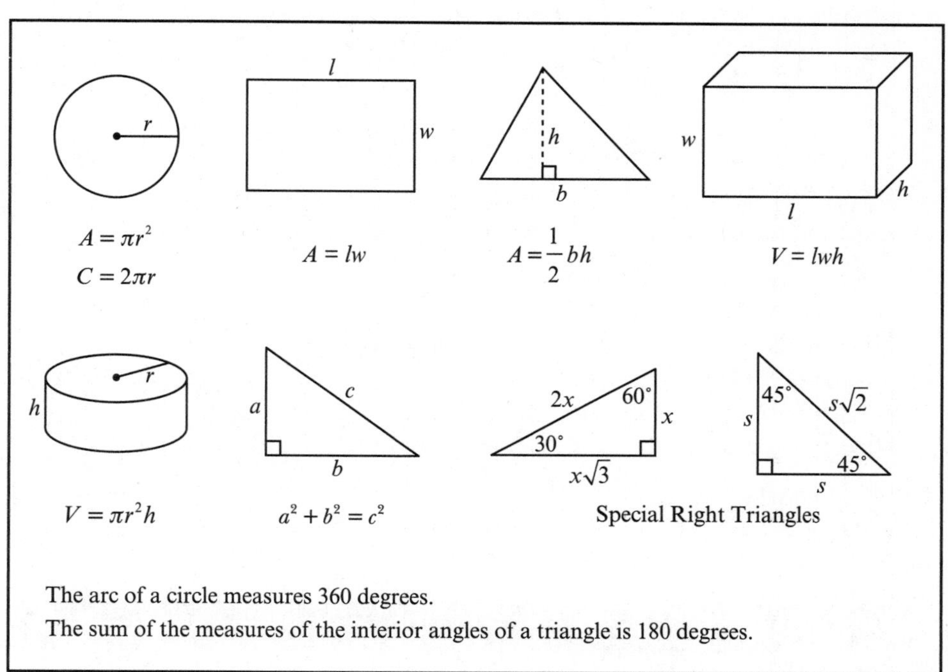

Section 1

Time Limit: 25 minutes

1 ⒶⒷⒸⒹⒺ	5 ⒶⒷⒸⒹⒺ	9 ⒶⒷⒸⒹⒺ	13 ⒶⒷⒸⒹⒺ	17 ⒶⒷⒸⒹⒺ
2 ⒶⒷⒸⒹⒺ	6 ⒶⒷⒸⒹⒺ	10 ⒶⒷⒸⒹⒺ	14 ⒶⒷⒸⒹⒺ	18 ⒶⒷⒸⒹⒺ
3 ⒶⒷⒸⒹⒺ	7 ⒶⒷⒸⒹⒺ	11 ⒶⒷⒸⒹⒺ	15 ⒶⒷⒸⒹⒺ	19 ⒶⒷⒸⒹⒺ
4 ⒶⒷⒸⒹⒺ	8 ⒶⒷⒸⒹⒺ	12 ⒶⒷⒸⒹⒺ	16 ⒶⒷⒸⒹⒺ	20 ⒶⒷⒸⒹⒺ

Section 2

Time Limit: 25 minutes

1 ⒶⒷⒸⒹⒺ	3 ⒶⒷⒸⒹⒺ	5 ⒶⒷⒸⒹⒺ	7 ⒶⒷⒸⒹⒺ
2 ⒶⒷⒸⒹⒺ	4 ⒶⒷⒸⒹⒺ	6 ⒶⒷⒸⒹⒺ	8 ⒶⒷⒸⒹⒺ

9. 10. 11. 12. 13.

	⊘	⊘	
⊙	⊙	⊙	⊙
	⓪	⓪	⓪
①	①	①	①
②	②	②	②
③	③	③	③
④	④	④	④
⑤	⑤	⑤	⑤
⑥	⑥	⑥	⑥
⑦	⑦	⑦	⑦
⑧	⑧	⑧	⑧
⑨	⑨	⑨	⑨

14. 15. 16. 17. 18.

Section 3

Time Limit: 20 minutes

1 ⒶⒷⒸⒹⒺ	5 ⒶⒷⒸⒹⒺ	9 ⒶⒷⒸⒹⒺ	13 ⒶⒷⒸⒹⒺ
2 ⒶⒷⒸⒹⒺ	6 ⒶⒷⒸⒹⒺ	10 ⒶⒷⒸⒹⒺ	14 ⒶⒷⒸⒹⒺ
3 ⒶⒷⒸⒹⒺ	7 ⒶⒷⒸⒹⒺ	11 ⒶⒷⒸⒹⒺ	15 ⒶⒷⒸⒹⒺ
4 ⒶⒷⒸⒹⒺ	8 ⒶⒷⒸⒹⒺ	12 ⒶⒷⒸⒹⒺ	16 ⒶⒷⒸⒹⒺ

Section 1

You have 25 minutes to complete this section of the test.

1. If the ratio of x to y is 3:4, what is the value of y when $x = 1$?

   (A) $\frac{1}{4}$

   (B) $\frac{1}{3}$

   (C) $\frac{3}{4}$

   (D) 1

   (E) $\frac{4}{3}$

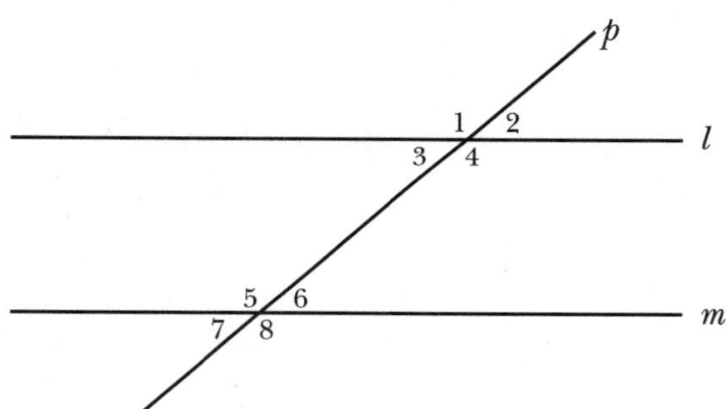

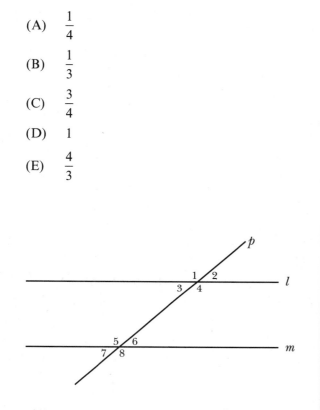

2. If lines l and m are parallel in the diagram above, what angle must be supplementary to angle 3?

   I. Angle 5
   II. Angle 7
   III. Angle 8

   (A) I only
   (B) II only
   (C) III only
   (D) I and II only
   (E) I and III only

3. Which of the following is a solution to the equation $|x - x^2| = 6$?

   I. $x = -3$
   II. $x = -2$
   III. $x = 3$

   (A) I only
   (B) II only
   (C) III only
   (D) II and III only
   (E) I, II, and III

4. If $2x - 6y = 18$, what is the value of x in terms of y?

   (A) $x = -\frac{3}{2}y$

   (B) $x = \frac{18}{x - 6y}$

   (C) $x = 3y + 9$

   (D) $x = 18y - 3$

   (E) $x = \frac{1}{3}y + 3$

$$2, 5, 8, x$$

5. If the average (arithmetic mean) of the four values listed above is 4.5, what is the value of x?

   (A) 2.5
   (B) 3
   (C) 3.5
   (D) 4
   (E) 4.5

6. If v is directly proportional to $\frac{3}{w+1}$ such that the constant of proportionality is m, which of the following expresses w in terms of m and v?

(A) $w = \frac{m}{v}$

(B) $w = \frac{m-1}{3v}$

(C) $w = \frac{v+1}{3m}$

(D) $w = \frac{3m-v}{v}$

(E) $w = \frac{3v}{m-1}$

7. The sum of three consecutive odd integers is 81. What is the least of the integers?

(A) 23
(B) 25
(C) 27
(D) 29
(E) 31

Note: Questions 8–9 refer to the information below.

Rectangle $ABCD$ in the coordinate plane has vertices $A = (x, 4)$, $B = (5, 4)$, $C = (5, -2)$, and $D = (x, -2)$. The length of one diagonal is $AC = 10$, and $x < 0$.

8. What is the value of x?

(A) –3
(B) –5
(C) –6
(D) –8
(E) –13

9. What is the perimeter of $ABCD$?

(A) 14
(B) 26
(C) 28
(D) 30
(E) 48

10. If $x = \frac{1}{\sqrt{y}}$ and $y^2 = \frac{z}{x}$, which of the following is equal to z?

(A) y^{-2}
(B) y^{-1}
(C) $y^{1/2}$
(D) $y^{3/2}$
(E) $y^{5/2}$

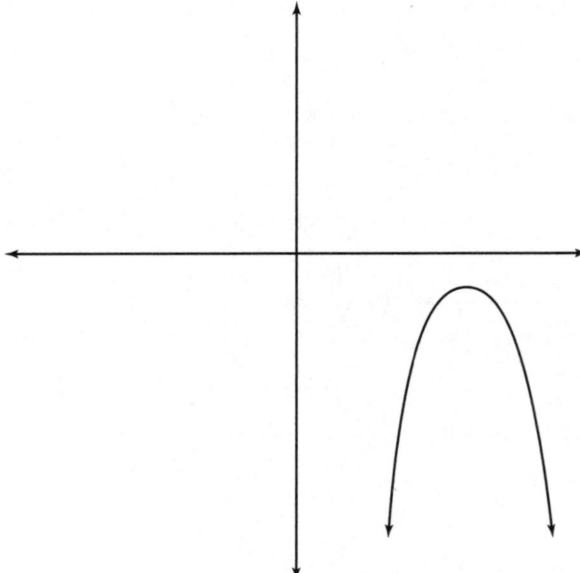

11. The function $f(x) = a(x + b)^2$ is graphed above. Which of the following statements must be true?

(A) $a = b$
(B) $a = -b$
(C) $a > 0$
(D) $b > 0$
(E) $ab > 0$

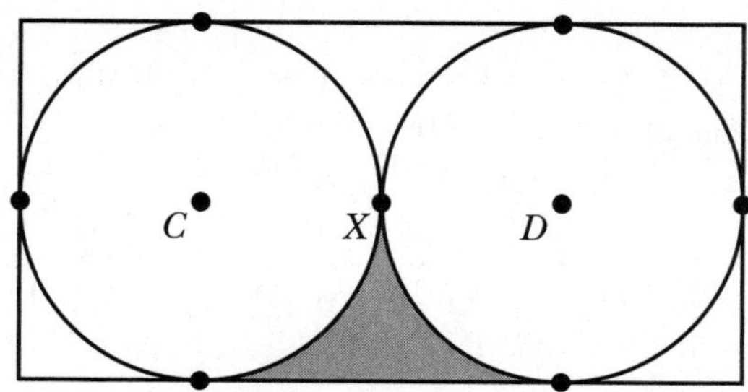

12. Congruent circles C and D above are tangent to each other at point X and tangent to the rectangle at all of the other marked points. If the area of the rectangle is 128, what is the area of the shaded region?

 (A) $4(3-2\pi)$
 (B) $8(4-\pi)$
 (C) $12(2-3\pi)$
 (D) $16(2-\pi)$
 (E) $32(1-\pi)$

13. A certain variety of seeds is sold in packets. Each packet contains s seeds and p% of those seeds will grow if planted. Miranda purchases n packets of seeds. Which of the following expressions represents the number of Miranda's seeds that will <u>*not*</u> grow when planted?

 (A) $100nps$
 (B) $\frac{100}{nps}$
 (C) $\frac{nps}{100}$
 (D) $\frac{100ns}{p}$
 (E) $ns-\frac{nps}{100}$

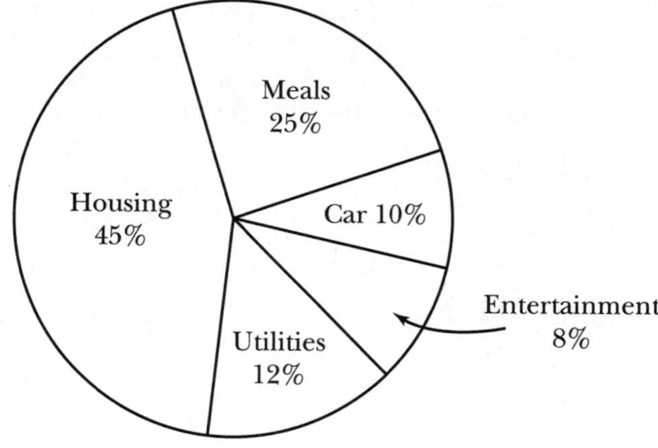

14. The circle graph above illustrates how Josh spends his monthly paycheck. He spends a total of $676.36 on utilities and meals each month. What is the difference between the amount he spends on housing and the amount he spends on utilities each month?

 (A) $146.24
 (B) $219.36
 (C) $457.00
 (D) $603.24
 (E) $822.60

15. Which of the following is the solution to the inequality $|2x|+3<11$?

 (A) $x<7$
 (B) $x<4$ or $x>7$
 (C) $x>4$ and $x<7$
 (D) $x<-4$ or $x>4$
 (E) $x>-4$ and $x<4$

16. Let the function $x \rrbracket\llbracket y$ be defined by $\frac{y-1}{x}$. If w is an integer less than –3, which of the following has the greatest value?

 (A) $w \rrbracket\llbracket w$
 (B) $w \rrbracket\llbracket w^2$
 (C) $w \rrbracket\llbracket -w$
 (D) $w \rrbracket\llbracket w^{-2}$
 (E) $w \rrbracket\llbracket 1$

17. Class A and Class B are two classes at a certain school. In Class A, there are 4 more girls than boys. In Class B, there are twice as many boys as girls. The total number of boys in both classes is 26. What is the total number of girls in Class B if there is a total of 49 boys and girls in both classes?

(A) 7
(B) 10
(C) 12
(D) 14
(E) 16

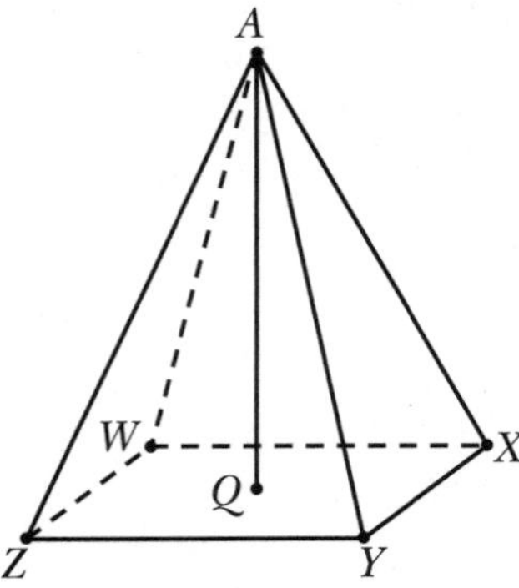

18. The right pyramid above has a square base $WXYZ$ with side length 4. If $h = AQ$ is the height of the pyramid, what is AX, the length of a lateral edge of the pyramid, in terms of h?

(A) $\sqrt{h^2+2}$
(B) $\sqrt{h^2+4}$
(C) $\sqrt{h^2+8}$
(D) $\sqrt{h^2+16}$
(E) $\sqrt{h^2+32}$

19. A certain phone company sells its most popular cell phones in two colors: black and white. In a typical week, it sells 400 more black phones than white phones. If B represents the number of black phones sold each week, which of the following represents the percentage of phones sold each week that are white, in terms of B?

(A) $\frac{50B}{B+200}$
(B) $\frac{50B}{B-200}$
(C) $\frac{B-400}{2B+800}$
(D) $\frac{50B-20,000}{B-200}$
(E) $\frac{B-400}{200(B-200)}$

20. Five friends write their names on separate pieces of paper and then three of those names are drawn at random. What is the probability that Nick and Justin—two of the friends—will be among the three names chosen?

(A) $\frac{1}{60}$
(B) $\frac{3}{20}$
(C) $\frac{1}{6}$
(D) $\frac{1}{5}$
(E) $\frac{3}{10}$

Section 2

You have 25 minutes to complete this section of the test.

1. If x is two less than y and y is three more than z, then what is the value of x when $z = -1$?

   (A) -6
   (B) -2
   (C) 0
   (D) 2
   (E) 4

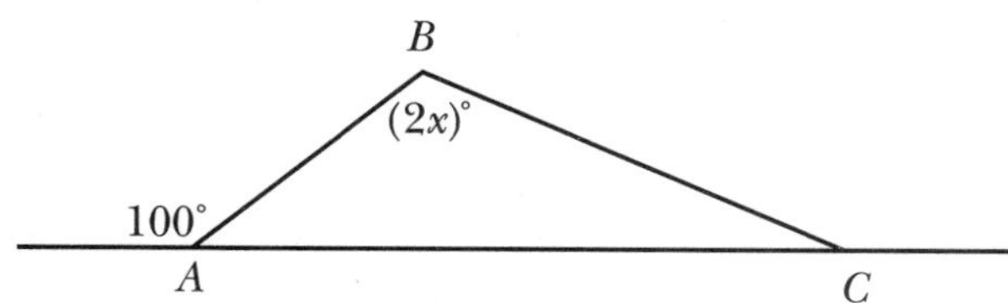

Note: Figure not drawn to scale.

2. In the figure above, $AB = BC$. What is the value of x?

   (A) 10
   (B) 20
   (C) 60
   (D) 80
   (E) 100

3. Points A, B, C, and D lie on line l. If C is the midpoint of AB and A is the midpoint of BD, which of the following is equal to AD?

   (A) $AB + AC$
   (B) $AB - AC$
   (C) $AC + BC$
   (D) $BD - BC$
   (E) $CD - AB$

4. A computer program selects an integer at random. The chosen number could be as low as 0 or as high as 20. What is the probability that a randomly selected integer is either even or divisible by 3?

   (A) $\frac{1}{2}$
   (B) $\frac{2}{3}$
   (C) $\frac{16}{21}$
   (D) $\frac{3}{4}$
   (E) $\frac{4}{5}$

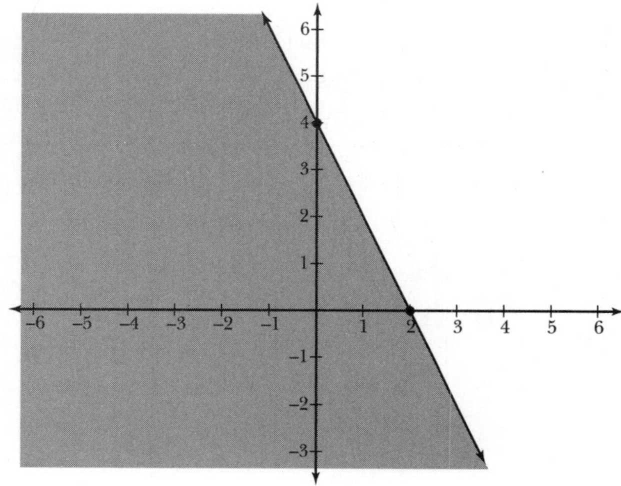

5. The graph above represents the solution region for the inequality $ax + by \leq 6$. What is the value of ab?

   (A) 1.5
   (B) 2.5
   (C) 4.5
   (D) 6.5
   (E) 8.5

6. If $16^{x+3} = 32^{x+2}$, what is the value of 3^{xy}?

(A) $\sqrt{3}$
(B) 2
(C) 3
(D) 9
(E) 27

7. If x and y are nonzero real numbers and z is the difference of the squares of the sum and the difference of x and y, what is the quotient of z and the product of x and y?

(A) xy
(B) $\frac{\sqrt{xy}}{2}$
(C) $\frac{x^2-y^2}{xy}$
(D) 0
(E) 4

8. If $a \div b$ has quotient c and remainder d, whereas $a \div c$ has quotient d and remainder b, which of the following statements must be true, assuming $c \neq 1$?

(A) $a = 2c$
(B) $b = d$
(C) $c = -b$
(D) $c = b - a$
(E) $d = \frac{1}{a}$

9. A personal identification number (PIN) for a certain banking website must contain four unique (non-repeating) digits between 0 and 9. How many unique PINs are possible? Assume that the order of the digits matters.

10. In a recent survey, students were asked if they participated in a spring or a fall sport in the preceding year. Of the 80 respondents, 12 participated in no sports during that time period, 29 students participated in a spring sport, and 15 students participated in both a spring and a fall sport. How many students participated in a fall sport?

$$-6, -1, 4, 9, \ldots$$

11. The first four terms of a sequence are listed above. Each term is five more than the previous term. What is the difference of the 44th term and the 4th term?

12. A full cylindrical water tank with radius $\frac{4}{\sqrt{\pi}}$ feet and height 6 feet begins leaking at a constant rate of 3 cubic feet per minute. How long, in minutes, will it take all of the water to leak out of the tank?

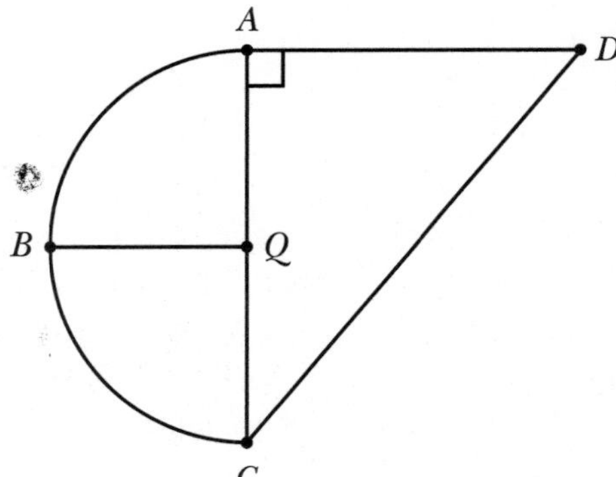

13. The figure above contains a semicircle $\overset{\frown}{ABC}$ and an isosceles right triangle ACD. If $CD = \sqrt{18}$, what is the radius of the semicircle?

Note: Problems 14–15 refer to the table below, which lists selected values of two functions, $f(x)$ and $g(x)$.

x	−3	−2	−1	0	1	2	3	4
$f(x)$	4	−1	0	4	6	2	−3	1
$g(x)$	−4	−3	−1	0	1	3	4	5

14. What is the value of $|(f-g)(3)|$?

15. What is the value of $g(f[g(-2)])$?

16. If the average of four different positive integers is 45, what is the greatest possible value for one of those integers?

Name	Average Words Per Minute
Bradley	50
Samantha	65
Lisa	70

17. The table above lists the average speed at which three administrative assistants can type, measured in words per minute. Bradley, Samantha, and Lisa need to work together to type a 35,000-word document in one afternoon, and they divide the document so that no two typists are typing the same content. All three typists begin typing at noon, and each is required to take a 5-minute break for every 30 minutes they spend typing. Bradley's workday ends at 3:00 p.m., Samantha's workday ends at 3:30 p.m., and Lisa's workday ends at 5:00 p.m. At what time will Lisa finish typing the document? Omit colons and "p.m." from your answer. For example, grid 3:45 p.m. as 345.

18. Circle C has circumference $a\pi$ and is inscribed in regular hexagon $KLMNOP$. If the perimeter of the hexagon is 24, what is the value of a?

Section 3

You have 20 minutes to complete this section of the test.

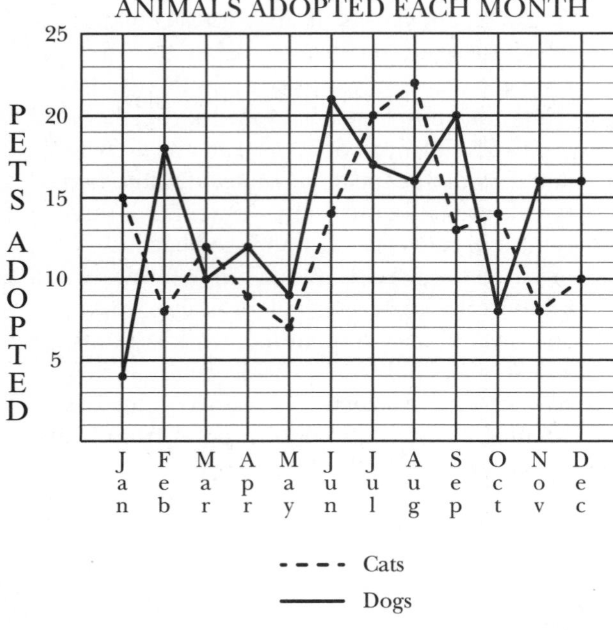

1. The line graph above reports the number of cats and dogs adopted at a certain rescue shelter each month for a one-year period. In how many months of that year were more cats than dogs adopted?

   (A) 3
   (B) 4
   (C) 5
   (D) 6
   (E) 7

2. 16 is 8% of what number?

   (A) 2
   (B) 12.8
   (C) 20
   (D) 128
   (E) 200

3. If $f(x) = x^3 - x^2 + x - 1$, what is the value of $f(-1)$?

   (A) 0
   (B) –1
   (C) –2
   (D) –3
   (E) –4

4. If point A has coordinates (–3,1) and point B has coordinates (7,–5), what are the coordinates of point M, the midpoint of $\overline{AB}$?

   (A) (10,8)
   (B) (10,–6)
   (C) (4,–4)
   (D) (2,–2)
   (E) (–5,–2)

5. A cube with side length s has surface area A. If the side length of the cube is increased by 2, what is the new surface area of the cube, in terms of s and A?

   (A) $12+A$
   (B) $(s+2)^3+A$
   (C) $24(s+1)+A$
   (D) $6(s^2+1)+A$
   (E) $\dfrac{12(s+2)+A}{A}$

6. If $\frac{x^2-y^2}{x-y}=3$ and $x-y\neq 0$, which of the following statements must be true?

(A) $xy=3$

(B) $\frac{x}{y}=3$

(C) $x=y+3$

(D) $x=y-3$

(E) $y=3-x$

7. The set R contains six real numbers with an average (arithmetic mean) of 9. The set S contains c real numbers with an average (arithmetic mean) of 4. Together, the sets have an average (arithmetic mean) of 6. What is the value of c?

(A) 4
(B) 6
(C) 8
(D) 9
(E) 10

8. If $BD=9$ and $CD=6$ in rectangle $ABCD$, what is the value of BC?

(A) 3
(B) 4
(C) 5
(D) $4\sqrt{3}$
(E) $3\sqrt{5}$

9. A particular family has three children: Maria, Manny, and Jose. Each child purchased a notebook for the first day of school, and each notebook was a different color. Given the information below, what is the order in which the children were born? In other words, what are the names of the children in order, from oldest to youngest?

1. The youngest child bought a blue notebook
2. Jose, who is not the oldest child, bought a red notebook
3. Maria's age is two-thirds Manny's age

(A) Maria, Jose, Manny
(B) Maria, Manny, Jose
(C) Manny, Jose, Maria
(D) Manny, Maria, Jose
(E) Maria, Manny, Jose

10. If x is an odd integer and y is an even integer such that $x<y$, which of the following statements must be true?

I. $x+y$ and xy are both odd

II. $\frac{x-y}{2}=-\frac{y-x}{2}$

III. $x^y<y^x$

(A) II only
(B) III only
(C) II and III only
(D) I, II, and III
(E) None of the above

11. If $3^a=\left(\frac{1}{9}\right)^b$, then which of the following expresses $-b$ in terms of a?

(A) $2a$

(B) $\frac{2}{a}$

(C) $\frac{a}{2}$

(D) a^2

(E) $\sqrt{a}$

12. If a triangle has sides with lengths x, $x - 3$, and $x + 4$, which of the following are possible values of x?

    I. 5
    II. 7
    III. 9

    (A) III only
    (B) I and II only
    (C) II and III only
    (D) I, II, and III
    (E) The answer cannot be determined based on the information given

13. If $\frac{Ax^2 - 9A}{2x+6} = Bx$ and $x > 0$, which of the following is the solution to the equation?

    (A) $x = \frac{A}{2B+3}$
    (B) $x = \frac{2A-2B}{3}$
    (C) $x = \frac{3A}{A-2B}$
    (D) $x = \frac{3A+2B}{A}$
    (E) $x = \frac{A-2B}{3A-2B}$

14. If $4w + 2x = y + 3z$ and $2y - 8w = 18 + 4x$, what is the value of z?

    (A) -2
    (B) -3
    (C) -4
    (D) -6
    (E) -8

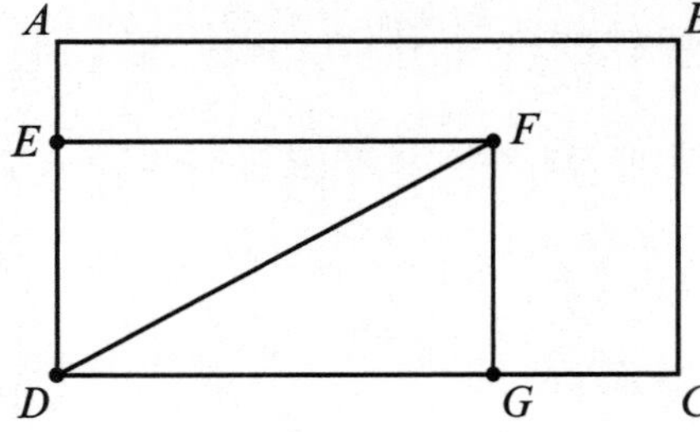

Note: Figure not drawn to scale.

15. In the diagram above, $ABCD$ and $DEFG$ are rectangles such that $DE = \frac{3}{4}AD$ and $CG = \frac{1}{6}CD$. If the area of triangle DEF is 18, what is the area of rectangle $ABCD$?

    (A) 40
    (B) 40.5
    (C) 57.6
    (D) 58.5
    (E) 81

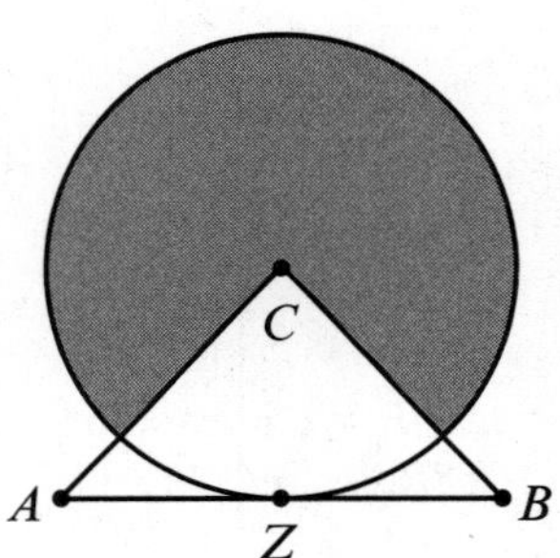

Note: Figure not drawn to scale.

16. In the diagram above, circle C and equilateral triangle ABC share point Z. If the radius of the circle is r, what is the ratio of the perimeter of triangle ABC to the area of the shaded region?

    (A) $\frac{2\sqrt{3}}{\pi r}$
    (B) $\frac{5\sqrt{3}}{\pi r}$
    (C) $\frac{12\sqrt{3}}{5\pi r}$
    (D) $\frac{18\sqrt{3}}{5\pi r}$
    (E) $\frac{36\sqrt{3}}{5\pi r}$

Solutions

Section I

1. **E**. Create a proportion based on the given information, substitute $x = 1$ into the proportion, and solve for y.

$$\frac{x}{y} = \frac{3}{4}$$
$$\frac{1}{y} = \frac{3}{4}$$
$$1(4) = 3(y)$$
$$4 = 3y$$
$$\frac{4}{3} = y$$

2. **E**. Lines l and m are parallel lines intersected by transversal p. Angles 3 and 5 are same-side interior angles, so they are supplementary. Thus, statement I is correct. Angles 3 and 7 are corresponding angles, so they are congruent, not supplementary. Statement II is incorrect.

See Problem 8.1 for more information.

Statement III is correct. Notice that angles 3 and 4 are supplementary because, combined, they form a straight angle. Furthermore, angles 4 and 8 are corresponding angles, so they are congruent. If angle 4 is a supplement of angle 3 and angle 8 has the same measure as angle 4, then angle 8 is also a supplement of angle 3.

3. **D**. Apply the plug and chug method, substituting each value of x into the equation to determine which produce true statements.

Test $x = -3$	Test $x = -2$	Test $x = 3$
$\left\|(-3)-(-3)^2\right\| = 6$	$\left\|(-2)-(-2)^2\right\| = 6$	$\left\|(3)-(3)^2\right\| = 6$
$\left\|-3-(9)\right\| = 6$	$\left\|-2-(4)\right\| = 6$	$\left\|3-(9)\right\| = 6$
$\left\|-12\right\| = 6$	$\left\|-6\right\| = 6$	$\left\|-6\right\| = 6$
$12 = 6$ **False**	$6 = 6$ **True**	$6 = 6$ **True**

Statements II and III are correct because $x = -2$ and $x = 3$ are solutions to the equation.

4. **C**. Begin by dividing each of the terms by 2, the greatest common factor, to simplify the equation.

$$\frac{2x}{2} - \frac{6y}{2} = \frac{18}{2}$$
$$x - 3y = 9$$

To express x in terms of y, you need to solve the equation for x by adding $3y$ to both sides.

$$x = 3y + 9$$

5. **B**. To calculate the average of four values, add them and divide by 4. The question states that the average is equal to 4.5.

$$\frac{2+5+8+x}{4}=4.5$$
$$\frac{15+x}{4}=4.5$$

Multiply both sides by 4 and solve for x.

$$(15+x)=4(4.5)$$
$$15+x=18$$
$$x=18-15$$
$$x=3$$

6. **D**. If a is directly proportional to b, then $a = kb$, where k is the constant of proportionality. In this question, v is directly proportional to $3/(w + 1)$ and the constant of proportionality is m.

$$v=m\cdot\frac{3}{w+1}$$
$$v=\frac{3m}{w+1}$$

Solve the equation for w.

$$v(w+1)=3m$$
$$w+1=\frac{3m}{v}$$
$$w=\frac{3m}{v}-1$$

This is not one of the answer choices, but if you combine the fractions using the least common denominator v, you do identify an answer choice.

$$w=\frac{3m}{v}-1\left(\frac{v}{v}\right)$$
$$w=\frac{3m-v}{v}$$

7. **B**. Notice that all of the answer choices are consecutive odd integers—the further down the list, the larger the number. In situations like this, you should start with Choice (C), 27. If 27 is the least of the three integers, then the sum of the three consecutive odd integers is 27 + 29 + 31 = 87. This number is too large, so select a smaller number as the least of the integers: Choice (B), 25. Note that 25 + 27 + 29 = 81, so the least of the three integers is 25.

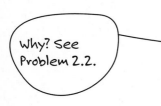

Why? See Problem 2.2.

8. **A**. Plot the points A, B, C, and D in the coordinate plane. You have both the x- and y-coordinates of points B and C, but you do not know the shared x-coordinate of A and D. The question states that $x < 0$, so A and D will appear left of the y-axis. Furthermore, because A and D share the same x-coordinate, they appear on the same vertical line, just as B and C appear on the vertical line $x = 5$.

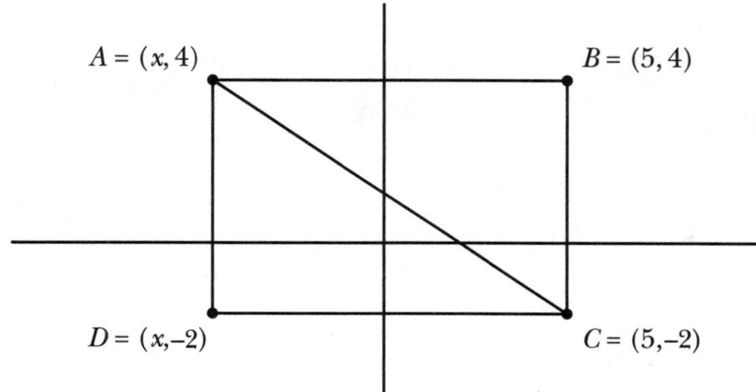

The width of the rectangle is the vertical distance between B and C. Calculate the width by subtracting the y-coordinates of B and C and, if necessary, take the absolute value.

$$\begin{aligned} BC &= |-2-(4)| \\ &= |-6| \\ &= 6 \end{aligned}$$

Notice that ABC is a right triangle with a side of length 6 and a hypotenuse of length 10. Recall that 6, 8, 10 is a Pythagorean triple, so $AB = 8$. Just like the vertical distance BC was equal to the absolute value of the y-coordinates of the points, the horizontal distance AB is equal to the absolute value of the x-coordinates of the points. Remember that 8 is the length of the rectangle: $|5 - x| = 8$. Apply the plug-and-chug method using the answer choices, noting that only $x = -3$ satisfies the equation.

$$\begin{aligned} |5-(-3)| &= 8 \\ |5+3| &= 8 \\ 8 &= 8 \quad \textbf{True} \end{aligned}$$

9. **C**. If you solved Question 8 correctly, this question is easy. In the solution to Question 8, you determined that the length and width of the rectangle are 8 and 6 respectively. The perimeter P of a rectangle is equal to twice the length plus twice the width.

$$\begin{aligned} P &= 2l + 2w \\ &= 2(8) + 2(6) \\ &= 16 + 12 \\ &= 28 \end{aligned}$$

10. **D**. Multiply both sides of the equation $y^2 = z/x$ by x to eliminate fractions and then solve for x.

$$\begin{aligned} y^2 &= \frac{z}{x} \\ xy^2 &= z \\ x &= \frac{z}{y^2} \end{aligned}$$

Substitute $x = z/y^2$ into the first equation and solve for z.

$$x = \frac{1}{\sqrt{y}}$$

$$\frac{z}{y^2} = \frac{1}{\sqrt{y}}$$

$$y^2\left(\frac{z}{y^2}\right) = y^2\left(\frac{1}{\sqrt{y}}\right)$$

$$z = \frac{y^2}{\sqrt{y}}$$

Recall that $\sqrt{y} = y^{1/2}$.

$$z = \frac{y^2}{y^{1/2}}$$

The quotient of two exponential expressions with the same base is the shared base raised to the difference of the exponents: $z = y^{2-1/2} = y^{3/2}$.

In other words, it's upside-down.

11. **E**. The function $f(x)$ is a transformed version of the function $y = x^2$. Compared to the graph of x^2, $f(x)$ is reflected across the x-axis and shifted to the right. That means a must be negative because of the reflection, as explained in Problem 6.83. Similarly, b must be negative because of the horizontal shift, as explained in Problem 6.81. If a and b are both negative, then their product is positive: $ab > 0$.

12. **B**. Notice that the length of the rectangle is equal to the two congruent diameters of the circles. In other words, the length is four times the radius of the circles. The width of the rectangle is one diameter (or two radii). The area of the rectangle is 128. Use this information to calculate r, the radius of the circles.

$$A = l \cdot w$$

$$128 = (4r)(2r)$$

$$128 = 8r^2$$

$$16 = r^2$$

$$\sqrt{16} = \sqrt{r^2}$$

$$4 = r$$

Consider the diagram below, which focuses on the left half of the rectangle. If you subtract the area of the circle (with radius r) from the area of the square (with side length $2r$), you get the area inside the square but outside the circle. This area is evenly divided into four congruent regions, one of which is shaded.

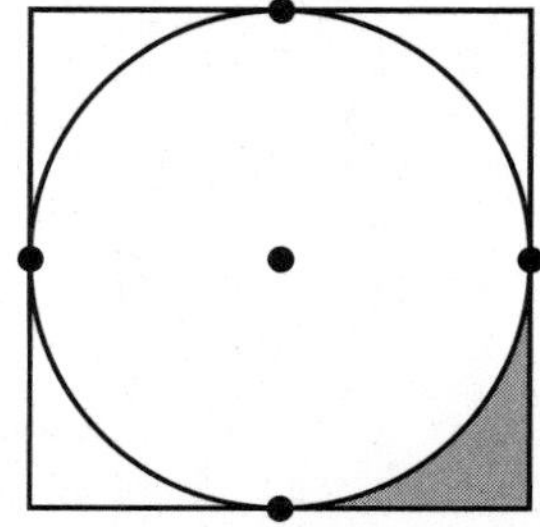

Calculate the area of this shaded region in this diagram. Recall that $r = 4$.

$$\begin{aligned} A &= \frac{\text{Area of square} - \text{Area of circle}}{4} \\ &= \frac{(2r)^2 - \pi(r)^2}{4} \\ &= \frac{(2\cdot 4)^2 - \pi(4)^2}{4} \\ &= \frac{(8)^2 - \pi(16)}{4} \\ &= \frac{64 - 16\pi}{4} \\ &= 16 - 4\pi \end{aligned}$$

The shaded region in the original question has an area twice as large as the area you just calculated, so its area is $2(16 - 4\pi) = 32 - 8\pi$. Factor 8 out of the expression to get $8(4 - \pi)$.

13. **E**. Apply the DIY strategy. If Miranda selects $n = 5$ packets of seeds that contain $s = 10$ seeds each, she purchases $ns = 5(10) = 50$ seeds. If $p = 80$ percent of the seeds will grow, then $50(0.80) = 40$ seeds will grow when planted. That means $50 - 40 = 10$ will *not* grow. If you substitute those values of n, s, and p into the formulas below, only Choice (E) produces the correct answer.

$$\begin{aligned} ns - \frac{nps}{100} &= 5(10) - \frac{5(80)(10)}{100} \\ &= 50 - \frac{4{,}000}{100} \\ &= 50 - 40 \\ &= 10 \end{aligned}$$

14. **D**. Utilities and meals, together, account for 37% of Josh's paycheck. If 37% of the paycheck is \$676.36, you can create a proportion to calculate the entire check.

$$\begin{aligned} \frac{\text{percentage of paycheck}}{\text{amount of that percentage}} &= \frac{\text{100 percent of paycheck}}{\text{amount of total paycheck}} \\ \frac{37}{676.36} &= \frac{100}{x} \\ 37x &= 67{,}636 \\ x &= \frac{67{,}636}{37} \\ x &= \$1{,}828 \end{aligned}$$

Josh's monthly paycheck is \$1,828. He spends 45% of that on housing, or (0.45)(\$1,828) = \$822.60. He spends 12%, or (0.12)(\$1,828) = \$219.36 on utilities. The difference between housing and utilities is \$822.60 – \$219.36 = \$603.24.

15. **E**. Apply the technique demonstrated in Problem 6.55, noting that 3 is outside of the absolute value bars. Thus, you should subtract 3 from both sides to isolate the absolute value expression before you create the compound inequality.

$$|2x| < 11 - 3$$
$$|2x| < 8$$
$$-8 < 2x < 8$$
$$-\frac{8}{2} < \frac{2x}{2} < \frac{8}{2}$$
$$-4 < x < 4$$

If a value x satisfies the compound inequality $-4 < x < 4$, then x is both greater than -4 and less than 4, Choice (E).

16. **A**. Apply the DIY strategy, substituting an integer value for x that is less than -3. In the expressions below, $w = -4$ is substituted into the expressions.

- Choice (A): $w]][[w = \frac{-4-1}{-4} = \frac{-5}{-4} = 1.25$. The more negative the value you choose for w, the closer this function gets to 1, but it is always greater than 1.
- Choice (B): $w]][[w^2 = \frac{(-4)^2 - 1}{-4} = \frac{15}{-4} = -3.75$. This will always be negative, because you divide a positive number by a negative number.
- Choice (C): $w]][[-w = \frac{-(-4)-1}{-4} = \frac{3}{-4} = -0.75$. This will always be negative, for the same reason as Choice (B).
- Choice (D): $w]][[w^{-2} = \frac{(-1/4)^2 - 1}{-4} = \frac{(1/16)-1}{-4} = 0.234375$. This will always be a small positive number less than 1.
- Choice (E): $w]][[1 = \frac{1-1}{-4} = \frac{0}{-4} = 0$. This always equals 0.

$(-4)^{-2} = \frac{1}{(-4)^2} = \frac{1}{16}$

The expression with the greatest value is Choice (A).

17. **A**. Construct a chart to organize the information given. Let x represent the number of boys in Class A and let y represent the number of girls in Class B. If there are 49 total students, then $49 - 26 = 23$ are girls.

	Class A	Class B	**Total**
Boys	x	$2y$	26
Girls	$x + 4$	y	23

Add the rows to create a system of equations: $x + 2y = 26$ and $x + y = 19$. Subtract the second equation from the first to solve for y.

This is the equation $(4 + x) + y = 23$ after you subtract 4 from both sides.

$$\begin{cases} x + 2y = 26 \\ -1(x + y = 19) \end{cases} \rightarrow \begin{cases} x + 2y = 26 \\ -x - y = -19 \end{cases}$$
$$\rightarrow (x - x) + (2y - y) = 26 - 19$$
$$\rightarrow 0x + y = 7$$

Adding corresponding terms produces the solution $y = 7$. There are 7 girls in Class B.

18. **C**. This is very similar to Problem 9.24. In fact, the bases are both squares with side length 4, so look at that problem for more information. Imagine right triangle AQX in the diagram, which has a leg of length $AQ = h$, a leg of length QX, and a hypotenuse of length AX, the side length you are asked to calculate in this problem. Consider the diagram below, a top-down view of the base $WXYZ$.

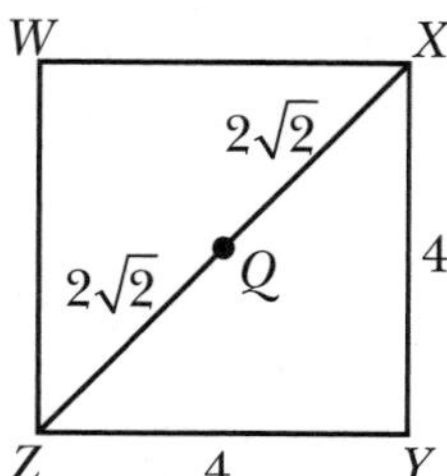

The diagonal $\overline{XZ}$ is the hypotenuse of a 45°–45°–90° triangle, so its length is $XZ = 4\sqrt{2}$. The point Q divides it in half, so $QX = QZ = 2\sqrt{2}$. Apply the Pythagorean theorem to the triangle AQX.

$$\begin{aligned}(AQ)^2 + (QX)^2 &= (AX)^2\\ h^2 + \left(2\sqrt{2}\right)^2 &= (AX)^2\\ h^2 + 4(2) &= (AX)^2\\ \sqrt{h^2 + 8} &= AX\end{aligned}$$

19. **D**. Apply the DIY strategy. If the company sells $B = 800$ black phones, then it sells 800 – 400 = 400 white phones. Of the 800 + 400 = 1,200 phones sold, $400 / 1{,}200 = 33.\bar{3}\%$ are white. Note that Choice (D) generates the correct percentage when $B = 800$.

$$\begin{aligned}33.\bar{3}\% &= \frac{50B - 20{,}000}{B - 200}\\ &= \frac{50(800) - 20{,}000}{800 - 200}\\ &= \frac{40{,}000 - 20{,}000}{600}\\ &= \frac{20{,}000}{600}\\ &= 33.\bar{3}\%\end{aligned}$$

20. **E**. Let the numbers 1, 2, 3, 4, and 5 represent the five names of the friends. List the different ways you can draw three of the five names, keeping in mind that the order of the names does not matter. In other words, the group 1 2 3 is the same as the group 2 1 3. There are 10 unique ways to select three names.

1 2 3	2 3 4	3 4 5
1 2 4	2 3 5	
1 2 5	2 4 5	
1 3 4		
1 3 5		
1 4 5		

Any two numbers—no matter which two numbers you choose—appear in exactly 3 of the 10 outcomes. For example, 1 and 2 both appear only in the first three rows of the leftmost column. Thus, no matter which numbers represent Nick and Justin, there is a 3/10 probability of selecting them randomly among the three names chosen.

Section 2

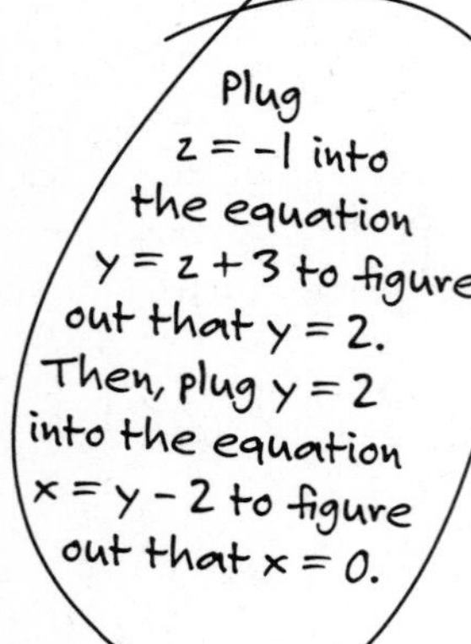

Plug
z = −1 into
the equation
y = z + 3 to figure
out that y = 2.
Then, plug y = 2
into the equation
x = y − 2 to figure
out that x = 0.

1. C. Translate the information into equations: $x = y - 2$ and $y = z + 3$. If $z = -1$, then $y = -1 + 3 = 2$. Therefore, $x = 2 - 2 = 0$.

2. A. Notice that angle *BAC* and the 100° angle are supplementary—they form a straight angle when combined. Therefore, the measure of angle *BAC* is 80°. If $AB = BC$, then *ABC* is an isosceles triangle and its base angles are congruent. In other words, angles *BAC* and *BCA* have the same measure. You now know two measures within triangle *ABC*, and their sum is 160°. Recall that a triangle's interior angles have a sum of 180°, so angle *B* must measure 20°. If $(2x)° = 20°$, then $x = 20/2 = 10$.

3. C. Construct a diagram that illustrates the information given. If *C* is the midpoint of *AB*, then $AC = BC$. If *A* is the midpoint of *BD*, then $AB = AD$.

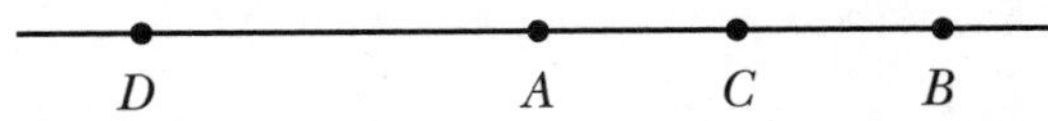

Only Choice (C) is equal to *AD*. The sum $AC + BC$ is equal to *AB*, which (as stated earlier) is equal to *AD*.

4. B. There are 21 integers from which the computer selects, listed below. Numbers that are even or divisible by 3 (or both) are boxed.

$$\boxed{0}\ 1\ \boxed{2}\ \boxed{3}\ \boxed{4}\ 5\ \boxed{6}\ 7\ \boxed{8}\ \boxed{9}\ \boxed{10}$$
$$11\ \boxed{12}\ 13\ \boxed{14}\ \boxed{15}\ \boxed{16}\ 17\ \boxed{18}\ 19\ \boxed{20}$$

There is a $14/21 = 2/3$ probability that the randomly selected number will be even or divisible by 3.

5. C. The graph of the line passes through points $(x,y) = (0,4)$ and $(x,y) = (2,0)$. Substitute each of these coordinate pairs into the equation $ax + by = 6$ to identify the values of *a* and *b*.

$$\begin{aligned} ax + by &= 6 \\ a(0) + b(4) &= 6 \\ 4b &= 6 \\ b &= \frac{6}{4} \\ b &= 1.5 \end{aligned} \qquad \begin{aligned} ax + by &= 6 \\ a(2) + b(0) &= 6 \\ 2a &= 6 \\ a &= \frac{6}{2} \\ a &= 3 \end{aligned}$$

Therefore, $ab = (1.5)(3) = 4.5$.

6. **D**. This is an exponential equation. It requires you to rewrite the bases 16 and 32 as powers of the same integer. Notice that $16 = 2^4$ and $32 = 2^5$; they are both powers of 2.

$$\left(2^4\right)^{x+3} = \left(2^5\right)^{x+2}$$

When an exponential expression (like 2^4) is raised to an exponent (like $x + 3$), you can multiply the exponents.

$$2^{4(x+3)} = 2^{5(x+2)}$$
$$2^{4x+12} = 2^{5x+10}$$
$$4x + 12 = 5x + 10$$
$$12 - 10 = 5x - 4x$$
$$2 = x$$

2 to the 4x + 12 power can only equal 2 to the 5x + 10 power if the powers are equal. Just ignore the base 2.

Be careful! The question does not ask for the value of x. It asks you to calculate 3^x, and $3^2 = 9$.

7. **E**. To answer this question correctly, you need to translate the mathematical statement carefully. The sum and difference of x and y are $x + y$ and $x - y$, respectively. The squares of those values are $(x + y)^2$ and $(x - y)^2$. The variable z is defined as the difference of those squares: $z = (x + y)^2 - (x - y)^2$. Simplify z by expanding the squared quantities.

$$\begin{aligned} z &= (x+y)(x+y) - \left[(x-y)(x-y)\right] \\ &= x^2 + 2xy + y^2 - \left[x^2 - 2xy + y^2\right] \\ &= x^2 + 2xy + y^2 - x^2 + 2xy - y^2 \\ &= 4xy \end{aligned}$$

The product of x and y is xy. The question asks you to calculate the quotient $4xy \div (xy)$. Reduce the fraction to lowest terms: $4xy/(xy) = 4$.

8. **B**. If $a \div b$ has quotient c and remainder d, then $bc + d = a$. Similarly, if $a \div c$ had quotient d and remainder b, then $cd + b = a$. Both expressions are equal to a, so you can set them equal to each other.

Use real numbers to help understand why this is true. If a = 9 is divided by b = 4, the quotient is c = 2 and the remainder is d = 1. Plug these values into the statement to see why it's true:

$$bc + d = a$$
$$4(2) + 1 = 9$$
$$8 + 1 = 9$$

$$bc + d = cd + b$$

Subtract b and d from both sides to group terms with like variables on the same side of the equation.

$$bc - b = cd - d$$

Factor the greatest common factor out of both expressions.

$$b(c - 1) = d(c - 1)$$

If $c \neq 1$, then $c - 1 \neq 0$, allowing you to divide both sides of the equation by $c - 1$.

$$b = d$$

9. **5040**. This is a fundamental counting principle problem. Because the order of the digits matters, you can multiply the numbers of ways to choose each digit to calculate the total number of possible PINs. There are 10 ways to choose the first digit but only 9 ways to choose the second, because the digits cannot repeat. By the same logic, there are 8 ways to choose the third digit and 7 ways to choose the fourth. Therefore, there are $10 \cdot 9 \cdot 8 \cdot 7 = 5{,}040$ possible unique PINs.

10. **54**. Construct a Venn diagram to organize the data given in the question. Note that the 29 students participating in a spring sport *includes* the 15 students who participated in a spring and a fall sport. Therefore, the number of students who participated *only* in a spring sport is $29 - 15 = 14$. In the following diagram, S represents the set of students who participated in a spring sport and F represents the set of students who participated in a fall sport.

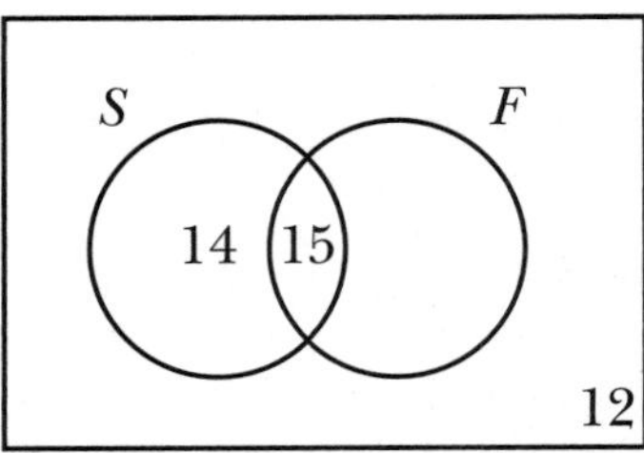

This diagram accounts for $14 + 15 + 12 = 41$ of the 80 students. The remaining 39 students must have participated in a fall sport only. Remember, 15 students were involved in a spring and fall sport, so a total of $39 + 15 = 54$ students participated in a fall sport.

11. **200**. You could apply the formula from Problem 4.50 to calculate the 44th term and then subtract the fourth term from it. Alternately, you could calculate each of the terms until you reach the 44th term and then subtract the first term. The solution below assumes that you cannot remember the formula and that you have no desire to calculate 44 terms. Draw a diagram that illustrates the relationships between consecutive terms.

First term	Second term	Third term	Fourth term	Fifth term
-6	$-6+5$	$-6+10$	$-6+15$	$-6+20$
-6	$-6+5(1)$	$-6+5(2)$	$-6+5(3)$	$-6+5(4)$

Notice that the a_n term is equal to $-6 + 5(n - 1)$. For example, the $n = 4$ term (a_4) is equal to $-6 + 5(4 - 1) = -6 + 5(3)$. Calculate the 44th term.

$$\begin{aligned} a_{44} &= -6+5(44-1) \\ &= -6+5(43) \\ &= -6+215 \\ &= 209 \end{aligned}$$

This is actually the formula from Problem 4.50, used to calculate the nth term of an arithmetic series in which a_1 is the first term and d is the common difference: $a_1 + (n - 1)d$. If you can't (or don't want to) memorize the formula, it is easy to re-create, as this solution demonstrates.

The difference between the 44th term and the 4th term is $a_{44} - a_4 = 209 - 9 = 200$.

12. **32**. Apply the formula for the volume of a cylinder, such that $r = 4/\sqrt{\pi}$ and height $h = 6$.

$$\begin{aligned} V &= \pi r^2 h \\ &= \pi\left(\frac{4}{\sqrt{\pi}}\right)^2 (6) \\ &= 6\not{\pi}\left(\frac{16}{\not{\pi}}\right) \\ &= 96 \text{ cubic feet} \end{aligned}$$

Each minute, the tank leaks 3 cubic feet of water. Divide 96 by 3 to calculate how many minutes it takes the tank to drain: 96 ÷ 3 = 32 minutes.

13. **3/2 or 1.5**. *ACD* is a 45°–45°–90° triangle. Note that $CD = \sqrt{18} = 3\sqrt{2}$. Recall that *CD* is $\sqrt{2}$ times the length of leg *AC*.

$$\begin{aligned} CD &= \sqrt{2}(AC) \\ 3\sqrt{2} &= \sqrt{2}(AC) \\ \frac{3\not{\sqrt{2}}}{\not{\sqrt{2}}} &= AC \\ 3 &= AC \end{aligned}$$

The semicircle has diameter 3, so the radius of the semicircle is 3/2 = 1.5.

14. **7**. This question asks you to evaluate the absolute value of $g(3) - f(3)$. According to the chart, $f(3) = -3$ and $g(3) = 4$.

$$\begin{aligned} |(f-g)(3)| &= |f(3) - g(3)| \\ &= |-3-4| \\ &= |-7| \\ &= 7 \end{aligned}$$

15. **5**. To calculate $g(f[g(-2)])$, begin with the innermost function: $g(-2) = -3$. Evaluate $f(x)$ for that value: $f[g(-2)] = f[-3] = 4$. Finally, evaluate $g(x)$ for that value: $g(f[g(-2)]) = g(4) = 5$.

The outermost function is g(x), so the final answer will be a value from the bottom row of the chart. There are no absolute value functions in the question, so the final answer cannot be negative or you wouldn't be able to grid it. Without doing any computations at all, you know the answer must be 0, 1, 3, 4, or 5.

16. **174**. If the average of four positive integers is 45, then the sum of those integers is 4(45) = 180. In order to create the largest possible value for one integer in that sum, you have to make the other three integers as small as possible. The question states that the integers are both positive and different, so the smallest possible values for the other three integers are the three smallest positive integers: 1, 2, and 3. Note that 1 + 2 + 3 = 6. The value of the fourth, and the largest, integer would be 180 – 6 = 174. The sum of these four integers is 180, and the average is 180 ÷ 4 = 45.

17. **417 or 418**. In 30 minutes, Bradley types 50(30) = 1,500 words, Samantha types 65(30) = 1,950 words, and Lisa types 2,100 words. Together, they type 5,550 words in 30 minutes. They start at 12:00 p.m., so after their 5-minute break ending at 12:35 p.m., they will have 35,000 – 5,550 = 29,450 words left to type. Similarly, at 1:10 there are 29,450 – 5,550 = 23,900 words left; at 1:45 there are 23,900 – 5,550 = 18,350 words left; at 2:20 there are 18,350 – 5,550 = 12,800 words left, and at 2:55 there are 12,800 – 5,550 = 7,250 words left.

Bradley will type 50(5) = 250 words in his last 5 minutes of work. Meanwhile, between 2:55 and 3:30 p.m., Samantha and Lisa will type 1,950 + 2,100 = 4,050 words. The trio will type 4,300 words by 3:30 p.m., which is the end of Samantha's workday. That leaves 7,250 – 4,300 = 2,950 words for Lisa to type. By 4:05, she will have 2,950 – 2,100 = 850 words left. She can type 70 words a minute, so it will take her 850 ÷ 70 ≈ 12.143 minutes beyond that time to finish the job. The final word will be typed slightly more than 12 minutes after 4:05, between 4:17 and 4:18.

The sum of the interior angles is 180°(n – 2), where n = 6. That gives you 720°. Divide by the six angles in the hexagon to figure out that each angle measures 120°.

18. **8**. Drag a diagram that illustrates the information given. Because the hexagon (6-sided polygon) is regular, the interior angles are congruent and the sides are congruent. Each side has length 24 ÷ 6 = 4. Each interior angle has measure 120°. The center of the hexagon and the circle are the same point, *C*.

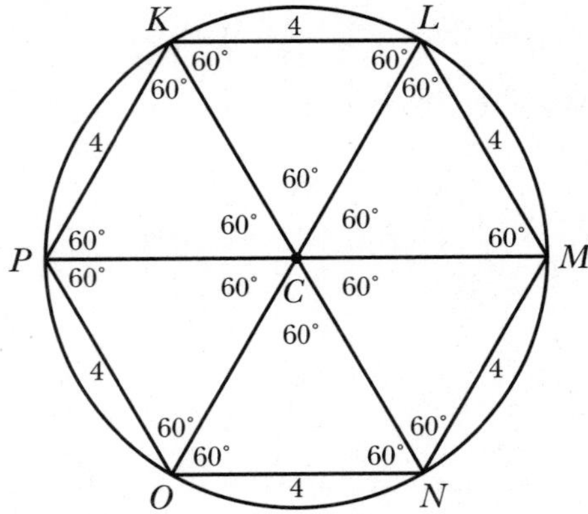

As the diagram illustrates, each of the 120° angles in the hexagon is bisected by the segments connecting *C* to the vertices. This forms equilateral triangles. Thus, radii of the circle ($\overline{CK}$, for example) are the same length as the sides of the hexagon, 4. Therefore, the circumference of the circle is $2\pi r = 2\pi(4) = 8\pi$. Thus, $a = 8$ if $a\pi$ is the circumference.

Section 3

1. **C**. More cats than dogs were adopted in the months of January, March, July, August, and October—a total of five months. In those months, the dotted line rises above the solid line in the graph.

2. **E**. Multiply the answer choices by 0.08, which is 8% expressed in decimal form, to determine which returns a value of 16. Notice that 200(0.08) = 16, so 16 is 8% of 200.

3. **E**. Substitute $x = -1$ into the function to evaluate $f(-1)$.

$$\begin{aligned} f(-1) &= (-1)^3 - (-1)^2 + (-1) - 1 \\ &= (-1) - (1) - 1 - 1 \\ &= -4 \end{aligned}$$

4. **D**. Apply the midpoint formula—the x-coordinate of the midpoint is the average of the endpoints' x-coordinates and the y-coordinate of the midpoint is the average of the endpoints' y-coordinates.

$$\begin{aligned} M &= \left(\frac{-3+7}{2}, \frac{1+(-5)}{2}\right) \\ &= \left(\frac{4}{2}, \frac{-4}{2}\right) \\ &= (2,-2) \end{aligned}$$

5. **C**. The original cube has side length s, so the area of one side (one face) of the original cube is s^2. The surface area of the original cube is 6 times that value, as there are six sides (six faces) on a cube: $6s^2$. The question states that the surface area of the original cube is A, so $A = 6s^2$.

When you increase the cube's side length by 2, the new side length is $s + 2$. The area of one face—one side—of the new cube is $(s + 2)^2 = s^2 + 4s + 4$. The surface area of the new cube is 6 times that value: $6(s^2 + 4s + 4) = 6s^2 + 24s + 24$. Recall that $A = 6s^2$. Therefore, the surface area of the new cube is $A + 24s + 24$. If you factor 24 out of the second and third terms of that expression, you get $A + 24(s + 1)$, which is equivalent to Choice (C).

Because $A = 6s^2$, you can replace $6s^2$ in the expression $6s^2 + 24s + 24$ with A.

6. **E**. The numerator of the fraction is a difference of perfect squares, which has a specific factoring pattern: $x^2 - y^2 = (x + y)(x - y)$. Reduce the fraction to lowest terms.

$$\begin{aligned} \frac{(x+y)\cancel{(x-y)}}{\cancel{x-y}} &= 3 \\ x+y &= 3 \end{aligned}$$

If you subtract x from both sides of the equation, you get Choice (E), $y = 3 - x$.

7. **D**. Calculate a weighted mean. There are 6 numbers in R that have an average of 9. Thus, the sum of those numbers is $6(9) = 54$. Similarly, there are c numbers in S that have an average of 4, so the sum of those numbers is $4c$. When combined, the $6 + c$ numbers in both sets have a total sum of $54 + 4c$. The average of that combined set is 6.

$$\frac{\text{sum of the values in a set}}{\text{number of values in the set}} = \text{average of the values in the set}$$

$$\frac{54+4c}{6+c} = 6$$

Solve the equation for c.

$$\begin{aligned} 6(6+c) &= 54+4c \\ 36+6c &= 54+4c \\ 6c-4c &= 54-36 \\ 2c &= 18 \\ c &= 9 \end{aligned}$$

8. **E**. Draw a diagram, noting that the side lengths listed are the lengths of the hypotenuse and leg of a right triangle, as illustrated below. In the diagram, x represents BC, the length you are asked to calculate.

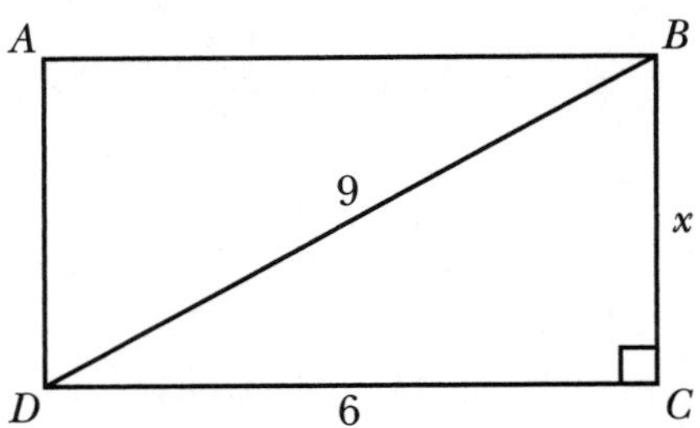

Apply the Pythagorean theorem and solve for x.

$$\begin{aligned} 6^2 + x^2 &= 9^2 \\ 36 + x^2 &= 81 \\ x^2 &= 81 - 36 \\ x &= \sqrt{45} \\ x &= 3\sqrt{5} \end{aligned}$$

9. **C**. According to statement 2, Jose is not the oldest child, and Jose bought a red notebook. Jose cannot be the youngest child either, because the youngest child bought a blue notebook according to statement 1. Jose must be the middle child, born second. According to statement 3, Maria is two-thirds as old as Manny, so Maria is younger than Manny. Maria cannot be the oldest child, so she must be the youngest. The correct order is Manny, Jose, Maria.

She can't be first in the birth order, but she can't be second either. You already figured out that Jose has to be second.

10. **A**. Statement I is false. The sum of an odd and an even integer is odd, but the product of an odd and even integer is always even. For example, if $x = 1$ and $y = 4$, $xy = 4$. Statement II is true. If you multiply both sides of the equation by 2, you get $x - y = -(y - x)$, a true statement for any values of x and y. If you are unconvinced, distribute the negative sign on the right side of the equation to get $x - y = -y + x$. Statement III is false. Though it is true for most odd values of x and even values of y, do not forget about negative values. For example, statement III is false when $x = -1$ and $y = 2$, as demonstrated below.

$$\begin{aligned} x^y &< y^x \\ (-1)^2 &< (2)^{-1} \\ 1 &< \frac{1}{2} \quad \textbf{False} \end{aligned}$$

Statement II is the only statement that is *always* true.

11. **C**. To express $-b$ in terms of a, solve the equation for $-b$.

$$3^a = \left(\frac{1}{9}\right)^b$$
$$3^a = \left(\frac{1}{3^2}\right)^b$$
$$3^a = \left(3^{-2}\right)^b$$
$$3^a = 3^{-2b}$$

If two exponential expressions are equal and they have the same base, the powers are equal: $a = -2b$. Divide both sides by 2; the result is $a/2 = -b$.

12. **A**. Apply the triangle inequality theorem, noting that the sum of any two side lengths must be greater than the length of the remaining side. Substituting $x = 5$ into the expressions gives you side lengths 5, $5 - 3 = 2$, and $5 + 4 = 9$. Because $2 + 5$ is not greater than 9, segments with lengths 2, 5, and 9 cannot form a triangle. Thus, $x = 5$ is not a possible value of x. Similarly, $x = 7$ produces side lengths of 7, 4, and 11. Because $7 + 4$ is not greater than 11, $x = 7$ is not a possible value of x. Only $x = 9$, which produces side lengths of 9, 6, and 13, is a valid value of x: $6 + 9 > 13$, $6 + 13 > 9$, and $9 + 13 > 6$.

13. **C**. Factor A out of the numerator and then factor the difference of perfect squares. Factor 2 out of the denominator.

$$\frac{A(x^2-9)}{2(x+3)} = Bx$$
$$\frac{A\cancel{(x+3)}(x-3)}{2\cancel{(x+3)}} = Bx$$
$$\frac{A(x-3)}{2} = Bx$$

The question states that $x > 0$, so x cannot equal -3, which would cause the denominator to equal 0. Because you're confident that you're not dividing by 0, you can reduce the fraction by canceling like terms $(x + 3)$ in the numerator and denominator.

Multiply both sides of the equation by 2 in order to eliminate the fraction. Then solve for x.

$$A(x-3) = 2Bx$$
$$Ax - 3A = 2Bx$$
$$Ax - 2Bx = 3A$$
$$x(A-2B) = 3A$$
$$x = \frac{3A}{A-2B}$$

14. **B**. If you divide each of the terms in the second equation by 2, many of those terms resemble terms in the first equation.

If you divide every term in an equation by a nonzero number, that equation is still true.

$$\frac{2y}{2} - \frac{8w}{2} = \frac{18}{2} + \frac{4x}{2}$$
$$y - 4w = 9 + 2x$$

Solve the new equation for $4w + 2x$, to match the left side of the first equation in the question.

$$y - 9 = 4w + 2x$$

Replace $4w + 2x$ in the first equation of the question with the equivalent value $y - 9$. Solve for z.

$$\begin{aligned}4w+2x&=y+3z\\y-9&=y+3z\\y-y-9&=3z\\-9&=3z\\-3&=z\end{aligned}$$

15. **C**. The area of triangle DEF is 18, so apply the area formula for a triangle: $18 = (1/2)(bh)$. The base and height of the triangle are DE and EF, respectively. Therefore, $18 = (1/2)(DE)(EF)$. The question states that $DE = (3/4)(AD)$. In other words, the width of rectangle $DEFG$ is 3/4 the width of rectangle $ABCD$. Now turn your attention to the lengths of the rectangles. If $CG = (1/6)(CD)$, then $DG = (5/6)(CD)$. Notice that $EF = DG$, because they are opposite sides of a rectangle. Thus, $EF = (5/6)(CD)$. Substitute the values of DE and EF that you have generated into the original triangle area formula.

If CG is only one-sixth the size of CD, then DG (the other part of CD) must represent the other five-sixths.

$$\begin{aligned}18&=\frac{1}{2}(DE)(EF)\\18&=\frac{1}{2}\left(\frac{3}{4}AD\right)\left(\frac{5}{6}CD\right)\\18&=\left(\frac{1}{2}\cdot\frac{3}{4}\cdot\frac{5}{6}\right)(AD)(CD)\\18&=\frac{15}{48}(AD)(CD)\\18&=\frac{5}{16}(AD)(CD)\end{aligned}$$

Multiply both sides of the equation by 16/5 to solve for $(AD)(CD)$, which is the area of rectangle $ABCD$.

$$\begin{aligned}\frac{16}{5}(18)&=(AD)(CD)\\57.6&=(AD)(CD)\end{aligned}$$

16. **C**. ABC is an equilateral triangle, a regular polygon, so its interior angles are congruent and each measures 60°. Notice that $\overline{CZ}$ is a radius of the circle, so $CZ = r$, as illustrated in the following diagram. Furthermore, that radius divides the equilateral triangle into two congruent 30°–60°–90° triangles.

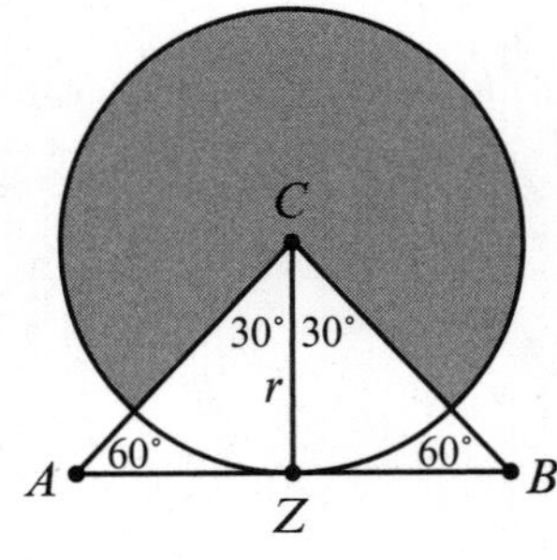

$CZ = r$ is the length of the long leg in both right triangles, so it is $\sqrt{3}$ times the length of both AZ and BZ, the lengths of the short legs: $AZ\left(\sqrt{3}\right) = r$, so $AZ = r / \sqrt{3}$. The hypotenuses of the right triangles represent the lengths of the sides of the equilateral triangles. Each is twice as long as the short leg, so each has length $2r / \sqrt{3}$. The perimeter of the triangle is $3\left(2r / \sqrt{3}\right) = 6r / \sqrt{3}$.

The shaded region of the circle is the entire area of the circle minus the area of a sector with central angle 60°. Calculate the area of the shaded region.

For more information, see Problem 8.59.

$$\begin{aligned}\text{area of shaded region} &= \text{area of circle} - \text{area of sector}\\ &= \pi r^2 - \frac{60}{360}\left(\pi r^2\right)\\ &= \pi r^2 - \frac{1}{6}\pi r^2\\ &= \frac{5}{6}\pi r^2\end{aligned}$$

Calculate the ratio of the perimeter of ABC to the area of the shaded region.

$$\begin{aligned}\frac{\frac{6r}{\sqrt{3}}}{\frac{5}{6}\pi r^2} &= \frac{\frac{6r}{\sqrt{3}}\cdot\frac{6}{5}}{\frac{\cancel{5}}{\cancel{6}}\pi r^2\cdot\frac{\cancel{6}}{\cancel{5}}}\\ &= \frac{\frac{36r}{5\sqrt{3}}}{\pi r^2}\\ &= \frac{36\cancel{r}}{\left(5\sqrt{3}\right)\left(\pi r\cdot\cancel{r}\right)}\\ &= \frac{36}{5\pi r\sqrt{3}}\end{aligned}$$

Dividing the fraction $\frac{A}{B}$ by C is equivalent to the fraction $\frac{A}{BC}$. You can move the number you're dividing by into the denominator of the fraction. In this case, you are dividing a fraction by πr^2, so you can move that expression into the denominator of $\frac{36r}{5\sqrt{3}}$.

Multiply the numerator and denominator by $\sqrt{3}$ to rationalize the denominator, which also transforms the expression into one of the answer choices.

$$\begin{aligned}\frac{36}{5\pi r\sqrt{3}}\cdot\frac{\sqrt{3}}{\sqrt{3}} &= \frac{36\sqrt{3}}{5\pi r(3)}\\ &= \frac{36}{3}\cdot\frac{\sqrt{3}}{5\pi r}\\ &= 12\cdot\frac{\sqrt{3}}{5\pi r}\\ &= \frac{12\sqrt{3}}{5\pi r}\end{aligned}$$

Chapter 13

SAT PRACTICE TEST 2

Remember to pace yourself—about one minute per problem.

This chapter is designed to simulate the mathematics portion of the SAT test. Record your answers on the student test booklet reproduced on the following page. The solutions are located at the end of the chapter. The following formulas are provided at the beginning of each SAT mathematics section, and you can refer to them during the practice test:

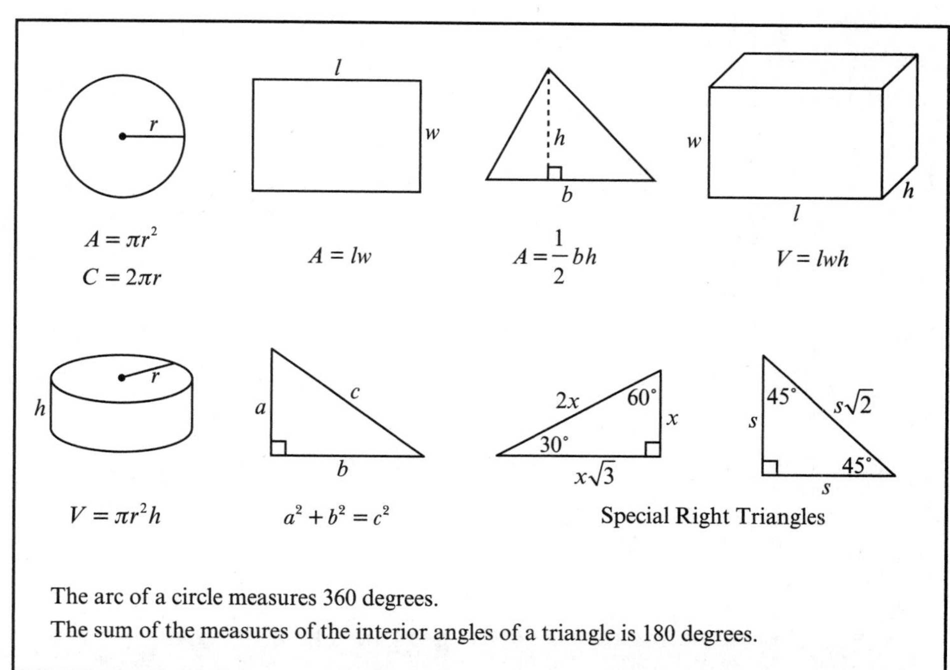

Section 1

Time Limit: 25 minutes

1 ⒶⒷⒸⒹⒺ 5 ⒶⒷⒸⒹⒺ 9 ⒶⒷⒸⒹⒺ 13 ⒶⒷⒸⒹⒺ 17 ⒶⒷⒸⒹⒺ
2 ⒶⒷⒸⒹⒺ 6 ⒶⒷⒸⒹⒺ 10 ⒶⒷⒸⒹⒺ 14 ⒶⒷⒸⒹⒺ 18 ⒶⒷⒸⒹⒺ
3 ⒶⒷⒸⒹⒺ 7 ⒶⒷⒸⒹⒺ 11 ⒶⒷⒸⒹⒺ 15 ⒶⒷⒸⒹⒺ 19 ⒶⒷⒸⒹⒺ
4 ⒶⒷⒸⒹⒺ 8 ⒶⒷⒸⒹⒺ 12 ⒶⒷⒸⒹⒺ 16 ⒶⒷⒸⒹⒺ 20 ⒶⒷⒸⒹⒺ

Section 2

Time Limit: 25 minutes

1 ⒶⒷⒸⒹⒺ 3 ⒶⒷⒸⒹⒺ 5 ⒶⒷⒸⒹⒺ 7 ⒶⒷⒸⒹⒺ
2 ⒶⒷⒸⒹⒺ 4 ⒶⒷⒸⒹⒺ 6 ⒶⒷⒸⒹⒺ 8 ⒶⒷⒸⒹⒺ

9. 10. 11. 12. 13.

14. 15. 16. 17. 18.

Section 3

Time Limit: 20 minutes

1 ⒶⒷⒸⒹⒺ 5 ⒶⒷⒸⒹⒺ 9 ⒶⒷⒸⒹⒺ 13 ⒶⒷⒸⒹⒺ
2 ⒶⒷⒸⒹⒺ 6 ⒶⒷⒸⒹⒺ 10 ⒶⒷⒸⒹⒺ 14 ⒶⒷⒸⒹⒺ
3 ⒶⒷⒸⒹⒺ 7 ⒶⒷⒸⒹⒺ 11 ⒶⒷⒸⒹⒺ 15 ⒶⒷⒸⒹⒺ
4 ⒶⒷⒸⒹⒺ 8 ⒶⒷⒸⒹⒺ 12 ⒶⒷⒸⒹⒺ 16 ⒶⒷⒸⒹⒺ

Section 1

You have 25 minutes to complete this section of the test.

1. If $2(x - 3) + 1 = 3x - 4$, what is the value of x?

   (A) –9
   (B) –3
   (C) –1
   (D) 3
   (E) 9

2. If $\overrightarrow{AX}$ is the angle bisector of $\angle WAY$, which of the following statements must be true?

   I. $m\angle XAY = m\angle WAX$
   II. $2(m\angle XAY) = m\angle WAY$
   III. $\Delta WAX \cong \Delta XAY$

   (A) I only
   (B) I and II only
   (C) I and III only
   (D) II and III only
   (E) I, II, and III

2, 4, 5, 13, 6, 1, 3, 2, 9

3. What is the mode of the data values above?

   (A) 1
   (B) 2
   (C) 4
   (D) 5
   (E) 11

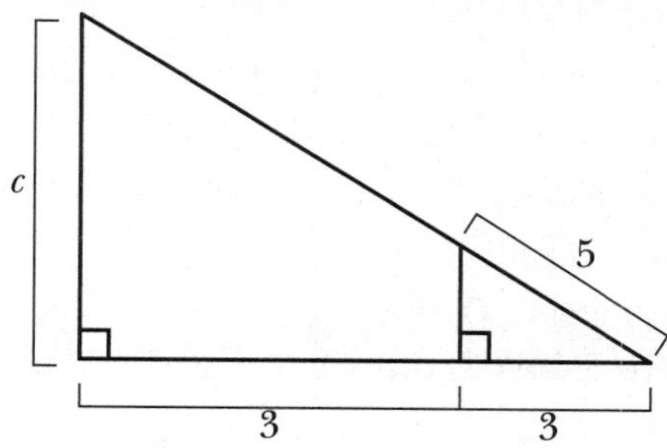

Note: Figure not drawn to scale.

4. Given the diagram above, what is the value of c?

   (A) 4
   (B) 6
   (C) $6\sqrt{2}$
   (D) 8
   (E) 10

5. Which of the following expressions represents the phrase "one-third of the difference of 2 and x"?

   (A) $\frac{1}{3(x-2)}$
   (B) $\frac{2-x}{3}$
   (C) $\frac{1}{3}(x-2)$
   (D) $\frac{1}{3}(2)-x$
   (E) $\frac{1}{3}x-\frac{2}{3}x$

6. If the prime factorization of a positive integer z is equal to $2^a \cdot 3^b \cdot 5^c$, such that a, b, and c are unique positive integers, which of the following statements must be true?

   I. z is even
   II. z is a multiple of 30
   III. z is divisible by 6

   (A) I only
   (B) II only
   (C) I and II only
   (D) I and III only
   (E) I, II, and III

7. If the slope of the line passing through points (–2,2) and (3,k) is 10, what is the value of k?

   (A) 1.9
   (B) 2.5
   (C) 48
   (D) 50
   (E) 52

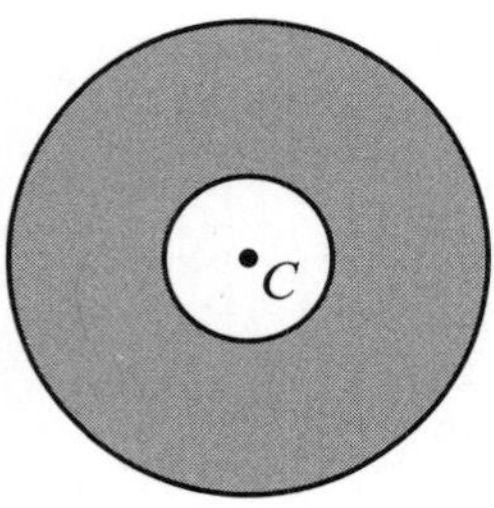

8. In the diagram above, concentric circles share center C. If the diameter of the large circle is 3 times the diameter of the small circle and the small circle has radius r, what is the area of the shaded region in terms of r?

   (A) $\frac{5}{4}\pi r^2$
   (B) $2\pi r^2$
   (C) $5\pi r^2$
   (D) $8\pi r^2$
   (E) $9\pi r^2$

9. The number of spiders in a certain home is inversely proportional to the number of flies. Last week, the home contained 2 spiders and 12 flies. This week, the combined total of spiders and flies is 10. How many more flies than spiders are present this week?

   (A) 1
   (B) 2
   (C) 3
   (D) 4
   (E) 6

2, 221, 222110, 222211100, 222221111000, …

10. In the sequence of integers above, the first term is $a_1 = 2$, the second term is $a_2 = 221$, and the third term is $a_3 = 222110$. Each of the subsequent terms has one additional 2, 1, and 0 digit than the term before it, such that the 2s precede the 1s and the 1s precede the 0s in each term. If x, y, and z represent the number of 2, 1, and 0 digits, respectively, in the a_{51} term of the sequence, what is the value of $x + y + z$?

   (A) 149
   (B) 150
   (C) 151
   (D) 152
   (E) 153

11. If Bob runs at a constant rate of 6 miles per hour, how far does he run in 11 minutes? (Note that 1 mile = 5,280 feet.)

   (A) 5,808 feet
   (B) 6,800 feet
   (C) 8,640 feet
   (D) 8,688 feet
   (E) 9,680 feet

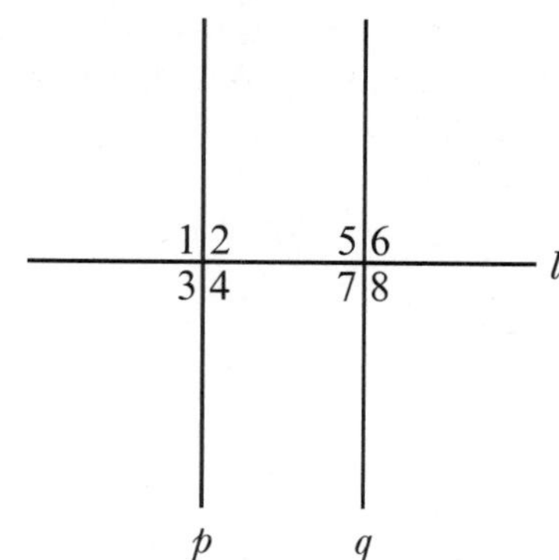

Note: Figure not drawn to scale.

12. Which of the following statements must be true only if $(m\angle 5) \neq (m\angle 6)$ in the figure above?

(A) $p \parallel q$
(B) $\angle 1 \cong \angle 5$
(C) $m\angle 4 + m\angle 7 = 180°$
(D) $m\angle 8 \neq 90°$
(E) $m\angle 5 + m\angle 6 + m\angle 7 + m\angle 8 = 360°$

13. Doni and her daughter, Asia, recently ate lunch at a certain restaurant where the tip is not included in the meal price. The price of Doni's lunch was $18, and she tipped 25% of that amount. Asia tipped 40% of the price of her own meal, and Asia's tip was $1 more than twice Doni's tip. What was the price of Asia's lunch, not including tip?

(A) $13.50
(B) $14.50
(C) $17.50
(D) $18.00
(E) $25.00

20	*a*
14	*b*
45	17
9	12

14. The table above lists the number of minutes eight teens spent sending electronic messages to friends in a 24-hour period. Two of the data values were misplaced and are represented by *a* and *b* in the table. If the complete set of all eight data values has a median of 15.5 and an average (arithmetic mean) of 20, which of the following are possible values of *a* and *b*?

(A) 1, 30
(B) 10, 21
(C) 11, 32
(D) 15, 16
(E) 20, 23

15. If $x \blacklozenge y$ is defined as $\frac{xy}{x-y}$ and $2 \blacklozenge 6 = z \blacklozenge 5$, what is the value of z?

(A) $-\frac{15}{2}$
(B) $-\frac{12}{5}$
(C) $\frac{5}{4}$
(D) $\frac{5}{3}$
(E) $\frac{15}{8}$

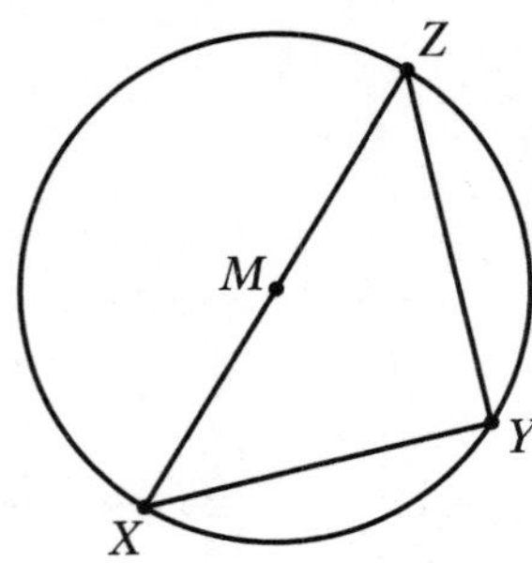

Note: Figure not drawn to scale.

16. In the diagram above, circle M has radius r. If $XM = XY$, what is the perimeter of triangle XYZ, in terms of r?

(A) $4r\sqrt{3}$

(B) $r(1+\sqrt{3})$

(C) $r\sqrt{3}(1+\sqrt{3})$

(D) $r(1+\sqrt{2})$

(E) $r\sqrt{2}(1+\sqrt{2})$

17. In a certain pet store, w% of the fish are goldfish and x% of the y total pets in the store are fish. If z pets in the store are not goldfish, which of the following expresses z in terms of w, x, and y?

(A) $xy(100w-1)$

(B) $\dfrac{x(100-w)}{y}$

(C) $\dfrac{x-w}{100y}$

(D) $\dfrac{100w(y-x)}{y}$

(E) $\dfrac{xy(100-w)}{10{,}000}$

18. Let $a = p + 1$, $b = p + 2$, and $c = p - 1$, such that p is an odd integer. Which of the following statements must be true?

I. $ab > ac$
II. $ac + 1 = p^2$
III. $(b^a)^c$ is even

(A) I only
(B) II only
(C) III only
(D) I and II only
(E) II and III only

19. Let the function k be defined as $k(x) = 2x + \dfrac{15-x^2}{3}$. If $k(3p) = p + 3$, which of the following could be a value of p?

(A) -2

(B) $-\dfrac{2}{5}$

(C) $-\dfrac{1}{3}$

(D) $\dfrac{2}{3}$

(E) $\dfrac{5}{3}$

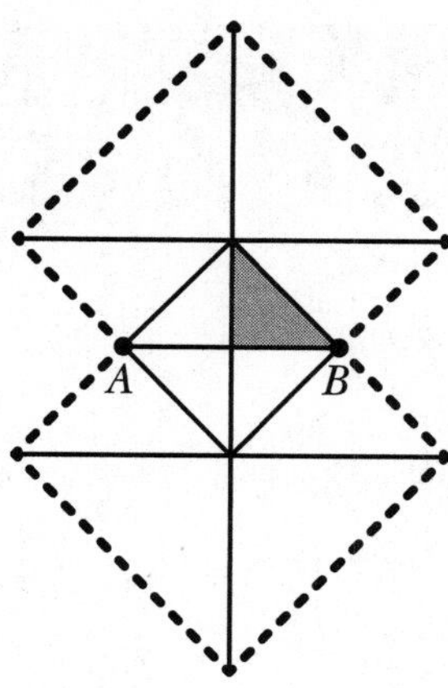

20. The company logo above contains two congruent, overlapping squares with nonzero side length s, such that points A and B are the midpoints of their respective sides on both squares. What is the ratio of the perimeter of the shaded region to the length of the dotted perimeter of the logo?

(A) $\dfrac{1+\sqrt{2}}{12}$

(B) $\dfrac{s+s\sqrt{2}}{2}$

(C) $\dfrac{s+s\sqrt{2}}{6}$

(D) $\dfrac{s\left(2+\sqrt{2}\right)}{6}$

(E) $\dfrac{s\left(2+\sqrt{2}\right)}{12}$

Section 2

You have 25 minutes to complete this section of the test.

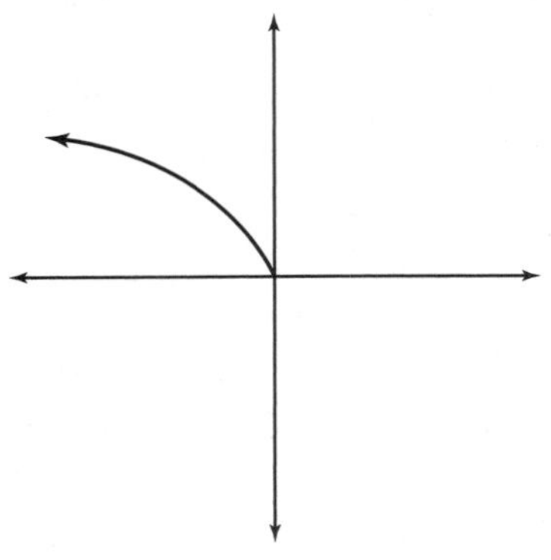

1. Which of the following is most likely the function $h(x)$ graphed above, if $a < 0$?

   (A) $h(x)=\sqrt{ax}$

   (B) $h(x)=\frac{a}{x}$

   (C) $h(x)=a|x|$

   (D) $h(x)=ax^2$

   (E) $h(x)=a(x-1)^2$

2. If a circle has radius r, such that r is an integer, which of the following could be the circumference of the circle?

   (A) π
   (B) 3π
   (C) 9π
   (D) 28π
   (E) 31π

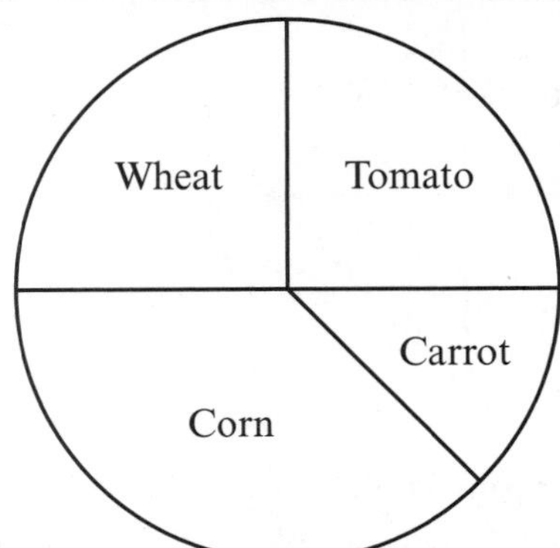

3. A farmer purchases a variety of seeds in quantities illustrated by the circle graph above. Approximately what percentage of the seeds she purchased were neither corn nor tomato seeds?

   (A) 13%
   (B) 25%
   (C) 38%
   (D) 63%
   (E) 75%

4. If half of three less than a number is equal to the quotient of double the number and 6, what is the value of the number?

   (A) 3
   (B) 5
   (C) 7
   (D) 9
   (E) 11

5. In a certain sequence containing only three terms, each term is two less than the square root of the previous term. If the third term is 0, what is the first term in the sequence?

   (A) 9
   (B) 25
   (C) 36
   (D) 49
   (E) 64

6. The measures of the interior angles A, B, and C of triangle ABC are $(3x + 4)°$, $(4x - 9)°$, and $\left[(x-3)^2 - 6\right]°$, respectively. Which of the angles has the least measure?

(A) Angle A
(B) Angle B
(C) Angle C
(D) The triangle is isosceles.
(E) The triangle is equilateral.

7. Coleman has 23 coins, a combination of quarters and nickels that is worth \$3.35. How many nickels does he have?

(A) 11
(B) 12
(C) 13
(D) 14
(E) 16

8. A square with side length x, such that x is a positive, odd integer, is divided completely into "unit squares," which are squares with side length 1. The unit squares in all four corners of the square (with side length x) are shaded, and so are as many of the other unit squares as possible according to this rule: No shaded unit square may be adjacent to any other shaded unit square. What is the ratio of shaded to unshaded unit squares in terms of n, if n is the area of the square with side length x?

(A) $1-\frac{1}{n}$

(B) $\frac{n+1}{2}$

(C) $\frac{n+1}{n}$

(D) $\frac{n+1}{n-1}$

(E) $\frac{n^2-1}{2n}$

9. What is the greatest value of z that satisfies the equation $\frac{|z-2|}{3}+1=5$?

10. If 10 hens can lay 12 eggs in one day, how many hens are needed to lay 30 eggs in one day?

Note: Problems 11–12 refer to the graphs of functions $f(x)$ and $g(x)$ below. The coordinates of all points on the graph are integers.

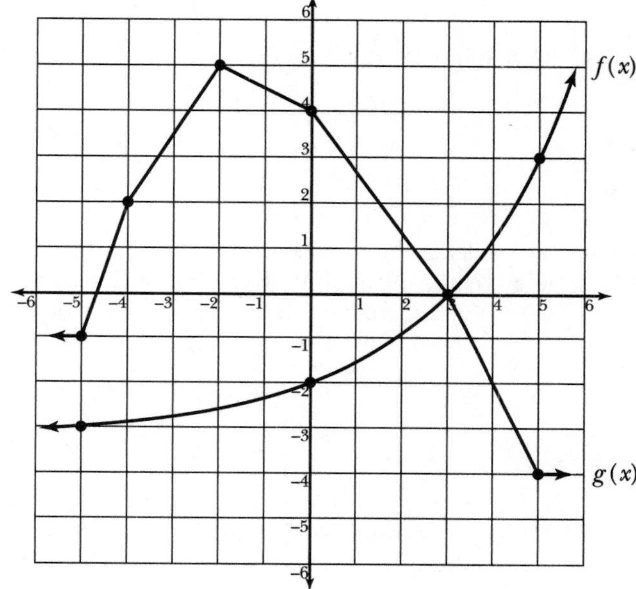

11. If $f(x) \le g(x)$ for all values of x such that $x \le c$, what is the value of c?

12. What is the value of $g\left[f(0)\right]$?

Note: Problems 13–14 refer to the following information.

A set of wooden blocks is placed into a bag, and a single positive integer is painted on each of the blocks. At least one of the following statements refers to each of the integers:

- Five integers are divisible by 2
- Three integers are divisible by 3
- Two integers are divisible by 5
- Three integers are divisible by 6
- One integers is divisible by 10
- No integers are divisible by 15

13. How many blocks does the bag contain?

14. Jodi selects two blocks randomly from the bag. What is the probability that the first cube she selects will feature a number that is divisible by 2 but is neither divisible by 3 nor divisible by 5?

15. If a positive integer z is divided by 9, the remainder is 6. Let r be the remainder when $z - 2$ is divided by 4. What is the sum of all possible unique values of r?

16. If the area of a regular hexagon is $\frac{75}{8}\sqrt{3}$, what is the length of one of its sides?

17. The expression $\frac{4x-1}{3}+\frac{x+8}{6}+\frac{x+3}{2}$ is how much more than $2x$?

18. A bucket contains only chicken eggs—some raw and some hardboiled, some brown and some white. There are six times as many raw eggs as hardboiled eggs. The only brown eggs in the bucket are raw, and there are three times as many raw brown eggs as there are raw white eggs. What is the probability of selecting a raw white egg from the bucket at random?

Section 3

You have 20 minutes to complete this section of the test.

1. Connor just bought food from a roadside farm stand that sells only apples, corn, and carrots. Which of the following statements must be true?

   (A) Connor did not buy carrots.
   (B) Connor bought more corn than apples.
   (C) Connor bought apples, corn, and carrots.
   (D) Connor did not buy peaches.
   (E) Connor bought apples but did not buy corn.

2. What is the height of a triangle with base 6 and area 18?

   (A) 1.5
   (B) 3
   (C) 6
   (D) 12
   (E) 24

3. If $2x + 5 = A$, $4x^2 - 25 = B$, and $A \neq 0$, then what is the value of $B \div A$?

   (A) $2x - 5$
   (B) $2x + 5$
   (C) $x^2 + 5$
   (D) $x^2 - 5$
   (E) $4x^2 + 25$

4. If $x = \frac{2-y}{5}$ and $y = \sqrt{p-1}$, what is the value of x when $p = 10$?

   (A) $-\frac{8}{5}$
   (B) $-\frac{7}{5}$
   (C) $-\frac{1}{5}$
   (D) 0
   (E) $\frac{2}{5}$

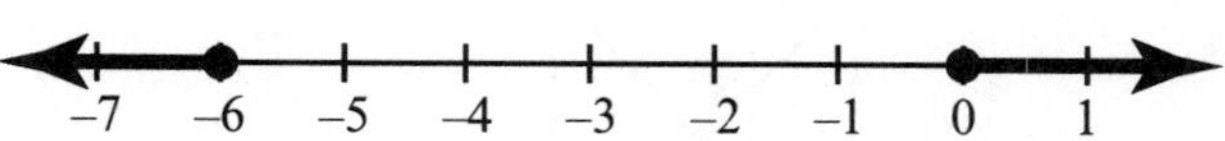

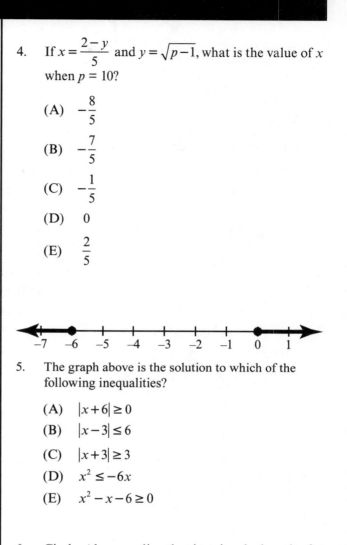

5. The graph above is the solution to which of the following inequalities?

   (A) $|x+6| \geq 0$
   (B) $|x-3| \leq 6$
   (C) $|x+3| \geq 3$
   (D) $x^2 \leq -6x$
   (E) $x^2 - x - 6 \geq 0$

6. Circle A has a radius that is twice the length of circle B's diameter. What is the ratio of circle B's circumference to circle A's circumference?

   (A) $\frac{1}{2}$
   (B) $\frac{1}{3}$
   (C) $\frac{1}{4}$
   (D) $\frac{1}{6}$
   (E) $\frac{1}{8}$

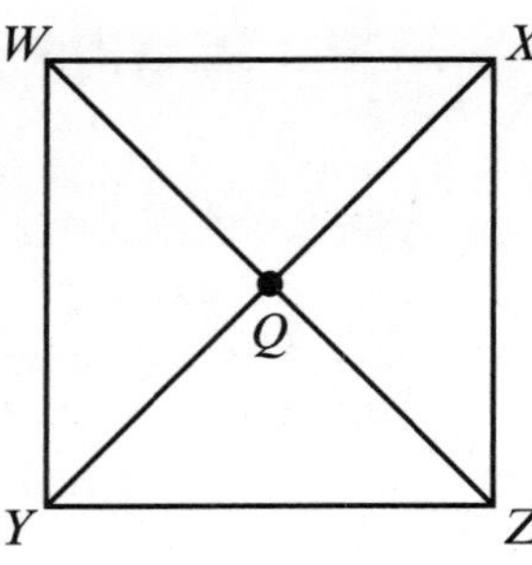

7. If square $WXZY$ above is rotated 270° in the counterclockwise direction around point Q, which of the following illustrates the result?

(A) 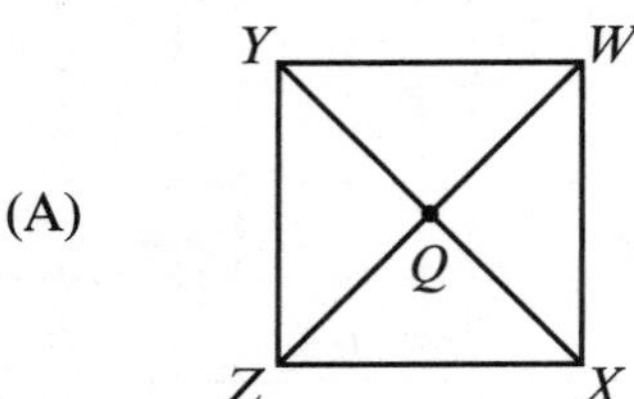

(B) 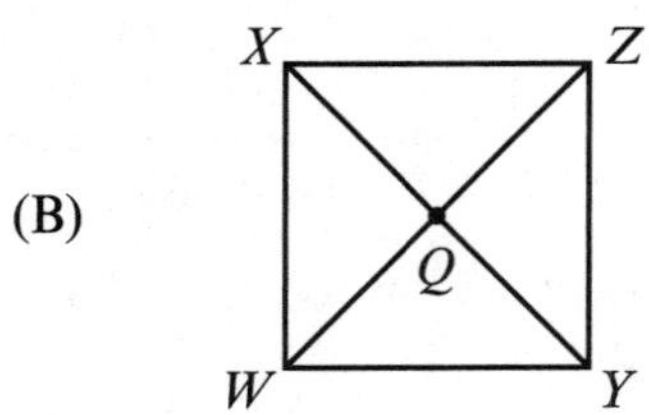

(C) 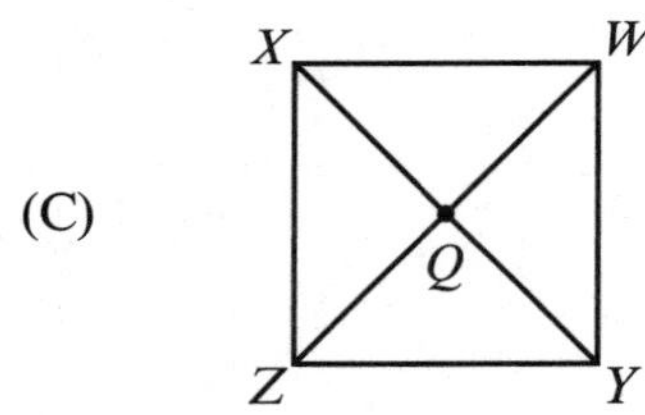

(D) 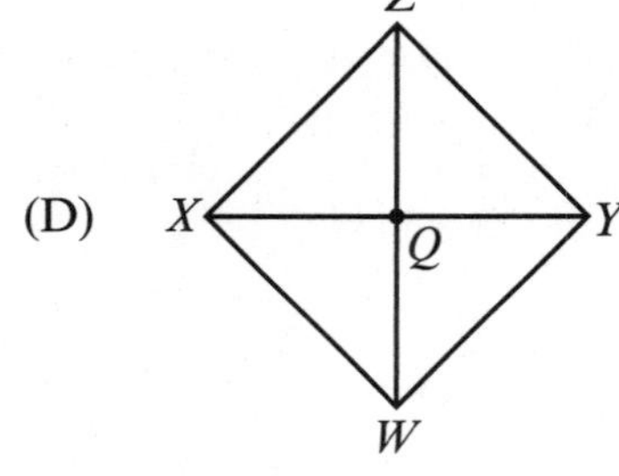

(E) 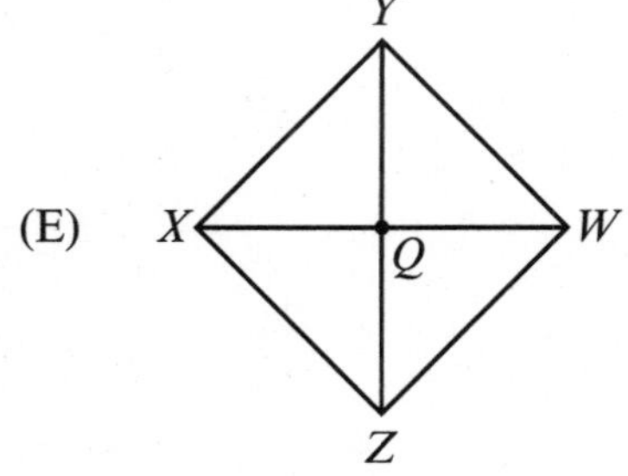

8. Carlie and Amanda can paint rooms with the same dimensions in 6 and 8 hours, respectively. Approximately how long would it take both of them to paint a single room with those dimensions if they worked together?

(A) 202 minutes
(B) 206 minutes
(C) 210 minutes
(D) 216 minutes
(E) 222 minutes

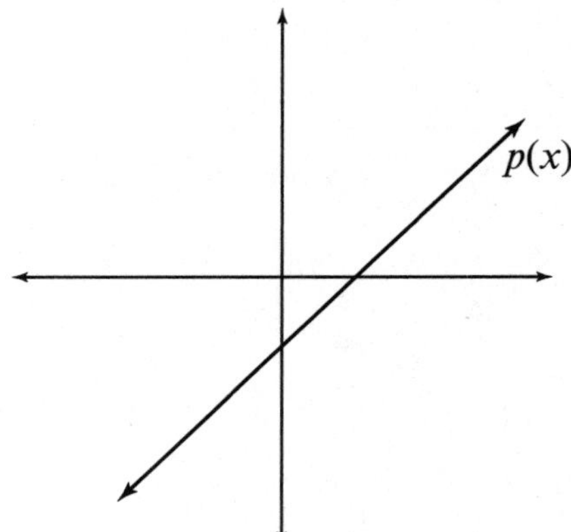

9. Which of the following statements must be true, given the graph of $p(x) = rx + s$ above?

I. $s < 0$

II. $p(x)$ has x-intercept $x = \frac{s}{r}$

III. $p(a) < p(b)$ if $b > a$

(A) I only
(B) I and II only
(C) I and III only
(D) I, II, and III
(E) Not enough information is given

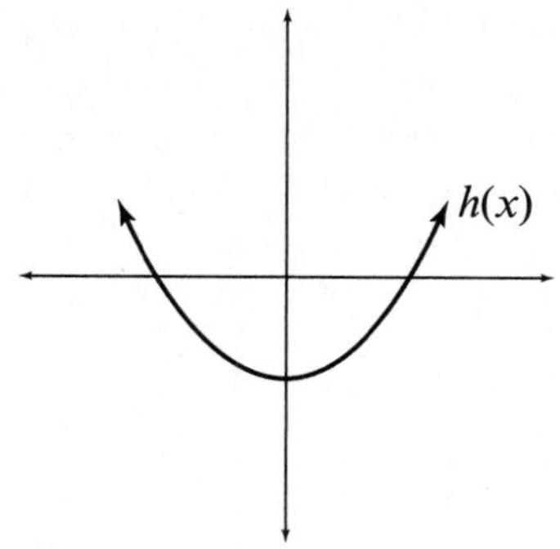

10. Given the graph of $h(x)$ above, which of the following is the graph of $|h(x)|$?

(A)
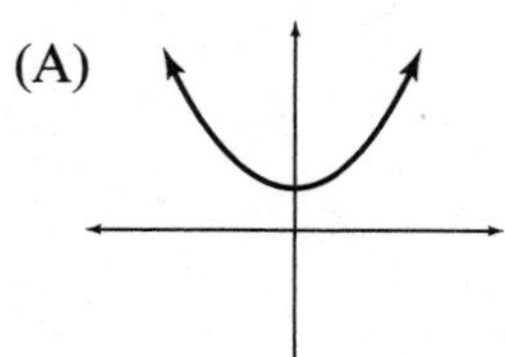

(B)
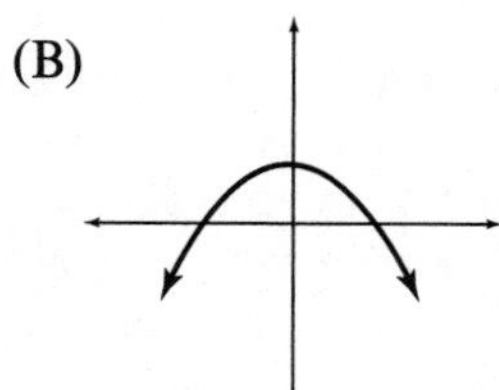

(C)
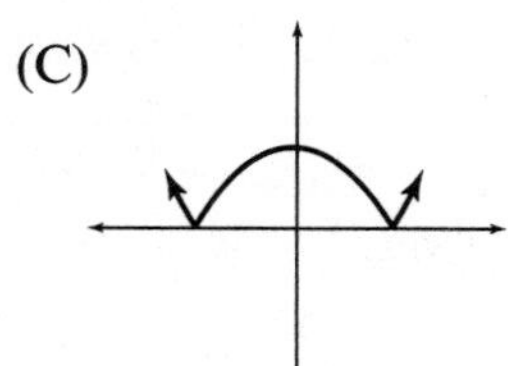

(D)
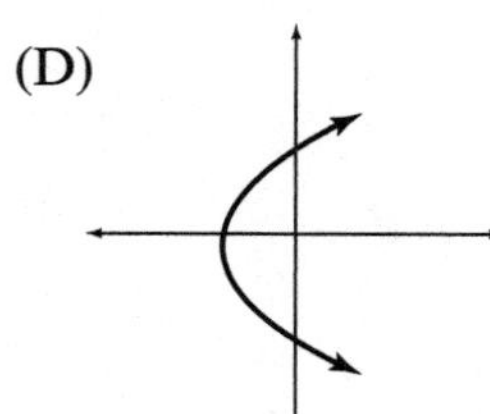

(E)
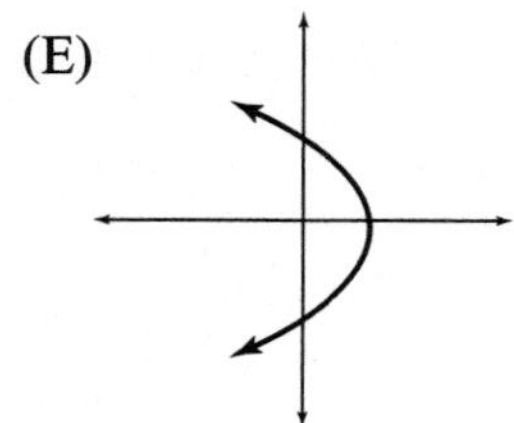

Note: Problems 11 and 12 refer to the following information.

A survey was conducted that asked people of different ages how many different cars they had driven in their lives. The results are recorded in the scatterplot below.

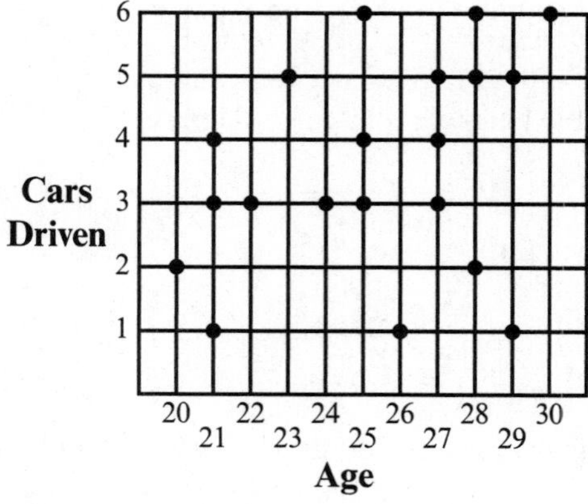

11. What was the median age of the people surveyed?

(A) 25
(B) 25.5
(C) 26
(D) 26.5
(E) 27

12. What was the average (arithmetic mean) of the number of cars driven per person?

(A) 3.6
(B) 3.7
(C) 3.8
(D) 3.9
(E) 4.2

13. The number of bacteria present in a Petri dish is modeled by the function $b(t)=\sqrt{t^2-t+c}$, where t represents the number of hours that have elapsed since the experiment began at time $t = 0$. After $t = 3$ hours have elapsed, 12 colonies were present. Which of the following best approximates the number of colonies that were present 6 hours after the experiment began?

    (A) 13
    (B) 24
    (C) 32
    (D) 144
    (E) 148

14. The price of a computer decreases every day for six consecutive days. Each day, the new price is 10 dollars less than 80% of the previous price. If the price of the computer is \$110 after the six-day period, which of the following best approximates the difference between the original and the final price?

    (A) \$381
    (B) \$389
    (C) \$432
    (D) \$450
    (E) \$603

15. Let the function $a\,\boxed{b}\,c$ be defined as $\dfrac{a-c}{b}$. Which of the following expressions is equal to $\left(5\,\boxed{4}\,2\right)\boxed{3}\,(1.25)$?

    (A) $\dfrac{1}{4}\,\boxed{\dfrac{1}{2}}\,\dfrac{1}{3}$
    (B) $\dfrac{1}{2}\,\boxed{1}\,\dfrac{1}{3}$
    (C) $\dfrac{1}{2}\,\boxed{6}\,1$
    (D) $\dfrac{3}{2}\,\boxed{-\dfrac{1}{3}}\,1$
    (E) $\dfrac{7}{6}\,\boxed{1}\,\dfrac{1}{3}$

16. A prom committee is having trouble agreeing on the color palette for this year's event. In order to resolve the conflict, the members reach a compromise: Any combination of three colors from the following list is acceptable: maroon, gold, black, silver, white, and pink. In order to finalize the color palette, they place one swatch representing each color into a bag and select three colors randomly. How many unique three-color combinations are possible, assuming that the order in which the colors are chosen does not matter?

    (A) 15
    (B) 20
    (C) 60
    (D) 120
    (E) 216

Solutions

Section I

1. **C**. Solve for x.

$$\begin{aligned}2(x)+2(-3)+1&=3x-4\\2x-6+1&=3x-4\\-6+1+4&=3x-2x\\-1&=x\end{aligned}$$

2. **B**. Draw a diagram of the information given. An angle bisector is a ray that divides an angle in half, into two congruent angles as illustrated below. Dotted lines are used to visualize triangles *WAX* and *XAY*, which are referenced in statement III.

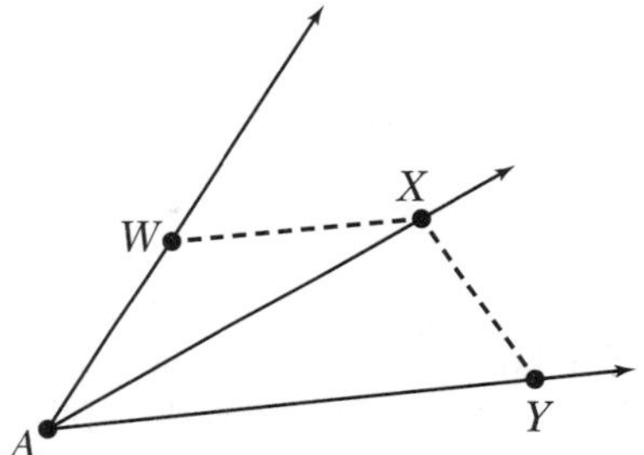

Statement I is true; these are the two congruent angles created by the angle bisector dividing the angle in half. Statement II is true; if a bisector divides an angle in half, then the original angle's measure is twice the measure of one of the halves. Statement III is false. The triangles share side $\overline{AX}$ and angles *WAX* and *XAY* are congruent. However, *WA* is not necessarily equal to *AY*, and *WX* is not necessarily equal to *XY*. Therefore, the triangles may not be congruent. Only statements I and II must be true, so the correct answer is (B).

3. **B**. The mode of the data is the value that occurs most frequently. Only one value in the data set repeats: 2.

See Problem 8.19.

4. **D**. The large right triangle with leg length c and the small right triangle with hypotenuse 5 are similar because they have two pairs of corresponding congruent angles. Therefore, corresponding legs are in the same proportion. The horizontal leg of the small triangle has length 3, and the corresponding horizontal leg of the large triangle has length 3 + 3 = 6. Therefore, every side in the large triangle is twice as long as the corresponding side of the small triangle.

The side with length c corresponds to the vertical leg of the small triangle. The smaller leg has length 4, because 3, 4, 5 is a Pythagorean triple. Thus, c is twice as large as 4, so $c = 2(4) = 8$.

Or you could get 4 using the Pythagorean theorem. Solve the equation $3^2 + x^2 = 5^2$ for x.

5. **B**. The word "difference" implies subtraction. The difference of 2 and x is $2 - x$, not $x - 2$. The order in which the terms are stated in the sentence is the order in which they appear in the expression: 2 comes before x. One-third of that difference is equal to that difference divided by 3, so the correct answer is choice (B).

6. **E**. Apply the DIY technique, selecting different positive integer values for a, b, and c. For example, $a = 3$, $b = 2$, and $c = 1$. The prime factorization would be $2^3 \cdot 3^2 \cdot 5^1 = 360$. All three statements are true. In fact, the statements are true no matter what positive integer values you select for a, b, and c. Notice that the factorization includes 2, so z is divisible by 2, which means it is even. The factorization includes factors of 2, 3, and 5, so $2 \cdot 3 \cdot 5 = 30$ is also a factor of z. If 30 is a factor of z, then z is a multiple of 30. Finally, 2 and 3 are factors, so $2 \cdot 3 = 6$ is also a factor of z, which means z is divisible by 6.

7. **E**. Apply the slope formula and solve for k.

$$\begin{aligned}\frac{k-2}{3-(-2)} &= 10\\ \frac{k-2}{5} &= 10\\ k-2 &= 10 \cdot 5\\ k &= 50+2\\ k &= 52\end{aligned}$$

8. **D**. If the diameter of the large circle is 3 times the diameter of the small circle, then the radius of the large circle is 3 times the radius of the small circle. If the radius of the small circle is r, then the radius of the large circle is $3r$. Calculate the area of the shaded region by subtracting the area of the small circle from the area of the large circle.

For example, if the large circle has diameter 12, and the small circle has diameter 4, then the radius of the large circle is 6 and the radius of the small circle is 2.

$$\begin{aligned}\text{shaded area} &= \text{area of large circle} - \text{area of small circle}\\ &= \pi(3r)^2 - \pi r^2\\ &= 9\pi r^2 - \pi r^2\\ &= 8\pi r^2\end{aligned}$$

9. **B**. If two quantities vary inversely, then their product is a fixed constant k called the constant of proportionality. In this question, the number of spiders s varies inversely with the number of flies f, so you conclude that $fs = k$. Substitute last week's values, $f = 12$ and $s = 2$, into the equation and calculate k.

$$\begin{aligned}fs &= k\\ 12 \cdot 2 &= k\\ 24 &= k\end{aligned}$$

You are told that this week's total sum of spiders and flies is 10, so $s + f = 10$. Solve the equation for one of the variables: $f = 10 - s$. This week, if there are s spiders, then there are $10 - s$ flies, and the product of those two numbers remains $k = 24$.

$$\begin{aligned}fs &= 24\\ (10-s)s &= 24\\ 10s - s^2 &= 24\end{aligned}$$

Solve the quadratic equation by factoring.

$$0 = s^2 - 10s + 24$$
$$0 = (s-4)(s-6)$$
$$s = 4 \quad \text{or} \quad s = 6$$

The question states that there are more flies than spiders, so $s = 4$ and $f = 10 - 4 = 6$. Be careful to respond to the question that is asked: How many more flies are there than spiders? Because there are 6 flies and 4 spiders, there are 2 more flies than spiders.

There are two possible values for the spider population, either 4 or 6. Remember, there is a total of 10 spiders and flies combined. Therefore, s = 6 cannot be the right answer, because there would be f = 10 - 6 = 4 flies. The spiders would outnumber the flies 6 to 4.

10. **B**. List a few terms of the series to identify a relationship between the term's number and how many digits are present. In the list below, groups of like digits are separated to help visualize the pattern.

$$a_3 = 222\ 11\ 0$$
$$a_4 = 2222\ 111\ 00$$
$$a_5 = 22222\ 1111\ 000$$
$$a_6 = 222222\ 11111\ 0000$$

The numbers of 2 digits, 1 digits, and 0 digits in the a_n term are n, $n-1$, and $n-2$, respectively. The number of 2 digits matches the subscript of the term. For example, the a_5 term contains five 2's. The number of 1 digits is one less than n and the number of 0 digits is two less than n. For example, in the a_5 term, there are four 1's and three 0's.

Therefore, the numbers of 2, 1, and 0 digits in the a_{51} term are $x = 51$, $y = 50$, and $z = 49$, respectively. You conclude that $x + y + z = 51 + 50 + 49 = 150$.

11. **A**. If Bob runs 6 miles per hour, then he runs 6(5,280) = 31,680 feet per hour. In other words, he runs 31,680 feet in 60 minutes. Set up a proportion to solve the problem, letting x represent the unknown distance in feet.

$$\frac{\text{distance}}{\text{time}} = \frac{\text{distance}}{\text{time}}$$
$$\frac{31{,}680 \text{ feet}}{60 \text{ minutes}} = \frac{x \text{ feet}}{11 \text{ minutes}}$$
$$(31{,}680)(11) = 60x$$
$$348{,}480 = 60x$$
$$\frac{348{,}480}{60} = x$$
$$5{,}808 \text{ feet} = x$$

He runs at a constant rate of 6 miles per hour. You now know that 31,680 feet per 60 minutes is equal to 6 miles per hour. The proportion calculates his speed in "feet per 11 minutes."

12. **D**. Angles 5 and 6 are supplementary. Only right angles are both congruent and supplementary, so if angles 5 and 6 do not have the same measure, they cannot be right angles. You must conclude that l and q are not perpendicular lines. Therefore, angle 8 cannot be a right angle, and choice (D) is correct.

Why are the other answer choices incorrect? Choices (B) and (C) are only correct if p and q are parallel lines. They *look* parallel. Line l *appears* to be perpendicular to p and q, but the drawing is not to scale, so neither of those assumptions may be true. Thus, choices (A), (B), and (C) should be eliminated. The question asks you to identify the statement that is true *only if* angles 5 and 6 have different measures. Because choice (E) is true no matter what the measures of angles 5 and 6 are, (E) should be eliminated also.

13. **E.** Doni's lunch cost \$18, and she tipped 25%, so Doni's tip was 0.25(\$18) = \$4.50. Asia's tip was \$1 more than twice Doni's tip, so Asia's tip was 2(\$4.50) + \$1 = \$9 + \$1 = \$10. This \$10 tip represents 40% of Asia's meal price, *p*. Therefore, 0.40(*p*) = \$10. Divide \$10 by 0.40 to solve the equation: *p* = \$10 ÷ 0.40 = \$25.

14. **C.** For the moment, ignore the unknown data values and list the known values in order from least to greatest.

9 12 14 17 20 45

The median of this set is 15.5, the average of 14 and 17. Note that the median of the complete set is also 15.5. If one missing data value is less than 14 and one is greater than 17, then the median of the new data set would still be 15.5. Now add the six known data values.

$$9 + 12 + 14 + 17 + 20 + 45 = 117$$

If the average of all eight data values (including the two unknown values *a* and *b*) is 20, then the sum of the values is 8(20) = 160. Therefore, the unknown values must have a sum of 160 – 117 = 43. Choice (C) is correct, because it is the only pair of numbers with a sum of 43 such that one data value is less than 14 and the other is greater than 17.

The numbers in Choice (E) also have a sum of 43, but if you include them in the data set, the median changes to (17 + 20)/2 = 18.5.

15. **E.** Calculate 2 ♦ 6 by substituting $x = 2$ and $y = 6$ into the function.

$$\begin{aligned} 2 \blacklozenge 6 &= \frac{2 \cdot 6}{2-6} \\ &= \frac{12}{-4} \\ &= -3 \end{aligned}$$

The question states that $z \blacklozenge 5$ is also equal to –3. Substitute $x = z$ and $y = 5$ into the function, set it equal to –3, and solve for z.

$$\begin{aligned} z \blacklozenge 5 &= -3 \\ \frac{z \cdot 5}{z-5} &= -3 \\ 5z &= -3(z-5) \\ 5z &= -3z+15 \\ 5z+3z &= 15 \\ 8z &= 15 \\ z &= \frac{15}{8} \end{aligned}$$

16. **C.** Notice that angle *Y* is an inscribed angle that subtends a semicircle, because *X* and *Z* are the endpoints of a diameter and vertex *Y* lies on the circle. A semicircle measures 180°. Recall that an inscribed angle is half the measure of the arc it subtends, so angle *Y* measures 180 ÷ 2 = 90°; it is a right angle. If *XM* = *XY*, then one leg of right triangle *XYZ* is half as long as the hypotenuse. This only occurs when that right triangle is a 30°–60°–90° triangle. Consider the following diagram, which redraws the triangle in the correct proportions, such that *XM* and *XY* are the same length.

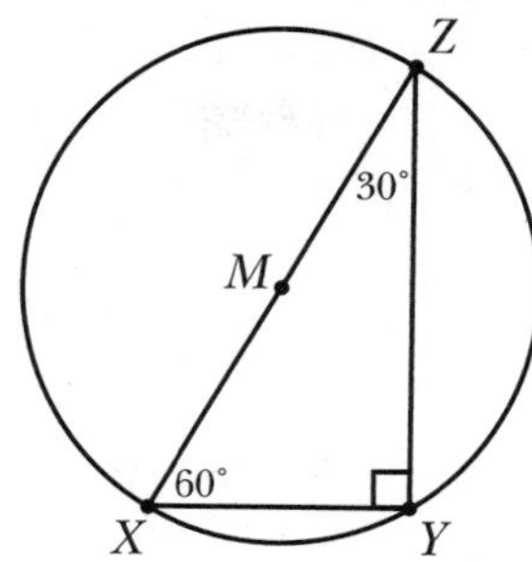

You know that $XY = r$, because it is the same length as radius $\overline{MX}$. $\overline{XZ}$ is a diameter, so its length is $2r$. The long leg of the triangle has length $r\sqrt{3}$. Thus, the perimeter of triangle XYZ is $r + 2r + r\sqrt{3} = 3r + r\sqrt{3}$. Note that this expression is equivalent to choice (C) if you distribute: $r\sqrt{3}\left(1+\sqrt{3}\right) = r\sqrt{3} + r\sqrt{9} = r\sqrt{3} + 3r$.

17. **E**. Apply the DIY technique. For example, assume $w = 10$ percent of the fish are goldfish and $x = 40$ percent of the $y = 200$ total pets in the store are fish. Using these values, there are $0.40(200) = 80$ fish in the store, and 90 percent of them are *not* goldfish: $0.90(80) = 72$.

If 10% of the 80 fish ARE goldfish, then 90% of the 80 fish ARE NOT goldfish.

Only choice (E) returns a value of $z = 72$ when you substitute $w = 10$, $x = 40$, and $y = 200$ into it.

$$\begin{aligned}\frac{xy(100-w)}{10{,}000} &= \frac{(40)(200)(100-10)}{10{,}000}\\ &= \frac{8{,}000(90)}{10{,}000}\\ &= \frac{720{,}000}{10{,}000}\\ &= 72\end{aligned}$$

18. **B**. If p is an odd integer—remember, it could be positive or negative—then a is the next consecutive integer (so a is even), b is the next consecutive integer after a (so b is odd), and c is the integer immediately preceding p (so c is even). Furthermore, you know that $c < a < b$. Now investigate the statements independently.

Statement I is false. The preceding paragraph concluded that $b > c$, but multiplying both sides of that inequality by a produces the true statement $ab > ac$ only when a is positive. Remember, if you multiply both sides of an inequality by a negative number, it reverses the inequality symbol. Therefore, if $a < 0$, $ab < ac$.

Statement II is true. If $a = p + 1$ and $c = p - 1$, then $ac = (p + 1)(p - 1)$. The expression right of the equal sign is a difference of perfect squares; simplify it to get $ac = p^2 - 1$. If you add 1 to both sides of this equation, the result is statement II: $ac + 1 = p^2$.

Statement III is false. The expression b^a is an odd integer raised to an even power. Such an expression is always odd. For example, the odd integer 3 raised to the even power 2 is equal to 9, an odd number. Raising this odd result to the c power, another even number, once again has an odd result.

Only statement II is true, so the correct answer is choice (B).

19. **C**. Evaluate the function $k(x)$ for $x = 3p$.

$$\begin{aligned} k(3p) &= 2(3p) + \frac{15-(3p)^2}{3} \\ &= 6p + \frac{15-9p^2}{3} \\ &= 6p + \frac{15}{3} - \frac{9p^2}{3} \\ &= 6p + 5 - 3p^2 \end{aligned}$$

The question states that $k(3p) = p + 3$, and now you know that $k(3p) = 6p + 5 - 3p^2$.

$$\begin{aligned} k(3p) &= p+3 \\ 6p+5-3p^2 &= p+3 \end{aligned}$$

Solve the quadratic equation by factoring.

$$\begin{aligned} 0 &= 3p^2 - 6p + p - 5 + 3 \\ 0 &= 3p^2 - 5p - 2 \\ 0 &= (3p+1)(p-2) \end{aligned}$$

Setting the factors $(3p + 1)$ and $(p - 2)$ equal to 0 produce possible solutions of $p = -1/3$ and $p = 2$. Only one of those solutions is listed among the answer choices: choice (C).

20. **A**. The segment extending from the center of the top square to point B has length $s/2$; it is half as long as a side of the square, as illustrated in the following diagram.

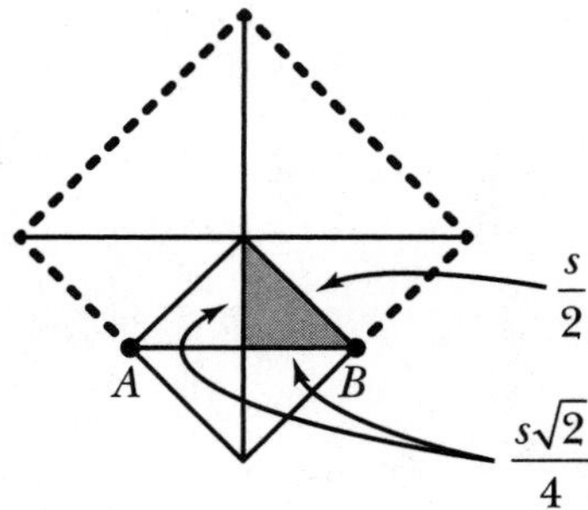

Note that the acute angles within the shaded region each measure 45°, so the shaded region is a 45°–45°–90° triangle. In a 45°–45°–90° triangle, the hypotenuse is $\sqrt{2}$ times the length of a leg. Calculate the length of the legs of the shaded triangle.

$$\text{hypotenuse length} = (\text{leg length})\sqrt{2}$$

$$\frac{s}{2} = (\text{leg length})\sqrt{2}$$

$$\frac{s}{2}\left(\frac{1}{\sqrt{2}}\right) = (\text{leg length})\left(\cancel{\sqrt{2}}\right)\left(\frac{1}{\cancel{\sqrt{2}}}\right)$$

$$\frac{s}{2\sqrt{2}} = \text{leg length}$$

Rationalize the expression by multiplying the numerator and denominator by $\sqrt{2}$.

$$\frac{s}{2\sqrt{2}}\left(\frac{\sqrt{2}}{\sqrt{2}}\right) = \frac{s\sqrt{2}}{2\sqrt{4}}$$

$$= \frac{s\sqrt{2}}{2 \cdot 2}$$

$$= \frac{s\sqrt{2}}{4}$$

The perimeter of the shaded triangle is twice the leg length calculated above plus the length of its hypotenuse.

Because the 45°-45°-90° triangle is isosceles and has two congruent legs

$$\text{perimeter of shaded triangle} = 2\left(\frac{s\sqrt{2}}{4}\right) + \frac{s}{2}$$

$$= \frac{2(s\sqrt{2})}{4} + \frac{s}{2}$$

$$= \frac{s\sqrt{2}}{2} + \frac{s}{2}$$

$$= \frac{s + s\sqrt{2}}{2}$$

Now calculate the dotted perimeter along the edge of the figure. As the diagram below illustrates, the perimeter is equal to $6s$.

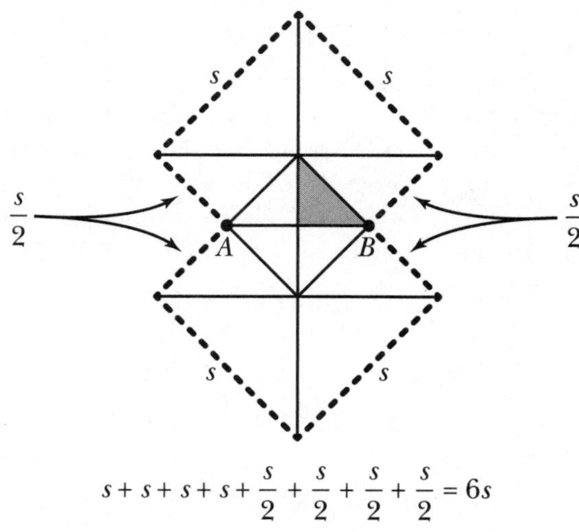

$$s + s + s + s + \frac{s}{2} + \frac{s}{2} + \frac{s}{2} + \frac{s}{2} = 6s$$

Finally (and I do mean *finally*), calculate the ratio of the perimeter of the shaded triangle to the dotted perimeter. In other words, divide the first perimeter by the second.

$$\begin{aligned}\frac{s+s\sqrt{2}}{2} \div 6s &= \frac{s+s\sqrt{2}}{2}\cdot\frac{1}{6s}\\ &= \frac{s+s\sqrt{2}}{12s}\\ &= \frac{\cancel{s}\left(1+\sqrt{2}\right)}{12\cancel{s}}\\ &= \frac{1+\sqrt{2}}{12}\end{aligned}$$

Section 2

1. **A**. This question requires you to recognize shapes of common graphs. If you have a graphing calculator, you can review those shapes quickly, which makes this much easier. For example, if you set $a = -4$ in choice (A), your graphing calculator plots something very similar to the graph in the question.

2. **D**. The circumference of a circle is defined by the formula $C = 2\pi r$, where r is the radius of the circle. If you set four of the answer choices equal to $2\pi r$, only one of them will result in an integer value for r. For example, set $2\pi r$ equal to 3π, the circumference in choice (B), and solve for r.

$$\begin{aligned}2\pi r &= 3\pi\\ r &= \frac{3\cancel{\pi}}{2\cancel{\pi}}\\ r &= \frac{3}{2}\end{aligned}$$

The circumference is 2 times the radius times π, and two times any integer is an even number. Choice (D) is the only even coefficient, so that's why it's the correct answer.

The question states that r must be an integer, so choice (B) is incorrect. Only 28π, choice (D), is a possible circumference of the circle. If you set $2\pi r = 28\pi$ and solve for r, you get $r = 14$, an integer.

3. **C**. Wheat and tomato seeds each appear to account for one-quarter, or 25%, of the seeds purchased. Carrots appear to be one-eighth, or 12.5%, of the seeds purchased. The question asks you to calculate the percentage of seeds that are neither corn nor tomatoes. In other words, add the percentages for wheat and carrots: 25% + 12.5% = 37.5%. These numbers are approximations, so you should choose the closest answer: choice (C), 38%.

The carrot sector looks like it has a central angle of 45°, which is half of the 90° angles in the wheat and tomato sectors. Therefore, it represents half of the percentage as well: 25% ÷ 2 = 12.5%.

4. **D**. Translate the sentence into an equation in which x represents the unknown number.

$$\begin{aligned}\frac{1}{2}(x-3) &= \frac{2x}{6}\\ \frac{x-3}{2} &= \frac{2x}{6}\end{aligned}$$

Cross-multiply and solve the proportion.

$$\begin{aligned} 6(x-3) &= 4x \\ 6x-18 &= 4x \\ 6x-4x &= 18 \\ 2x &= 18 \\ x &= 9 \end{aligned}$$

5. **C**. If each term is two less than the square root of the previous term, then each term is also two *more* than the *square* of the following term. In other words, add 2 to a term and then square it to move backwards in the sequence. If the third term is 0, then the second term is $(0 + 2)^2 = 4$. The first term is $(4 + 2)^2 = 36$.

6. **D**. The sum of the measures of the interior angles of a triangle is 180°. Add the angle measures and solve for x in the equation below to calculate the actual measures of the angles.

$$\begin{aligned} (3x+4)+(4x-9)+\left[(x-3)^2-6\right] &= 180 \\ 7x-5+\left[x^2-6x+9-6\right] &= 180 \\ x^2+(7x-6x)+(-5+9-6) &= 180 \\ x^2+x-2 &= 180 \\ x^2+x-182 &= 0 \end{aligned}$$

Solve the equation by applying the quadratic formula or by factoring. The solution below applies the factoring method.

$$\begin{aligned} (x-13)(x+14) &= 0 \\ x &= -14, 13 \end{aligned}$$

Ask yourself this question: What two numbers have a sum of +1 and a product of −182? One of them must be positive and the other negative; the larger is positive because the sum is positive. Try pairs of numbers like −8 and 9, −9 and 10, −10 and 11, which have a difference of 1. Eventually, you reach −13 and 14, which have a product of −182.

Note that x must be positive or the angle measurements will be negative. Thus, $x = 13$ is the only valid solution. Substitute $x = 13$ into each of the angle measures.

$$\begin{aligned} 3(13)+4 &= 39+4 \\ &= 43 \end{aligned} \qquad \begin{aligned} 4(13)-9 &= 52-9 \\ &= 43 \end{aligned} \qquad \begin{aligned} (13-3)^2-6 &= 10^2-6 \\ &= 100-6 \\ &= 94 \end{aligned}$$

Two of the angles are congruent, and they share the least measure of 43°. If a triangle has two congruent angles, the sides opposite those angles are congruent. Therefore, ABC is an isosceles triangle.

7. **B**. You can answer this question using trial and error, or you can create a surprisingly simple equation to solve it. Let n represent the number of nickels Coleman has. That means he has $23 - n$ quarters. Each quarter is worth 25 cents and each nickel is worth 5 cents. When you add their values, the total is 335 cents.

$$5¢(\text{number of nickels}) + 25¢(\text{number of quarters}) = 335¢$$
$$5n + 25(23 - n) = 335$$
$$5n + 575 - 25n = 335$$
$$-25n + 5n = 335 - 575$$
$$-20n = -240$$
$$n = \frac{-240}{-20}$$
$$n = 12$$

Coleman has 12 nickels and 23 – 12 = 11 quarters, for a total of 12($0.05) + 11($0.25) = $3.35.

8. **D**. Draw two examples of squares with an odd side length and their shaded unit squares. Remember that the unit squares in the corners must be shaded. The diagram below illustrates squares with side lengths $x = 3$ and $x = 5$. Following the rule about adjacency of shaded unit squares produces a checkerboard.

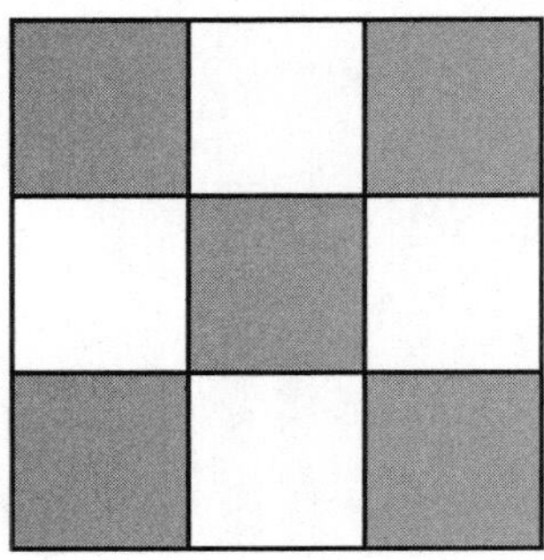

$$\frac{\text{shaded}}{\text{unshaded}} = \frac{5}{4}$$

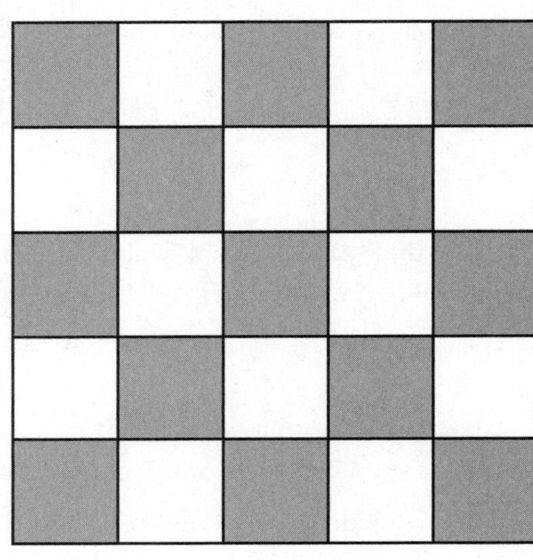

$$\frac{\text{shaded}}{\text{unshaded}} = \frac{13}{12}$$

Apply the DIY technique with $x = 3$. A square with side length $x = 3$ has area $n = 3^2 = 9$. The ratio of shaded to unshaded unit squares is 5/4. Only choice (D) produces the correct ratio.

$$\frac{n+1}{n-1} = \frac{9+1}{9-1}$$
$$= \frac{10}{8}$$
$$= \frac{5}{4}$$

9. **14**. Isolate the absolute value expression left of the equal sign. Begin by multiplying all of the terms by 3 to eliminate the fraction.

$$\left(\frac{3}{1}\right)\left(\frac{|z-2|}{3}\right)+(3)(1)=(3)(5)$$
$$|z-2|+3=15$$
$$|z-2|=12$$

There are two solutions to this absolute value equation. To calculate them, set the contents of the absolute value expression equal to 12 and –12 and solve those two equations.

$$\begin{aligned} z-2&=12 & z-2&=-12 \\ z&=12+2 & z&=-12+2 \\ z&=14 & z&=-10 \end{aligned}$$

The greater of the two solutions is 14.

If this were a multiple-choice question, you'd just plug all of the choices into the equation, determine which choices were actually solutions, and then pick the biggest one. However, this is a grid-in question, which means the test writers expect you to actually solve the absolute value equation.

10. **25**. Create a proportion, where x represents the number of hens needed.

$$\frac{\text{number of hens}}{\text{number of eggs laid by the hens}}=\frac{\text{number of hens}}{\text{number of eggs laid by the hens}}$$
$$\frac{10}{12}=\frac{x}{30}$$

Cross-multiply and solve for x.

$$10(30)=12x$$
$$300=12x$$
$$\frac{300}{12}=x$$
$$25=x$$

11. **3**. The function $f(x)$ is less than or equal to the function $g(x)$ when the graph of $f(x)$ is below or overlapping the graph of $g(x)$. Notice that the graphs overlap at $x = 3$, but left of that x-value, the graph of $f(x)$ is below the graph of $g(x)$. Therefore, $c = 3$.

12. **5**. Begin by evaluating $f(0)$. Notice that $f(x)$ passes through the point (0,–2). Therefore, $f(0) = -2$. Substitute this value into the expression given: $g[f(0)] = g(-2)$. The graph of $g(x)$ passes through the point (–2,5), so $g(-2) = 5$. You conclude that $g[f(0)] = 5$.

13. **6**. Use a Venn diagram to visualize the data. You need three sets: one representing integers divisible by 2, one representing integers divisible by 3, and one representing integers divisible by 5. In the following diagram, 5 integers must appear in the upper-left circle, three integers must appear in the upper-right circle, and 2 integers must appear in the lower circle. However, they may appear in different parts of the circles, including the overlapping regions.

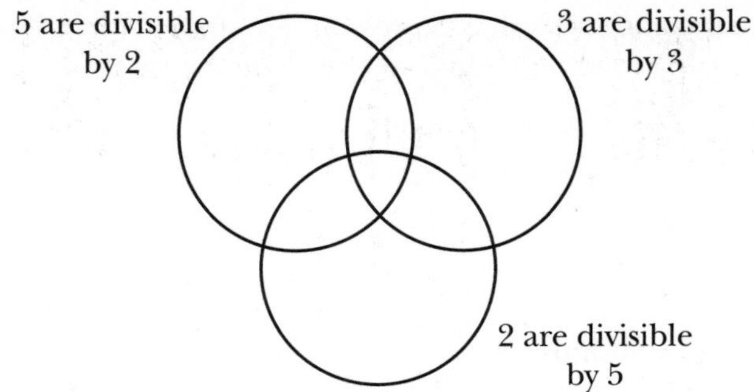

No integers are divisible by 15. Notice that $3 \cdot 5 = 15$, so place a 0 in the region where "divisible by 3" and "divisible by 5" overlap. If no integer is divisible by 3 and 5, then no integer is divisible by all 3 numbers (2, 3, and 5). Visually, this means that the region in which all three sets overlap—the center of the diagram—contains a 0.

Exactly 1 integer is divisible by 10. Because $10 = 2 \cdot 5$, that single integer is divisible by 2 and 5 but not by 3, because no integers are divisible by 2, 3, and 5. Indicate this with a 1 in the region where those sets overlap. Similarly, three of the integers are divisible by 6, so they are divisible by 2 and 3 (the factors of 6) but not by 5. Therefore, you should write a 3 in the appropriate intersection.

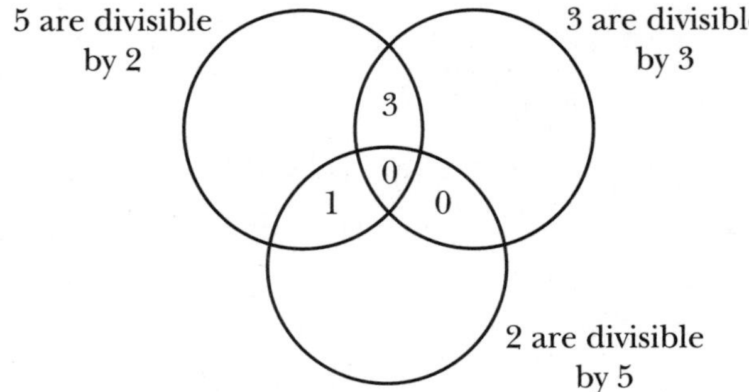

Now complete the rest of the diagram. The upper-left set must contain five integers. It already contains $1 + 0 + 3 = 4$ integers, so the portion of that set that does not overlap any other set must contain $5 - 4 = 1$ integer. The upper-right set already contains the 3 integers allotted to it, so the non-overlapping portion of that set contains 0 additional integers. In the lower set, one integer has been identified, which leaves $2 - 1 = 1$ integer left in the non-overlapping portion of the set.

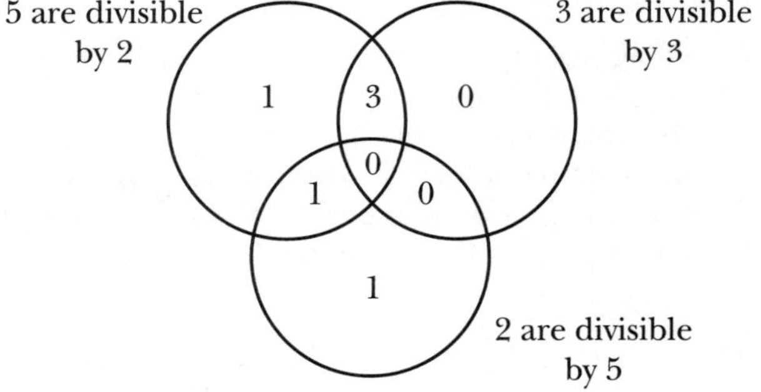

The total number of integers is $1 + 3 + 1 + 1 = 6$.

14. **.166, .167, or 1/6**. If you used a Venn diagram to complete the previous question, this question is very simple. If an integer is divisible by 2 but is neither divisible by 3 nor divisible by 5, then it appears in the shaded region of the following diagram.

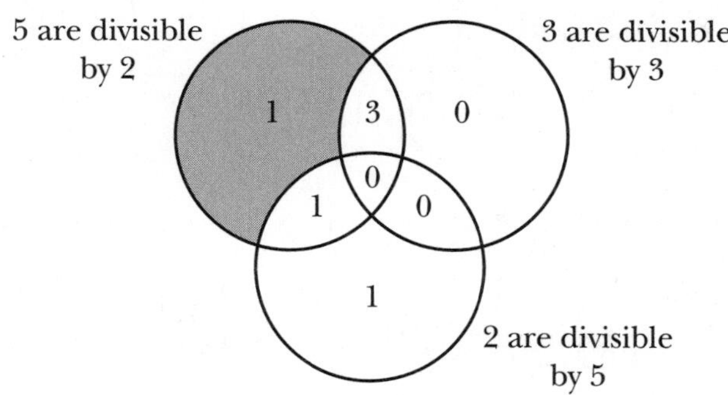

Only one numbered block satisfies this condition. According to the previous question, there are 6 blocks in the bag, so Jodi has a 1/6 or $0.1\overline{6}$ probability of drawing that block randomly.

15. **6**. Identify values of z that have a remainder of 6 when divided by 9, such as 15, 24, 33, and 42. Calculate the remainder of each when you subtract 2 and divide by 4.

$13 \div 4 = 3 \text{ r } 1 \qquad 22 \div 4 = 5 \text{ r } 2 \qquad 31 \div 4 = 7 \text{ r } 3 \qquad 40 \div 4 = 10 \text{ r } 0$

There are only four possible remainders when you divide by 4, and they are all possible values of r: 0, 1, 2, and 3. Therefore, the sum of all possible unique remainders is $0 + 1 + 2 + 3 = 6$.

16. **2.5 or 5/2**. Draw a regular hexagon with side length s. Note that the interior angles of a regular hexagon each measure 120°. You can divide the hexagon into a rectangle and two isosceles triangles. Note that the large angles in the isosceles triangles are interior angles of the hexagon, so they measure 120°.

Use the formula $(180)(n - 2)$ where n is the number of sides. In this case, $n = 6$. That's the sum of the interior angles. Divide by 6 to calculate one angle.

In this question, $180(6 - 2) = 720$, and $720 \div 6 = 120$.

The sum of the measures of the interior angles of a triangle is 180°, and the base angles of an isosceles triangle are congruent, so the small angles in the isosceles triangles each measure 30°. Divide the isosceles triangles into 30°–60°–90° triangles by extending sides perpendicular to the bases that bisect the 120° vertex angles. The hypotenuses of those right triangles each have length s, so the height of the isosceles triangles is $s/2$ and their bases have length $s\sqrt{3}$, as illustrated below.

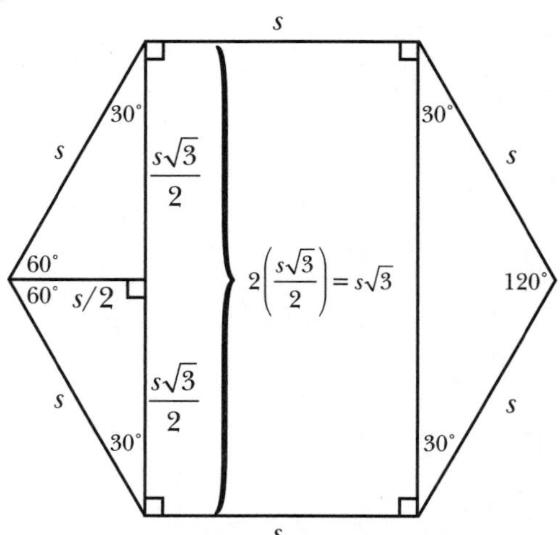

Calculate the area of the rectangle and the isosceles triangles. Remember, there are two triangles in the hexagon, so double the area of one triangle.

Area of two triangles

$$2\left(\frac{1}{2}bh\right)=bh$$
$$=\left(s\sqrt{3}\right)\left(\frac{s}{2}\right)$$
$$=\frac{s^2\sqrt{3}}{2}$$

Area of rectangle

$$l\cdot w=\left(s\sqrt{3}\right)(s)$$
$$=s^2\sqrt{3}$$

Add these areas and set the sum equal to the area stated in the question. Use common denominators to simplify the left side of the equation.

$$\frac{s^2\sqrt{3}}{2}+s^2\sqrt{3}=\frac{75}{8}\sqrt{3}$$
$$\frac{s^2\sqrt{3}}{2}+\frac{2s^2\sqrt{3}}{2}=\frac{75}{8}\sqrt{3}$$
$$\frac{3s^2\sqrt{3}}{2}=\frac{75}{8}\sqrt{3}$$

Solve for s.

$$\left(\frac{\cancel{2}}{\cancel{3}}\right)\frac{\cancel{3}s^2\sqrt{3}}{\cancel{2}}=\left(\frac{2}{3}\right)\frac{75}{8}\sqrt{3}$$
$$s^2\cancel{\sqrt{3}}=\frac{150}{24}\cancel{\sqrt{3}}$$
$$s^2=\frac{25}{4}$$
$$s=\sqrt{\frac{25}{4}}$$
$$s=\frac{5}{2}$$

The hexagon has side length $s=5/2$.

17. **2.5, 5/2, or 15/6**. Combine the expressions using the least common denominator, 6.

$$\frac{4x-1}{3}+\frac{x+8}{6}+\frac{x+3}{2}=\frac{2}{2}\left(\frac{4x-1}{3}\right)+\frac{x+8}{6}+\left(\frac{3}{3}\right)\left(\frac{x+3}{2}\right)$$
$$=\frac{8x-2}{6}+\frac{x+8}{6}+\frac{3x+9}{6}$$
$$=\frac{(8x+x+3x)+(-2+8+9)}{6}$$
$$=\frac{12x+15}{6}$$

Express the fraction as a sum and reduce to lowest terms.

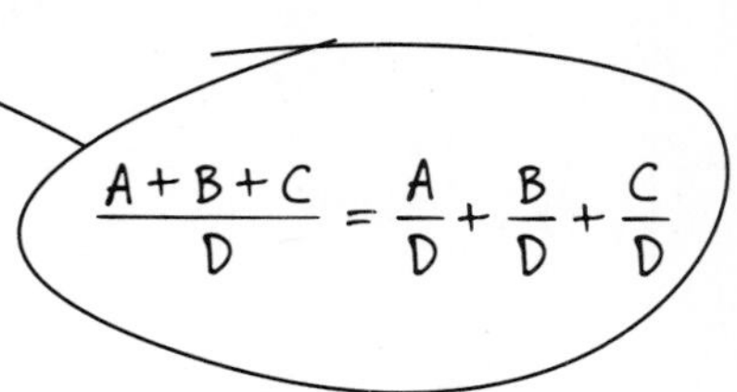

$$= \frac{12x}{6} + \frac{15}{6}$$
$$= 2x + \frac{5}{2}$$

The expression is 5/2 larger than $2x$.

18. **.214 or 3/14**. Draw a diagram to illustrate the information given. There are six times as many raw eggs as hardboiled eggs. In other words, for every hardboiled egg, there are six raw eggs. In the visual representation below, H represents hardboiled eggs and R represents raw eggs.

R R R R R R H

All of the hardboiled eggs are white, because the only brown eggs are raw. More information is given about the raw eggs: There are three times as many raw brown eggs as there are raw white eggs. Ideally, you could manipulate the preceding diagram, rewriting the Rs as RW and RB to designate raw white and raw brown eggs, respectively. However, with only 6 Rs, there is no way to do this.

Add another row of eggs, matching the first row. In the new diagram, there are 12 raw eggs and 2 hardboiled eggs, so there are still six times as many raw eggs.

R R R R R R H

R R R R R R H

Now that you have 12 raw eggs, you can make three of them raw brown eggs (RB) for every raw white egg (RW).

RW RW RW RB RB RB H

RB RB RB RB RB RB H

In this diagram, there are six times as many raw eggs as hardboiled eggs, and there are three times as many raw brown eggs as raw white eggs. In the diagram, there are 14 total outcomes of a random draw. In three of them you select a raw white egg, so the probability is 3/14.

You could alter the diagram if there were an equal number of raw white and raw brown eggs:

RW RW RW RB RB RB H

You could even make it work if there were twice as many raw browns as raw whites:

RW RW RB RB RB RB H

Section 3

1. **D**. You cannot determine what Connor bought based on the information given, other than the food he bought must have been apples, corn, or carrots. Therefore, Choice (D) must be true. Connor did not buy peaches, because he bought food from a farm stand that does not sell peaches.

2. **C**. The area A of a triangle with base b and height h is $A = (1/2)bh$. Substitute $A = 18$ and $b = 6$ into the equation and solve for h.

$$\begin{aligned} A &= \frac{1}{2}bh \\ 18 &= \frac{1}{2}(6)h \\ 18 &= 3h \\ \frac{18}{3} &= h \\ 6 &= h \end{aligned}$$

3. **A**. Notice that $4x^2 - 25$ is a difference of perfect squares: $4x^2 - 25 = (2x + 5)(2x - 5)$. According to the question, $A = 2x + 5$ and $B = 4x^2 - 25$. Substitute A and B into the factored equation.

$$\begin{aligned} 4x^2 - 25 &= (2x+5)(2x-5) \\ B &= A(2x-5) \end{aligned}$$

Divide both sides of the equation by A to get $B/A = 2x - 5$, so Choice (A) is correct.

4. **C**. Substitute $p = 10$ into the equation solved for y to get $y = \sqrt{10-1} = \sqrt{9} = 3$. Next, substitute $y = 3$ into the equation solved for x.

$$\begin{aligned} x &= \frac{2-y}{5} \\ &= \frac{2-3}{5} \\ &= \frac{-1}{5} \\ &= -\frac{1}{5} \end{aligned}$$

5. **C**. To solve the inequality $|x+3| \geq 3$, you should convert it into the two linear inequalities below and solve each for x.

$$\begin{aligned} x+3 &\geq 3 & \qquad x+3 &\leq -3 \\ x &\geq 3-3 & x &\leq -3-3 \\ x &\geq 0 & x &\leq -6 \end{aligned}$$

These two inequalities, together, form the graph given in the question.

6. **C**. Draw a diagram to illustrate the information given. If circle A has a radius that is twice the length of circle B's diameter, then two full diameters of circle B are equal to one radius of circle A.

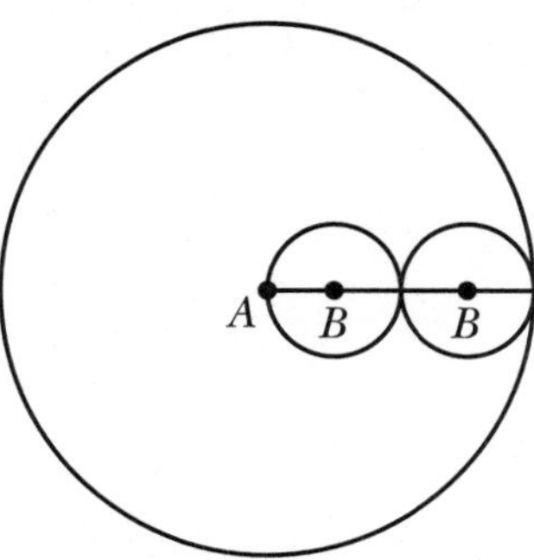

If circle B has diameter d, then circle A has diameter $4d$. Recall that the circumference of a circle is equal to π times its diameter. Calculate the ratio of circle B's circumference to circle A's circumference.

$$\begin{aligned}\frac{\text{circumference of circle } B}{\text{circumference of circle } A} &= \frac{\pi d}{\pi(4d)} \\ &= \frac{\cancel{\pi}\,\cancel{d}}{4\cancel{\pi}\,\cancel{d}} \\ &= \frac{1}{4}\end{aligned}$$

The formula at the beginning of the chapter states $C = 2\pi r$. Because $2r$ is the length of the diameter, you can replace $2r$ in that formula with d to get $C = \pi d$.

7. **A**. Rotating the square 270° counterclockwise around point Q is $360° - 270° = 90°$ less than a full rotation around Q. If you had rotated 360°, all of the points would return to their original locations and the square's appearance would be unchanged. Instead, you rotate $270/360 = 3/4$ of a full rotation. The diagram below illustrates the square as you rotate it 90° at a time around Q. In each step, the vertices move one position in the counterclockwise direction.

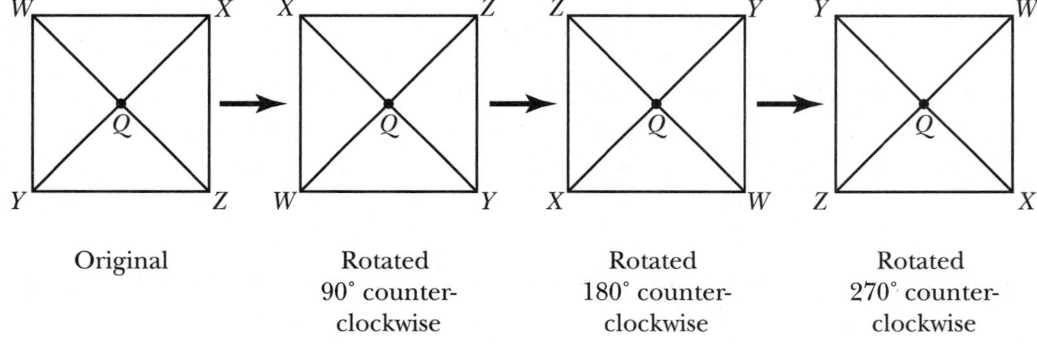

8. **B.** It takes Carlie 6 hours to paint a specific room. Therefore, in one hour she can paint 1/6 of that room. Similarly, in one hour, Amanda can paint 1/8 of that room. Add the fractions to calculate how much of the room they can paint if they work together. Use the least common denominator 24 to combine the fractions.

$$\begin{aligned}\frac{1}{6}+\frac{1}{8}&=\frac{1}{6}\left(\frac{4}{4}\right)+\frac{1}{8}\left(\frac{3}{3}\right)\\&=\frac{4}{24}+\frac{3}{24}\\&=\frac{7}{24}\end{aligned}$$

Together, the two girls can paint 7/24 of the room in one hour. In two hours, they could paint 7/24 + 7/24 = 14/24 = 7/12 of the room—just over half of it. In three hours, they could paint 21/24 = 7/8 of the room. To summarize, multiplying 7/24 by 2 hours means 7/12 of the room is painted; multiplying 7/24 by 3 means 7/8 of the room is painted. Your job is to figure out how many hours it takes until 1 entire room is painted. Let x equal that unknown number in the following equation.

$$\frac{7}{24}x=1$$

Solve for x.

$$\begin{aligned}\left(\frac{\cancel{24}}{\cancel{7}}\right)\left(\frac{\cancel{7}}{\cancel{24}}x\right)&=\left(\frac{24}{7}\right)(1)\\x&=\frac{24}{7}\end{aligned}$$

The room will be finished in 24/7 ≈ 3.429 hours. To convert into minutes, multiply by 60.

$$3.429(60)\approx 205.7 \text{ minutes}$$

Choice (B) best approximates 205.7 minutes.

9. **C.** Notice that $p(x) = rx + s$ is the slope-intercept form of a line, using r instead of m to refer to the slope and s instead of b to refer to the y-intercept. Because $p(x)$ crosses the y-axis below the x-axis, the y-intercept is negative: $s < 0$. Therefore, statement I is true.

Statement II is false. To calculate the x-intercept of the function, set $p(x) = 0$ and solve for x.

$$\begin{aligned}rx+s&=0\\rx&=-s\\x&=-\frac{s}{r}\end{aligned}$$

The x-intercept of $p(x)$ is $x = -s/r$, not s/r.

Statement III claims that smaller x-values result in smaller function values. In other words, if $x = a$ is less than $x = b$, then $p(a) < p(b)$. This is a true statement. Choose any two points on the graph of $p(x)$. Whichever point is further left (and therefore has a lesser x-value) will also be lower (and therefore has a lesser function value).

You conclude that statements I and III must be true.

10. **C.** The entirety of the graph of $|h(x)|$ must be above the x-axis, because taking the absolute value of a function changes any negative function values to positive values. Therefore, choices (B), (D), and (E) can be eliminated immediately. Note that taking the absolute value of a positive number does not change that number. Therefore, positive portions of the graph of $h(x)$ are unchanged when you take the absolute value. Choice (C) must be correct, because choice (A) increases the positive values of the function by shifting them upward.

In short, the graph of the absolute value of a function reflects any portion of the graph below the x-axis across the x-axis, as illustrated in choice (C).

11. **B.** List the ages of the people surveyed in numerical order, from least to greatest. All dots appearing on vertical lines represent one person surveyed of that age. For example, one dot appears on the vertical line 20, so one 20-year-old person was surveyed. However, three dots appear on the vertical line 21, so three 21-year-olds were surveyed. There are 20 data values.

20 21 21 21 22 23 24 25 25 **25 26** 27 27 27 28 28 28 29 29 30

The median is the average of the two middle data values, 25 and 26.

$$\begin{aligned}\text{median} &= \frac{25+26}{2}\\ &= \frac{51}{2}\\ &= 25.5\end{aligned}$$

12. **A.** In Question 11, you pay specific attention to the vertical lines on which each of the dots lie. In this problem, you examine the horizontal lines for each data point—the number of cars each person drove. The average of these values is the solution to this problem.

For example, three dots lie on the horizontal line 1, so three people surveyed had driven only 1 car in their lives. Five dots appear on the horizontal line 3, so five people had driven 3 cars. Calculate a weighted average, multiplying each number of cars (1 through 6) by the number of people who drove that many cars. Divide that sum by 20, the total number of people surveyed.

$$\begin{aligned}\text{average cars driven} &= \frac{3(1)+2(2)+5(3)+3(4)+4(5)+3(6)}{20}\\ &= \frac{3+4+15+12+20+18}{20}\\ &= \frac{72}{20}\\ &= 3.6\end{aligned}$$

The population surveyed drove an average of 3.6 cars per person.

13. **A**. According to the question, $b(3) = 12$. Substitute $t = 3$ into the function and set it equal to 12 in order to calculate the value of c.

$$\begin{aligned} b(t) &= \sqrt{t^2 - t + c} \\ 12 &= \sqrt{3^2 - 3 + c} \\ 12 &= \sqrt{9 - 3 + c} \\ 12 &= \sqrt{6 + c} \end{aligned}$$

To solve for c, square both sides of the equation.

$$\begin{aligned} 12^2 &= \left(\sqrt{6+c}\right)^2 \\ 144 &= 6 + c \\ 144 - 6 &= c \\ 138 &= c \end{aligned}$$

Knowing the value of c, you can more accurately express the function: $b(t) = \sqrt{t^2 - t + 138}$. To calculate the number of colonies present after $t = 6$ hours, evaluate $b(6)$.

$$\begin{aligned} b(6) &= \sqrt{6^2 - 6 + 138} \\ &= \sqrt{36 - 6 + 138} \\ &= \sqrt{168} \\ &\approx 12.9615 \end{aligned}$$

The answer choice that best approximates $b(6)$ is choice (A), 13.

14. **D**. Think of the decreasing prices as a sequence in which a_0 is the original price, a_1 is the price after one daily discount, a_2 is the price after two discounts, etc. The question states that $a_6 = 110$. This price is 10 less than 80% of a_5, so to calculate a_5, you must first add 10 to a_6: $110 + 10 = 120$. Ask yourself this question: "120 is 80% of what number?" You can construct the following proportion to answer that question, such that x is the unknown number.

$$\frac{120}{x} = \frac{80}{100}$$

Cross-multiply and solve for x.

$$\begin{aligned} 80x &= 120(100) \\ 80x &= 12{,}000 \\ x &= \frac{12{,}000}{80} \\ x &= 150 \end{aligned}$$

Thus, $a_5 = 150$. To check your answer, calculate 80% of 150 and subtract 10.

$$\begin{aligned} 0.80(150) - 10 &= 120 - 10 \\ &= 110 \end{aligned}$$

Creating a proportion each time you need to calculate the previous number in the sequence is a bit tedious. Consider this shortcut: Dividing 120 by 0.80 calculates the number of which 80% is 120.

$$\frac{120}{0.80} = 150$$

Apply this shortcut to calculate a_4. Remember, $a_5 = 150$ is 10 less than 80% of a_4, so add 10 to get 150 + 10 = 160. Then, divide 160 by 0.80.

$$\frac{160}{0.80} = 200$$

You now know that $a_4 = 200$. Repeat the process four more times to calculate a_0, the original price of the computer.

$$\begin{aligned} a_6 &= 110 \\ a_5 &= \frac{110+10}{0.80} &&= \frac{120}{0.80} &&= 150 \\ a_4 &= \frac{150+10}{0.80} &&= \frac{160}{0.80} &&= 200 \\ a_3 &= \frac{200+10}{0.80} &&= \frac{210}{0.80} &&= 262.5 \\ a_2 &= \frac{262.5+10}{0.8} &&= \frac{272.5}{0.8} &&= 340.625 \\ a_1 &= \frac{340.625+10}{0.8} &&= \frac{350.625}{0.8} &&= 438.28125 \\ a_0 &= \frac{438.28125+10}{0.8} &&= \frac{448.28125}{0.8} &&= 560.3515625 \end{aligned}$$

The original price is approximately \$560.35 and the final discounted price is \$110; the difference between those prices is \$560.35 – \$110 = \$450.35. The answer choice that most closely approximates that difference is choice (D), \$450.

15. **A**. The question contains a composition of functions. Begin by calculating $5 \boxed{4} 2$.

$$\begin{aligned} 5 \boxed{4} 2 &= \frac{5-2}{4} \\ &= \frac{3}{4} \end{aligned}$$

Substitute the value you just calculated into the expression $\left(5 \boxed{4} 2\right) \boxed{3} (1.25)$. It is helpful to express all values either as fractions or decimals. The solution below uses fractions.

In other words, 1.25 = 5/4.

$$\left(5 \boxed{4} 2\right) \boxed{3} (1.25) = \left(\frac{3}{4}\right) \boxed{3} \left(\frac{5}{4}\right)$$

Evaluate the function.

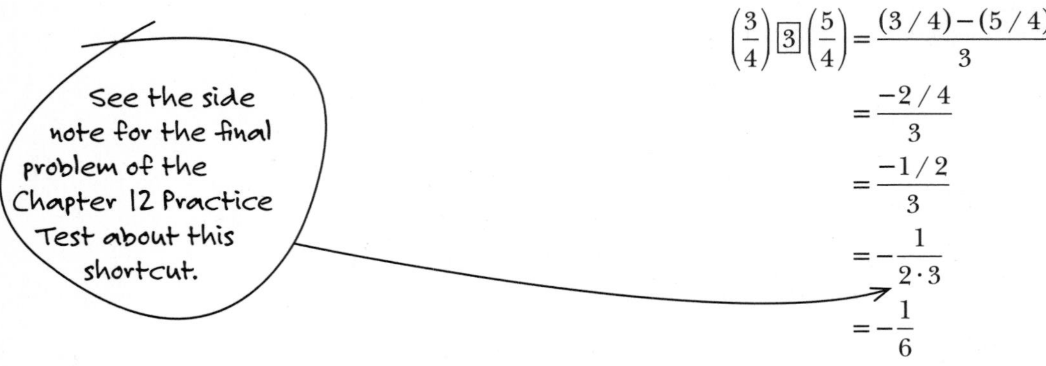

See the side note for the final problem of the Chapter 12 Practice Test about this shortcut.

$$\left(\frac{3}{4}\right)\boxed{3}\left(\frac{5}{4}\right)=\frac{(3/4)-(5/4)}{3}$$
$$=\frac{-2/4}{3}$$
$$=\frac{-1/2}{3}$$
$$=-\frac{1}{2\cdot 3}$$
$$=-\frac{1}{6}$$

Only choice (A) has a value of –1/6.

$$\frac{1}{4}\boxed{\frac{1}{2}}\frac{1}{3}=\frac{(1/4)-(1/3)}{1/2}$$
$$=\frac{(3/12)-(4/12)}{1/2}$$
$$=\frac{-1/12}{1/2}$$

Dividing –1/12 by 1/2 is equivalent to multiplying –1/12 by 2.

$$=-\frac{1}{12}\cdot 2$$
$$=-\frac{2}{12}$$
$$=-\frac{1}{6}$$

16. **B**. When the order does not matter, the three-color combinations black-white-silver and white-silver-black count as the same combination. Use the first letter of each color (M = maroon, G = gold, B = black, S = silver, W = white, and P = pink) to represent that color and list the possible three-color combinations in an orderly fashion.

M G B	M B W	G B S	G W P
M G S	M B P	G B W	B S W
M G W	M S W	G B P	B S P
M G P	M S P	G S W	B W P
M B S	M W P	G S P	S W P

There are 20 different unique combinations of three colors.

SHORTCUT ALERT!

This is a mathematical combination problem because order does not matter. (If order did matter, it would be a permutation problem.) Your calculator may have a combination function, which makes this question incredibly easy. On many Texas Instruments graphing calculators, you can press the Math button and then select the Probability menu to find it. It looks like this: "nCr." If you type "6," select "nCr," and then type "3," you are asking the calculator to compute the number of different ways to choose 3 out of 6 possible things when the order of those things doesn't matter. Press Enter and poof! The answer is 20.

Chapter 14

SAT PRACTICE TEST 3

Make sure to time yourself when you take this practice test

This chapter is designed to simulate the mathematics portion of the SAT test. Record your answers on the student test booklet reproduced on the following page. The solutions are located at the end of the chapter. The following formulas are provided at the beginning of each SAT mathematics section, and you can refer to them during the practice test:

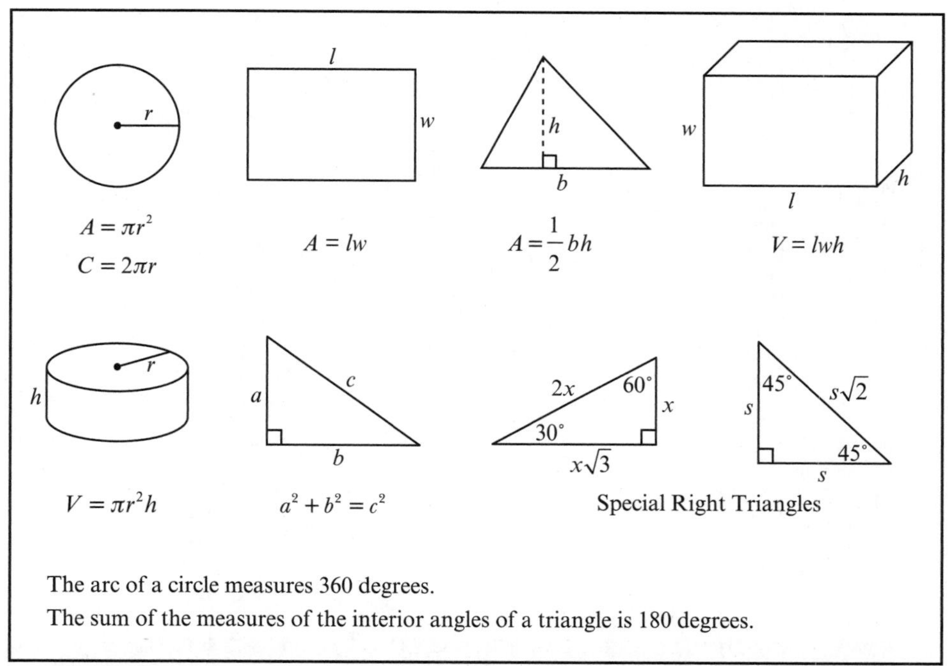

Section 1

Time Limit: 25 minutes

1 ⒶⒷⒸⒹⒺ 5 ⒶⒷⒸⒹⒺ 9 ⒶⒷⒸⒹⒺ 13 ⒶⒷⒸⒹⒺ 17 ⒶⒷⒸⒹⒺ
2 ⒶⒷⒸⒹⒺ 6 ⒶⒷⒸⒹⒺ 10 ⒶⒷⒸⒹⒺ 14 ⒶⒷⒸⒹⒺ 18 ⒶⒷⒸⒹⒺ
3 ⒶⒷⒸⒹⒺ 7 ⒶⒷⒸⒹⒺ 11 ⒶⒷⒸⒹⒺ 15 ⒶⒷⒸⒹⒺ 19 ⒶⒷⒸⒹⒺ
4 ⒶⒷⒸⒹⒺ 8 ⒶⒷⒸⒹⒺ 12 ⒶⒷⒸⒹⒺ 16 ⒶⒷⒸⒹⒺ 20 ⒶⒷⒸⒹⒺ

Section 2

Time Limit: 25 minutes

1 ⒶⒷⒸⒹⒺ 3 ⒶⒷⒸⒹⒺ 5 ⒶⒷⒸⒹⒺ 7 ⒶⒷⒸⒹⒺ
2 ⒶⒷⒸⒹⒺ 4 ⒶⒷⒸⒹⒺ 6 ⒶⒷⒸⒹⒺ 8 ⒶⒷⒸⒹⒺ

9. 10. 11. 12. 13.

14. 15. 16. 17. 18.

Section 3

Time Limit: 20 minutes

1 ⒶⒷⒸⒹⒺ 5 ⒶⒷⒸⒹⒺ 9 ⒶⒷⒸⒹⒺ 13 ⒶⒷⒸⒹⒺ
2 ⒶⒷⒸⒹⒺ 6 ⒶⒷⒸⒹⒺ 10 ⒶⒷⒸⒹⒺ 14 ⒶⒷⒸⒹⒺ
3 ⒶⒷⒸⒹⒺ 7 ⒶⒷⒸⒹⒺ 11 ⒶⒷⒸⒹⒺ 15 ⒶⒷⒸⒹⒺ
4 ⒶⒷⒸⒹⒺ 8 ⒶⒷⒸⒹⒺ 12 ⒶⒷⒸⒹⒺ 16 ⒶⒷⒸⒹⒺ

Section 1

You have 25 minutes to complete this section of the test.

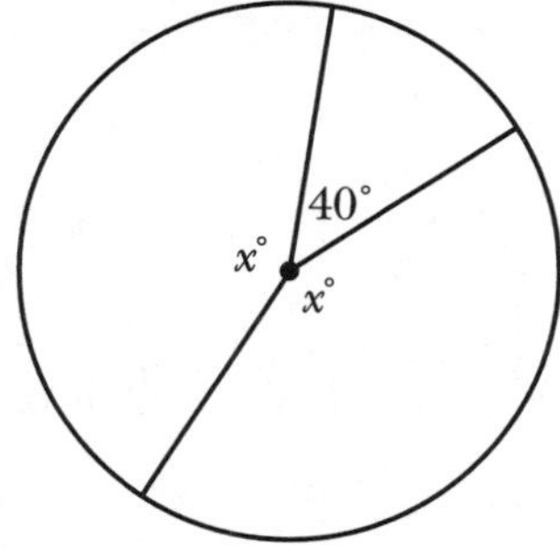

1. Given the circle above with central angles $x°$, $x°$, and 40°, what is the value of x?

   (A) 70
   (B) 80
   (C) 120
   (D) 140
   (E) 160

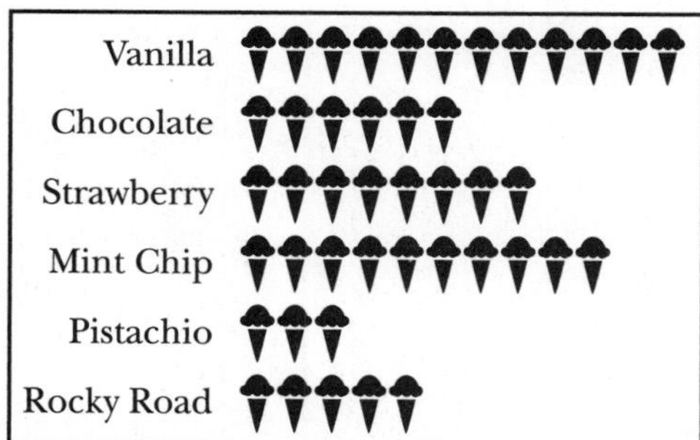

2. The diagram above illustrates the number of ice cream cones of different flavors that were sold at a certain ice cream shop during a 12-hour period. If the shop sold a combined total of 78 mint chip and pistachio ice cream cones during this time period, how many strawberry cones did it sell?

   (A) 40
   (B) 48
   (C) 50
   (D) 54
   (E) 56

3. If $x - 4y = -2$, what is the value of $3x - 12y$?

   (A) –36
   (B) –24
   (C) –9
   (D) –6
   (E) 8

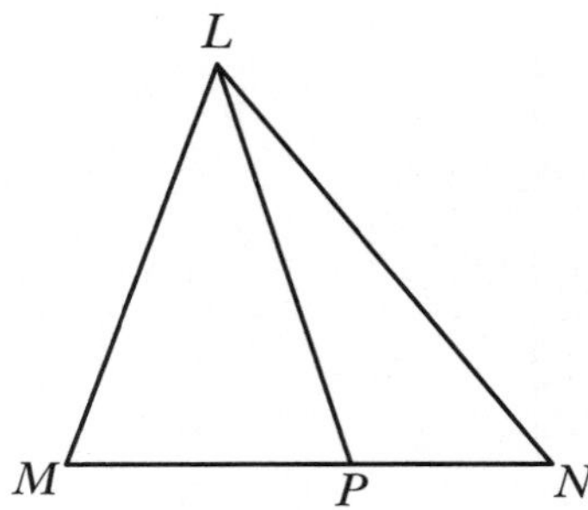

Note: Figure not drawn to scale.

4. In the diagram above, if angles LPM and LPN are congruent, which of the following statements must be true?

   I. $LM = LN$
   II. $MP = NP$
   III. $\overline{LP} \perp \overline{MN}$

   (A) I only
   (B) II only
   (C) III only
   (D) I and II only
   (E) I, II, and III

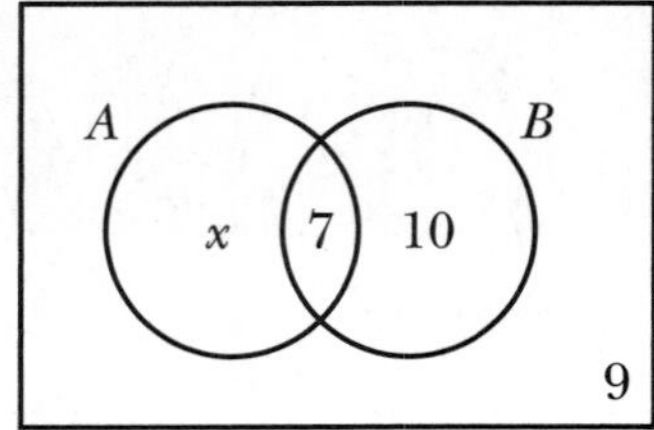

5. If the diagram above describes 30 unique elements, how many of the elements belong to set A?

(A) 4
(B) 11
(C) 13
(D) 17
(E) 26

6. The domain of function $f(s)=\sqrt{s+2}$ is the set of all real numbers *except* for which of the following values of s?

(A) $s<-2$
(B) $s\le-2$
(C) $s\ne-2$
(D) $s>2$
(E) $s\ge 2$

7. Which of the following is equal to $\sqrt{x}+\sqrt{x^3}$, if $x\ge 0$?

(A) x^2
(B) $2x\sqrt{x}$
(C) $x^{1/2}+x^{2/3}$
(D) $\sqrt{x}(x+1)$
(E) $\sqrt{x(1+x^2)}$

Note: Problems 8 and 9 refer to the following information.

A company's logo, illustrated below, is composed of a square, a semicircle, and an equilateral triangle. The solid lines in the illustration represent the perimeter of the figure. Let s be equal to the length of a side of the square.

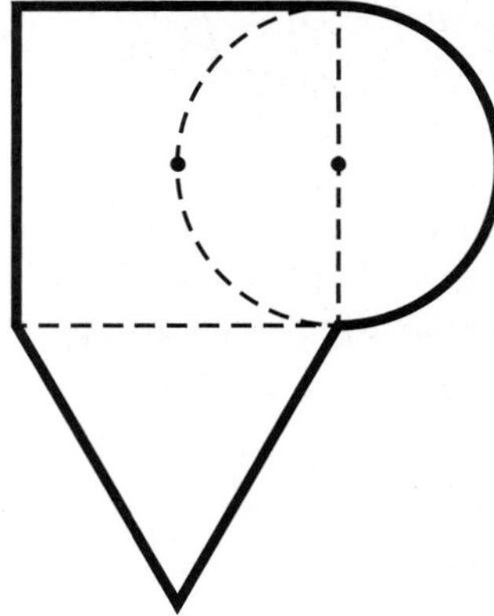

8. What is the perimeter of the logo, in terms of s?

(A) $\frac{s}{2}(8s+\pi)$
(B) $s\left(s+2+\frac{\pi}{2}\right)$
(C) $2s(2s+\pi)$
(D) $s\left(4+\frac{\pi}{2}\right)$
(E) $s(2\pi+7)$

9. What is the area of the logo, in terms of s?

(A) $\frac{s^2}{8}\left(\pi+2\sqrt{3}+8\right)$
(B) $\frac{s^2}{4}\left(2\pi+\sqrt{3}+4\right)$
(C) $\frac{s^2}{2}\left(\pi+\frac{\sqrt{3}}{2}+2\right)$
(D) $s^2\left(\frac{\pi+\sqrt{3}}{4}+1\right)$
(E) $s\left(s+\frac{\pi s}{2}+\frac{3}{4}\right)$

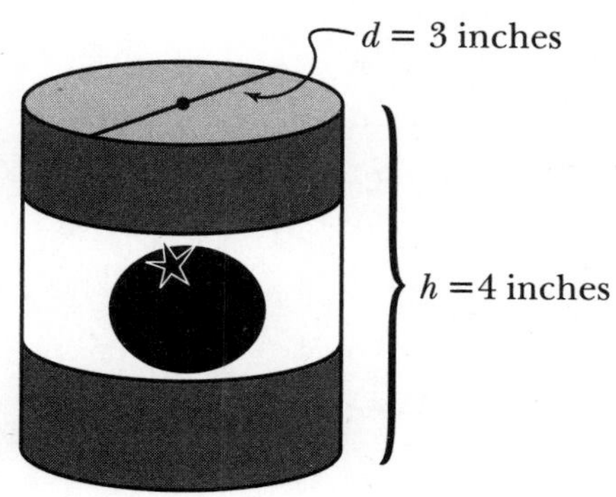

10. The unopened can of soup illustrated above is a right circular cylinder with height h and base diameter d. What is the total surface area of the can, measured in square inches?

(A) 10.5π
(B) 12π
(C) 16.5π
(D) 24π
(E) 30π

11. On a recent quiz, the lowest and highest possible scores were 0 and 100, respectively. If 9 students earned an average (arithmetic mean) score of 75, what is the maximum number of students who could receive a score of 40?

(A) 2
(B) 3
(C) 4
(D) 5
(E) 6

12. Scott is 8 years older than Dennis. In exactly 5 years, Scott's age will be 9 less than twice Dennis' age at that time. What is Scott's current age?

(A) 12
(B) 13
(C) 17
(D) 20
(E) 25

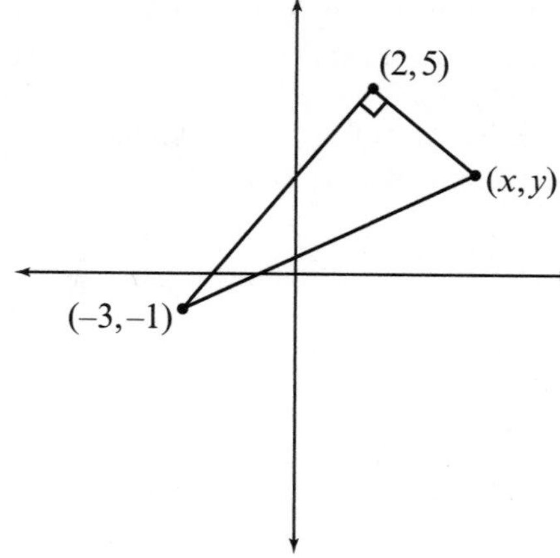

13. Given the right triangle in the diagram above, which of the following expresses x in terms of y?

(A) $x = 8 - 1.2y$
(B) $x = 0.8y + 0.6$
(C) $x = 0.8\overline{3}y - 2.1\overline{6}$
(D) $x = 0.8\overline{3}y - 2.8\overline{3}$
(E) $x = -0.8\overline{3}y + 6.1\overline{6}$

14. The students in a certain class are arranged in order according to their ages, from oldest to youngest. Elizabeth is the fifth oldest student in her class. Janice is the eighth youngest. If Janice is older than Elizabeth and two other students separate them, how many students are in the class?

(A) 7
(B) 8
(C) 9
(D) 10
(E) 11

15. Let x, y, and z be nonzero real numbers such that x varies proportionally with y but varies inversely proportionally with the sum of y and z. If the constants of proportionality are k_1 and k_2, respectively, which of the following is equal to z?

(A) $k_1 y + \frac{k_2}{y}$

(B) $\frac{1}{x}\left(k_2 y - \frac{k_1}{y}\right)$

(C) $\frac{1}{y}\left(\frac{k_2}{k_1} - y^2\right)$

(D) $\frac{k_1}{k_2 y} - \frac{k_2 x}{k_1}$

(E) $k_1 k_2 \left(\frac{y}{y + 1/x}\right)$

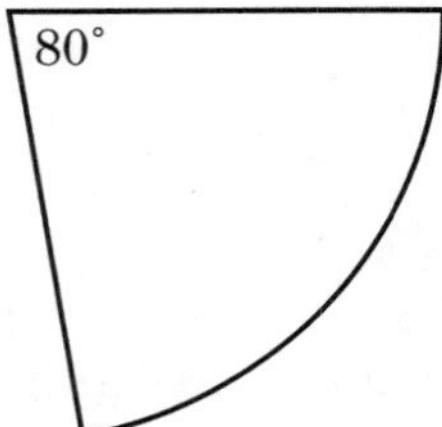

16. If the sector in the diagram above has central angle 80° and an area of $\frac{25\pi}{288}$, what is the length of the arc of the sector?

(A) $\frac{5}{9}\pi$

(B) $\frac{25}{64}\pi$

(C) $\frac{5}{16}\pi$

(D) $\frac{5}{18}\pi$

(E) $\frac{2}{9}\pi$

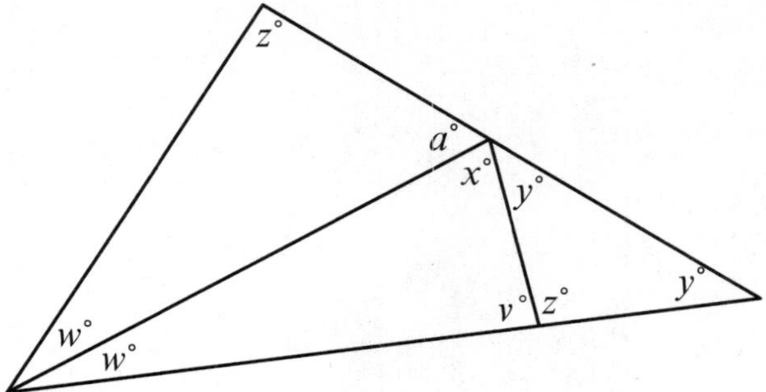

Note: Figure not drawn to scale.

17. Given the diagram above, what is the value of a in terms of v?

(A) $\frac{v}{2}$

(B) $\frac{3v}{4}$

(C) $\frac{5v}{4}$

(D) $\frac{3v}{2}$

(E) $\frac{5v}{2}$

18. If $\sqrt{a^2 + b^2} = b - 3a$ and $b > a > 0$, what is the value of $\frac{2a}{3b}$?

(A) $\frac{1}{2}$

(B) $\frac{2}{3}$

(C) $\frac{3}{4}$

(D) $\frac{4}{3}$

(E) No values of a and b satisfy the equation

Starch	couscous	potato	rice	stuffing
Vegetable	corn	peas	radishes	spinach
Meat	chicken	pork	roast	steak
Dessert	cheesecake	pie	raisin bars	soufflé

19. Tonya has learned how to prepare four different foods in each of the following four categories: starches, vegetables, meats, and desserts. The foods are listed in the table above. She notices that each category contains exactly one food that begins with the letters C, P, R, and S. How many different meals can she prepare under the following conditions?

    - The meal contains exactly one food from each of the four categories.
    - No foods in the meal begin with the same letter.
    - The meal contains spinach and cheesecake.

    (A) 2
    (B) 4
    (C) 6
    (D) 8
    (E) 12

20. If p and q are real numbers and the expressions $x^2 + px - 5$ and $(x - 1)(x + q)$ are equivalent, what is the value of pq?

(A) 20

(B) 10

(C) 5

(D) –5

(E) –15

Section 2

You have 25 minutes to complete this section of the test.

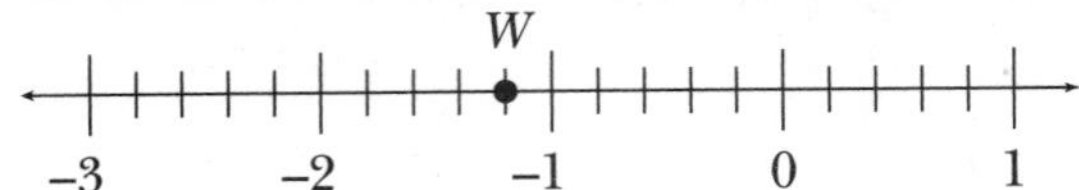

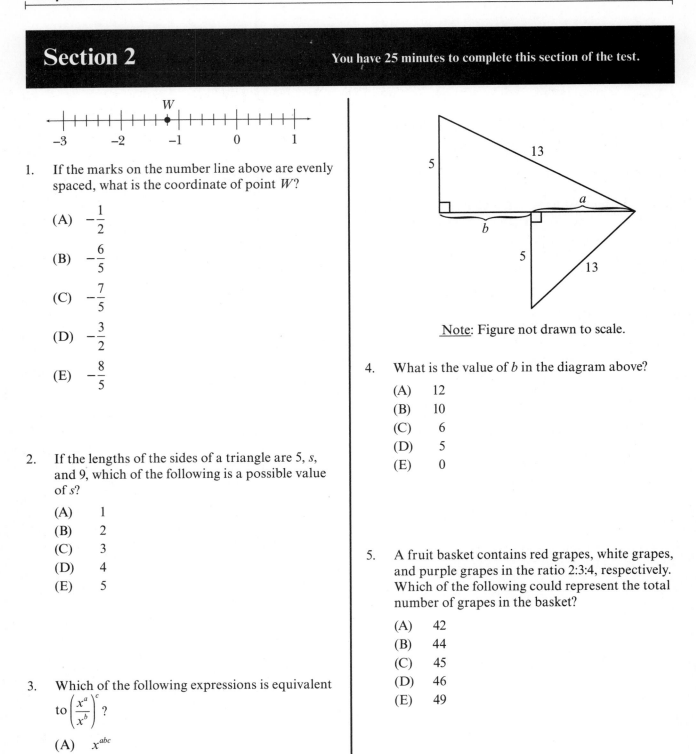

1. If the marks on the number line above are evenly spaced, what is the coordinate of point W?

(A) $-\frac{1}{2}$

(B) $-\frac{6}{5}$

(C) $-\frac{7}{5}$

(D) $-\frac{3}{2}$

(E) $-\frac{8}{5}$

2. If the lengths of the sides of a triangle are 5, s, and 9, which of the following is a possible value of s?

(A) 1
(B) 2
(C) 3
(D) 4
(E) 5

3. Which of the following expressions is equivalent to $\left(\frac{x^a}{x^b}\right)^c$?

(A) x^{abc}

(B) $x^{ac/b}$

(C) x^{ca-cb}

(D) $x^{(a+b)/c}$

(E) $x^{(a/c)+(b/c)}$

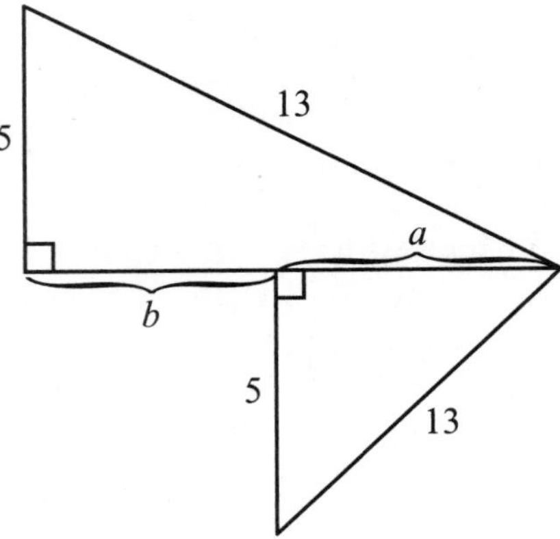

Note: Figure not drawn to scale.

4. What is the value of b in the diagram above?

(A) 12
(B) 10
(C) 6
(D) 5
(E) 0

5. A fruit basket contains red grapes, white grapes, and purple grapes in the ratio 2:3:4, respectively. Which of the following could represent the total number of grapes in the basket?

(A) 42
(B) 44
(C) 45
(D) 46
(E) 49

6. Jeanette writes the letters of the alphabet on separate index cards and then places the 26 cards into a bag. If she selects 3 random cards at the same time, what is the probability that she can spell the "WIN" with the cards she draws?

(A) $\frac{1}{17,576}$

(B) $\frac{27}{17,576}$

(C) $\frac{9}{5,200}$

(D) $\frac{3}{8,788}$

(E) $\frac{1}{2,600}$

$$\begin{cases} x-4y+3z=1 \\ 3x+y=4 \\ z-2x=5 \end{cases}$$

7. If (x, y, z) is the solution to the above system of equations, what is the value of $x + y + z$?

(A) 6
(B) 7
(C) 8
(D) 9
(E) 10

8. Let b, p, and m be positive integers. If b boxes, each containing p pencils, cost a total of m dollars, and the cost per pencil is 25 cents, which of the following statements must be true?

I. $p \geq m$

II. b is even

III. $\frac{m}{4} = \frac{p}{b}$

(A) I only
(B) II only
(C) III only
(D) I and II only
(E) II and III only

9. Marcus is 2 inches taller than Stan, who is 5 inches shorter than Barry. If Barry is 5 feet, 9 inches tall, what is Marcus' height, in inches?

10. A square has side length s, area a, and perimeter p, such that s, a, and p are positive integers that are greater than 1 and less than 20. What is one possible value of ap?

11. Let X and Y be points on opposite bases of a right circular cylinder that has height 12 and bases of radius 3. If the greatest possible value of XY is $3\sqrt{c}$, what is the value of c?

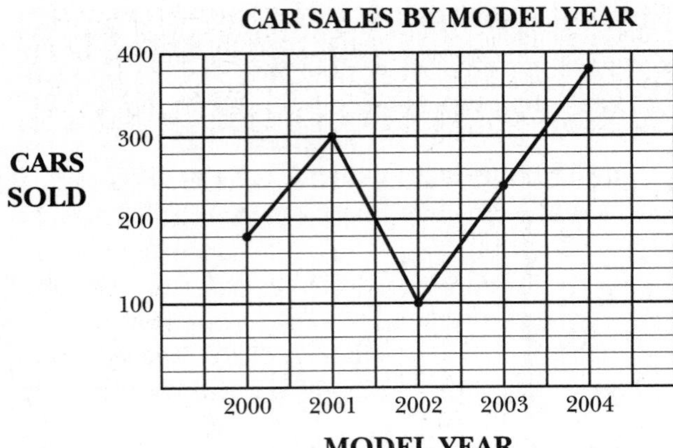

12. If you were to express the data in the line graph above as a circle graph, what central angle represents the sales of model year 2000? Omit the degree symbol from your answer.

13. Let $s \otimes t$ be defined by the function $s \otimes t = s^2 - 2st + t^2$. If $a \otimes b = 100$, what is one possible value of $a - b$?

$a_1 = 2 \qquad a_2 = -6$

14. The first two terms of a geometric sequence are listed above. What is the sum of the fourth and fifth terms?

15. Renée is studying positive integers whose prime factorizations are the product of two unique squared prime factors. How many such numbers exist that are less than 2,000?

16. If $\frac{2x-5y}{7} = \frac{y}{4}$, what is the value of $\frac{y}{x}$?

17. Jason needs to order t-shirts to promote his business. Company A will charge him a one-time setup fee of \$750 and then \$10 per shirt he orders. Company B will charge him \$15 for orders between 1 and 100 shirts, \$13 for orders between 101 and 250 shirts, and \$12 for orders of 251 shirts or more. What is the least number of shirts Jason can order for which Company A is less expensive than Company B?

x	0	1	2	3	4	5	6
$f(x)$	0	2	3	2	−6	−8	−6

18. The chart above lists a few values of the odd function $f(x)$, which is defined for all real numbers. Given the points $A = (-4, f(-4))$, $B = (3, f(3))$, $C = (4, f(4))$, and $D = (-3, f(-3))$, what is the area of quadrilateral $ABCD$?

Section 3

You have 20 minutes to complete this section of the test.

1. If the sum of four consecutive even integers is 380, what is the least of the four integers?

   (A) 90
   (B) 92
   (C) 94
   (D) 96
   (E) 98

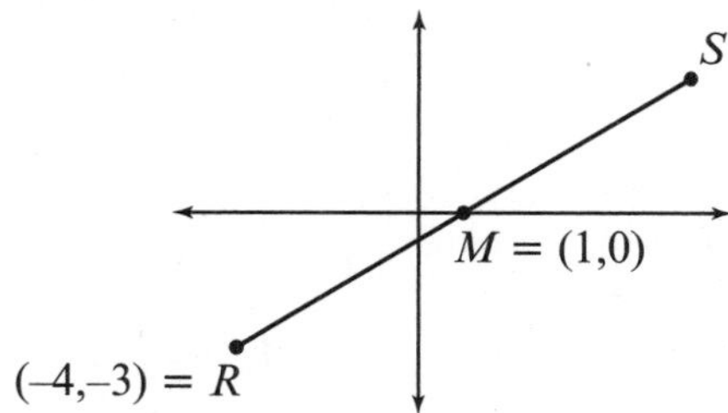

2. Given the diagram above, in which M is the midpoint of $\overline{RS}$, what are the coordinates of S?

   (A) $(3,4)$
   (B) $(4,3)$
   (C) $(5,2)$
   (D) $(5,4)$
   (E) $(6,3)$

3. Two cars, driving at an identical and constant rate, leave from the same location at the same time. Car A travels due south, and car B travels due west. At the precise moment car A has traveled $6\sqrt{2}$ miles, what is the distance between the cars, in miles?

   (A) $3\sqrt{2}$
   (B) 6
   (C) 12
   (D) $6\sqrt{6}$
   (E) $12\sqrt{2}$

4. A certain data set contains n elements that are unique real numbers. In other words, no element appears more than once in the set. Assume $n > 0$. If the median of the set is not an element of the set, which of the following statements must be true? (Note that the word "average" in the choices below refers to the arithmetic mean.)

   (A) The number n is even.
   (B) The number n is odd.
   (C) The average is less than the median.
   (D) The average is equal to the median.
   (E) The average is greater than the median.

5. Which of the following is a solution to the equation $\frac{x+1}{x-1} = c$, if c is an integer?

   I. $x = 0$
   II. $x = 1$
   III. $x = \frac{1}{2}$

   (A) II only
   (B) III only
   (C) I and II only
   (D) I and III only
   (E) II and III only

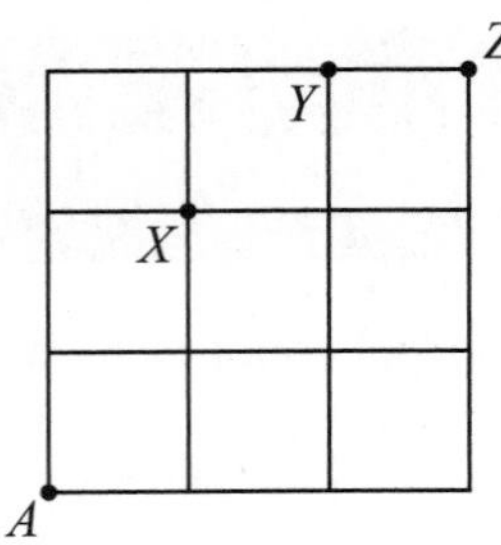

6. In the figure above, you may travel either up or right along the linear segments. How many paths exist from point A to point Z that pass either through point X or point Y but <u>not</u> through both points X and Y?

   (A) 4
   (B) 6
   (C) 8
   (D) 10
   (E) 12

7. Ruth, Mel, and David are presented with equal gifts of d dollars each. Ruth gives half of her gift to Mel. After receiving the additional money, Mel gives two-thirds of her total to David. Once these transactions are complete, David's total is what percentage of his original gift, d?

   (A) 100%
   (B) $166.\overline{6}\%$
   (C) 200%
   (D) $233.\overline{3}\%$
   (E) 250%

8. If point A is the vertex of function $g(x) = cx^2$, and point B is the vertex of function $h(x) = -g(x)$, which of the following best describes the relationship between A and B on the coordinate plane?

   (A) B is above A
   (B) B is below A
   (C) B is left of A
   (D) B is right of A
   (E) A and B are the same point

9. A catering company is preparing b boxed lunches and places the same number of cookies in each lunch. If the company has c cookies to distribute and n cookies remain when the lunches are complete (such that $n < b$), each lunch contains how many cookies?

   (A) $bc - n$
   (B) $\frac{c-n}{b}$
   (C) $\frac{c+bn}{b}$
   (D) $\frac{n}{b} + c$
   (E) $\frac{c}{b} + n$

$$\begin{array}{r} C\ \ 3\ \ C\ \ B \\ +\quad D\ \ B\ \ 5 \\ \hline B\ \ 0\ \ A\ \ 3 \end{array}$$

10. If A, B, C, and D are positive digits in the sum above, what is the value of $A + D$?

    (A) 12
    (B) 13
    (C) 14
    (D) 15
    (E) 16

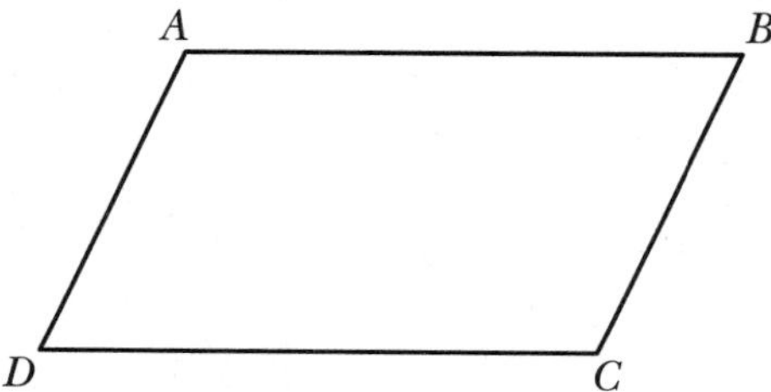

11. In the parallelogram above, $AB = 4x$, $BC = x$, and the measure of angle A is 120°. If a right prism with height 4 has this parallelogram as its base, what is the volume of the prism?

    (A) $2\sqrt{3}x^2$
    (B) $6\sqrt{3}x^2$
    (C) $7\sqrt{3}x^2$
    (D) $8\sqrt{3}x^2$
    (E) $9\sqrt{3}x^2$

12. If a and b are nonzero numbers and $ab^2 < 0$, which of the following statements must be true?

    I. $\frac{a}{b^2} < 0$

    II. $a < 0$

    III. $\sqrt{2} - a > 0$

    (A) I only
    (B) II only
    (C) I and II only
    (D) II and III only
    (E) I, II, and III

13. If a circle with center $(1, 2)$ and diameter 30 passes through point $(p, 2p)$, what is the value of p?

    (A) $1+\sqrt{6}$
    (B) $1+3\sqrt{5}$
    (C) $1+5\sqrt{3}$
    (D) $1+6\sqrt{5}$
    (E) $1+10\sqrt{3}$

14. If $\frac{1}{xy} - \frac{1}{y} = \frac{1}{x+y}$ and $\frac{x}{y} = 2$, what is the value of $x + y$?

    (A) $\frac{3}{8}$
    (B) $\frac{1}{2}$
    (C) $\frac{3}{4}$
    (D) $\frac{9}{8}$
    (E) 3

15. The average (arithmetic mean) of p, q, r, s, t, and u is a. If the average of p, r, t, and u is b, what is the average of q and s, in terms of a and b?

(A) $\frac{a}{6}+\frac{b}{2}$

(B) $\frac{a}{2}+\frac{b}{2}$

(C) $a-\frac{1}{3}b$

(D) $3a-2b$

(E) $6a-4b$

16. If $y-\frac{1}{3y}=4$ and $y\neq 0$, what is the value of $\frac{9y^4+1}{9y^2}$?

(A) 16

(B) $\frac{50}{3}$

(C) 12

(D) $\frac{14}{3}$

(E) $\frac{8}{3}$

Solutions

Section I

1. **E.** The sum of the central angles within a circle is equal to 360°. This circle contains exactly three central angles that have measures $x°$, $x°$, and 40°. The sum of those three values must equal 360°. Set up an equation and solve it for x.

$$x + x + 40 = 360$$
$$2x + 40 = 360$$
$$2x = 360 - 40$$
$$2x = 320$$
$$x = \frac{320}{2}$$
$$x = 160$$

2. **B.** The ice cream cone symbols in the pictograph do not each represent one cone sold, and no key is given to interpret the graph. However, the question states that 78 mint chip and pistachio ice cream cones were sold. There are 10 ice cream cone symbols representing mint chip sales and 3 ice cream cone symbols representing pistachio sales, a total of 10 + 3= 13 symbols representing combined sales of the two flavors.

Let x represent the value of the symbol and solve the equation $13x = 78$ to determine the value of the symbol in the pictograph.

$$13x = 78$$
$$x = \frac{78}{13}$$
$$x = 6$$

Each ice cream cone symbol represents 6 cones sold. The question asks how many strawberry cones were sold, and the pictograph lists 8 symbols in the strawberry row. Therefore, $8(6) = 48$ strawberry cones were sold.

3. **D.** The question states that $x - 4y = -2$. Notice that the second expression, $3x - 12y$, is exactly three times as large as $x - 4y$.

$$3(x - 4y) = 3x - 12y$$

If $3x - 12y$ is three times as large as $x - 4y$, then $3x - 12y$ is three times as large as -2. In other words, $3x - 12y$ is equivalent to $3(-2) = -6$.

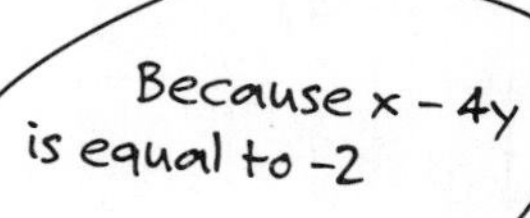
Because x - 4y
is equal to -2

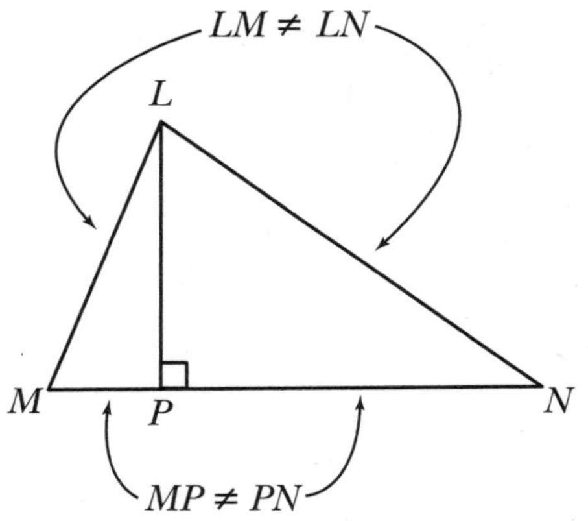

Think of it this way: Imagine that some student X is at least 18 years old. I have no other information about the age of student X. Is it accurate to say that student X's age cannot be 18 or lower? No! student X might be 18.

To calculate the square root of 16, you ask yourself, "What times itself equals 16?" There are two answers to that question: +4 and −4. However, the square root of 16 is always equal to +4, not −4. This positive answer is called the "principal square root." When you simplify the expression $\sqrt{x^2} = |x|$, you use absolute value signs to make the answer positive, to ensure that you're reporting the principal square root.

4. **C.** Notice that angles *LPM* and *LPN*, together, form a straight angle. Thus, the sum of their measures is 180°. The question states that the angles are congruent, so each measures $180° \div 2 = 90°$. If both angles measure 90°, then the segments forming those angles are perpendicular, and statement III is true. Neither statement I nor statement II must be true if $\overline{LP} \perp \overline{MN}$, as the following diagram demonstrates.

$LM \neq LN$

L

M P N

$MP \neq PN$

5. **B.** According to the diagram, 10 elements belong only to set *B*, 7 elements belong to both sets *A* and *B*, and 9 elements belong to neither set. That accounts for $10 + 7 + 9 = 26$ of the 30 elements. The remaining 4 must belong only to set *A*. In other words, $x = 4$. The question asks you to calculate the number of elements in set *A*, which includes the $x = 4$ elements only in set *A* and the 7 elements shared by sets *A* and *B*. Therefore, set *A* contains $4 + 7 = 11$ elements.

6. **A.** The function $f(s) = \sqrt{s+2}$ is only defined for values of *s* that are greater than or equal to −2. Notice that $f(-2) = \sqrt{-2+2} = 0$. Any value of *s* less than −2 results in a negative value within the radical symbol, and the square root of a negative number is not a real number value. Therefore, Choice (A) is correct.

Note that Choice (B) is similar but incorrect. The function *is* defined when $s = -2$, so it is not undefined when *s* is less than *or equal to* −2.

7. **D.** Notice that you can rewrite x^3 within the square root to simplify the expression.

$$\begin{aligned}\sqrt{x} + \sqrt{x^3} &= \sqrt{x} + \sqrt{x^2 \cdot x} \\ &= \sqrt{x} + \sqrt{x^2} \cdot \sqrt{x} \\ &= \sqrt{x} + x\sqrt{x}\end{aligned}$$

Factor $\sqrt{x}$ out of both terms and you get choice (D).

$$\begin{aligned}&= \sqrt{x}(1+x) \\ &= \sqrt{x}(x+1)\end{aligned}$$

This question skillfully dodges an important mathematical concept about which you should be aware. Technically, $\sqrt{x^2} = |x|$, not just *x*. Why? Consider the expression $(-4)^2$. While this is equal to 16, it would be incorrect to state that $\sqrt{16} = -4$. While it is true that a square and a square root "cancel each other out," the square root of a real number cannot be negative. Luckily, the question states that $x \geq 0$, so this is not a concern. You can conclude that $\sqrt{x^2} = x$, without worrying that *x* might be negative.

8. **D**. Two sides of the square and two sides of the equilateral triangle—each with length s—form the straight portions of the logo. That part of the perimeter is equal to $4s$. The semicircle has diameter s, so the length of the curved portion of the logo is equal to half the circumference of a circle with diameter s: $(s\pi/2)$.

The circumference of a circle is π times the diameter, in this case $s\pi$. The logo only contains a semicircle—not the whole circle—so divide the circumference by 2.

The perimeter of the logo is equal to $4s + (s\pi)/2$, which is not one of the answer choices. However, notice that choice (D) has the same value if you distribute s through the parentheses.

$$s\left(4+\frac{\pi}{2}\right)=4s+\frac{s\pi}{2}$$

9. **A**. The area of the logo is equal to the sum of the areas of the square, semicircle, and triangle. The square has area s^2. The semicircle has an area equal to half the area of a circle with radius $r = s/2$.

$$\begin{aligned}\text{area of semicircle} &= \frac{1}{2}(\pi r^2)\\ &= \frac{1}{2}\left[\pi\left(\frac{s}{2}\right)^2\right]\\ &= \frac{1}{2}\left(\frac{s^2\pi}{4}\right)\\ &= \frac{s^2\pi}{8}\end{aligned}$$

To calculate the area of the equilateral triangle, you need to divide it into two 30°–60°–90° triangles and apply the properties of those triangles, as illustrated below.

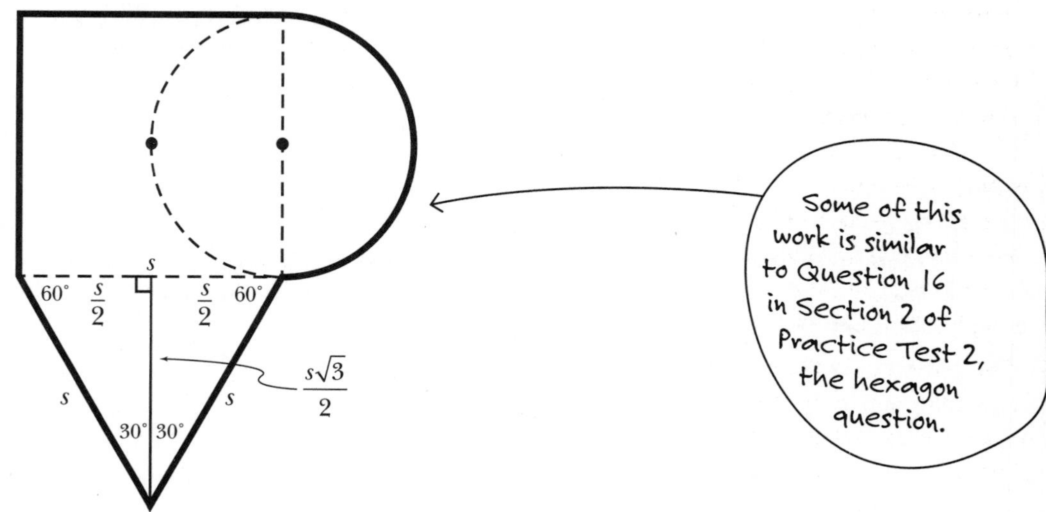

Calculate the area of the equilateral triangle.

$$\begin{aligned}\text{area of triangle} &= \frac{1}{2}bh\\ &= \frac{1}{2}s\left(\frac{s\sqrt{3}}{2}\right)\\ &= \frac{s^2\sqrt{3}}{4}\end{aligned}$$

Calculate the total area of the logo.

$$\begin{aligned}\text{logo area} &= \text{square area} + \text{semicircle area} + \text{triangle area}\\ &= s^2 + \frac{s^2\pi}{8} + \frac{s^2\sqrt{3}}{4}\end{aligned}$$

This is not one of the answer choices, but if you distribute $s^2/8$ through the parentheses in choice (A), you get this answer.

$$\begin{aligned}\frac{s^2}{8}\left(\pi + 2\sqrt{3} + 8\right) &= \frac{s^2\pi}{8} + \frac{s^2 \cdot \cancel{2}\sqrt{3}}{\cancel{2}\cdot 4} + \frac{\cancel{8}s^2}{\cancel{8}}\\ &= \frac{s^2\pi}{8} + \frac{s^2\sqrt{3}}{4} + s^2\end{aligned}$$

10. **C.** An unopened soup can has three surfaces: two lids and the lateral surface (which is covered by the label). Each lid is a circle with radius $d \div 2 = 3/2$. Calculate the surface area of one lid using the formula for the area of a circle.

$$\begin{aligned}A_{\text{lid}} &= \pi r^2\\ &= \pi\left(\frac{3}{2}\right)^2\\ &= \pi\left(\frac{9}{4}\right)\\ &= \frac{9\pi}{4}\end{aligned}$$

There are two lids, so multiply that area by 2.

$$\begin{aligned}A_{\text{two lids}} &= 2\left(\frac{9\pi}{4}\right)\\ &= \frac{18\pi}{4}\\ &= \frac{9\pi}{2}\end{aligned}$$

See Problem 8.67.

The lateral surface of a right circular cylinder is a rectangle with width h and length equal to the circumference of the base. The base has diameter $d = 3$, so the circumference of the base is $C = \pi d = 3\pi$. Calculate the lateral surface area.

$$\begin{aligned}A_{\text{lateral}} &= l \cdot w\\ &= (3\pi)(4)\\ &= 12\pi\end{aligned}$$

Add the individual areas to calculate the total surface area.

$$\begin{aligned}A_{\text{total}} &= A_{\text{two lids}} + A_{\text{lateral}}\\ &= \frac{9\pi}{2} + 12\pi\\ &= \frac{9\pi}{2} + \frac{24\pi}{2}\\ &= \frac{33\pi}{2}\\ &= 16.5\pi\end{aligned}$$

11. **B.** If 9 students earned an average of 75, then the sum of the 9 student scores must be 9(75) = 675. To answer the question, try the solutions suggested by the answer choices to determine which is true. It is helpful to begin with choice (C), because the choices are listed in order.

If 4 students received a score of 40, then their scores have a sum of 4(40) = 160. Therefore, the remaining 5 student scores have a sum of 675 – 160 = 515. Unfortunately, the greatest possible test score is 100, so 5 students can score a maximum of 5(100) = 500 points, which is less than the necessary 515. Fewer than 4 students must have scored 40 on the test. Try the next lowest answer choice, (B).

If 3 students received a score of 40, then their scores have a sum of 3(40) = 120, and the remaining 6 student scores have a sum of 675 – 120 = 555. It is possible for 6 students scores to have a sum of 555, so (B) is the correct answer.

12. **D.** Let s represent Scott's current age and let d represent Dennis' current age. You know that $s = d + 8$, because Scott is 8 years older than Dennis. In 5 years, Scott's age will be $s + 5$ and Dennis' age will be $d + 5$. At that point, Scott's age will be 9 less than twice Dennis' age.

$$\begin{aligned}\text{Scott's age in 5 years} &= 2(\text{Dennis' age in 5 years}) - 9\\ s+5 &= 2(d+5)-9\\ s+5 &= 2d+10-9\\ s+5 &= 2d+1\end{aligned}$$

Recall that $s = d + 8$. Substitute this value into the equation and solve for d.

$$\begin{aligned}(d+8)+5 &= 2d+1\\ d+13 &= 2d+1\\ 13-1 &= 2d-d\\ 12 &= d\end{aligned}$$

Dennis is currently 12 years old. Scott is 8 years older, so Scott's current age is 12 + 8 = 20.

13. **A.** A right triangle has perpendicular legs, so the slopes of the legs must be opposite reciprocals. Calculate the slope of the leg passing through points (–3,–1) and (2,5).

$$\begin{aligned}m &= \frac{5-(-1)}{2-(-3)}\\ &= \frac{5+1}{2+3}\\ &= \frac{6}{5}\end{aligned}$$

The slope of the leg passing through points (2,5) and (x,y) must be –5/6.

$$\begin{aligned}-\frac{5}{6} &= \frac{y-5}{x-2}\\ -5(x-2) &= 6(y-5)\\ -5x+10 &= 6y-30\end{aligned}$$

Solve the equation for x.

$$-5x = 6y - 30 - 10$$
$$-5x = 6y - 40$$
$$\frac{-5}{-5}x = \frac{6}{-5}y - \frac{40}{-5}$$
$$x = -\frac{6}{5}y + 8$$

Note that $-6/5 = -1.2$. Therefore, $x = -1.2y + 8$, choice (A).

14. **C.** Draw a diagram to illustrate that Elizabeth is the fifth oldest in her class. In the following diagram, the further left the student, the older he or she is, so Elizabeth is in the fifth blank from the left. Notice that four blanks appear left of Elizabeth, because four students are older than she is.

Near the end of the question, you find that Janice is older than Elizabeth; two other students separate them. Update the diagram.

____ Janice ____ ____ Elizabeth

Janice is the eighth youngest in the class, so she should be in the eighth blank counting from the right. This requires you to add four blanks to the right of Elizabeth in the diagram.

____ Janice ____ ____ Elizabeth ____ ____ ____ ____

Janice is eight positions from the right and Elizabeth is five positions from the left. Janice is older and two blanks separate the students. You have satisfied all of the conditions described in the question, so count the blanks to calculate the number of students in class: 9.

15. **C.** Apply the DIY technique. If $x = 4$ varies proportionally with $y = 2$, solve the following equation to calculate the constant of proportionality k_1.

$$x = k_1 y$$
$$4 = k_1(2)$$
$$\frac{4}{2} = k_1$$
$$2 = k_1$$

Let $z = 6$. If $x = 4$ varies inversely proportionally with $y + z = 2 + 6 = 8$, solve the following equation to calculate the constant of proportionality k_2.

$$x(y+z) = k_2$$
$$4(8) = k_2$$
$$32 = k_2$$

Substitute $x = 4$, $y = 2$, $k_1 = 2$, and $k_2 = 32$ into the answer choices to determine which is equivalent to $z = 6$. Only choice (C) generates the correct answer.

$$\begin{aligned}\frac{1}{y}\left(\frac{k_2}{k_1} - y^2\right) &= \frac{1}{2}\left(\frac{32}{2} - 2^2\right)\\ &= \frac{1}{2}(16-4)\\ &= \frac{1}{2}(12)\\ &= 6\end{aligned}$$

16. **D.** The sector has central angle 80°, so it represents 80/360 = 2/9 of a circle. Therefore, its area is 2/9 the area A of a circle with the same radius. Construct a proportion to calculate A.

$$\begin{aligned}\frac{\text{area of sector}}{\text{central angle of sector}} &= \frac{\text{area of circle}}{360}\\ \frac{25\pi/288}{80} &= \frac{A}{360}\\ \frac{25\pi}{288}\cdot\frac{360}{1} &= 80A\\ \frac{9000\pi}{288} &= 80A\\ \frac{125\pi}{4} &= 80A\end{aligned}$$

Solve the equation for A.

$$\begin{aligned}\frac{1}{80}\left(\frac{125\pi}{4}\right) &= \frac{1}{\cancel{80}}\left(\cancel{80}A\right)\\ \frac{125\pi}{320} &= A\\ \frac{25\pi}{64} &= A\end{aligned}$$

If the area of the circle is $25\pi/64$, you can apply the area formula for a circle to calculate the radius.

$$\begin{aligned}A &= \pi r^2\\ \frac{25\cancel{\pi}}{64} &= \cancel{\pi}r^2\\ \sqrt{\frac{25}{64}} &= \sqrt{r^2}\\ \frac{5}{8} &= r\end{aligned}$$

Divide both sides by π to simplify the equation.

Calculate the circumference of a circle with radius $r = 5/8$.

$$\begin{aligned} C &= 2\pi r \\ &= 2\pi\left(\frac{5}{8}\right) \\ &= \frac{10\pi}{8} \\ &= \frac{5\pi}{4} \end{aligned}$$

The length of the arc of the sector is 2/9 the circumference of the circle with radius $r = 5/8$, just like the area of the sector is 2/9 the area of the circle with radius $r = 5/8$.

$$\begin{aligned} \text{arc length} &= \left(\frac{2}{9}\right)\left(\frac{5\pi}{4}\right) \\ &= \frac{10\pi}{36} \\ &= \frac{5\pi}{18} \end{aligned}$$

17. **B.** There are a number of different ways to answer this question. Your solution may be longer or it may be shorter, but the key to answering this question correctly is keeping your information organized. You can identify many relationships between the variables, noting that some angles are supplementary while others are interior angles of a triangle. Your final goal is to express everything in terms of v. Keep that in mind, and you will eventually reach the answer. For clarity, the following solution is expressed in steps.

Step 1: Notice that $a° + x° + y° = 180°$, because the angles form a straight angle. Additionally, $v° + x° + w° = 180°$, because they are the interior angles of a triangle. Both expressions are equal to 180°, so they are also equal to each other.

$$a + x + y = v + x + w$$

Subtract x from both sides of the equation and solve for a, because the question asks you to calculate the value of a.

$$\begin{aligned} a + y &= v + w \\ a &= v + w - y \end{aligned}$$

One of the terms on the right side of the equation is v. The other two terms, w and $-y$, must be expressed in terms of v as well. That is the focus of steps 2 through 4.

Step 2: The angle measures of the smallest triangle have a sum of 180°.

$$\begin{aligned} y + y + z &= 180 \\ 2y + z &= 180 \end{aligned}$$

Furthermore, the angles measuring $v°$ and $z°$ must be supplementary, because they form a straight angle.

$$v + z = 180$$

Like in step 1, you have two expressions that are equal to 180, so they must be equal to each other.

$$2y + z = v + z$$

Subtract z from both sides of the equation and solve for y.

$$2y = v$$

$$y = \frac{v}{2}$$

You have now expressed y in terms of v.

Step 3: Recall that $2y + z = 180$. Now add the angle measures that form the largest triangle and set that equal to 180°.

The huge triangle you get when you combine the three, smaller triangles

$$2w + y + z = 180$$

Both expressions are equal to 180, so set them equal to each other. Solve for w.

$$\begin{aligned} 2y+z &= 2w+y+z \\ 2y+z-z &= 2w+y+z-z \\ 2y &= 2w+y \\ 2y-y &= 2w \\ y &= 2w \\ \frac{y}{2} &= w \end{aligned}$$

Step 4: In step 2, you determined that $y = v/2$. Substitute this into the equation you created in step 3.

$$\begin{aligned} \frac{y}{2} &= w \\ \frac{v/2}{2} &= w \\ \frac{v}{4} &= w \end{aligned}$$

You have now expressed w in terms of v.

Step 5: You now know that $y = v/2$ and $w = v/4$. Substitute these values into the equation you generated in step 1.

$$\begin{aligned} a &= v+w-y \\ &= v+\frac{v}{4}-\frac{v}{2} \\ &= \frac{4v}{4}+\frac{v}{4}-\frac{2v}{4} \\ &= \frac{4v+v-2v}{4} \\ &= \frac{3v}{4} \end{aligned}$$

18. **A.** Square both sides of the radical equation and simplify.

$$a^2+b^2=(b-3a)^2$$
$$a^2+\cancel{b^2}=\cancel{b^2}-6ab+9a^2$$
$$a^2=-6ab+9a^2$$
$$6ab=9a^2-a^2$$
$$6ab=8a^2$$

Divide both sides by a.

$$\frac{6\cancel{a}b}{\cancel{a}}=\frac{8a\cdot\cancel{a}}{\cancel{a}}$$
$$6b=8a$$

Solve for a/b, which is the variable portion of the expression in the question.

$$\frac{6\cancel{b}}{\cancel{b}}=\frac{8a}{b}$$
$$\left(\frac{1}{8}\right)6=\left(\frac{1}{\cancel{8}}\right)\frac{\cancel{8}a}{b}$$
$$\frac{6}{8}=\frac{a}{b}$$
$$\frac{3}{4}=\frac{a}{b}$$

To evaluate $(2a)/(3b)$, multiply both sides of the equation by $2/3$.

$$\left(\frac{2}{3}\right)\left(\frac{3}{4}\right)=\left(\frac{2}{3}\right)\left(\frac{a}{b}\right)$$
$$\frac{6}{12}=\frac{2a}{3b}$$
$$\frac{1}{2}=\frac{2a}{3b}$$

19. **A.** Create diagrams that illustrate all possible meals. The diagram below, for example, illustrates all possible meals that include couscous as a starch. Each branch represents a different meal, so there are six possible meals.

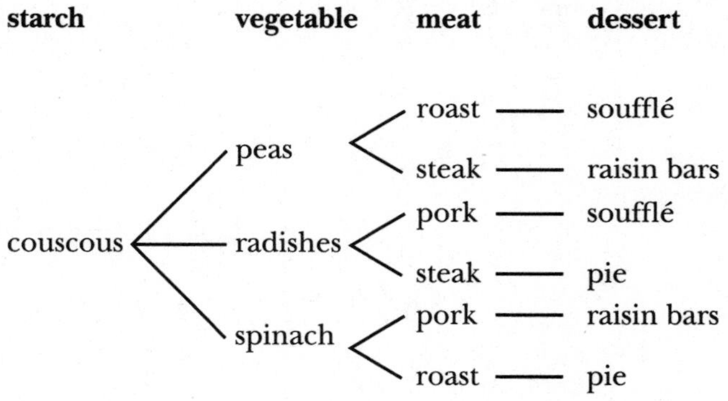

starch	**vegetable**	**meat**	**dessert**
couscous	peas	roast	soufflé
		steak	raisin bars
	radishes	pork	soufflé
		steak	pie
	spinach	pork	raisin bars
		roast	pie

None of these meals include spinach and cheesecake. To generate the other diagrams quickly, use only the first letters of the dishes and keep the order the same. In other words, from left to right, the categories are starch, vegetable, meat, and dessert. Arrows identify the meals that include spinach and cheesecake.

P — C — R—S
P — C — S—R
P — R — C—S
P — R — S—C
P — S — C—R
P — S — R—C ←

R — C — P—S
R — C — S—P
R — P — C—S
R — P — S—C
R — S — C—P
R — S — P—C ←

S — C — P—R
S — C — R—P
S — P — C—R
S — P — R—C
S — R — C—P
S — R — P—C

Tonya can make 24 unique meals. Only two include spinach and cheesecake.

20. **A.** Notice that $x^2 + px - 5$ is a quadratic expression. It is equivalent to $(x - 1)(x + q)$. Expand the product by multiplying $(x - 1)$ and $(x + q)$.

$$\begin{aligned}(x-1)(x+q) &= x^2 + qx - 1x - q \\ &= x^2 + x(q-1) - q\end{aligned}$$

Both expressions, $x^2 + px - 5$ and $x^2 + x(q - 1) - q$, contain x^2. Both contain an x-term, and their coefficients must be equal: $p = q - 1$. Both expressions also contain a constant: $-5 = -q$. Therefore, $q = 5$. Substitute this value into the coefficient equation you just generated.

$$\begin{aligned}p &= q - 1 \\ &= 5 - 1 \\ &= 4\end{aligned}$$

You conclude that $pq = (4)(5) = 20$.

You may have found this question to be slightly easier than 18 and 19. It's still a hard question, but just because it's the LAST question in the section, that doesn't make it the HARDEST question in the section. Remember, questions are clumped by difficulty, but the hard questions are not necessarily in order from "sort of hard" to "ridiculously hard." That's why you should skip questions when you get stuck, especially at the end of the section. You might be able to answer some of the later questions.

Section 2

1. **B.** The marks on the number line divide the space between each integer into five congruent segments. Therefore, each mark represents one-fifth of the number line. Point W is located 6 marks left of 0, which is six of the one-fifth lengths in the negative direction. Therefore, the coordinate of $W = -6/5$.

2. **E.** This is a triangle inequality theorem question. Remember that the sum of the lengths of any two sides of a triangle must be greater than the length of the remaining side. In this question that means $s + 5 > 9$, $s + 9 > 5$, and $5 + 9 > s$. The only value of s that satisfies all three of those equations is choice (E), $s = 5$. Note that $5 + 5 > 9$, $5 + 9 > 5$, and $5 + 9 > 5$. The other four answer choices fail to satisfy the first inequality: $s + 5 > 9$.

3. **C.** Apply properties of exponents to rewrite the expression. Recall that a quotient of two exponential expressions with the same base is equal to the shared base raised to the difference of the powers.

$$\left(\frac{x^a}{x^b}\right)^c = \left(x^{a-b}\right)^c$$

When an exponential expression is, itself, raised to another exponent, you can rewrite the expression as the base raised to the product of the exponents.

$$\left(x^{a-b}\right)^c = x^{(a-b)c}$$
$$= x^{ca-cb}$$

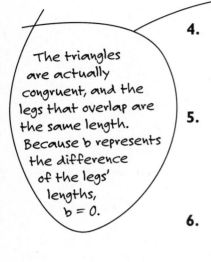

The triangles are actually congruent, and the legs that overlap are the same length. Because b represents the difference of the legs' lengths, b = 0.

4. **E.** The diagram is very misleading. Both right triangles have hypotenuse 13 and one leg of length 5. According to the Pythagorean triple 5, 12, 13, the other leg in both triangles must have length 12. In other words, in the triangle that appears smaller, $a = 12$. The leg in the triangle that appears larger has length $a + b = 12$. If $a = 12$, then b equals 0.

5. **C.** The grapes are in the ratio 2:3:4, so for every 2 red grapes, there are 3 white grapes and 4 purple grapes. In order to maintain the ratio, the total number of grapes must be a multiple of 2 + 3 + 4 = 9. For example, you could multiply each number in the ratio by 5, and the basket would contain 5(2) = 10 red grapes, 5(3) = 15 white grapes, and 5(4) = 20 purple grapes. The total number of grapes would be 10 + 15 + 20 = 45, which is choice (C).

6. **E.** This is a fundamental counting principle question. To make the question easier, assume she draws the letters one at a time without replacing them in the bag before she draws the next card.

In order to spell the word "WIN," with her first draw, she must draw either the W, the I, or the N index card at random from the 26 cards in the bag. There is a 3/26 probability that she will select one of those cards. There are only 25 cards left in the bag for the second draw, because the cards are not replaced. There are two letters remaining in the word that she must select, so the probability of selecting one of them is 2/25. With her final draw, she must select the single letter she has not yet selected in the word WIN from among the 24 cards in the bag. This probability is 1/24.

For example, if she drew the letter N with the first selection, she'd need to draw the letter W or I with the second selection—only 2 of the 25 remaining letters will work.

Apply the multiplication principle of probability to calculate the probability of selecting the letters W, I, and N at random in three consecutive draws.

$$\left(\frac{3}{26}\right)\left(\frac{2}{25}\right)\left(\frac{1}{24}\right) = \frac{6}{15{,}600}$$
$$= \frac{1}{2{,}600}$$

7. **D.** Apply the substitution method. Begin by solving the second equation for y and the third equation for z.

$$y = 4 - 3x \qquad z = 2x + 5$$

Substitute these values of y and z into the first equation. The equation is now entirely in terms of x.

$$x - 4y + 3z = 1$$
$$x - 4(4 - 3x) + 3(2x + 5) = 1$$
$$x - 16 + 12x + 6x + 15 = 1$$
$$x + 12x + 6x = 1 + 16 - 15$$
$$19x = 2$$
$$x = \frac{2}{19}$$

Now substitute this value of x into the equations you previously solved for y and z.

$$\begin{aligned} y &= 4-3x \\ &= 4-3\left(\frac{2}{19}\right) \\ &= 4-\frac{6}{19} \\ &= 4\left(\frac{19}{19}\right)-\frac{6}{19} \\ &= \frac{76}{19}-\frac{6}{19} \\ &= \frac{70}{19} \end{aligned}$$

$$\begin{aligned} z &= 2x+5 \\ &= 2\left(\frac{2}{19}\right)+5 \\ &= \frac{4}{19}+5 \\ &= \frac{4}{19}+5\left(\frac{19}{19}\right) \\ &= \frac{4}{19}+\frac{95}{19} \\ &= \frac{99}{19} \end{aligned}$$

The solution to the system is $(x, y, z) = (2/19, 70/19, 99/19)$. Calculate the sum of x, y, and z.

$$\begin{aligned} x+y+z &= \frac{2}{19}+\frac{70}{19}+\frac{99}{19} \\ &= \frac{171}{19} \\ &= 9 \end{aligned}$$

8. **A.** You can apply the DIY technique to answer the question, but your answers may feel specific to the numbers you chose. You may not be comfortable concluding that the statements are or are not always true. In this case, it may be helpful to manipulate the expressions algebraically.

One pencil costs 25 cents, which is 1/4 of a dollar. If you buy b boxes that each contain p pencils, then you buy a total of bp pencils. Calculate the total cost, m, in dollars.

$$\begin{aligned} \text{total cost} &= (\text{total number of pencils})(\text{cost per pencil}) \\ m &= (bp)\left(\frac{1}{4}\right) \\ m &= \frac{bp}{4} \end{aligned}$$

Multiply both sides of the equation by 4 to get $4m = bp$. Use this equation to test each of the statements in the question. Statement I is true. The number of pencils is always higher than the total cost in dollars, because each pencil costs 25 cents. For every $m = 1$ dollar, you can purchase $p = 4$ pencils, regardless of the number of pencils per box or the number of boxes you purchase.

Statement II is false. Recall that $4m = bp$. Whether m is even or odd, $4m$ is an even number, so the equivalent value bp must also be even. However, b does not have to be even. For example, the product of $b = 3$ and $p = 6$ is 18, an even number. If p is even, b could be an even or an odd number.

Not convinced? Think of it this way: The proportion 1/2 = 2/4 is true, but if you swap the denominators, it's false: 1/4 ≠ 2/2.

Statement III is false. You can divide both sides of $4m = bp$ by 4 and b to get numerators that match the numerators of statement III, but the denominators do not match.

$$\frac{\cancel{4}m}{\cancel{4}b} = \frac{\cancel{b}p}{4\cancel{b}}$$

$$\frac{m}{b} = \frac{p}{4}$$

The proportion $m/b = p/4$ is not equivalent to $m/4 = p/b$.

9. **66.** Barry's height is 5 feet, 9 inches. One foot contains 12 inches, so Barry's height is $5(12) + 9 = 69$ inches. Stan is 5 inches shorter than Barry, so Stan's height is $69 - 5 = 64$ inches. Marcus is 2 inches taller than Stan, so Marcus is $64 + 2 = 66$ inches tall.

10. **32, 108, or 256.** Create a table that calculates the area a and perimeter p of squares with integer side lengths. Start with $s = 2$, because the question states that $s > 1$.

side length	area (a)	perimeter (p)
2	$2^2 = 4$	$4(2) = 8$
3	$3^2 = 9$	$4(3) = 12$
4	$4^2 = 16$	$4(4) = 16$
5	~~$5^2 = 25$~~	~~$4(5) = 20$~~

Notice that squares of side length 5 or greater do not satisfy the conditions described in the question because they have areas that are greater than 20. Therefore, the square must have side length 2, 3, or 4, and any of the following answers are valid: $4(8) = 32$, $9(12) = 108$, or $16(16) = 256$.

11. **20.** The greatest value of XY corresponds with the greatest distance between X and Y. Consider the following diagram, which illustrates this distance. In it, X and Y are opposite endpoints of identical diameters on both bases. In other words, if the height of the cylinder was 0 and the bases overlapped, the diameters would lie along the same line. The distance between X and Y is the hypotenuse of the shaded right triangle, which has leg length 6 (the diameter of the bottom circular base) and leg length 12 (the height of the cylinder).

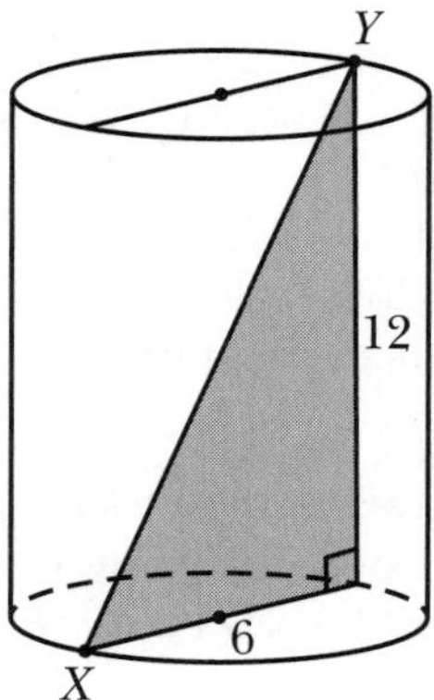

Apply the Pythagorean theorem to calculate *XY*.

$$\begin{aligned} 6^2+12^2 &= (XY)^2 \\ 36+144 &= (XY)^2 \\ 180 &= (XY)^2 \\ \sqrt{180} &= \sqrt{(XY)^2} \\ \sqrt{9\cdot 20} &= XY \\ 3\sqrt{20} &= XY \end{aligned}$$

According to the question, $XY = 3\sqrt{c}$, so $c = 20$.

12. **54.** Five points of data are provided in the line graph. The respective sales for model years 2000–2004 are 180, 300, 100, 240, and 380. Calculate the total sales by adding those numbers.

$$180 + 300 + 100 + 240 + 380 = 1{,}200$$

In model year 2000, there were 180 car sales, which represents 180/1,200 = 0.15 (or 15%) of the total sales. Therefore, the central angle for model year 2000 should represent 15% of the 360° in the circle graph.

$$0.15(360°) = 54°$$

The central angle for model year 2000 is 54°.

13. **10.** Factor the quadratic expression.

$$\begin{aligned} s^2-2st+t^2 &= (s-t)(s-t) \\ &= (s-t)^2 \end{aligned}$$

Therefore, $a \otimes b = (a-b)^2$. The question states that $a \otimes b = 100$.

$$\begin{aligned} a\otimes b &= 100 \\ (a-b)^2 &= 100 \end{aligned}$$

Take the square root of both sides of the equation.

$$\begin{aligned} \sqrt{(a-b)^2} &= \sqrt{100} \\ a-b &= \pm 10 \end{aligned}$$

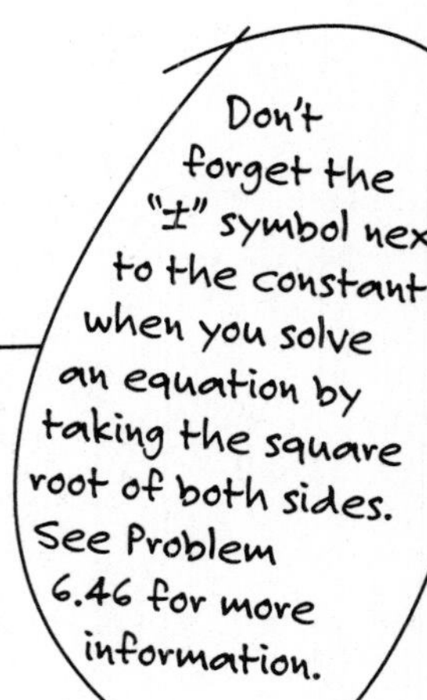

Don't forget the "±" symbol next to the constant when you solve an equation by taking the square root of both sides. See Problem 6.46 for more information.

The expression $a-b$ is equal to –10 or +10. Because you cannot grid a negative number on the SAT, the only valid solution is 10.

14. **108.** Each term of a geometric sequence is equal to the preceding term times some *r* called the common ratio. In other words, $a_2 = r \cdot a_1$. Substitute a_2 and a_1 into the equation to calculate *r*.

$$\begin{aligned} -6 &= r(2) \\ \frac{-6}{2} &= r \\ -3 &= r \end{aligned}$$

The common ratio is $r = -3$, so multiply a_2 by -3 to calculate a_3, multiply a_3 by -3 to calculate a_4, and multiply a_4 by -3 to calculate a_5.

$$\begin{aligned} a_3 &= r(a_2) \\ &= (-3)(-6) \\ &= 18 \end{aligned} \qquad \begin{aligned} a_4 &= r(a_3) \\ &= (-3)(18) \\ &= -54 \end{aligned} \qquad \begin{aligned} a_5 &= r(a_4) \\ &= (-3)(-54) \\ &= 162 \end{aligned}$$

Add the fourth and fifth terms.

$$\begin{aligned} a_4 + a_5 &= -54 + 162 \\ &= 108 \end{aligned}$$

15. **12.** Renée is looking for numbers whose prime factorizations are of the form $a^2 \cdot b^2$, where a and b are different prime numbers. To identify all prime factorizations of this form, begin with the smallest prime number, 2. In other words, if $a = 2$, identify all prime values of b such that $a^2 \cdot b^2 < 2{,}000$.

$$\begin{aligned} 2^2 \cdot 3^2 &= 4 \cdot 9 &&= 36 \\ 2^2 \cdot 5^2 &= 4 \cdot 25 &&= 100 \\ 2^2 \cdot 7^2 &= 4 \cdot 49 &&= 196 \\ 2^2 \cdot 11^2 &= 4 \cdot 121 &&= 484 \\ 2^2 \cdot 13^2 &= 4 \cdot 169 &&= 676 \\ 2^2 \cdot 17^2 &= 4 \cdot 289 &&= 1{,}156 \\ 2^2 \cdot 19^2 &= 4 \cdot 361 &&= 1{,}444 \\ \cancel{2^2 \cdot 23^2} &\cancel{= 4 \cdot 529} &&\cancel{= 2{,}116} \end{aligned}$$

Notice that $2^2 \cdot 23^2 = 2{,}116$ does not meet Renée's requirements, because $2{,}116 > 2{,}000$. So far, you have identified 7 of the numbers Renée is seeking. Now list the numbers that have a least squared prime factor of 3. Notice that $3^2 \cdot 2^2$ is not included in the following list, because it is equal to $2^2 \cdot 3^2$, which appears in the previous list.

$$\begin{aligned} 3^2 \cdot 5^2 &= 9 \cdot 25 &&= 225 \\ 3^2 \cdot 7^2 &= 9 \cdot 49 &&= 441 \\ 3^2 \cdot 11^2 &= 9 \cdot 121 &&= 1{,}089 \\ 3^2 \cdot 13^2 &= 9 \cdot 169 &&= 1{,}521 \\ \cancel{3^2 \cdot 17^2} &\cancel{= 9 \cdot 289} &&\cancel{= 2{,}601} \end{aligned}$$

You are now up to 11 of the numbers Renée is seeking. Only one remains.

$$5^2 \cdot 7^2 = 25 \cdot 49 = 1{,}225$$

There are a total of 12 numbers whose prime factorizations are the product of two unique squared prime factors.

16. **8/27 or .296.** Cross-multiply the proportion and solve for y/x.

$$\begin{aligned}(2x-5y)(4)&=(7)(y)\\ 8x-20y&=7y\\ 8x&=20y+7y\\ 8x&=27y\\ \frac{8\cancel{x}}{\cancel{x}}&=\frac{27y}{x}\\ \frac{8}{27}&=\frac{\cancel{27}y}{\cancel{27}x}\\ \frac{8}{27}&=\frac{y}{x}\end{aligned}$$

17. **376.** Company A's price begins at \$750 and then increases by \$10 for every shirt ordered. If x represents the number of shirts ordered, then the price is y = \$750 + \$10(x). Note that this is a linear equation in slope-intercept form.

Company B's price is more complicated, because the cost per shirt varies. The graph is a piecewise-defined function, with breaks at 100 shirts ordered and 250 shirts ordered, because the pricing model changes at those values. Consider the graph below, which compares the two pricing models.

The graph is composed of separate pieces.

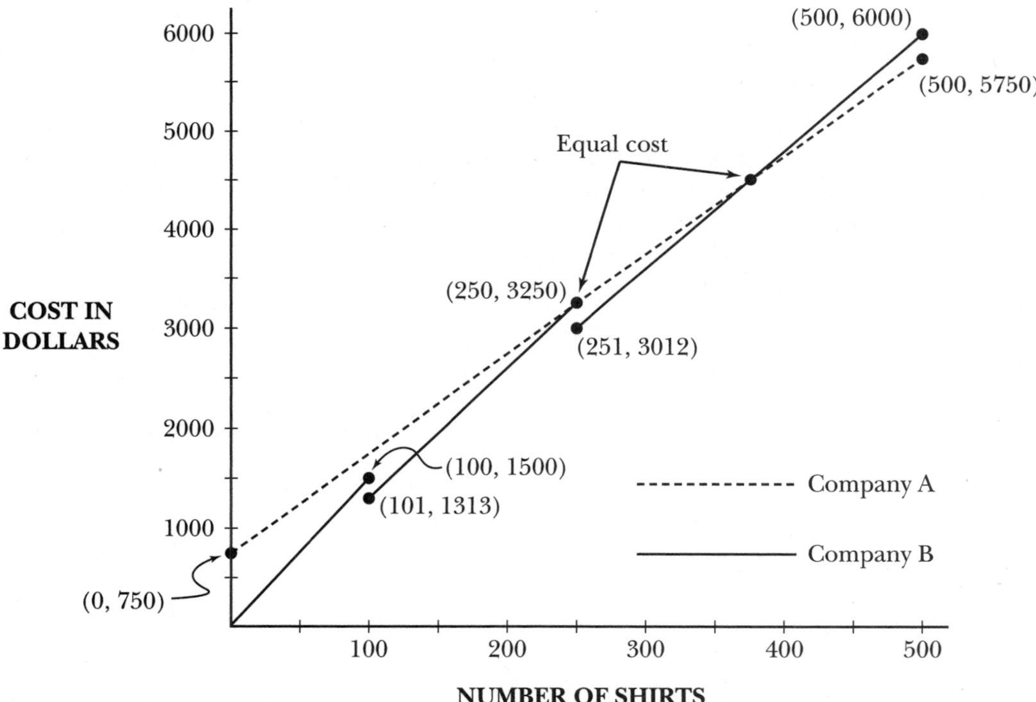

To create the graph, calculate prices for different shirt orders. For example, company A charges \$750 + \$10(0) = \$750 for 0 shirts because of the setup fee. Therefore, (0,750) is a point on its graph. Company A charges \$750 + \$10(500) = \$5,750 for x = 500 shirts, so the point (500, 5750) is also on the dotted cost graph of company A.

Use the same process to plot the segments of company B's graph; calculate the boundaries of each pricing tier by computing the prices for 0, 100, 101, 250, 251, and 500 shirts. For example, the middle segment of the graph represents the \$13 price range. Thus, the cost for 101 shirts is \$13(101) = \$1,313 and the cost for 250 shirts is \$13(250) = \$3,250. The endpoints for the middle segment of the graph are (101,1313) and (250,3250).

The solid graph is beneath the dotted graph from 0 to 250 along the horizontal axis.

Note that Company B charges less per shirt until you order 250 shirts. For that one specific order number, both companies charge the same amount (\$3,250), so the graphs intersect. However, if you order just one additional shirt, the Company B price is cheaper once again. This changes somewhere between 250 shirts and 500 shirts—the solid line intersects and finally overtakes the dotted line. Where does this happen? To find the answer, you need to write the equations of both lines.

You already know the equation of the dotted line: $y = 750 + 10x$, if x is the total number of shirts ordered and y is the price of the order. The equation of the rightmost segment of the solid graph—the portion that intersects the dotted graph—is very simple: $y = 12x$. The price of x shirts is \$12 each when $x \geq 251$.

Substitute $y = 12x$ into the equation $y = 750 + 10x$ and solve for x.

$$\begin{aligned} 12x &= 750+10x \\ 12x-10x &= 750 \\ 2x &= 750 \\ x &= \frac{750}{2} \\ x &= 375 \end{aligned}$$

The graphs intersect when $x = 375$, because both companies charge \$4,500 for that quantity of shirts. Company A is finally less expensive than company B if you order 1 more than 375 shirts, so the correct answer is 376.

18. **52.** Because $f(x)$ is an odd function, $f(-x) = -f(x)$. In other words, if you substitute the opposite of a number from the top row into the function, the result will be the opposite of the corresponding number in the bottom row. Specifically, if $f(4) = -6$, then $f(-4) = 6$. Similarly, if $f(3) = 2$, then $f(-3) = -2$. The question uses these values to define four points: $A = (-4,6)$, $B = (3,2)$, $C = (4,-6)$, and $D = (-3,-2)$. Plot these points in the coordinate plane.

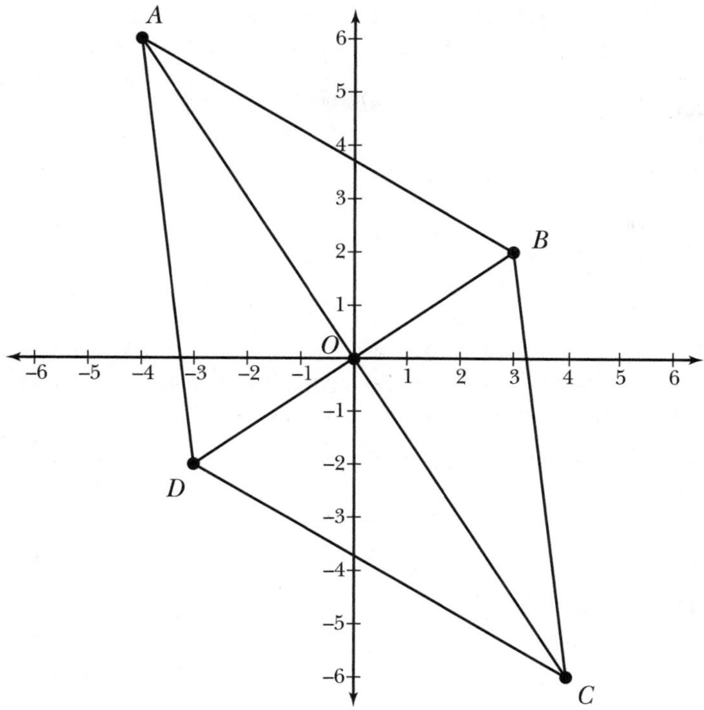

The diagonals of quadrilateral *ABCD* are perpendicular to each other, because their slopes are opposite reciprocals.

$$\text{slope of } \overline{AC} = \frac{-6-6}{4-(-4)} = \frac{-12}{8} = -\frac{3}{2} \qquad \text{slope of } \overline{BD} = \frac{-2-2}{-3-3} = \frac{-4}{-6} = \frac{2}{3}$$

Perpendicular lines form right angles, so *ABCD* can be divided into four right triangles. Let *O* be the origin, point (0,0) on the coordinate plane. You can form right triangles *AOB*, *BOC*, *COD*, and *AOD*. Furthermore, you can verify that the slopes of opposite sides of the quadrilateral are equal (by applying the slope formula once again). Thus, *ABCD* is a parallelogram, and its diagonals bisect each other: *AO* = *CO* and *BO* = *DO*.

Apply the distance formula to calculate the lengths of *AO* and *BO*.

$$AO = \sqrt{[0-(-4)]^2+[0-(6)]^2} = \sqrt{[0+4]^2+[0-6]^2} = \sqrt{16+36} = \sqrt{52}$$

$$BO = \sqrt{(0-3)^2+(0-2)^2} = \sqrt{(-3)^2+(-2)^2} = \sqrt{9+4} = \sqrt{13}$$

Because the triangle's legs are perpendicular, one leg can represent the base of triangle *AOB* and the other can represent the height of the triangle. Calculate the area of triangle *AOB*.

See Problem 8.52.

$$\begin{aligned}\text{area}(\text{triangle } AOB) &= \frac{1}{2}bh \\ &= \frac{1}{2}\left(\sqrt{52}\right)\left(\sqrt{13}\right) \\ &= \frac{1}{2}\left(\sqrt{676}\right) \\ &= \frac{1}{2}(26) \\ &= 13\end{aligned}$$

Not only is 13 the area of triangle *AOB*, it is the area of all four triangles. Remember that $AO = CO$ and $BO = DO$, so each right triangle has legs with lengths $\sqrt{52}$ and $\sqrt{13}$, just like triangle *AOB*. The area of *ABCD* is equal to the sum of the four congruent areas: 13 + 13 + 13 + 13 = 52.

Section 3

1. **B.** If the smallest of the integers is x, then the other consecutive even integers are $x + 2$, $x + 4$, and $x + 6$. Add those expressions, set the sum equal to 380, and solve for x.

$$\begin{aligned}x + (x+2) + (x+4) + (x+6) &= 380 \\ 4x + 12 &= 380 \\ 4x &= 368 \\ x &= 92\end{aligned}$$

You can also use the plug and chug method to solve this. Beginning with choice (C), 94, you could test whether it and the next three consecutive even integers have a sum of 380. You would get a sum that is too high (388), so you would move to choice (B) and test those values.

Some students prefer to use an equation to answer questions like these; others prefer the plug and chug experimental technique. Use the approach that works most quickly and reliably for you.

2. **E.** In order to reach the midpoint *M* from endpoint *R*, you travel 5 units to the right and 3 units up. In other words, adding 5 to the x-coordinate of *R* and 3 to the y-coordinate of *R* gives you the coordinates of midpoint *M*.

$$(-4 + 5, -3 + 3) = (1, 0)$$

To reach endpoint *S* from midpoint *M*, travel the same horizontal and vertical distances. Add 5 to the x-coordinate of *M* and add 3 to the y-coordinate.

$$(1 + 5, 0 + 3) = (6, 3)$$

3. **C.** The cars are traveling at the same rate, so if car A has traveled $6\sqrt{2}$ miles, so has car B. Furthermore, the cars are traveling along perpendicular paths, so the distance between them is the hypotenuse of an isosceles right triangle, as illustrated by the following diagram.

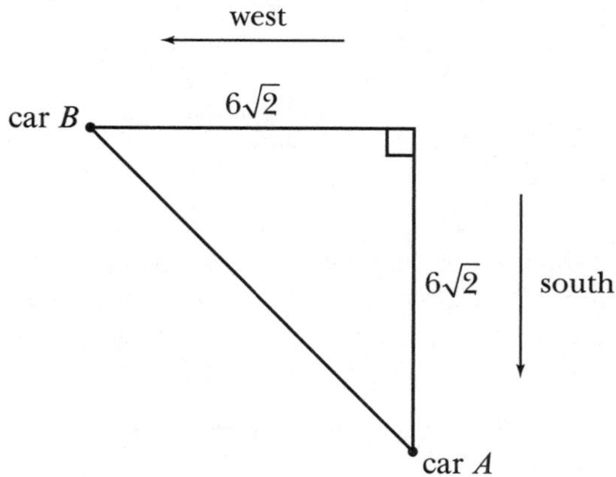

An isosceles right triangle is a 45°–45°–90° right triangle, so its hypotenuse is $\sqrt{2}$ times as long as a leg.

$$\begin{aligned}\text{hypotenuse length} &= \sqrt{2}\left(\text{leg length}\right)\\ &= \sqrt{2}\left(6\sqrt{2}\right)\\ &= 6\sqrt{4}\\ &= 6(2)\\ &= 12\end{aligned}$$

The hypotenuse has length 12, so the cars are 12 miles apart.

4. **A.** The median is the middle value of a data set, once that set is arranged in order from least to greatest. For example, the median of the set $\{1, 2, 3, 4, 5\}$ is 3. However, when a data set contains an even number of data values, the median is the average of the *two* middle data values. For example, the median of $\{1, 2, 3, 4\}$ is 2.5, the average of 2 and 3. Notice that 2.5 is not a data value in the set $\{1, 2, 3, 4\}$.

Or greatest to least

As this example demonstrates, when a data set contains an even number of values, the median may not be one of the values in the data set. The correct answer is (A).

5. **D.** Substitute each value of x into the fraction.

$\boxed{x=0}$

$$\frac{0+1}{0-1}=\frac{1}{-1}=-1$$

$\boxed{x=1}$

$$\frac{1+1}{1-1}=\frac{2}{0}=\text{undefined}$$

$\boxed{x=\frac{1}{2}}$

$$\frac{(1/2)+1}{(1/2)-1}=\frac{3/2}{-1/2}=-\frac{3}{1}=-3$$

The numerator and denominator of this fraction are fractions! Multiply the big fraction by 2/2 to eliminate the smaller fractions.

Substituting $x = 0$ and $x = 1/2$ into the fraction produces integer values –1 and –3, respectively. However, you cannot substitute $x = 1$ into the fraction, because it results in division by 0. You conclude that statements I and III are true.

6. **B.** There are 6 paths that begin at *A*, pass either through *X* or *Y* (but not both), and end at *Z*. Three of the paths pass through *X*, and three of the paths pass through *Y*.

Paths through *X*

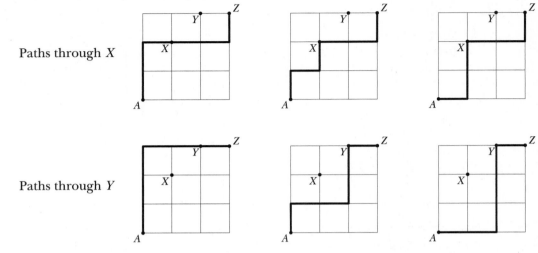

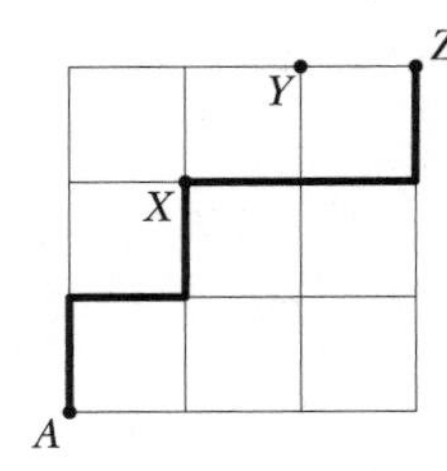

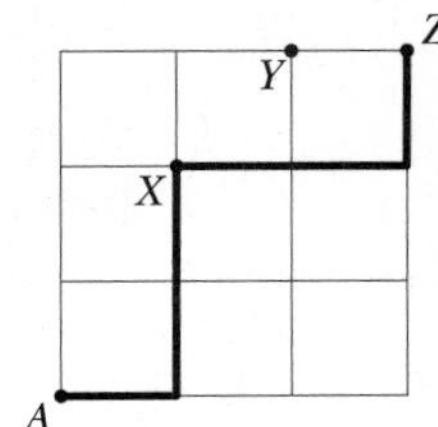

Paths through *Y*

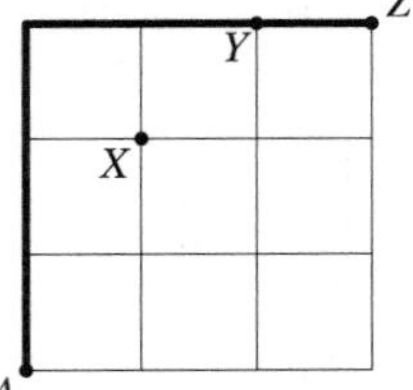

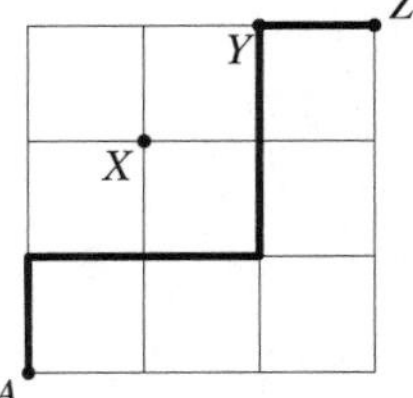

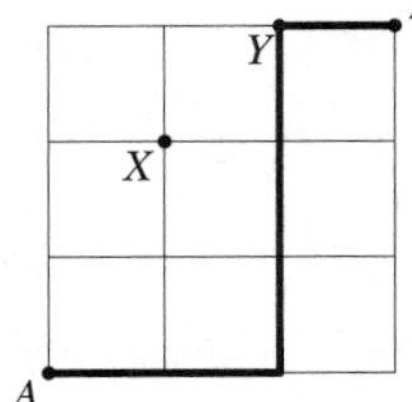

7. **C.** Construct a diagram that records how much money each individual has throughout the transactions. They begin with *d* dollars each.

Ruth	Mel	David
d	d	d

Ruth gives half of her gift ($d/2$) to Mel, which means Ruth has $d/2$ dollars left and Mel has $d + (d/2)$ dollars.

Ruth	Mel	David
$\frac{d}{2}$	$d+\frac{d}{2}=\frac{2d}{2}+\frac{d}{2}=\frac{3d}{2}$	d

Mel currently has $(3d)/2$ dollars, but she gives $2/3$ of that amount to David, leaving her with $1/3$ of that amount. Calculate these amounts in terms of *d*.

$$\text{Mel donates: } \frac{3d}{2}\left(\frac{2}{3}\right)=\frac{\cancel{6}d}{\cancel{6}}=d \text{ dollars}$$

$$\text{Mel is left with: } \frac{3d}{2}\left(\frac{1}{3}\right)=\frac{3d}{6}=\frac{d}{2} \text{ dollars}$$

Update the diagram using the values you calculated.

Ruth	Mel	David
$\frac{d}{2}$	$\frac{d}{2}$	$d+d=2d$

David's final total is $2d$, which is twice his original total, *d*. Therefore, his total is 200% of his original gift.

If you picked (A), you made a classic percentage mistake. David doubled his money, which means two things are true: (1) he increased his money by 100%, and (2) his total is 200% of the original amount. If the question had asked "By what percentage did David increase his money?" the answer would have been (A). However, this question asks you to express his final total as a percentage of his original gift. If David's final total is 100% of his original gift, then he starts and ends with the same amount, *d* dollars.

8. **E.** The value of c affects the wideness or narrowness of the parabola in the graph. However, all parabolas of the form $g(x) = cx^2$ will pass through the origin, because $g(0) = 0$. Furthermore, the origin represents the vertex of the graph, the maximum or minimum height of the function, depending on the direction in which the ends of the parabola point.

Consider the graphs below, in which $c = 1$. The vertex of $g(x)$, point A, represents the minimum value of the function, and it lies on the origin. The vertex of $h(x)$, point B, represents the maximum value of the function, and it also lies on the origin.

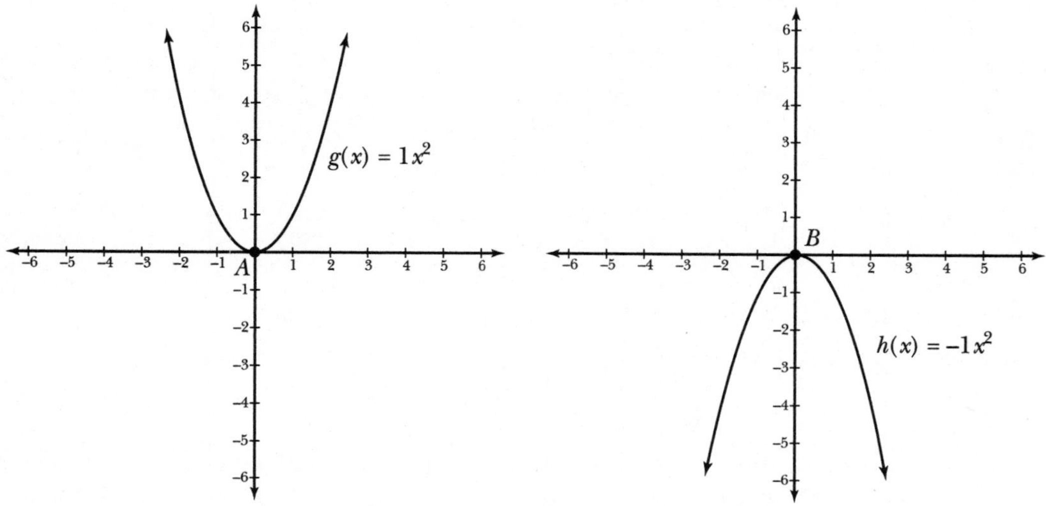

A and B are the same point in both graphs.

9. **B.** Apply the DIY technique. For example, you can prepare $b = 10$ boxed lunches using $c = 47$ cookies. If each lunch has 4 cookies in it, then $n = 7$ cookies remain once the lunches are complete. Only choice (B) returns the correct answer of 4 cookies per lunch if you substitute $b = 10$, $c = 47$, and $n = 7$ into the expression.

$$\begin{aligned}\frac{c-n}{b} &= \frac{47-7}{10} \\ &= \frac{40}{10} \\ &= 4\end{aligned}$$

10. **A.** Start with the rightmost column, which is the ones column. You cannot add 5 to a positive digit B and get 3, because 3 is less than 5. Therefore, the 3 in the ones column must represent 13. The 1 ten is carried to the tens column. If $B + 5 = 13$, then $B = 8$.

Now consider the hundreds column. Because 3 hundreds added to some quantity of D hundreds cannot be equal to 0 hundreds, it must be equal to 10 hundreds, or a thousand. Carry that 1 thousand to the thousands column, and you conclude that $C + 1 = B$. You know that $B = 8$, so $C = 7$.

Rewrite the addition problem, substituting the known values of B and C.

$$\begin{array}{r} 7\ 3\ 7\ 8 \\ +\quad D\ 8\ 5 \\ \hline 8\ 0\ A\ 3 \end{array}$$

They are same-side interior angles formed by two parallel lines cut by a transversal.

This allows you to calculate A and D. Add the digits in the tens column, and remember that you carried 1 ten from the ones column: $1 + 7 + 8 = 16$. A cannot equal 16 tens; it has to be a single digit. Therefore, $A = 6$ tens. That leaves 10 tens, which is equal to 1 hundred. Carry that 1 to the hundreds column: $1 + 3 + D = 10$. Thus, $D = 6$.

The question asks you to calculate $A + D$: $6 + 6 = 12$.

11. **D.** Opposite angles of a parallelogram are congruent; adjacent angles are supplementary. Opposite sides of a parallelogram are congruent. Use this information to update the diagram.

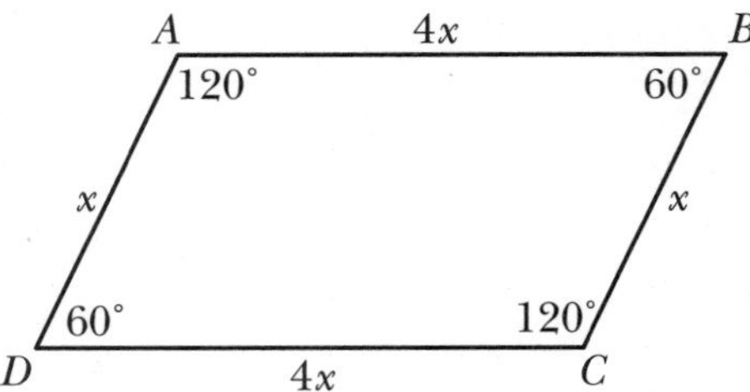

The volume of a prism is equal to the area of the base B multiplied by the height h of the prism. The question states that $h = 4$. The area of the parallelogram base is equal to the base of the parallelogram ($4x$) multiplied by the height of the parallelogram, the perpendicular distance between the bases. Consider the diagram below, in which perpendicular segments extend from two vertices and intersect the opposite side. This forms two 30°–60°–90° right triangles.

You know two of the angles in the right triangle: 60° and 90°. The small angle must measure 30° because all three angles together have a sum of 180°.

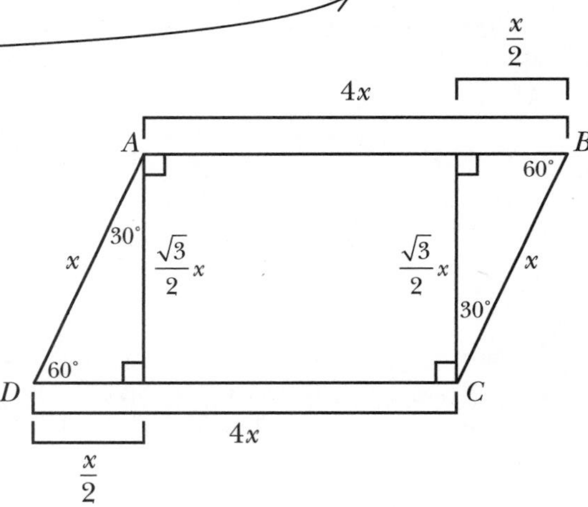

According to the question, the hypotenuses of the right triangles have length x. Apply properties of 30°–60°–90° triangles to calculate the lengths of the legs of the right triangles: $x/2$ and $\left(\sqrt{3}/2\right)x$. The area B of the parallelogram is the product of its base and height.

$$\begin{aligned} B &= (4x)\left(\frac{\sqrt{3}}{2}x\right) \\ &= \frac{4}{2}\sqrt{3}\cdot x^2 \\ &= 2\sqrt{3}\,x^2 \end{aligned}$$

Now calculate the volume V of the prism with base B and height $h = 4$.

$$\begin{aligned} V &= B\cdot h \\ &= \left(2\sqrt{3}\,x^2\right)(4) \\ &= 8\sqrt{3}\,x^2 \end{aligned}$$

12. **E.** If $ab^2 < 0$, then the product ab^2 is negative. Note that the square of any real number is positive, so b^2 must be positive. Therefore, a must be negative in order for the product of a and b^2 to be negative. Statement II is true. Notice that statement I is also true: A negative number divided by a positive number equals a negative number.

Statement III is also true. If a is a negative number, then subtracting the negative number from $\sqrt{2}$ is the same thing as adding it to $\sqrt{2}$. For example, assume that $a = -3$. The expression $\sqrt{2}-(-3)$ is equivalent to $\sqrt{2}+3$; you are adding a positive number to the positive number $\sqrt{2}$, so the result is positive. All three of the statements are true, so the correct answer is choice (E).

13. **B.** The diameter of the circle is 30, so the radius of the circle is $30 \div 2 = 15$. Apply the distance formula, noting that the distance between the center and any point on the circle is equal to the radius, 15.

$$\sqrt{(p-1)^2+(2p-2)^2} = 15$$

Factor the greatest common factor out of $(2p-2)^2$, noting that the factor is squared as well.

$$\begin{aligned} \sqrt{(p-1)^2+\left[2(p-1)\right]^2} &= 15 \\ \sqrt{(p-1)^2+2^2(p-1)^2} &= 15 \\ \sqrt{(p-1)^2+4(p-1)^2} &= 15 \\ \sqrt{5(p-1)^2} &= 15 \end{aligned}$$

Square both sides of the equation and simplify.

$$\begin{aligned} \left[\sqrt{5(p-1)^2}\right]^2 &= (15)^2 \\ 5(p-1)^2 &= 225 \\ (p-1)^2 &= \frac{225}{5} \\ (p-1)^2 &= 45 \end{aligned}$$

Take the square root of both sides of the equation and solve for p.

$$\sqrt{(p-1)^2} = \pm\sqrt{45}$$
$$p-1 = \pm 3\sqrt{5}$$
$$p = 1 \pm 3\sqrt{5}$$

There are two values of p that satisfy the equation, and choice (B) is one of them.

14. **D.** Apply the substitution method, solving the equation $x/y = 2$ for x.

$$\frac{x}{y} = 2$$
$$x = 2y$$

Substitute $x = 2y$ into the other equation of the system.

$$\frac{1}{xy} - \frac{1}{y} = \frac{1}{x+y}$$
$$\frac{1}{(2y)y} - \frac{1}{y} = \frac{1}{2y+y}$$
$$\frac{1}{2y^2} - \frac{1}{y} = \frac{1}{3y}$$

Combine the fractions on the left side of the equation using the least common denominator, $2y^2$.

$$\frac{1}{2y^2} - \frac{1}{y}\left(\frac{2y}{2y}\right) = \frac{1}{3y}$$
$$\frac{1}{2y^2} - \frac{2y}{2y^2} = \frac{1}{3y}$$
$$\frac{1-2y}{2y^2} = \frac{1}{3y}$$

Cross-multiply and solve for y.

$$(1-2y)(3y) = (2y^2)(1)$$
$$3y - 6y^2 = 2y^2$$
$$3y - 8y^2 = 0$$
$$y(3-8y) = 0$$
$$y = 0 \text{ and } y = \frac{3}{8}$$

Notice that y is a denominator in the original equation, so y cannot equal 0. Therefore, $y = 3/8$. Remember, the question asks you to calculate $x + y$. Substitute $y = 3/8$ into the equation you originally solved for x to calculate y.

$$\begin{aligned} x &= 2y \\ &= 2\left(\frac{3}{8}\right) \\ &= \frac{6}{8} \\ &= \frac{3}{4} \end{aligned}$$

Calculate $x + y$.

$$\begin{aligned} x + y &= \frac{3}{4} + \frac{3}{8} \\ &= \frac{3}{4}\left(\frac{2}{2}\right) + \frac{3}{8} \\ &= \frac{6}{8} + \frac{3}{8} \\ &= \frac{9}{8} \end{aligned}$$

15. **D.** The average of p, q, r, s, t, and u is a, so the sum of the six terms is equal to $6a$.

$$p + q + r + s + t + u = 6a$$

The average of p, r, t, and u is b, so the sum of those four terms is $4b$.

$$p + r + t + u = 4b$$

Now return your attention to the first equation you created. Four of the terms on the left side of that equation have a sum of $4b$. Solve for $q + s$.

$$\begin{aligned} p+q+r+s+t+u &= 6a \\ (p+r+t+u)+(q+s) &= 6a \\ 4b+(q+s) &= 6a \\ q+s &= 6a-4b \end{aligned}$$

To calculate the average of q and s, divide the sum of q and s by 2.

$$\begin{aligned} \frac{q+s}{2} &= \frac{6a-4b}{2} \\ &= \frac{6a}{2} - \frac{4b}{2} \\ &= 3a - 2b \end{aligned}$$

16. **B.** Simplify the fraction $(9y^4 + 1)/(9y^2)$ by rewriting it as a sum of two fractions with the denominator $9y^2$.

$$\frac{9y^4+1}{9y^2} = \frac{\cancel{9}y^4}{\cancel{9}y^2} + \frac{1}{9y^2}$$
$$= y^2 + \frac{1}{9y^2}$$

Compare this to the left side of the equation in the question: $y - 1/(3y)$. One expression contains y and the other contains y^2. One expression contains $3y$ and the other contains $(3y)^2 = 9y^2$. Square both sides of the original equation to introduce these terms, which are present in the simplified form of the expression you need to evaluate.

However, one expression IS NOT the square of the other! Remember, $(x + y)^2$ is NOT equal to $x^2 + y^2$.

$$\left(y - \frac{1}{3y}\right)^2 = (4)^2$$
$$y^2 + \left(-\frac{1}{3y}\right)(y) + \left(-\frac{1}{3y}\right)(y) + \left(-\frac{1}{3y}\right)^2 = 4^2$$
$$y^2 - \frac{\cancel{y}}{3\cancel{y}} - \frac{\cancel{y}}{3\cancel{y}} + \frac{1}{9y^2} = 16$$
$$y^2 - \frac{1}{3} - \frac{1}{3} + \frac{1}{9y^2} = 16$$
$$y^2 - \frac{2}{3} + \frac{1}{9y^2} = 16$$

If you add 2/3 to both sides of the equation, the expression left of the equal sign is the (simplified form of the) expression you are asked to evaluate.

$$y^2 + \frac{1}{9y^2} = \frac{2}{3} + 16$$
$$= \frac{2}{3} + \frac{48}{3}$$
$$= \frac{50}{3}$$

You conclude that $\frac{9y^4+1}{9y^2} = y^2 + \frac{1}{9y^2} = \frac{50}{3}$.

Index

ALPHABETICAL LIST OF CONCEPTS WITH PROBLEM NUMBERS

This comprehensive index organizes the concepts and skills discussed within the book alphabetically. Each entry is accompanied by one or more problem numbers, in which the topics are most prominently featured.

Problems in Chapters 1–11 are numbered according to chapter and problem. Chapters 12–14 are practice SAT exams. Each of those chapters is divided into three sections, and each of those sections is numbered independently. Therefore, problems in Chapters 12–14 are indexed by chapter, section, and number.

If the index reference contains two numbers, they refer to chapter and problem, respectively. In other words, 8.1 refers to Chapter 8 Problem 1 and 3.11 refers to Chapter 3 Problem 11.

If there are three numbers in the reference, the problem appears in a practice test. Each practice test contains three sections that are numbered separately, which is why a different numbering system is required. The numbers identify the chapter, seciton, and question number. For example, 12.3.4 refers to Question 4 in Section 3 of Chapter 12.

Numbers & Symbols

A

B

C

D

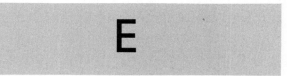

E

F

G

H

I

L

M

N–O

P

Q

R

S

T

U–V

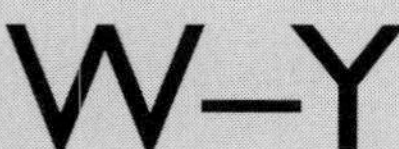

W–Y